The Scientist and Engineer's Guide to
Digital Signal Processing

Pictures on the cover

Physics
A rare and spectacular collision is seen in the *Cartwheel galaxy*, located 500 million light-years away in the constellation Sculptor. The striking ring-like feature was formed when a smaller intruder galaxy careened through the core of the host galaxy about 200 million years ago, sending gas and dust into space at 200,000 miles per hour. The intruder is not known, but is thought to be one of the two galaxies at the right of the image. This picture was taken with the Hubble Space Telescope on October 16, 1994.

Earth Science
Water surface height measurement of the Pacific Ocean, acquired with microwave imaging from the TOPEX/Poseidon satellite on August 21, 1997. The oblong mass near the equator is the weather-disrupting phenomenon known as *El Nino*. Green represents the normal sea level; white indicates that the sea surface is greater than 18 cm (7 inches) higher than normal; purple indicates it is more than 18 cm below normal. The excess water contained in this El Nino is about 30 times the volume of all the U.S. great lakes combined.

Medicine
Magnetic Resonance Image (MRI) of a living human's head, using pseudo color to delineate the brain from other soft tissue and bone. MRI has revolutionized medical diagnosis with its ability to visualize the body's internal anatomy. Digital Signal Processing has been at the forefront of MRI and other medical imaging techniques.

Engineering
Cut away view of a turbine engine, such as used on commercial and military aircraft. Air enters the engine on the left and is compressed by a multistage turbine. In the center, fuel is injected into the air flow and ignited. The hot expanding gases exit the engine toward the right, passing through another turbine that extracts power from the high-velocity gas flow. Large turbine engines can generate more than 100,000 pounds of thrust.

The Scientist and Engineer's Guide to

Digital Signal Processing

by
Steven W. Smith

California Technical Publishing
San Diego, California

The Scientist and Engineer's Guide to
Digital Signal Processing

by
Steven W. Smith

ISBN 0-9660176-3-3
LCCN 97-80293

California Technical Publishing
P.O. Box 502407
San Diego, CA 92150-2407

To contact the author or publisher through the internet:
    website: **DSPguide.com**
    e-mail: **Smith@DSPguide.com**

Printed in the United States of America

Contents at a Glance

Table of Contents

FOUNDATIONS

FUNDAMENTALS

APPLICATIONS

x

COMPLEX TECHNIQUES

Visit our website at **www.DSPguide.com** for free software, homework problems, suggested reading, biography of the author, and other up-to-date DSP information.

Preface

How this book is different

The technical world is changing very rapidly. In only 15 years, the power of personal computers has increased by a factor of nearly *one-thousand*. By all accounts, it will increase by *another* factor of one-thousand in the next 15 years. This tremendous power has changed the way science and engineering is done, and there is no better example of this than Digital Signal Processing.

In the early 1980s, DSP was taught as a graduate level course in electrical engineering. A decade later, DSP had become a standard part of the undergraduate curriculum. Today, DSP is a *basic skill* needed by scientists and engineers in many fields. Unfortunately, DSP education has been slow to adapt to this change. Nearly all DSP textbooks are still written in the traditional electrical engineering style of detailed and rigorous mathematics. DSP is incredibly powerful, but if you can't understand it, you can't use it!

This book was written for scientists and engineers in a wide variety of fields: physics, bioengineering, geology, oceanography, mechanical and electrical engineering, to name just a few. The goal is to present practical techniques while avoiding the barriers of detailed mathematics and abstract theory. To achieve this goal, three strategies were employed in writing this book:

First, the techniques are *explained*, not simply proven to be true through mathematical derivations. While much of the mathematics is included, it is not used as the primary means of conveying the information. Nothing beats a few well written paragraphs supported by good illustrations!

Second, *complex numbers are treated as an advanced topic*, something to be learned after the fundamental principles are understood. Chapters 1-27 explain all the basic techniques using only algebra, and in rare cases, a small amount of elementary calculus. Chapters 28-31 show how complex math extends the power of DSP, presenting techniques that cannot be implemented with real numbers alone. Many would view this approach as heresy! Traditional DSP textbooks are full of complex math, often starting right from the first chapter.

Third, *very simple computer programs* are used. Most DSP programs are written in C, Fortran, or a similar language. However, *learning* DSP has different requirements than *using* DSP. The student needs to concentrate on the algorithms and techniques, without being distracted by the quirks of a particular language. Power and flexibility aren't important; simplicity is critical. The programs in this book are written to teach DSP in the most straightforward way, with all other factors being treated as secondary. Good programming style is disregarded if it makes the program logic more clear. For instance:

- a simplified version of BASIC is used
- line numbers are included
- the only control structure used is the FOR-NEXT loop
- there are no I/O statements

This is the simplest programming style I could find. Some may think that this book would be better if the programs had been written in C. I couldn't disagree more.

The Intended Audience

This book is primarily intended for a one year course in practical DSP, with the students being drawn from a wide variety of science and engineering fields. The suggested prerequisites are:

- A course in practical electronics: (op amps, RC circuits, etc.)
- A course in computer programming (Fortran or similar)
- One year of calculus

This book was also written with the practicing professional in mind. Many everyday DSP applications are discussed: digital filters, neural networks, data compression, audio and image processing, etc. As much as possible, these chapters stand on their own, not requiring the reader to review the entire book to solve a specific problem.

Be sure to visit our Website!

Anything you see in a book will be at least two years out-of-date, and usually much more. For this reason, we have placed much of the supplementary material you would expect in a textbook on our website. This includes: homework problems, computer exercises, references, suggested reading, information about the author, etc. We have also included free computer programs and links to other DSP sites. Visit this site often for up-to-date information about Digital Signal Processing.

Acknowledgements

A special thanks to the many reviewers who provided comments and suggestions on this book. Their generous donation of time and skill has made this a better work: **Magnus Aronsson** (Department of Electrical Engineering, University of Utah); **Vernon L. Chi** (Department of Computer Science, University of North Carolina); **Manohar Das, Ph.D.** (Department of Electrical and Systems Engineering, Oakland University); **Fred DePiero, Ph.D.** (Department of Electrical Engineering, CalPoly State University); **Kenneth H. Jacker**, (Department of Computer Science, Appalachian State University); **Rajiv Kapadia, Ph.D.** (Department of Electrical Engineering, Mankato State University); **A. Dale Magoun, Ph.D.** (Department of Computer Science, Northeast Louisiana University); **Bernard J. Maxum, Ph.D.** (Department of Electrical Engineering, Lamar University); **Paul Morgan, Ph.D.** (Department of Geology, Northern Arizona University); **Dale H. Mugler, Ph.D.** (Department of Mathematical Science, University of Akron); **Christopher L. Mullen, Ph.D.** (Department of Civil Engineering, University of Mississippi); **Cynthia L. Nelson, Ph.D.** (Sandia National Laboratories); **Branislava Perunicic-Drazenovic, Ph.D.** (Department of Electrical Engineering, Lamar University); **John Schmeelk, Ph.D.** (Department of Mathematical Science, Virginia Commonwealth University); **Richard R. Schultz, Ph.D.** (Department of Electrical Engineering, University of North Dakota); **Jay L. Smith, Ph.D.** (Center for Aerospace Technology, Weber State University); **Jeffrey Smith, Ph.D.** (Department of Computer Science, University of Georgia); **Oscar Yanez Suarez, Ph.D.** (Department of Electrical Engineering, Metropolitan University, Iztapalapa campus, Mexico City), and other reviewers who wish to remain anonymous.

This book is now in the hands of the final reviewer, you. Please take the time to give me your comments and suggestions. This will allow future reprints and editions to serve your needs even better. All it takes is a two minute e-mail message to: Smith@DSPguide.com. Thanks; I hope you enjoy the book.

Steve Smith
October 1997

CHAPTER 1

The Breadth and Depth of DSP

Digital Signal Processing is one of the most powerful technologies that will shape science and engineering in the twenty-first century. Revolutionary changes have already been made in a *broad* range of fields: communications, medical imaging, radar & sonar, high fidelity music reproduction, and oil prospecting, to name just a few. Each of these areas has developed a *deep* DSP technology, with its own algorithms, mathematics, and specialized techniques. This combination of breath and depth makes it impossible for any one individual to master all of the DSP technology that has been developed. DSP education involves two tasks: learning general concepts that apply to the field as a whole, and learning specialized techniques for your particular area of interest. This chapter starts our journey into the world of Digital Signal Processing by describing the dramatic effect that DSP has made in several diverse fields. The revolution has begun.

The Roots of DSP

Digital Signal Processing is distinguished from other areas in computer science by the unique type of data it uses: *signals*. In most cases, these signals originate as sensory data from the real world: seismic vibrations, visual images, sound waves, etc. DSP is the mathematics, the algorithms, and the techniques used to manipulate these signals after they have been converted into a digital form. This includes a wide variety of goals, such as: enhancement of visual images, recognition and generation of speech, compression of data for storage and transmission, etc. Suppose we attach an analog-to-digital converter to a computer and use it to acquire a chunk of real world data. DSP answers the question: *What next?*

The roots of DSP are in the 1960s and 1970s when digital computers first became available. Computers were expensive during this era, and DSP was limited to only a few critical applications. Pioneering efforts were made in four key areas: *radar & sonar*, where national security was at risk; *oil exploration*, where large amounts of money could be made; *space exploration*, here the data are irreplaceable; and *medical imaging*, where lives could be saved. The personal computer revolution of the 1980s and 1990s caused DSP to exploded with new applications. Rather than being motivated by military and government needs, DSP was suddenly driven by the commercial marketplace. Anyone who thought they could make money in the rapidly expanding field was suddenly a DSP vender. DSP reached the public in such products as: mobile telephones, compact disc players, and electronic voice mail. Figure 1-1 illustrates a few of these varied applications.

This technological revolution occurred from the top-down. In the early 1980s, DSP was taught as a *graduate* level course in electrical engineering. A decade later, DSP had become a standard part of the *undergraduate* curriculum. Today, DSP is a *basic skill* needed by scientists and engineers

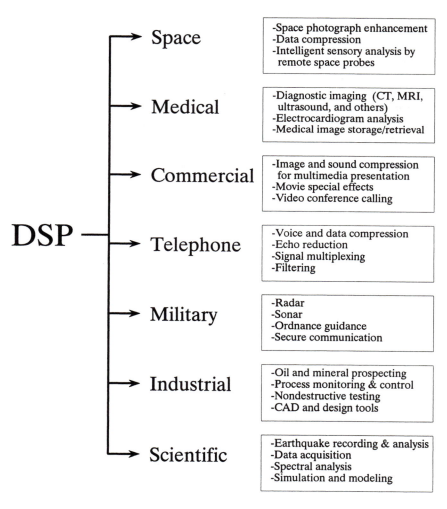

FIGURE 1-1
DSP has revolutionized many areas in science and engineering.
A few of these diverse applications are shown here.

in many fields.   As an analogy, DSP can be compared to a previous technological revolution: *electronics*.   While still the realm of electrical engineering, nearly every scientist and engineer has some background in basic circuit design.   Without it, they would be lost in the technological world. DSP has the same future.

This recent history is more than a curiosity; it has a tremendous impact on *your* ability to learn and use DSP. Suppose you encounter a DSP problem, and turn to textbooks or other publications to find a solution.   What you will typically find is page after page of equations, obscure mathematical symbols, and unfamiliar terminology. It's a nightmare! Much of the DSP literature is baffling even to those experienced in the field. It's not that there is anything wrong with this material, it is just intended for a very specialized audience.   State-of-the-art researchers need this kind of detailed mathematics to understand the theoretical implications of the work.

A basic premise of this book is that most practical DSP techniques can be learned and used without the traditional barriers of detailed mathematics and theory. *The Scientist and Engineers Guide to Digital Signal Processing* is written for those who want to use DSP as a *tool*, not a new *career*.

The remainder of this chapter illustrates areas where DSP has produced revolutionary changes. As you go through each application, notice that DSP is very *interdisciplinary*, relying on the technical work in many adjacent fields. As Fig. 1-2 suggests, the borders between DSP and other technical disciplines are not sharp and well defined, but rather fuzzy and overlapping. If you want to specialize in DSP, these are the allied areas you will also need to study.

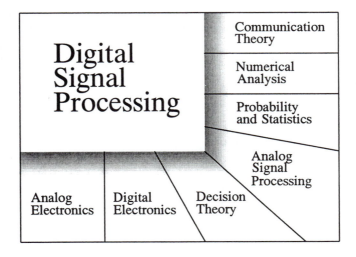

FIGURE 1-2
Digital Signal Processing has fuzzy and overlapping borders with many other areas of science, engineering and mathematics.

Telecommunications

Telecommunications is about transferring information from one location to another. This includes many forms of information: telephone conversations, television signals, computer files, and other types of data. To transfer the information, you need a *channel* between the two locations. This may be a wire pair, radio signal, optical fiber, etc. Telecommunications companies receive *payment* for transferring their customer's information, while they must *pay* to establish and maintain the channel. The financial bottom line is simple: the more information they can pass through a single channel, the more money they make. DSP has revolutionized the telecommunications industry in many areas: signaling tone generation and detection, frequency band shifting, filtering to remove power line hum, etc. Three specific examples from the telephone network will be discussed here: multiplexing, compression, and echo control.

Multiplexing

There are approximately *one billion* telephones in the world. At the press of a few buttons, switching networks allow any one of these to be connected to any other in only a few seconds. The immensity of this task is mind boggling! Until the 1960s, a connection between two telephones required passing the analog voice signals through mechanical switches and amplifiers. One connection required one pair of wires. In comparison, DSP converts audio signals into a stream of serial digital data. Since bits can be easily intertwined and later separated, many telephone conversations can be transmitted on a single channel. For example, a telephone standard known as the *T-carrier system* can simultaneously transmit 24 voice signals. Each voice signal is sampled 8000 times per second using an 8 bit companded (logarithmic compressed) analog-to-digital conversion. This results in each voice signal being represented as 64,000 bits/sec, and all 24 channels being contained in 1.544 megabits/sec. This signal can be transmitted about 6000 feet using ordinary telephone lines of 22 gauge copper wire, a typical interconnection distance. The financial advantage of digital transmission is enormous. Wire and analog switches are expensive; digital logic gates are cheap.

Compression

When a voice signal is digitized at 8000 samples/sec, most of the digital information is *redundant*. That is, the information carried by any one sample is largely duplicated by the neighboring samples. Dozens of DSP algorithms have been developed to convert digitized voice signals into data streams that require fewer bits/sec. These are called **data compression** algorithms. Matching **uncompression** algorithms are used to restore the signal to its original form. These algorithms vary in the amount of compression achieved and the resulting sound quality. In general, reducing the data rate from 64 kilobits/sec to 32 kilobits/sec results in no loss of sound quality. When compressed to a data rate of 8 kilobits/sec, the sound is noticeably affected, but still usable for long distance telephone networks. The highest achievable compression is about 2 kilobits/sec, resulting in sound that is highly distorted, but usable for some applications such as military and undersea communications.

Echo control

Echoes are a serious problem in long distance telephone connections. When you speak into a telephone, a signal representing your voice travels to the connecting receiver, where a portion of it returns as an echo. If the connection is within a few hundred miles, the elapsed time for receiving the echo is only a few milliseconds. The human ear is accustomed to hearing echoes with these small time delays, and the connection sounds quite normal. As the distance becomes larger, the echo becomes increasingly noticeable and irritating. The delay can be several hundred milliseconds for intercontinental communications, and is particularity objectionable. Digital Signal Processing attacks this type of problem by measuring the returned signal and generating an appropriate *antisignal* to cancel the offending echo. This same technique allows speakerphone users to hear and speak at the same time without fighting audio feedback (squealing). It can also be used to reduce environmental noise by canceling it with digitally generated *antinoise*.

Audio Processing

The two principal human senses are vision and hearing. Correspondingly, much of DSP is related to image and audio processing. People listen to both *music* and *speech*. DSP has made revolutionary changes in both these areas.

Music

The path leading from the musician's microphone to the audiophile's speaker is remarkably long. Digital data representation is important to prevent the degradation commonly associated with analog storage and manipulation. This is very familiar to anyone who has compared the musical quality of cassette tapes with compact disks. In a typical scenario, a musical piece is recorded in a sound studio on multiple channels or tracks. In some cases, this even involves recording individual instruments and singers separately. This is done to give the sound engineer greater flexibility in creating the final product. The complex process of combining the individual tracks into a final product is called *mix down*. DSP can provide several important functions during mix down, including: filtering, signal addition and subtraction, signal editing, etc.

One of the most interesting DSP applications in music preparation is *artificial reverberation*. If the individual channels are simply added together, the resulting piece sounds frail and diluted, much as if the musicians were playing outdoors. This is because listeners are greatly influenced by the echo or reverberation content of the music, which is usually minimized in the sound studio. DSP allows artificial echoes and reverberation to be added during mix down to simulate various ideal listening environments. Echoes with delays of a few hundred milliseconds give the impression of cathedral like locations. Adding echoes with delays of 10-20 milliseconds provide the perception of more modest size listening rooms.

Speech generation

Speech generation and recognition are used to communicate between humans and machines. Rather than using your hands and eyes, you use your mouth and ears. This is very convenient when your hands and eyes should be doing something else, such as: driving a car, performing surgery, or (unfortunately) firing your weapons at the enemy. Two approaches are used for computer generated speech: *digital recording* and *vocal tract simulation*. In digital recording, the voice of a human speaker is digitized and stored, usually in a compressed form. During playback, the stored data are uncompressed and converted back into an analog signal. An entire hour of recorded speech requires only about three megabytes of storage, well within the capabilities of even small computer systems. This is the most common method of digital speech generation used today.

Vocal tract simulators are more complicated, trying to mimic the physical mechanisms by which humans create speech. The human vocal tract is an acoustic cavity with resonate frequencies determined by the size and shape of the chambers. Sound originates in the vocal tract in one of two basic ways, called *voiced* and *fricative* sounds. With voiced sounds, vocal cord vibration produces near periodic pulses of air into the vocal cavities. In comparison, fricative sounds originate from the noisy air turbulence at narrow constrictions, such as the teeth and lips. Vocal tract simulators operate by generating digital signals that resemble these two types of excitation. The characteristics of the resonate chamber are simulated by passing the excitation signal through a digital filter with similar resonances. This approach was used in one of the very early DSP success stories, the *Speak & Spell*, a widely sold electronic learning aid for children.

Speech recognition

The automated recognition of human speech is immensely more difficult than speech generation. Speech recognition is a classic example of things that the human brain does well, but digital computers do poorly. Digital computers can store and recall vast amounts of data, perform mathematical calculations at blazing speeds, and do repetitive tasks without becoming bored or inefficient. Unfortunately, present day computers perform very poorly when faced with raw sensory data. Teaching a computer to send you a monthly electric bill is easy. Teaching the same computer to understand your voice is a major undertaking.

Digital Signal Processing generally approaches the problem of voice recognition in two steps: *feature extraction* followed by *feature matching*. Each word in the incoming audio signal is isolated and then analyzed to identify the type of excitation and resonate frequencies. These parameters are then compared with previous examples of spoken words to identify the closest match. Often, these systems are limited to only a few hundred words; can only accept speech with distinct pauses between words; and must be retrained for each individual speaker. While this is adequate for many commercial applications, these limitations are humbling when compared to the abilities of human hearing. There is a great deal of work to be done in this area, with tremendous financial rewards for those that produce successful commercial products.

Echo Location

A common method of obtaining information about a remote object is to bounce a *wave* off of it. For example, radar operates by transmitting pulses of radio waves, and examining the received signal for echoes from aircraft. In sonar, sound waves are transmitted through the water to detect submarines and other submerged objects. Geophysicists have long probed the earth by setting off explosions and listening for the echoes from deeply buried layers of rock. While these applications have a common thread, each has its own specific problems and needs. Digital Signal Processing has produced revolutionary changes in all three areas.

Radar

Radar is an acronym for *RAdio Detection And Ranging*. In the simplest radar system, a radio transmitter produces a pulse of radio frequency energy a few microseconds long. This pulse is fed into a highly directional antenna, where the resulting radio wave propagates away at the speed of light. Aircraft in the path of this wave will reflect a small portion of the energy back toward a receiving antenna, situated near the transmission site. The distance to the object is calculated from the elapsed time between the transmitted pulse and the received echo. The direction to the object is found more simply; you known *where* you pointed the directional antenna when the echo was received.

The operating range of a radar system is determined by two parameters: how much energy is in the initial pulse, and the noise level of the radio receiver. Unfortunately, increasing the energy in the pulse usually requires making the pulse *longer*. In turn, the longer pulse reduces the accuracy and precision of the elapsed time measurement. This results in a conflict between two important parameters: the ability to detect objects at long range, and the ability to accurately determine an object's distance.

DSP has revolutionized radar in three areas, all of which relate to this basic problem. First, DSP can *compress* the pulse after it is received, providing better distance determination without reducing the operating range. Second, DSP can filter the received signal to decrease the noise. This increases the range, without degrading the distance determination. Third, DSP enables the rapid selection and generation of different pulse shapes and lengths. Among other things, this allows the pulse to be optimized for a particular detection problem. Now the impressive part: much of this is done at a sampling rate comparable to the radio frequency used, at high as several hundred megahertz! When it comes to radar, DSP is as much about high-speed hardware design as it is about algorithms.

Sonar

Sonar is an acronym for *SOund NAvigation and Ranging*. It is divided into two categories, *active* and *passive*. In active sonar, sound pulses between 2 kHz and 40 kHz are transmitted into the water, and the resulting echoes detected and analyzed. Uses of active sonar include: detection & localization of undersea bodies, navigation, communication, and mapping the sea floor. A maximum operating range of 10 to 100 kilometers is typical. In comparison, passive sonar simply *listens* to underwater sounds, which includes: natural turbulence, marine life, and mechanical sounds from submarines and surface vessels. Since passive sonar emits no energy, it is ideal for covert operations. You want to detect *the other guy*, without him detecting *you*. The most important application of passive sonar is in military surveillance systems that detect and track submarines. Passive sonar typically uses lower frequencies than active sonar because they propagate through the water with less absorption. Detection ranges can be thousands of kilometers.

DSP has revolutionized sonar in many of the same areas as radar: pulse generation, pulse compression, and filtering of detected signals. In one view, sonar is *simpler* than radar because of the lower frequencies involved. In another view, sonar is more *difficult* than radar because the environment is much less uniform and stable. Sonar systems usually employ extensive arrays of transmitting and receiving elements, rather than just a single channel. By properly controlling and mixing the signals in these many elements, the sonar system can steer the emitted pulse to the desired location and determine the direction that echoes are received from. To handle these multiple channels, sonar systems require the same massive DSP computing power as radar.

Reflection seismology

As early as the 1920s, geophysicists discovered that the structure of the earth's crust could be probed with sound. Prospectors could set off an explosion and record the echoes from boundary layers more than ten kilometers below the surface. These echo seismograms were interpreted by the raw eye to map the subsurface structure. The reflection seismic method rapidly became the primary method for locating petroleum and mineral deposits, and remains so today.

In the ideal case, a sound pulse sent into the ground produces a single echo for each boundary layer the pulse passes through. Unfortunately, the situation is not usually this simple. Each echo returning to the surface must pass through all the other boundary layers above where it originated. This can result in the echo bouncing between layers, giving rise to *echoes of echoes* being detected at the surface. These secondary echoes can make the detected signal very complicated and difficult to interpret. Digital Signal Processing has been widely used since the 1960s to isolate the primary from the secondary echoes in reflection seismograms. How did the early geophysicists manage without DSP? The answer is simple: they looked in *easy* places, where multiple reflections were minimized. DSP allows oil to be found in *difficult* locations, such as under the ocean.

Image Processing

Images are signals with special characteristics. First, they are a measure of a parameter over *space* (distance), while most signals are a measure of a parameter over *time*. Second, they contain a great deal of information. For example, more than 10 megabytes can be required to store one second of television video. This is more than a thousand times greater than for a similar length voice signal. Third, the final judge of quality is often a subjective human evaluation, rather than an objective criteria. These special characteristics have made image processing a distinct subgroup within DSP.

Medical

In 1895, Wilhelm Conrad Röntgen discovered that x-rays could pass through substantial amounts of matter. Medicine was revolutionized by the ability to look inside the living human body. Medical x-ray systems spread throughout the world in only a few years. In spite of its obvious success, medical x-ray imaging was limited by four problems until DSP and related techniques came along in the 1970s. First, overlapping structures in the body can hide behind each other. For example, portions of the heart might not be visible behind the ribs. Second, it is not always possible to distinguish between similar tissues. For example, it may be able to separate bone from soft tissue, but not distinguish a tumor from the liver. Third, x-ray images show *anatomy*, the body's structure, and not *physiology*, the body's operation. The x-ray image of a living person looks exactly like the x-ray image of a dead one! Forth, x-ray exposure can cause cancer, requiring it to be used sparingly and only with proper justification.

The problem of overlapping structures was solved in 1971 with the introduction of the first **computed tomography** scanner (formerly called computed axial tomography, or *CAT* scanner). Computed tomography (CT) is a classic example of Digital Signal Processing. X-rays from many directions are passed through the section of the patient's body being examined. Instead of simply forming images with the detected x-rays, the signals are converted into digital data and stored in a computer. The information is then used to *calculate* images that appear to be *slices through the body*. These images show much greater detail than conventional techniques, allowing significantly better diagnosis and treatment. The impact of CT was nearly as large as the original introduction of x-ray imaging itself. Within only a few years, every major hospital in the world had access to a CT scanner. In 1979, two of CT's principle contributors, Godfrey N. Hounsfield and Allan M. Cormack, shared the Nobel Prize in Medicine. *That's good DSP!*

The last three x-ray problems have been solved by using penetrating energy other than x-rays, such as radio and sound waves. DSP plays a key role in all these techniques. For example, Magnetic Resonance Imaging (MRI) uses magnetic fields in conjunction with radio waves to probe the interior of the human body. Properly adjusting the strength and frequency of the fields cause the atomic nuclei in a localized region of the body to resonate between quantum energy states. This resonance results in the emission of a secondary radio wave, detected with an antenna placed near the body. The strength and other characteristics of this detected signal provide information about the localized region in resonance. Adjustment of the magnetic field allows the resonance region to be scanned throughout the body, mapping the internal structure. This information is usually presented as images, just as in computed tomography. Besides providing excellent discrimination between different types of soft tissue, MRI can provide information about physiology, such as blood flow through arteries. MRI relies totally on Digital Signal Processing techniques, and could not be implemented without them.

Space

Sometimes, you just have to make the most out of a bad picture. This is frequently the case with images taken from unmanned satellites and space exploration vehicles. No one is going to send a repairman to Mars just to tweak the knobs on a camera! DSP can improve the quality of images taken under extremely unfavorable conditions in several ways: brightness and contrast adjustment, edge detection, noise reduction, focus adjustment, motion blur reduction, etc. Images that have spatial distortion, such as encountered when a flat image is taken of a spherical planet, can also be *warped* into a correct representation. Many individual images can also be combined into a single database, allowing the information to be displayed in unique ways. For example, a video sequence simulating an aerial flight over the surface of a distant planet.

Commercial Imaging Products

The large information content in images is a problem for systems sold in mass quantity to the general public. Commercial systems must be *cheap*, and this doesn't mesh well with large memories and high data transfer rates. One answer to this dilemma is *image compression*. Just as with voice signals, images contain a tremendous amount of redundant information, and can be run through algorithms that reduce the number of bits needed to represent them. Television and other moving pictures are especially suitable for compression, since most of the image remain the same from frame-to-frame. Commercial imaging products that take advantage of this technology include: video telephones, computer programs that display moving pictures, and digital television.

CHAPTER 2

Statistics, Probability and Noise

Statistics and probability are used in Digital Signal Processing to characterize signals and the processes that generate them. For example, a primary use of DSP is to reduce interference, noise, and other undesirable components in acquired data. These may be an inherent part of the signal being measured, arise from imperfections in the data acquisition system, or be introduced as an unavoidable byproduct of some DSP operation. Statistics and probability allow these disruptive features to be measured and classified, the first step in developing strategies to remove the offending components. This chapter introduces the most important concepts in statistics and probability, with emphasis on how they apply to acquired signals.

Signal and Graph Terminology

A *signal* is a description of how one parameter depends on another parameter. For example, the most common type of signal in analog electronics is a *voltage* that varies with *time*. Since both parameters can assume a continuous range of values, we will call this a **continuous signal**. In comparison, passing this signal through an analog-to-digital converter forces each of the two parameters to be *quantized*. For instance, imagine the conversion being done with 12 bits at a sampling rate of one kilohertz. The voltage is curtailed to 4096 possible binary levels, and the time is only defined at one millisecond increments. Signals formed from parameters that are quantized in this manner are said to be **discrete signals** or **digitized signals**. For the most part, continuous signals exist in nature, while discrete signals exist inside computers (although you can find exceptions to both cases). It is also possible to have signals where one parameter is continuous and the other is discrete. Since these mixed signals are quite uncommon, they do not have special names given to them, and the nature of the two parameters must be explicitly stated.

Figure 2-1 shows two discrete signals, such as might be acquired with a digital data acquisition system. The **vertical axis** may represent voltage,

11

light intensity, sound pressure, or an infinite number of other parameters. Since we don't know what it represents in this particular case, we will give it the generic label: **amplitude**. This parameter is also called several other names: the **y-axis**, the **dependent variable**, the **range**, and the **ordinate**.

The **horizontal axis** represents the other parameter of the signal, going by such names as: the **x-axis**, the **independent variable**, the **domain**, and the **abscissa**. *Time* is the most common parameter to appear on the horizontal axis of acquired signals; however, other parameters are used in specific applications. For example, a geophysicist might acquire measurements of rock density at equally spaced *distances* along the surface of the earth. To keep things general, we will simply label the horizontal axis: **sample number**. If this were a continuous signal, another label would have to be used, such as: *time, distance, x*, etc.

The two parameters that form a signal are generally not interchangeable. The parameter on the y-axis (the dependent variable) is said to be a **function** of the parameter on the x-axis (the independent variable). In other words, the independent variable describes *how* or *when* each sample is taken, while the dependent variable is the actual measurement. Given a specific value on the x-axis, we can always find the corresponding value on the y-axis, but usually not the other way around.

Pay particular attention to the word: *domain*, a very widely used term in DSP. For instance, a signal that uses time as the independent variable (i.e., the parameter on the horizontal axis), is said to be in the **time domain**. Another common signal in DSP uses frequency as the independent variable, resulting in the term, **frequency domain**. Likewise, signals that use distance as the independent parameter are said to be in the **spatial domain** (distance is a measure of space). The type of parameter on the horizontal axis *is* the domain of the signal; it's that simple. What if the x-axis is labeled with something very generic, such as *sample number*? Authors commonly refer to these signals as being in the *time* domain. This is because sampling at equal intervals of time is the most common way of obtaining signals, and they don't have anything more specific to call it.

Although the signals in Fig. 2-1 are discrete, they are displayed in this figure as continuous lines. This is because there are too many samples to be distinguishable if they were displayed as individual markers. In graphs that portray shorter signals, say less than 100 samples, the individual markers are usually shown. Continuous lines may or may not be drawn to connect the markers, depending on how the author wants you to view the data. For instance, a continuous line could imply what is happening *between* samples, or simply be an aid to help the reader's eye follow a trend in noisy data. The point is, examine the labeling of the horizontal axis to find if you are working with a discrete or continuous signal. Don't rely on an illustrator's ability to draw dots.

The variable, N, is widely used in DSP to represent the total number of samples in a signal. For example, $N = 512$ for the signals in Fig. 2-1. To

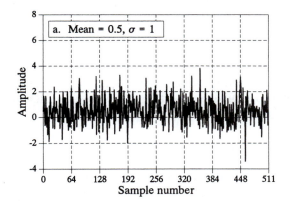
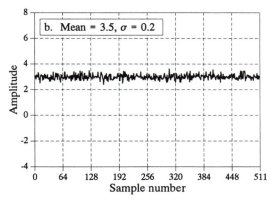

FIGURE 2-1
Examples of two digitized signals with different means and standard deviations.

keep the data organized, each sample is assigned a **sample number** or **index**. These are the numbers that appear along the horizontal axis. Two notations for assigning sample numbers are commonly used. In the first notation, the sample indexes run from 1 to N (e.g., 1 to 512). In the second notation, the sample indexes run from 0 to $N-1$ (e.g., 0 to 511). Mathematicians often use the first method (1 to N), while those in DSP commonly uses the second (0 to $N-1$). In this book, we will use the second notation. Don't dismiss this as a trivial problem. It *will* confuse you sometime during your career. Look out for it!

Mean and Standard Deviation

The **mean**, indicated by μ (a lower case Greek *mu*), is the statistician's jargon for the average value of a signal. It is found just as you would expect: add all of the samples together, and divide by N. It looks like this in mathematical form:

EQUATION 2-1
Calculation of a signal's mean. The signal is contained in x_0 through x_{N-1}, i is an index that runs through these values, and μ is the mean.

$$\mu = \frac{1}{N} \sum_{i=0}^{N-1} x_i$$

In words, sum the values in the signal, x_i, by letting the index, i, run from 0 to $N-1$. Then finish the calculation by dividing the sum by N. This is identical to the equation: $\mu = (x_0+x_1+x_2+\cdots+x_{N-1})/N$. If you are not already familiar with Σ (upper case Greek *sigma*) being used to indicate *summation*, study these equations carefully, and compare them with the computer program in Table 2-1. Summations of this type are abundant in DSP, and you need to understand this notation fully.

In electronics, the *mean* is commonly called the **DC** (direct current) value. Likewise, **AC** (alternating current) refers to how the signal fluctuates around the mean value. If the signal is a simple repetitive waveform, such as a sine or square wave, its excursions can be described by its peak-to-peak amplitude. Unfortunately, most acquired signals do not show a well defined peak-to-peak value, but have a random nature, such as the signals in Fig. 2-1. A more generalized method must be used in these cases, called the **standard deviation**, denoted by σ (a lower case Greek *sigma*).

As a starting point, the expression, $|x_i - \mu|$, describes how far the ith sample *deviates* (differs) from the mean. The *average deviation* of a signal is found by summing the deviations of all the individual samples, and then dividing by the number of samples, N. Notice that we take the absolute value of each deviation before the summation; otherwise the positive and negative terms would average to zero. The average deviation provides a single number representing the typical distance that the samples are from the mean. While convenient and straightforward, the average deviation is almost never used in statistics. This is because it doesn't fit well with the physics of how signals operate. In most cases, the important parameter is not the *deviation* from the mean, but the *power* represented by the deviation from the mean. For example, when random noise signals combine in an electronic circuit, the resultant noise is equal to the combined *power* of the individual signals, not their combined *amplitude*.

The *standard deviation* is similar to the *average deviation*, except the averaging is done with power instead of amplitude. This is achieved by squaring each of the deviations before taking the average (remember, power $\propto$ voltage^2). To finish, the *square root* is taken to compensate for the initial squaring. In equation form, the standard deviation is calculated:

EQUATION 2-2
Calculation of the standard deviation of a signal. The signal is stored in x_i, μ is the mean found from Eq. 2-1, N is the number of samples, and σ is the standard deviation.

$$\sigma^2 = \frac{1}{N-1} \sum_{i=0}^{N-1} (x_i - \mu)^2$$

In the alternative notation: $\sigma = \sqrt{(x_0 - \mu)^2 + (x_1 - \mu)^2 + \cdots + (x_{N-1} - \mu)^2 / (N-1)}$. Notice that the average is carried out by dividing by $N-1$ instead of N. This is a subtle feature of the equation that will be discussed in the next section. The term, σ^2, occurs frequently in statistics and is given the name **variance**. The standard deviation is a measure of how far the signal fluctuates from the mean. The variance represents the power of this fluctuation. Another term you should become familiar with is the **rms (root-mean-square)** value, frequently used in electronics. By definition, the standard deviation only measures the AC portion of a signal, while the rms value measures both the AC and DC components. If a signal has no DC component, its rms value is identical to its standard deviation. Figure 2-2 shows the relationship between the standard deviation and the peak-to-peak value of several common waveforms.

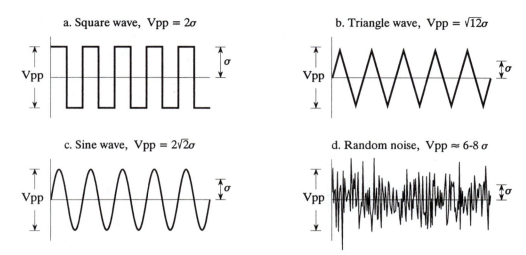

FIGURE 2-2
Ratio of the peak-to-peak amplitude to the standard deviation for several common waveforms. For the square wave, this ratio is 2; for the triangle wave it is $\sqrt{12} = 3.46$; for the sine wave it is $2\sqrt{2} = 2.83$. While random noise has no *exact* peak-to-peak value, it is *approximately* 6 to 8 times the standard deviation.

Table 2-1 lists a computer routine for calculating the mean and standard deviation using Eqs. 2-1 and 2-2. The programs in this book are intended to convey *algorithms* in the most straightforward way; all other factors are treated as secondary. Good programming techniques are disregarded if it makes the program logic more clear. For instance: a simplified version of BASIC is used, line numbers are included, the only control structure allowed is the FOR-NEXT loop, there are no I/O statements, etc. Think of these programs as an alternative way of understanding the equations used

```
100 CALCULATION OF THE MEAN AND STANDARD DEVIATION
110 '
120 DIM X[511]                    'The signal is held in X[0] to X[511]
130 N% = 512                      'N% is the number of points in the signal
140 '
150 GOSUB XXXX                    'Mythical subroutine that loads the signal into X[ ]
160 '
170 MEAN = 0                      'Find the mean via Eq. 2-1
180 FOR I% = 0 TO N%-1
190   MEAN = MEAN + X[I%]
200 NEXT I%
210 MEAN = MEAN/N%
220 '
230 VARIANCE = 0                  'Find the standard deviation via Eq. 2-2
240 FOR I% = 0 TO N%-1
250   VARIANCE = VARIANCE + ( X[I%] - MEAN )^2
260 NEXT I%
270 VARIANCE = VARIANCE/(N%-1)
280 SD = SQR(VARIANCE)
290 '
300 PRINT  MEAN  SD               'Print the calculated mean and standard deviation
310 '
320 END
```

TABLE 2-1

in DSP. If you can't grasp one, maybe the other will help. In BASIC, the % character at the end of a variable name indicates it is an integer. All other variables are floating point. Chapter 4 discusses these variable types in detail.

This method of calculating the mean and standard deviation is adequate for many applications; however, it has two limitations. First, if the mean is much larger than the standard deviation, Eq. 2-2 involves subtracting two numbers that are very close in value. This can result in excessive round-off error in the calculations, a topic discussed in more detail in Chapter 4. Second, it is often desirable to recalculate the mean and standard deviation as new samples are acquired and added to the signal. We will call this type of calculation: **running statistics.** While the method of Eqs. 2-1 and 2-2 can be used for running statistics, it requires that *all* of the samples be involved in each new calculation. This is a very inefficient use of computational power and memory.

A solution to these problems can be found by manipulating Eqs. 2-1 and 2-2 to provide another equation for calculating the standard deviation:

EQUATION 2-3
Calculation of the standard deviation using running statistics. This equation provides the same result as Eq. 2-2, but with less round-off noise and greater computational efficiency. The signal is expressed in terms of three accumulated parameters: N, the total number of samples; *sum*, the sum of these samples; and *sum of squares*, the sum of the squares of the samples. The mean and standard deviation are then calculated from these three accumulated parameters.

$$\sigma^2 = \frac{1}{N-1}\left[\sum_{i=0}^{N-1} x_i^2 - \frac{1}{N}\left(\sum_{i=0}^{N-1} x_i\right)^2\right]$$

or using a simpler notation,

$$\sigma^2 = \frac{1}{N-1}\left[sum\ of\ squares - \frac{sum^2}{N}\right]$$

While moving through the signal, a running tally is kept of three parameters: (1) the number of samples already processed, (2) the sum of these samples, and (3) the sum of the squares of the samples (that is, square the value of each sample and add the result to the accumulated value). After any number of samples have been processed, the mean and standard deviation can be efficiently calculated using only the current value of the three parameters. Table 2-2 shows a program that reports the mean and standard deviation in this manner as each new sample is taken into account. This is the method used in hand calculators to find the statistics of a sequence of numbers. Every time you enter a number and press the Σ (summation) key, the three parameters are updated. The mean and standard deviation can then be found whenever desired, without having to recalculate the entire sequence.

```
100 'MEAN AND STANDARD DEVIATION USING RUNNING STATISTICS
110 '
120 DIM X[511]                       'The signal is held in X[0] to X[511]
130 '
140 GOSUB XXXX                       'Mythical subroutine that loads the signal into X[ ]
150 '
160 N% = 0                          'Zero the three running parameters
170 SUM = 0
180 SUMSQUARES = 0
190 '
200 FOR I% = 0 TO 511               'Loop through each sample in the signal
210   '
220   N% = N%+1                     'Update the three parameters
230   SUM = SUM + X(I%)
240   SUMSQUARES = SUMSQUARES + X(I%)^2
250   '
260   MEAN = SUM/N%                 'Calculate mean and standard deviation via Eq. 2-3
270   VARIANCE = (SUMSQUARES -  SUM^2/N%) / (N%-1)
280   SD = SQR(VARIANCE)
290   '
300   PRINT  MEAN  SD               'Print the running mean and standard deviation
310   '
320 NEXT I%
330 '
340 END
```

TABLE 2-2

Before ending this discussion on the mean and standard deviation, two other terms need to be mentioned. In some situations, the *mean* describes what is being measured, while the *standard deviation* represents noise and other interference. In these cases, the standard deviation is not important in itself, but only in *comparison* to the mean. This gives rise to the term: **signal-to-noise ratio** (**SNR**), which is equal to the mean divided by the standard deviation. Another term is also used, the **coefficient of variation** (**CV**). This is defined as the standard deviation divided by the mean, multiplied by 100 percent. For example, a signal (or other group of measure values) with a CV of 2%, has an SNR of 50. Better data means a *higher* value for the SNR and a *lower* value for the CV.

Signal vs. Underlying Process

Statistics is the science of interpreting *numerical data*, such as acquired signals. In comparison, **probability** is used in DSP to understand the *processes* that generate signals. Although they are closely related, the distinction between the **acquired signal** and the **underlying process** is key to many DSP techniques.

For example, imagine creating a 1000 point signal by flipping a coin 1000 times. If the coin flip is heads, the corresponding sample is made a value of one. On tails, the sample is set to zero. The *process* that created this signal has a mean of exactly 0.5, determined by the relative probability of each possible outcome: 50% heads, 50% tails. However, it is unlikely that the actual 1000 point signal will have a mean of exactly 0.5. Random chance will make the number of ones and zeros slightly different each time the signal is generated. The *probabilities* of the underlying process are constant, but the *statistics* of the acquired signal change each time the experiment is repeated. This random irregularity found in actual data is called by such names as: **statistical variation, statistical fluctuation,** and **statistical noise.**

This presents a bit of a dilemma. When you see the terms: *mean* and *standard deviation*, how do you know if the author is referring to the statistics of an actual signal, or the probabilities of the underlying process that created the signal? Unfortunately, the only way you can tell is by the context. This is not so for all terms used in statistics and probability. For example, the *histogram* and *probability mass function* (discussed in the next section) are matching concepts that are given separate names.

Now, back to Eq. 2-2, calculation of the standard deviation. As previously mentioned, this equation divides by N-1 in calculating the average of the squared deviations, rather than simply by N. To understand why this is so, imagine that you want to find the mean and standard deviation of some *process* that generates signals. Toward this end, you acquire a signal of N samples from the process, and calculate the mean of the signal via Eq. 2.1. You can then use this as an *estimate* of the mean of the underlying process; however, you know there will be an error due to statistical noise. In particular, for random signals, the typical error between the mean of the N points, and the mean of the underlying process, is given by:

EQUATION 2-4
Typical error in calculating the mean of an underlying process by using a finite number of samples, N. The parameter, σ, is the standard deviation.

$$Typical\ error = \frac{\sigma}{N^{1/2}}$$

If N is small, the statistical noise in the calculated mean will be very large. In other words, you do not have access to enough data to properly characterize the process. The larger the value of N, the smaller the expected error will become. A milestone in probability theory, the *Strong Law of Large Numbers*, guarantees that the error becomes zero as N approaches infinity.

In the next step, we would like to calculate the standard deviation of the acquired signal, and use it as an estimate of the standard deviation of the underlying process. Herein lies the problem. Before you can calculate the standard deviation using Eq. 2-2, you need to already know the mean, μ. However, you don't know the mean of the underlying process, only the mean of the N point signal, *which contains an error due to statistical noise*. This error tends to reduce the calculated value of the standard deviation. To compensate for this, N is replaced by N-1. If N is large, the difference doesn't matter. If N is small, this replacement provides a more accurate

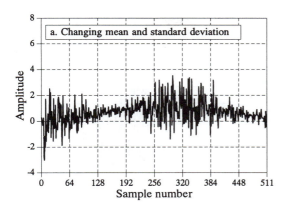
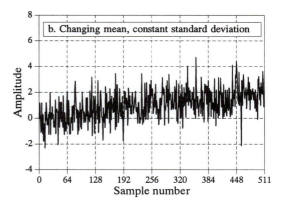

FIGURE 2-3
Examples of signals generated from nonstationary processes. In (a), both the mean and standard deviation change. In (b), the standard deviation remains a constant value of one, while the mean changes from a value of zero to two. It is a common analysis technique to break these signals into short segments, and calculate the statistics of each segment individually.

estimate of the standard deviation of the underlying process. In other words, Eq. 2-2 is an *estimate* of the standard deviation of the *underlying process*. If we divided by *N* in the equation, it would provide the standard deviation of the *acquired signal*.

As an illustration of these ideas, look at the signals in Fig. 2-3, and ask: are the variations in these signals a result of statistical noise, or is the underlying process changing? It probably isn't hard to convince yourself that these changes are too large for random chance, and must be related to the underlying process. Processes that change their characteristics in this manner are called **nonstationary**. In comparison, the signals previously presented in Fig. 2-1 were generated from a stationary process, and the variations result completely from statistical noise. Figure 2-3b illustrates a common problem with nonstationary signals: the slowly changing *mean* interferes with the calculation of the *standard deviation*. In this example, the standard deviation of the signal, over a short interval, is *one*. However, the standard deviation of the entire signal is 1.16. This error can be nearly eliminated by breaking the signal into short sections, and calculating the statistics for each section individually. If needed, the standard deviations for each of the sections can be averaged to produce a single value.

The Histogram, Pmf and Pdf

Suppose we attach an 8 bit analog-to-digital converter to a computer, and acquire 256,000 samples of some signal. As an example, Fig. 2-4a shows 128 samples that might be a part of this data set. The value of each sample will be one of 256 possibilities, 0 through 255. The **histogram** displays the *number of samples* there are in the signal that have each of these *possible values*. Figure (b) shows the histogram for the 128 samples in (a). For

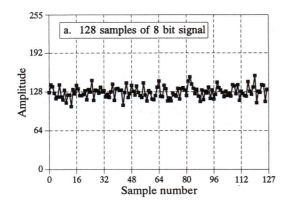

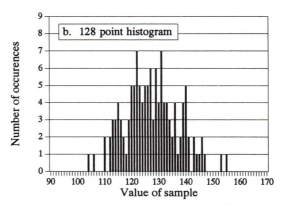

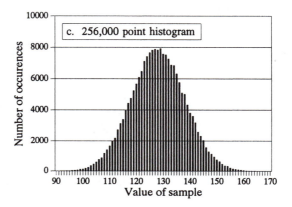

FIGURE 2-4

Examples of histograms. Figure (a) shows 128 samples from a very long signal, with each sample being an integer between 0 and 255. Figures (b) and (c) shows histograms using 128 and 256,000 samples from the signal, respectively. As shown, the histogram is smoother when more samples are used.

example, there are 2 samples that have a value of 110, 8 samples that have a value of 131, 0 samples that have a value of 170, etc. We will represent the histogram by H_i, where i is an index that runs from 0 to M-1, and M is the number of possible values that each sample can take on. For instance, H_{50} is the number of samples that have a value of 50. Figure (c) shows the histogram of the signal using the full data set, all 256k points. As can be seen, the larger number of samples results in a much smoother appearance. Just as with the mean, the statistical noise (roughness) of the histogram is inversely proportional to the square root of the number of samples used.

From the way it is defined, the sum of all of the values in the histogram must be equal to the number of points in the signal:

EQUATION 2-5
The sum of all of the values in the histogram is equal to the number of points in the signal. In this equation, H_i is the histogram, N is the number of points in the signal, and M is the number of points in the histogram.

$$N = \sum_{i=0}^{M-1} H_i$$

The histogram can be used to efficiently calculate the mean and standard deviation of very large data sets. This is especially important for *images*, which can contain millions of samples. The histogram groups samples together that have the same value. This allows the statistics to be calculated by working with a few groups, rather than a large number of individual samples. Using this approach, the mean and standard deviation are calculated from the histogram by the equations:

EQUATION 2-6
Calculation of the mean from the histogram. This can be viewed as combining all samples having the same value into groups, and then using Eq. 2-1 on each group.

$$\mu = \frac{1}{N} \sum_{i=0}^{M-1} i H_i$$

EQUATION 2-7
Calculation of the standard deviation from the histogram. This is the same concept as Eq. 2-2, except that all samples having the same value are operated on at once.

$$\sigma^2 = \frac{1}{N-1} \sum_{i=0}^{M-1} (i-\mu)^2 H_i$$

Table 2-3 contains a program for calculating the histogram, mean, and standard deviation using these equations. Calculation of the histogram is very fast, since it only requires indexing and incrementing. In comparison,

```
100 'CALCULATION OF THE HISTOGRAM, MEAN, AND STANDARD DEVIATION
110 '
120 DIM X%[25000]          'X%[0] to X%[25000] holds the signal being processed
130 DIM H%[255]            'H%[0] to H%[255] holds the histogram
140 N% = 25001             'Set the number of points in the signal
150 '
160 FOR I% = 0 TO 255      'Zero the histogram, so it can be used as an accumulator
170   H%[I%] = 0
180 NEXT I%
190 '
200 GOSUB XXXX             'Mythical subroutine that loads the signal into X%[ ]
210 '
220 FOR I% = 0 TO 25000    'Calculate the histogram for 25001 points
230   H%[ X%[I%] ] = H%[ X%[I%] ] + 1
240 NEXT I%
250 '
260 MEAN = 0               'Calculate the mean via Eq. 2-6
270 FOR I% = 0 TO 255
280   MEAN = MEAN + I% * H%[I%]
290 NEXT I%
300 MEAN = MEAN / N%
310 '
320 VARIANCE = 0           'Calculate the standard deviation via Eq. 2-7
330 FOR I% = 0 TO 255
340   VARIANCE = VARIANCE + H[I%] * (I%-MEAN)^2
350 NEXT I%
360 VARIANCE = VARIANCE / (N%-1)
370 SD = SQR(VARIANCE)
380 '
390 PRINT MEAN SD          'Print the calculated mean and standard deviation.
400 '
410 END
```

TABLE 2-3

calculating the mean and standard deviation requires the time consuming operations of addition and multiplication. The strategy of this algorithm is to use these slow operations only on the few numbers in the histogram, not the many samples in the signal. This makes the algorithm much faster than the previously described methods. Think a factor of ten for very long signals with the calculations being performed on a general purpose computer.

The notion that the acquired signal is a noisy version of the underlying process is very important; so important that some of the concepts are given different names. The *histogram* is what is formed from an acquired signal. The corresponding curve for the underlying process is called the **probability mass function (pmf)**. A histogram is always calculated using a finite number of samples, while the pmf is what *would be* obtained with an infinite number of samples. The pmf can be estimated (inferred) from the histogram, or it may be deduced by some mathematical technique, such as in the coin flipping example.

Figure 2-5 shows an example pmf, and one of the possible histograms that could be associated with it. The key to understanding these concepts rests in the units of the vertical axis. As previously described, the vertical axis of the histogram is the *number of times* that a particular value occurs in the signal. The vertical axis of the pmf contains similar information, except expressed on a *fractional basis*. In other words, each value in the histogram is divided by the total number of samples to approximate the pmf. This means that each value in the pmf must be between zero and one, and that the sum of all of the values in the pmf will be equal to one.

The pmf is important because it describes the *probability* that a certain value will be generated. For example, imagine a signal generated by the process described by Fig. 2-5b, such as previously shown in Fig. 2-4a. What is the probability that a sample taken from this signal will have a value of 120? Figure 2-5b provides the answer, 0.03, or about 1 chance in 34. What is the probability that a randomly chosen sample will have a value greater than 150? Adding up the values in the pmf for: 151, 152, 153,···, 255, provides the answer, 0.0122, or about 1 chance in 82. Thus, the signal would be expected to have a value exceeding 150 on an average of every 82 points. What is the probability that any one sample will be between 0 to 255? Summing all of the values in the histogram produces the probability of 1.00, a certainty that this will occur.

The histogram and pmf can only be used with discrete data, such as a digitized signal residing in a computer. A similar concept applies to continuous signals, such as voltages appearing in analog electronics. The **probability density function (pdf)**, also called the **probability distribution function**, is to continuous signals what the probability mass function is to discrete signals. For example, imagine an analog signal passing through an analog-to-digital converter, resulting in the digitized signal of Fig. 2-4a. For simplicity, we will assume that voltages between 0 and 255 millivolts become digitized into digital numbers between 0 and 255. The pmf of this digital

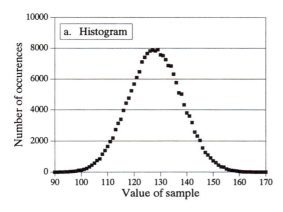

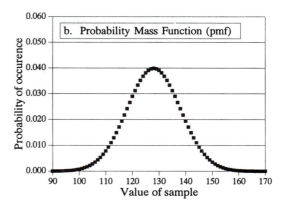

FIGURE 2-5
The relationship between (a) the histogram, (b) the probability mass function (pmf), and (c) the probability density function (pdf). The histogram is calculated from a finite number of samples. The pmf describes the probabilities of the underlying process. The pdf is similar to the pmf, but is used with continuous rather than discrete signals. Even though the vertical axis of (b) and (c) have the same values (0 to 0.06), this is only a coincidence of this example. The amplitude of these three curves is determined by: (a) the sum of the values in the histogram being equal to the number of samples in the signal; (b) the sum of the values in the pmf being equal to one, and (c) the area under the pdf curve being equal to one.

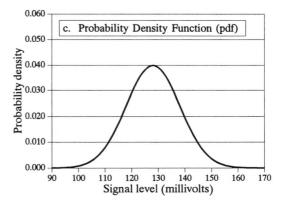

signal is shown by the *markers* in Fig. 2-5b. Similarly, the pdf of the analog signal is shown by the *continuous line* in (c), indicating the signal can take on a continuous range of values, such as the voltage in an electronic circuit.

The vertical axis of the pdf is in units of **probability density**, rather than just probability. For example, a pdf of 0.03 at 120.5 *does not* mean that the a voltage of 120.5 millivolts will occur 3% of the time. In fact, the probability of the continuous signal being exactly 120.5 millivolts is infinitesimally small. This is because there are an infinite number of possible values that the signal needs to divide its time between: 120.49997, 120.49998, 120.49999, etc. The chance that the signal happens to be exactly 120.50000··· is very remote indeed!

To calculate a *probability*, the *probability density* is multiplied by a *range* of values. For example, the probability that the signal, at any given instant, will be between the values of 120 and 121 is: 121-120 × 0.03 = 0.03. The probability that the signal will be between 120.4 and 120.5 is: 120.5-120.4 × 0.03 = 0.003, etc. If the pdf is not constant over the range of interest, the multiplication becomes the integral of the pdf over that range. In other words, the area under the pdf bounded by the specified values. Since the value of the signal must always be *something*, the total area under the pdf curve, the integral from −∞ to +∞, will always be equal to one. This is analogous to the sum of all of the pmf values being equal to one, and the sum of all of the histogram values being equal to N.

The histogram, pmf, and pdf are very similar concepts. Mathematicians always keep them straight, but you will frequently find them used interchangeably (and therefore, incorrectly) by many scientists and

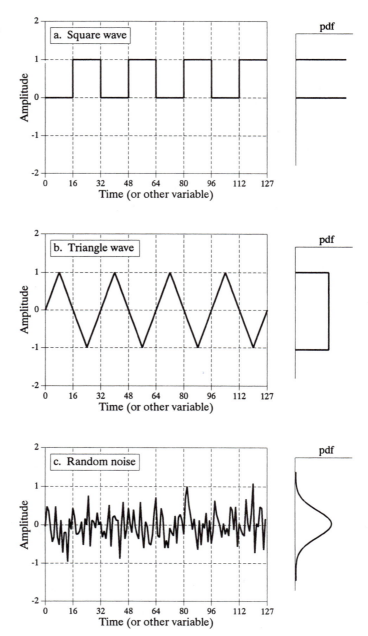

FIGURE 2-6
Three common waveforms and their probability density functions. As in these examples, the pdf graph is often rotated one-quarter turn and placed at the side of the signal it describes. The pdf of a square wave, shown in (a), consists of two infinitesimally narrow spikes, corresponding to the signal only having two possible values. The pdf of the triangle wave, (b), has a constant value over a range, and is often called a *uniform* distribution. The pdf of random noise, as in (c), is the most interesting of all, a bell shaped curve known as a *Gaussian*.

engineers. Figure 2-6 shows three *continuous* waveforms and their *pdfs*. If these were *discrete signals*, signified by changing the horizontal axis labeling to "sample number", *pmfs* would be used.

A problem occurs in calculating the histogram when the number of levels each sample can take on is much larger than the number of samples in the signal. This is always true for signals represented in *floating point* notation, where each sample is stored as a fractional value. For example, integer representation might require the sample value to be 3 or 4, while floating point allows millions of possible fractional values *between* 3 and 4. The previously described approach for calculating the histogram involves counting the number of samples that have each of the possible quantization levels. This is not possible with floating point data because there are *billions* of possible levels that would have to be taken into account. Even worse, nearly all of these possible levels would have no samples that correspond to them. For example, imagine a 10,000 sample signal, with each sample having one billion possible values. The conventional histogram would consist of one billion data points, with all but about 10,000 of them having a value of zero.

The solution to these problems is a technique called **binning**. This is done by arbitrarily selecting the length of the histogram to be some convenient number, such as 1000 points, often called **bins**. The value of each bin represent the total number of samples in the signal that have a value within a *certain range*. For example, imagine a floating point signal that contains values from 0.0 to 10.0, and a histogram with 1000 bins. Bin 0 in the histogram is the number of samples in the signal with a value between 0 and 0.01, bin 1 is the number of samples with a value between 0.01 and 0.02, and so forth, up to bin 999 containing the number of samples with a value between 9.99 and 10.0. Table 2-4 presents a program for calculating a binned histogram in this manner.

```
100 'CALCULATION OF BINNED HISTOGRAM
110 '
120 DIM X[25000]              'X[0] to X[25000] holds the floating point signal,
130 '                         'with each sample being in the range: 0.0 to 10.0
140 DIM H%[999]               'H%[0] to H%[999] holds the binned histogram
150 '
160 FOR I% = 0 TO 999         'Zero the binned histogram for use as an accumulator
170   H%[I%] = 0
180 NEXT I%
190 '
200 GOSUB XXXX                'Mythical subroutine that loads the signal into X%[ ]
210 '
220 FOR I% = 0 TO 25000       'Calculate the binned histogram for 25001 points
230   BINNUM% = INT( X[I%] * .01 )
240   H%[ BINNUM%] = H%[ BINNUM%] + 1
250 NEXT I%
260 '
270 END
```
TABLE 2-4

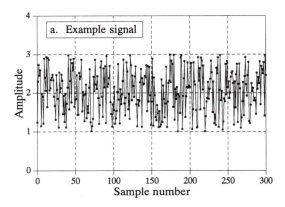

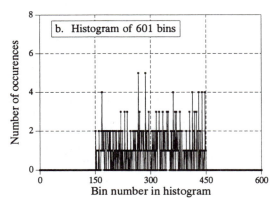

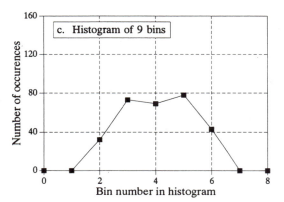

FIGURE 2-7
Example of binned histograms. As shown in (a), the signal used in this example is 300 samples long, with each sample a floating point number uniformly distributed between 1 and 3. Figures (b) and (c) show binned histograms of this signal, using 601 and 21 bins, respectively. As shown, a large number of bins results in poor resolution along the *vertical axis*, while a small number of bins provides poor resolution along the *horizontal axis*. Using more samples makes the resolution better in both directions.

How many bins should be used? This is a compromise between two problems. As shown in Fig. 2-7, too many bins makes it difficult to estimate the *amplitude* of the underlying pmf. This is because only a few samples fall into each bin, making the statistical noise very high. At the other extreme, too few of bins makes it difficult to estimate the underlying pmf in the *horizontal* direction. In other words, the number of bins controls a tradeoff between resolution in along the y-axis, and resolution along the x-axis.

The Normal Distribution

Signals formed from random processes usually have a bell shaped pdf. This is called a **normal distribution**, a **Gauss distribution**, or a **Gaussian**, after the great German mathematician, Karl Friedrich Gauss (1777-1855). The reason why this curve occurs so frequently in nature will be discussed shortly in conjunction with *digital noise generation*. The basic shape of the curve is generated from a *negative squared exponent*:

$$y(x) = e^{-x^2}$$

This raw curve can be converted into the complete Gaussian by adding an adjustable mean, μ, and standard deviation, σ. In addition, the equation must be normalized so that the total area under the curve is equal to *one*, a requirement of all probability distribution functions. This results in the general form of the normal distribution, one of the most important relations in statistics and probability:

EQUATION 2-8
Equation for the *normal distribution*, also called the *Gauss distribution*, or simply a *Gaussian*. In this relation, $P(x)$ is the probability distribution function, μ is the mean, and σ is the standard deviation.

$$P(x) = \frac{1}{\sqrt{2\pi}\,\sigma}\, e^{-(x-\mu)^2/2\sigma^2}$$

Figure 2-8 shows several examples of Gaussian curves with various means and standard deviations. The *mean* centers the curve over a particular value, while the *standard deviation* controls the width of the bell shape.

An interesting characteristic of the Gaussian is that the *tails* drop toward zero very rapidly, much faster than with other common functions such as decaying exponentials or $1/x$. For example, at two, four, and six standard

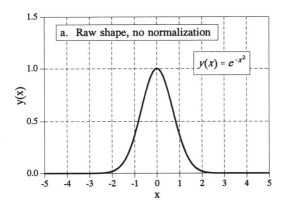

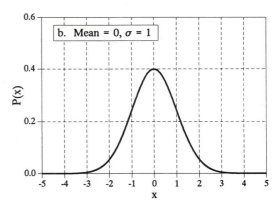

FIGURE 2-8
Examples of Gaussian curves. Figure (a) shows the shape of the raw curve without normalization or the addition of adjustable parameters. In (b) and (c), the complete Gaussian curve is shown for various means and standard deviations.

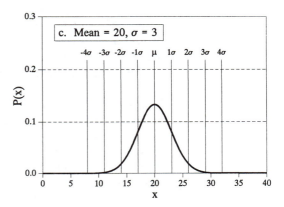

deviations from the mean, the value of the Gaussian curve has dropped to about 1/19, 1/7563, and 1/166,666,666, respectively. This is why normally distributed signals, such as illustrated in Fig. 2-6c, *appear* to have an approximate peak-to-peak value. In principle, signals of this type can experience excursions of unlimited amplitude. In practice, the sharp drop of the Gaussian pdf dictates that these extremes almost never occur. This results in the waveform having a relatively bounded appearance with an apparent peak-to-peak amplitude of about 6-8σ.

As previously shown, the integral of the pdf is used to find the probability that a signal will be within a certain range of values. This makes the integral of the pdf important enough that it is given its own name, the **cumulative distribution function (cdf)**. An especially obnoxious problem with the Gaussian is that it cannot be integrated using elementary methods. To get around this, the integral of the Gaussian can be calculated by *numerical integration*. This involves sampling the continuous Gaussian curve very finely, say, a few million points between -10σ and +10σ. The samples in this discrete signal are then *added* to simulate *integration*. The discrete curve resulting from this simulated integration is then stored in a table for use in calculating probabilities.

The cdf of the normal distribution is shown in Fig. 2-9, with its numeric values listed in Table 2-5. Since this curve is used so frequently in probability, it is given its own symbol: $\Phi(x)$ (upper case Greek *phi*). For example, $\Phi(-2)$ has a value of 0.0228. This indicates that there is a 2.28% probability that the value of the signal will be between $-\infty$ and two standard deviations below the mean, at any randomly chosen time. Likewise, the value: $\Phi(1) = 0.8413$, means there is an 84.13% chance that the value of the signal, at a randomly selected instant, will be between $-\infty$ and one standard deviation above the mean. To calculate the probability that the signal will be will be *between* two values, it is necessary to subtract the appropriate numbers found in the $\Phi(x)$ table. For example, the probability that the value of the signal, at some randomly chosen time, will be between two standard deviations below the mean and one standard deviation above the mean, is given by: $\Phi(1) - \Phi(-2) = 0.8185$, or 81.85%

Using this method, samples taken from a normally distributed signal will be within $\pm 1\sigma$ of the mean about 68% of the time. They will be within $\pm 2\sigma$ about 95% of the time, and within $\pm 3\sigma$ about 99.75% of the time. The probability of the signal being more than 10 standard deviations from the mean is so minuscule, it would be expected to occur for only a few *microseconds* since the beginning of the universe, about 10 billion years!

Equation 2-8 can also be used to express the probability mass function of normally distributed *discrete* signals. In this case, x is restricted to be one of the quantized levels that the signal can take on, such as one of the 4096 binary values exiting a 12 bit analog-to-digital converter. Ignore the $1/\sqrt{2\pi}\sigma$ term, it is only used to make the total area under the pdf curve equal to *one*. Instead, you must include whatever term is needed to make the sum of all the values in the pmf equal to *one*. In most cases, this is done by

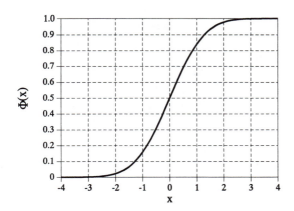

x	Φ(x)	x	Φ(x)
-3.4	.0003	0.0	.5000
-3.3	.0005	0.1	.5398
-3.2	.0007	0.2	.5793
-3.1	.0010	0.3	.6179
-3.0	.0013	0.4	.6554
-2.9	.0019	0.5	.6915
-2.8	.0026	0.6	.7257
-2.7	.0035	0.7	.7580
-2.6	.0047	0.8	.7881
-2.5	.0062	0.9	.8159
-2.4	.0082	1.0	.8413
-2.3	.0107	1.1	.8643
-2.2	.0139	1.2	.8849
-2.1	.0179	1.3	.9032
-2.0	.0228	1.4	.9192
-1.9	.0287	1.5	.9332
-1.8	.0359	1.6	.9452
-1.7	.0446	1.7	.9554
-1.6	.0548	1.8	.9641
-1.5	.0668	1.9	.9713
-1.4	.0808	2.0	.9772
-1.3	.0968	2.1	.9821
-1.2	.1151	2.2	.9861
-1.1	.1357	2.3	.9893
-1.0	.1587	2.4	.9918
-0.9	.1841	2.5	.9938
-0.8	.2119	2.6	.9953
-0.7	.2420	2.7	.9965
-0.6	.2743	2.8	.9974
-0.5	.3085	2.9	.9981
-0.4	.3446	3.0	.9987
-0.3	.3821	3.1	.9990
-0.2	.4207	3.2	.9993
-0.1	.4602	3.3	.9995
0.0	.5000	3.4	.9997

FIGURE 2-9 & TABLE 2-5
$\Phi(x)$, the cumulative distribution function of the normal distribution. These values are calculated by numerically integrating the normal distribution shown in Fig. 2-8b. In words, $\Phi(x)$ is the probability that the value of a normally distributed signal, at some randomly chosen time, will be less than x. The parameter, x, is expressed in standard deviations referenced to the mean.

generating the curve without worrying about normalization, summing all of the unnormalized values, and then dividing all of the values by the sum.

Digital Noise Generation

Random noise is an important topic in both electronics and DSP. For example, it limits how small of signals an instrument will measure, the distance a radio system can communication, and how much radiation is required to produce an x-ray image. A common need in DSP is to generate signals that resemble various types of random noise. This is required to test the performance of algorithms that must *work* in the presence of noise.

The heart of digital noise generation is the **random number generator**. Most programming languages have this as a standard function. The *BASIC* statement: $X = RND$, loads the variable, X, with a new random number each time the command is encountered. Each random number has a value between zero and one, with an equal probability of being anywhere between these two extremes. Figure 2-10a shows a signal formed by taking 128 samples from this type of random number generator. The mean of the underlying process that generated this signal is 0.5, the standard deviation is $1/\sqrt{12} = 0.29$, and the distribution is uniform between zero and one.

Algorithms need to be tested using the same kind of data they will encounter in actual operation. This creates the need to generate digital noise with a *Gaussian* pdf. There are two methods for generating such signals using a random number generator. Figure 2-10 illustrates the first method. Figure (b) shows a signal obtained by adding two random numbers to form each sample, i.e., $X = RND + RND$. Since each of the random numbers can run from zero to one, the sum can run from zero to two. The mean is now *one*, and the standard deviation is $1/\sqrt{6}$ (remember, when random signals are added, the variances also add). As shown, the pdf has changed from a *uniform* distribution to a *triangular* distribution. That is, the signal spends more of its time around a value of *one*, with less time spent near *zero* or *two*.

Figure (c) takes this idea a step further by adding twelve random numbers to produce each sample. The mean is now *six*, and the standard deviation is *one*. What is most important, the pdf has virtually become a *Gaussian*. This procedure can be used to create a normally distributed noise signal with an arbitrary mean and standard deviation. For each sample in the signal: (1) add twelve random numbers, (2) subtract six to make the mean equal to zero, (3) multiply by the standard deviation desired, and (4) add the desired mean.

The mathematical basis for this algorithm is contained in the **Central Limit Theorem**, one of the most important concepts in probability. In its simplest form, the Central Limit Theorem states that a *sum* of random numbers becomes normally distributed as more and more of the random numbers are added together. The Central Limit Theorem *does not* require the individual random numbers be from any particular distribution, or even that the random numbers be from the *same* distribution. The Central Limit Theorem provides the reason why normally distributed signals are seen so widely in nature. Whenever many different random forces are interacting, the resulting pdf becomes a Gaussian.

In the second method for generating normally distributed random numbers, the random number generator is invoked *twice*, to obtain R_1 and R_2. A normally distributed random number, X, can then be found:

EQUATION 2-9
Generation of normally distributed random numbers. R_1 and R_2 are random numbers with a uniform distribution between zero and one. This results in X being normally distributed with a mean of zero, and a standard deviation of one. The log is base e, and the cosine is in radians.

$$X = (-2 \log R_1)^{1/2} \cos(2\pi R_2)$$

Just as before, this approach can generate normally distributed random signals with an arbitrary mean and standard deviation. Take each number generated by this equation, multiply it by the desired standard deviation, and add the desired mean.

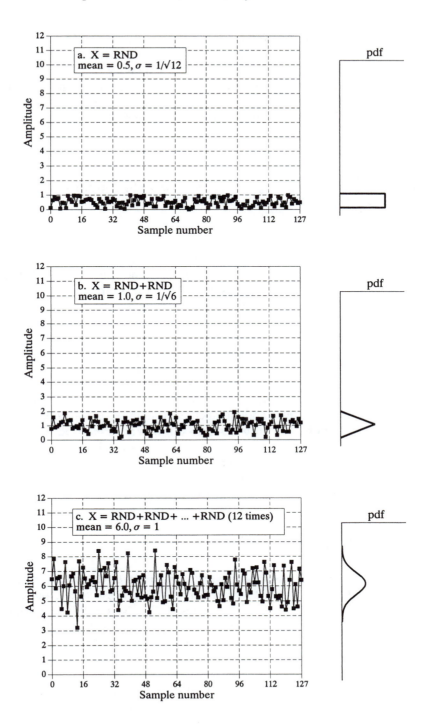

FIGURE 2-10

Converting a uniform distribution to a Gaussian distribution. Figure (a) shows a signal where each sample is generated by a random number generator. As indicated by the pdf, the value of each sample is uniformly distributed between zero and one. Each sample in (b) is formed by adding *two* values from the random number generator. In (c), each sample is created by adding *twelve* values from the random number generator. The pdf of (c) is very nearly Gaussian, with a mean of *six*, and a standard deviation of *one*.

Random number generators operate by starting with a **seed**, a number between zero and one. When the random number generator is invoked, the seed is passed through a fixed algorithm, resulting in a new number between zero and one. This new number is reported as the *random number*, and is then internally stored to be used as the seed the next time the random number generator is called. The algorithm that transforms the seed into the new random number is often of the form:

EQUATION 2-10
Common algorithm for generating uniformly distributed random numbers between zero and one. In this method, S is the seed, R is the new random number, and a,b,& c are appropriately chosen constants. In words, the quantity $aS+b$ is divided by c, and the remainder is taken as R.

$$R = (aS + b) \text{ modulo } c$$

In this manner, a continuous sequence of random numbers can be generated, all starting from the same seed. This allows a program to be run multiple times using exactly the same random number sequences. If you want the random number sequence to change, most languages have a provision for **reseeding** the random number generator, allowing you to choose the number first used as the seed. A common technique is to use the *time* (as indicated by the system's clock) as the seed, thus providing a new sequence each time the program is run.

From a pure mathematical view, the numbers generated in this way cannot be absolutely random since each number is fully determined by the *previous* number. The term **pseudo-random** is often used to describe this situation. However, this is not something you should be concerned with. The sequences generated by random number generators are statistically random to an exceedingly high degree. It is very unlikely that you will encounter a situation where they are not adequate.

Precision and Accuracy

Precision and accuracy are terms used to describe systems and methods that *measure*, *estimate*, or *predict*. In all these cases, there is some parameter you wish to know the value of. This is called the **true value**, or simply, **truth**. The method provides a **measured value**, that you want to be as close to the true value as possible. *Precision* and *accuracy* are ways of describing the error that can exist between these two values.

Unfortunately, precision and accuracy are used interchangeably in nontechnical settings. In fact, dictionaries define them by referring to each other! In spite of this, science and engineering has very specific definitions for each. You should make a point of using the terms correctly, and quietly tolerate others when they used them incorrectly.

As an example, consider an oceanographer measuring water depth using a *sonar* system. Short bursts of sound are transmitted from the ship, reflected from the ocean floor, and received at the surface as an echo. Sound waves travel at a relatively constant velocity in water, allowing the depth to be found from the elapsed time between the transmitted and received pulses. As with all empirical measurements, a certain amount of error exists between the measured and true values. This particular measurement could be affected by many factors: random noise in the electronics, waves on the ocean surface, plant growth on the ocean floor, variations in the water temperature causing the sound velocity to change, etc.

To investigate these effects, the oceanographer takes many successive readings at a location known to be *exactly* 1000 meters deep (the true value). These measurements are then arranged as the histogram shown in Fig. 2-11. As would be expected from the Central Limit Theorem, the acquired data are normally distributed. The *mean* occurs at the center of the distribution, and represents the best estimate of the depth based on all of the measured data. The *standard deviation* defines the width of the distribution, describing how much variation occurs between successive measurements.

This situation results in two general types of error that the system can experience. First, the mean may be shifted from the true value. The amount of this shift is called the **accuracy** of the measurement. Second, individual measurements may not agree well with each other, as indicated by the width of the distribution. This is called the **precision** of the measurement, and is expressed by quoting the standard deviation, the signal-to-noise ratio, or the CV.

Consider a measurement that has good accuracy, but poor precision; the histogram is centered over the true value, but is very broad. Although the measurements are correct *as a group*, each individual reading is a poor measure of the true value. This situation is said to have poor *repeatability;* measurements taken in succession don't agree well. Poor precision results from **random errors**. This is the name given to errors that change each

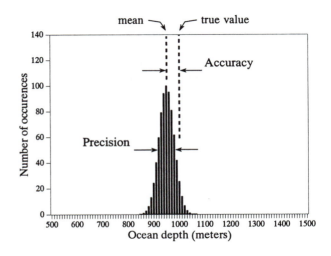

FIGURE 2-11
Definitions of accuracy and precision. Accuracy is the difference between the true value and the mean of the underlying process that generates the data. Precision is the spread of the values, specified by the standard deviation, the signal-to-noise ratio, or the CV.

time the measurement is repeated. Averaging several measurements will *always* improve the precision. In short, *precision is a measure of random noise.*

Now, imagine a measurement that is very precise, but has poor accuracy. This makes the histogram very slender, but not centered over the true value. Successive readings are close in value; however, they *all* have a large error. Poor accuracy results from **systematic errors.** These are errors that become repeated in exactly the same manner each time the measurement is conducted. Accuracy is usually dependent on how you *calibrate* the system. For example, in the ocean depth measurement, the parameter directly measured is elapsed time. This is converted into depth by a calibration procedure that relates *milliseconds* to *meters*. This may be as simple as multiplying by a fixed velocity, or as complicated as dozens of second order corrections. Averaging individual measurements does nothing to improve the accuracy. In short, *accuracy is a measure of calibration.*

In actual practice there are many ways that precision and accuracy can become intertwined. For example, imagine building an electronic amplifier from 1% resistors. This tolerance indicates that the value of each resistor will be within 1% of the stated value over a wide range of conditions, such as temperature, humidity, age, etc. This error in the resistance will produce a corresponding error in the gain of the amplifier. Is this error a problem of accuracy or precision?

The answer depends on how you take the measurements. For example, suppose you build *one* amplifier and test it several times over a few minutes. The error in gain remains constant with each test, and you conclude the problem is *accuracy*. In comparison, suppose you build *one thousand* of the amplifiers. The gain from device to device will fluctuate randomly, and the problem appears to be one of *precision*. Likewise, any one of these amplifiers will show gain fluctuations in response to temperature and other environmental changes. Again, the problem would be called *precision*.

When deciding which name to call the problem, ask yourself two questions. First: Will averaging successive readings provide a better measurement? If yes, call the error precision; if no, call it accuracy. Second: Will calibration correct the error? If yes, call it accuracy; if no, call it precision. This may require some thought, especially related to how the device will be calibrated, and how often it will be done.

ADC and DAC

Most of the signals directly encountered in science and engineering are *continuous*: light intensity that changes with distance; voltage that varies over time; a chemical reaction rate that depends on temperature, etc. Analog-to-Digital Conversion (ADC) and Digital-to-Analog Conversion (DAC) are the processes that allow digital computers to interact with these everyday signals. Digital information is different from its continuous counterpart in two important respects: it is *sampled*, and it is *quantized*. Both of these restrict how much information a digital signal can contain. This chapter is about *information management*: understanding what information you need to retain, and what information you can afford to lose. In turn, this dictates the selection of the sampling frequency, number of bits, and type of analog filtering needed for converting between the analog and digital realms.

Quantization

First, a bit of trivia. As you know, it is a *digital* computer, not a *digit* computer. The information processed is called *digital* data, not *digit* data. Why then, is analog-to-digital conversion generally called: *digitize and digitization*, rather than *digitalize and digitalization*? The answer is nothing you would expect. When electronics got around to inventing digital techniques, the preferred names had already been snatched up by the medical community nearly a century before. *Digitalize* and *digitalization* mean to administer the heart stimulant *digitalis*.

Figure 3-1 shows the electronic waveforms of a typical analog-to-digital conversion. Figure (a) is the analog signal to be digitized. As shown by the labels on the graph, this signal is a *voltage* that varies over *time*. To make the numbers easier, we will assume that the voltage can vary from 0 to 4.095 volts, corresponding to the digital numbers between 0 and 4095 that will be produced by a 12 bit digitizer. Notice that the block diagram is broken into two sections, the sample-and-hold (S/H), and the analog-to-digital converter (ADC). As you probably learned in electronics classes, the sample-and-hold is required to keep the voltage entering the ADC constant while the conversion is taking place. However, this is *not* the reason it is shown here; breaking the digitization into these two stages is an important theoretical model for understanding digitization. The fact that it happens to look like common electronics is just a fortunate bonus.

As shown by the difference between (a) and (b), the output of the sample-and-hold is allowed to change only at periodic intervals, at which time it is made identical to the instantaneous value of the input signal. Changes in the input signal that occur between these sampling times are completely ignored. That is, **sampling** converts the *independent variable* (time in this example) from continuous to discrete.

As shown by the difference between (b) and (c), the ADC produces an integer value between 0 and 4095 for each of the flat regions in (b). This introduces an error, since each plateau can be *any* voltage between 0 and 4.095 volts. For example, both 2.56000 volts and 2.56001 volts will be converted into digital number 2560. In other words, **quantization** converts the *dependent variable* (voltage in this example) from continuous to discrete.

Notice that we carefully avoid comparing (a) and (c), as this would lump the sampling and quantization together. It is important that we analyze them separately because they degrade the signal in different ways, as well as being controlled by different parameters in the electronics. There are also cases where one is used without the other. For instance, sampling without quantization is used in switched capacitor filters.

First we will look at the effects of quantization. Any one sample in the digitized signal can have a maximum error of ±½ **LSB** (**Least Significant Bit**, jargon for the distance between adjacent quantization levels). Figure (d) shows the quantization error for this particular example, found by subtracting (b) from (c), with the appropriate conversions. In other words, the digital output (c), is equivalent to the continuous input (b), *plus* a quantization error (d). An important feature of this analysis is that the quantization error appears very much like *random noise*.

This sets the stage for an important model of quantization error. In most cases, *quantization results in nothing more than the addition of a specific amount of random noise to the signal*. The additive noise is uniformly distributed between ±½ LSB, has a mean of zero, and a standard deviation of $1/\sqrt{12}$ LSB (~0.29 LSB). For example, passing an analog signal through an 8 bit digitizer adds an rms noise of: 0.29/256, or about 1/900 of the full scale value. A 12 bit conversion adds a noise of: $0.29/4096 \approx 1/14,000$, while a 16 bit conversion adds: $0.29/65536 \approx 1/227,000$. Since quantization error is a random noise, the *number of bits* determines the *precision* of the data. For example, you might make the statement: "We increased the precision of the measurement from 8 to 12 bits."

This model is extremely powerful, because the random noise generated by quantization will simply add to whatever noise is already present in the

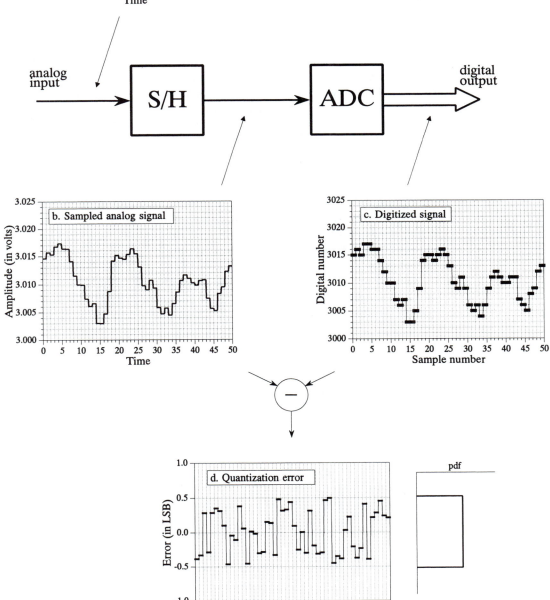

FIGURE 3-1
Waveforms illustrating the digitization process. The conversion is broken into two stages to allow the effects of *sampling* to be separated from the effects of *quantization*. The first stage is the sample-and-hold (S/H), where the only information retained is the instantaneous value of the signal when the periodic sampling takes place. In the second stage, the ADC converts the voltage to the nearest integer number. This results in each sample in the digitized signal having an error of up to ±½ LSB, as shown in (d). As a result, quantization can usually be modeled as simply adding noise to the signal.

analog signal. For example, imagine an analog signal with a maximum amplitude of 1.0 volts, and a random noise of 1.0 millivolts rms. Digitizing this signal to 8 bits results in 1.0 volts becoming digital number 255, and 1.0 millivolts becoming 0.255 LSB. As discussed in the last chapter, random noise signals are combined by adding their *variances*. That is, the signals are added in quadrature: $\sqrt{A^2 + B^2} = C$. The total noise on the digitized signal is therefore given by: $\sqrt{0.255^2 + 0.29^2} = 0.386$ LSB. This is an increase of about 50% over the noise already in the analog signal. Digitizing this same signal to 12 bits would produce virtually no increase in the noise, and *nothing* would be lost due to quantization. When faced with the decision of how many bits are needed in a system, ask two questions: (1) How much noise is *already* present in the analog signal? (2) How much noise can be *tolerated* in the digital signal?

When isn't this model of quantization valid? Only when the quantization error cannot be treated as random. The only common occurrence of this is when the analog signal remains at about the same value for many consecutive samples, as is illustrated in Fig. 3-2a. The output remains *stuck* on the same digital number for many samples in a row, even though the analog signal may be changing up to ±½ LSB. Instead of being an additive random noise, the quantization error now looks like a thresholding effect or weird distortion.

Dithering is a common technique for improving the digitization of these slowly varying signals. As shown in Fig. 3-2b, a small amount of random noise is added to the analog signal. In this example, the added noise is normally distributed with a standard deviation of 2/3 LSB, resulting in a peak-to-peak amplitude of about 3 LSB. Figure (c) shows how the addition of this dithering noise has affected the digitized signal. Even when the original analog signal is changing by less than ±½ LSB, the added noise causes the digital output to randomly toggle between adjacent levels.

To understand how this improves the situation, imagine that the input signal is a constant analog voltage of 3.0001 volts, making it one-tenth of the way between the digital levels 3000 and 3001. Without dithering, taking 10,000 samples of this signal would produce 10,000 identical numbers, all having the value of 3000. Next, repeat the thought experiment with a small amount of dithering noise added. The 10,000 values will now oscillate between two (or more) levels, with about 90% having a value of 3000, and 10% having a value of 3001. Taking the average of all 10,000 values results in something close to 3000.1. Even though a single measurement has the inherent ±½ LSB limitation, the statistics of a large number of the samples can do much better. This is quite a strange situation: *adding noise provides more information*.

Circuits for dithering can be quite sophisticated, such as using a computer to generate random numbers, and then passing them through a DAC to produce the added noise. After digitization, the computer can *subtract*

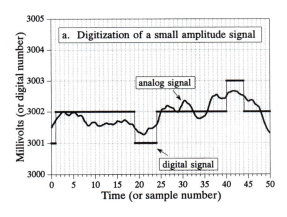

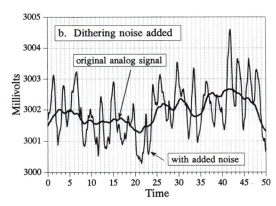

FIGURE 3-2
Illustration of dithering. Figure (a) shows how an analog signal that varies less than ±½ LSB can become *stuck* on the same quantization level during digitization. Dithering improves this situation by adding a small amount of random noise to the analog signal, such as shown in (b). In this example, the added noise is normally distributed with a standard deviation of 2/3 LSB. As shown in (c), the added noise causes the digitized signal to toggle between adjacent quantization levels, providing more information about the original signal.

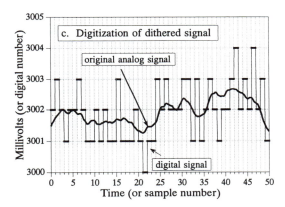

the random numbers from the digital signal using floating point arithmetic. This elegant technique is called **subtractive dither**, but is only used in the most elaborate systems. The simplest method, although not always possible, is to use the noise already present in the analog signal for dithering.

The Sampling Theorem

The definition of *proper sampling* is quite simple. Suppose you sample a continuous signal in some manner. If you can exactly *reconstruct* the analog signal from the samples, you must have done the sampling *properly*. Even if the sampled data appears confusing or incomplete, the key information has been captured if you can reverse the process.

Figure 3-3 shows several sinusoids before and after digitization. The continious line represents the analog signal entering the ADC, while the square markers are the digital signal leaving the ADC. In (a), the analog signal is a constant DC value, a cosine wave of *zero* frequency. Since the analog signal is a series of straight lines between each of the samples, all of the information needed to reconstruct the analog signal is contained in the digital data. According to our definition, this is *proper sampling*.

The sine wave shown in (b) has a frequency of 0.09 of the sampling rate. This might represent, for example, a 90 cycle/second sine wave being sampled at 1000 samples/second. Expressed in another way, there are 11.1 samples taken over each complete cycle of the sinusoid. This situation is more complicated than the previous case, because the analog signal cannot be reconstructed by simply drawing straight lines between the data points. Do these samples properly represent the analog signal? The answer is yes, because no other sinusoid, or combination of sinusoids, will produce this pattern of samples (within the reasonable constraints listed below). These samples correspond to only one analog signal, and therefore the analog signal can be exactly reconstructed. Again, an instance of *proper sampling*.

In (c), the situation is made more difficult by increasing the sine wave's frequency to 0.31 of the sampling rate. This results in only 3.2 samples per sine wave cycle. Here the samples are so sparse that they don't even appear to follow the general trend of the analog signal. Do these samples properly represent the analog waveform? Again, the answer is yes, and for exactly the same reason. The samples are a unique representation of the analog signal. All of the information needed to reconstruct the continuous waveform is contained in the digital data. How you go about doing this will be discussed later in this chapter. Obviously, it must be more sophisticated than just drawing straight lines between the data points. As strange as it seems, this is *proper sampling* according to our definition.

In (d), the analog frequency is pushed even higher to 0.95 of the sampling rate, with a mere 1.05 samples per sine wave cycle. Do these samples properly represent the data? *No, they don't!* The samples represent a *different* sine wave from the one contained in the analog signal. In particular, the original sine wave of 0.95 frequency misrepresents itself as a sine wave of 0.05 frequency in the digital signal. This phenomenon of sinusoids changing frequency during sampling is called **aliasing**. Just as a criminal might take on an assumed name or identity (an *alias*), the sinusoid assumes another frequency that is not its own. Since the digital data is no longer uniquely related to a particular analog signal, an unambiguous reconstruction is impossible. There is nothing in the sampled data to suggest that the original analog signal had a frequency of 0.95 rather than 0.05. The sine wave has hidden its true identity completely; the perfect crime has been committed! According to our definition, this is an example of *improper sampling*.

This line of reasoning leads to a milestone in DSP, the **sampling theorem**. Frequently this is called the *Shannon* sampling theorem, or the *Nyquist* sampling theorem, after the authors of 1940s papers on the topic. The sampling theorem indicates that a continuous signal can be properly sampled, *only if it does not contain frequency components above one-half of the sampling rate*. For instance, a sampling rate of 2,000 samples/second requires the analog signal to be composed of frequencies below 1000 cycles/second. If frequencies above this limit *are* present in the signal, they will be aliased to frequencies between 0 and 1000 cycles/second, combining with whatever information that was legitimately there.

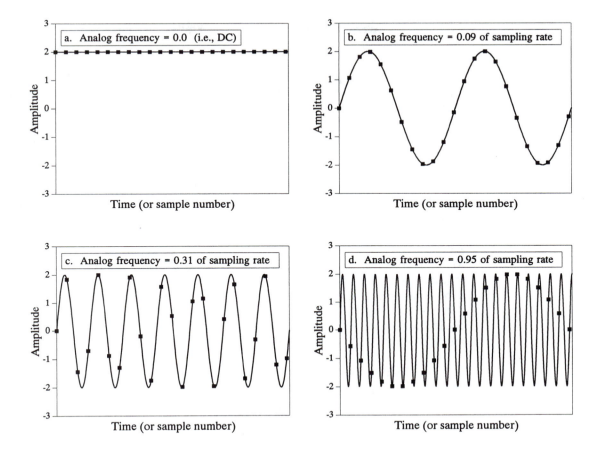

FIGURE 3-3
Illustration of proper and improper sampling. A continuous signal is sampled *properly* if the samples contain all the information needed to recreated the original waveform. Figures (a), (b), and (c) illustrate *proper sampling* of three sinusoidal waves. This is certainly not obvious, since the samples in (c) do not even appear to capture the shape of the waveform. Nevertheless, each of these continuous signals forms a unique one-to-one pair with its pattern of samples. This guarantees that reconstruction can take place. In (d), the frequency of the analog sine wave is greater than the Nyquist frequency (one-half of the sampling rate). This results in *aliasing*, where the frequency of the sampled data is different from the frequency of the continuous signal. Since aliasing has corrupted the information, the original signal cannot be reconstructed from the samples.

Two terms are widely used when discussing the sampling theorem: the **Nyquist frequency** and the **Nyquist rate**. Unfortunately, their meaning is not standardized. To understand this, consider an analog signal composed of frequencies between DC and 3 kHz. To properly digitize this signal it must be sampled at 6,000 samples/sec (6 kHz) or higher. Suppose we choose to sample at 8,000 samples/sec (8 kHz), allowing frequencies between DC and 4 kHz to be properly represented. In this situation their are four important frequencies: (1) the highest frequency in the signal, 3 kHz; (2) twice this frequency, 6 kHz; (3) the sampling rate, 8 kHz; and (4) one-half the sampling rate, 4 kHz. Which of these four is the *Nyquist frequency* and which is the *Nyquist rate*? It depends who you ask! All of the possible combinations are used. Fortunately, most authors are careful to define how they are using the terms. In this book, they are both used to mean *one-half the sampling rate*.

Figure 3-4 shows how frequencies are changed during aliasing. The key point to remember is that a digital signal *cannot* contain frequencies above one-half the sampling rate (i.e., the Nyquist frequency/rate). When the frequency of the continuous wave is below the Nyquist rate, the frequency of the sampled data is a match. However, when the continuous signal's frequency is above the Nyquist rate, aliasing *changes* the frequency into something that *can* be represented in the sampled data. As shown by the zigzagging line in Fig. 3-4, every continuous frequency above the Nyquist rate has a corresponding digital frequency between zero and one-half the sampling rate. It there happens to be a sinusoid already at this lower frequency, the aliased signal will add to it, resulting in a loss of information. Aliasing is a double curse; information can be lost about the higher *and* the lower frequency. Suppose you are given a digital signal containing a frequency of 0.2 of the sampling rate. If this signal were obtained by proper sampling, the original analog signal *must* have had a frequency of 0.2. If aliasing took place during sampling, the digital frequency of 0.2 could have come from any one of an infinite number of frequencies in the analog signal: 0.2, 0.8, 1.2, 1.8, 2.2, ⋯ .

Just as aliasing can change the frequency during sampling, it can also change the *phase*. For example, look back at the aliased signal in Fig. 3-3d. The aliased digital signal is *inverted* from the original analog signal; one is a sine wave while the other is a negative sine wave. In other words, aliasing has changed the frequency *and* introduced a 180° phase shift. Only two phase shifts are possible: 0° (no phase shift) and 180° (inversion). The zero phase shift occurs for analog frequencies of 0 to 0.5, 1.0 to 1.5, 2.0 to 2.5, etc. An inverted phase occurs for analog frequencies of 0.5 to 1.0, 1.5 to 2.0, 3.5 to 4.0, and so on.

Now we will dive into a more detailed analysis of sampling and how aliasing occurs. Our overall goal is to understand what happens to the information when a signal is converted from a continuous to a discrete form. The problem is, these are very different things; one is a *continuous waveform* while the other is an *array of numbers*. This "apples-to-oranges" comparison makes the analysis very difficult. The solution is to introduce a theoretical concept called the **impulse train**.

Figure 3-5a shows an example analog signal. Figure (c) shows the signal sampled by using an *impulse train*. The impulse train is a continuous signal consisting of a series of narrow spikes (impulses) that match the original signal at the sampling instants. Each impulse is infinitesimally narrow, a concept that will be discussed in Chapter 13. Between these sampling times the value of the waveform is zero. Keep in mind that the impulse train is a *theoretical* concept, not a waveform that can exist in an electronic circuit. Since both the original analog signal and the impulse train are continuous waveforms, we can make an "apples-apples" comparison between the two.

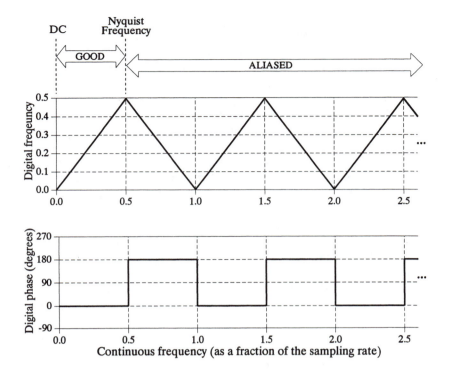

FIGURE 3-4
Conversion of analog frequency into digital frequency during sampling. Continuous signals with a frequency less than one-half of the sampling rate are directly converted into the corresponding digital frequency. Above one-half of the sampling rate, aliasing takes place, resulting in the frequency being misrepresented in the digital data. Aliasing always changes a higher frequency into a lower frequency between 0 and 0.5. In addition, aliasing may also change the phase of the signal by 180 degrees.

Now we need to examine the relationship between the impulse train and the discrete signal (an array of numbers). This one is easy; in terms of *information content*, they are *identical*. If one is known, it is trivial to calculate the other. Think of these as different ends of a bridge crossing between the analog and digital worlds. This means we have achieved our overall goal once we understand the consequences of changing the waveform in (a) into the waveform in (c).

Three continuous waveforms are shown in the left-hand column in Fig. 3-5. The corresponding *frequency spectra* of these signals are displayed in the right-hand column. This should be a familiar concept from you knowledge of electronics; every waveform can be viewed as being composed of sinusoids of varying amplitude and frequency. Later chapters will discuss the frequency domain in detail. (You may want to revisit this discussion after becoming more familiar with frequency spectra).

Figure (a) shows an analog signal we wish to sample. As indicated by its frequency spectrum in (b), it is composed only of frequency components between 0 and about 0.33 f_s, where f_s is the sampling frequency we intend to use. For example, this might be a speech signal that has been filtered to remove all frequencies above 3.3 kHz. Correspondingly, f_s would be 10 kHz (10,000 samples/second), our intended sampling rate.

Sampling the signal in (a) by using an impulse train produces the signal shown in (c), and its frequency spectrum shown in (d). This spectrum is a *duplication* of the spectrum of the original signal. Each multiple of the sampling frequency, f_s, $2f_s$, $3f_s$, $4f_s$, etc., has received a *copy* and a *left-for-right flipped copy* of the original frequency spectrum. The copy is called the **upper sideband**, while the flipped copy is called the **lower sideband**. Sampling has generated *new* frequencies. Is this proper sampling? The answer is yes, because the signal in (c) can be transformed back into the signal in (a) by eliminating all frequencies above ½f_s. That is, an analog low-pass filter will convert the impulse train, (b), back into the original analog signal, (a).

If you are already familiar with the basics of DSP, here is a more technical explanation of why this spectral duplication occurs. (Ignore this paragraph if you are new to DSP). In the time domain, sampling is achieved by multiplying the original signal by an impulse train of *unity amplitude* spikes. The frequency spectrum of this unity amplitude impulse train is also a unity amplitude impulse train, with the spikes occurring at multiples of the sampling frequency, f_s, $2f_s$, $3f_s$, $4f_s$, etc. When two time domain signals are multiplied, their frequency spectra are convolved. This results in the original spectrum being duplicated to the location of each spike in the impulse train's spectrum. Viewing the original signal as composed of both positive and negative frequencies accounts for the upper and lower sidebands, respectively. This is the same as amplitude modulation, discussed in Chapter 10.

Figure (e) shows an example of *improper sampling*, resulting from too low of sampling rate. The analog signal still contains frequencies up to 3.3 kHz, but the sampling rate has been lowered to 5 kHz. Notice that f_s, $2f_s$, $3f_s$ ··· along the horizontal axis are spaced closer in (f) than in (d). The frequency spectrum, (f), shows the problem: the duplicated portions of the spectrum have invaded the band between zero and one-half of the sampling frequency. Although (f) shows these overlapping frequencies as retaining their separate identity, in actual practice they add together forming a single confused mess. Since there is no way to separate the overlapping frequencies, information is lost, and the original signal cannot be reconstructed. This overlap occurs when the analog signal contains frequencies greater than one-half the sampling rate, that is, we have proven the sampling theorem.

Digital-to-Analog Conversion

In theory, the simplest method for digital-to-analog conversion is to pull the samples from memory and convert them into an *impulse train*. This is

Time Domain

Frequency Domain

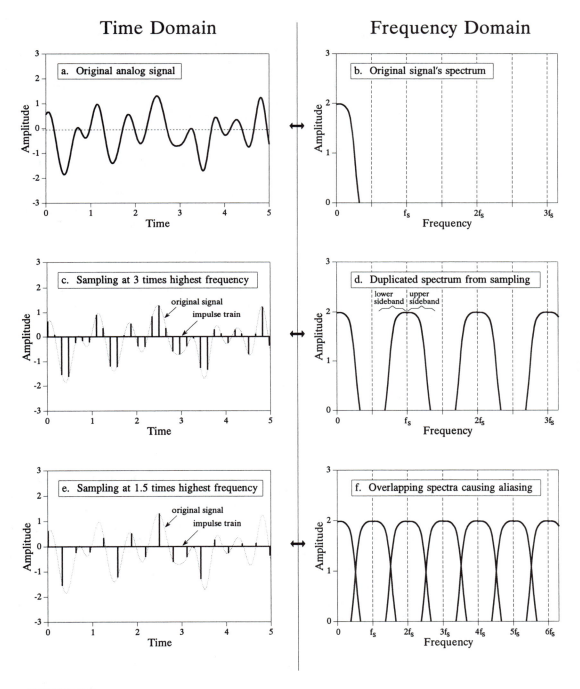

FIGURE 3-5
The sampling theorem in the time and frequency domains. Figures (a) and (b) show an analog signal composed of frequency components between zero and 0.33 of the sampling frequency, f_s. In (c), the analog signal is sampled by converting it to an impulse train. In the frequency domain, (d), this results in the spectrum being duplicated into an infinite number of upper and lower sidebands. Since the original frequencies in (b) exist undistorted in (d), proper sampling has taken place. In comparison, the analog signal in (e) is sampled at 0.66 of the sampling frequency, a value exceeding the Nyquist rate. This results in aliasing, indicated by the sidebands in (f) overlapping.

illustrated in Fig. 3-6a, with the corresponding frequency spectrum in (b). As just described, the original analog signal can be perfectly reconstructed by passing this impulse train through a low-pass filter, with the cutoff frequency equal to one-half of the sampling rate. In other words, the original signal and the impulse train have identical frequency spectra below the Nyquist frequency (one-half the sampling rate). At higher frequencies, the impulse train contains a duplication of this information, while the original analog signal contains nothing (assuming aliasing did not occur).

While this method is mathematically pure, it is difficult to generate the required narrow pulses in electronics. To get around this, nearly all DACs operate by holding the last value until another sample is received. This is called a **zeroth-order hold**, the DAC equivalent of the sample-and-hold used during ADC. (A first-order hold is straight lines between the points, a second-order hold uses parabolas, etc.). The zeroth-order hold produces the staircase appearance shown in (c).

In the frequency domain, the zeroth-order hold results in the spectrum of the impulse train being *multiplied* by the dark curve shown in (d), given by the equation:

EQUATION 3-1
High frequency amplitude reduction due to the zeroth-order hold. This curve is plotted in Fig. 3-6d. The sampling frequency is represented by f_s. For $f = 0$, $H(f) = 1$.

$$H(f) = \left| \frac{\sin(\pi f / f_s)}{\pi f / f_s} \right|$$

This is of the general form: $\sin(\pi x)/(\pi x)$, called the **sinc function** or **sinc(x)**. The sinc function is very common in DSP, and will be discussed in more detail in later chapters. If you already have a background in this material, the zeroth-order hold can be understood as the convolution of the impulse train with a rectangular pulse, having a width equal to the sampling period. This results in the frequency domain being *multiplied* by the Fourier transform of the rectangular pulse, i.e., the sinc function. In Fig. (d), the light line shows the frequency spectrum of the impulse train (the "correct" spectrum), while the dark line shown the sinc. The frequency spectrum of the zeroth order hold signal is equal to the product of these two curves.

The analog filter used to convert the zeroth-order hold signal, (c), into the reconstructed signal, (f), needs to do two things: (1) remove all frequencies above one-half of the sampling rate, and (2) boost the frequencies by the reciprocal of the zeroth-order hold's effect, i.e., *1/sinc(x)*. This amounts to an amplification of about 36% at one-half of the sampling frequency. Figure (e) shows the ideal frequency response of this analog filter.

This *1/sinc(x)* frequency boost can be handled in four ways: (1) ignore it and accept the consequences, (2) design an analog filter to include the *1/sinc(x)*

Time Domain

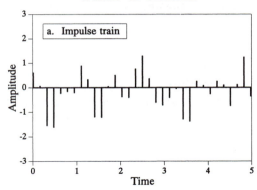

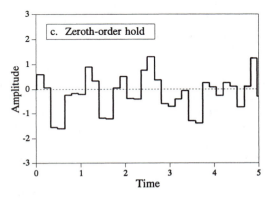

FIGURE 3-6
Analysis of digital-to-analog conversion. In (a), the
digital data are converted into an impulse train, with the
spectrum in (b). This is changed into the reconstructed
signal, (f), by using an electronic low-pass filter to
remove frequencies above one-half the sampling rate
[compare (b) and (g)]. However, most electronic DACs
create a zeroth-order hold waveform, (c), instead of an
impulse train. The spectrum of the zeroth-order hold is
equal to the spectrum of the impulse train multiplied by
the sinc function shown in (d). To convert the zeroth-
order hold into the reconstructed signal, the analog filter
must remove all frequencies above the Nyquist rate, *and*
correct for the sinc, as shown in (e).

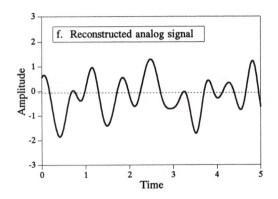

Frequency Domain

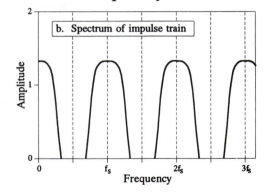

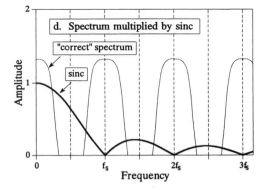

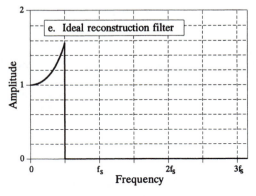

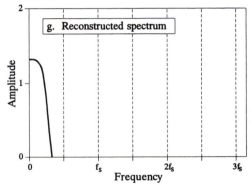

response, (3) use a fancy *multirate* technique described later in this chapter, or (4) make the correction in software before the DAC (see Chapter 24).

Before leaving this section on sampling, we need to dispel a common myth about analog versus digital signals. As this chapter has shown, the amount of information carried in a digital signal is limited in two ways: First, the number of bits per sample limits the resolution of the *dependent* variable. That is, small changes in the signal's amplitude may be lost in the quantization noise. Second, the sampling rate limits the resolution of the *independent variable*, i.e., closely spaced events in the analog signal may be lost between the samples. This is another way of saying that frequencies above one-half the sampling rate are lost.

Here is the myth: "Since analog signals use continuous parameters, they have infinitely good resolution in both the independent and the dependent variables." Not true! Analog signals are limited by the same two problems as digital signals: *noise* and *bandwidth* (the highest frequency allowed in the signal). The noise in an analog signal limits the measurement of the waveform's amplitude, just as quantization noise does in a digital signal. Likewise, the ability to separate closely spaced events in an analog signal depends on the highest frequency allowed in the waveform. To understand this, imagine an analog signal containing two closely spaced pulses. If we place the signal through a low-pass filter (removing the high frequencies), the pulses will blur into a single blob. For instance, an analog signal formed from frequencies between DC and 10 kHz will have *exactly* the same resolution as a digital signal sampled at 20 kHz. It must, since the sampling theorem guarantees that the two contain the same information.

Analog Filters for Data Conversion

Figure 3-7 shows a block diagram of a DSP system, *as the sampling theorem dictates it should be.* Before encountering the analog-to-digital converter,

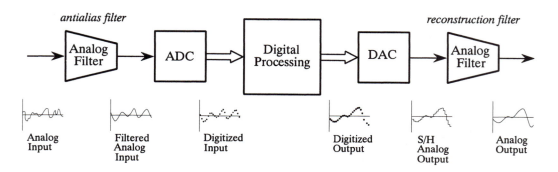

FIGURE 3-7

Analog electronic filters used to comply with the sampling theorem. The electronic filter placed before an ADC is called an *antialias filter.* It is used to remove frequency components above one-half of the sampling rate that would alias during the sampling. The electronic filter placed after a DAC is called a *reconstruction filter.* It also eliminates frequencies above the Nyquist rate, and may include a correction for the zeroth-order hold.

the input signal is processed with an electronic low-pass filter to remove all frequencies above the Nyquist frequency (one-half the sampling rate). This is done to prevent aliasing during sampling, and is correspondingly called **an antialias filter**. On the other end, the digitized signal is passed through a digital-to-analog converter and another low-pass filter set to the Nyquist frequency. This output filter is called a **reconstruction filter**, and may include the previously described frequency boost. Unfortunately, there is a serious problem with this simple model: the limitations of electronic filters can be as bad as the problems they are trying to prevent.

If your main interest is in software, you are probably thinking that you don't need to read this section. *Wrong!* Even if you have vowed never to touch an oscilloscope, an understanding of the properties of analog filters is important for successful DSP. First, the characteristics of every digitized signal you encounter will depend on what type of antialias filter was used when it was acquired. If you don't understand the nature of the antialias filter, you cannot understand the nature of the digital signal. Second, the future of DSP is to replace *hardware* with *software*. For example, the *multirate* techniques presented later in this chapter reduce the need for antialias and reconstruction filters by fancy software tricks. If you don't understand the hardware, you cannot design software to replace it. Third, much of DSP is related to digital filter design. A common strategy is to start with an equivalent *analog filter*, and convert it into software. Later chapters assume you have a basic knowledge of analog filter techniques.

Three types of analog filters are commonly used: **Chebyshev**, **Butterworth**, and **Bessel** (also called a Thompson filter). Each of these is designed to optimize a different performance parameter. The complexity of each filter can be adjusted by selecting the number of **poles**, a mathematical term that will be discussed in later chapters. The more poles in a filter, the more electronics it requires, and the better it performs. Each of these names describe what the filter *does*, not a particular arrangement of resistors and capacitors. For example, a six pole Bessel filter can be implemented by many different types of circuits, all of which have the same overall characteristics. For DSP purposes, the characteristics of these filters are more important than how they are constructed. Nevertheless, we will start with a short segment on the electronic design of these filters to provide an overall framework.

Figure 3-8 shows a common building block for analog filter design, the modified Sallen-Key circuit. This is named after the authors of a 1950s paper describing the technique. The circuit shown is a two pole low-pass filter that can be configured as any of the three basic types. Table 3-1 provides the necessary information to select the appropriate resistors and capacitors. For example, to design a 1 kHz, 2 pole Butterworth filter, Table 3-1 provides the parameters: $k_1 = 0.1592$ and $k_2 = 0.586$. Arbitrarily selecting $R_1 = 10K$ and $C = 0.01uF$ (common values for op amp circuits), R and R_f can be calculated as 15.95K and 5.86K, respectively. Rounding these last two values to the nearest 1% standard resistors, results in R = 15.8K and $R_f = 5.90K$ All of the components should be 1% precision or

FIGURE 3-8
The modified Sallen-Key circuit, a building block for active filter design. The circuit shown implements a 2 pole low-pass filter. Higher order filters (more poles) can be formed by cascading stages. Find k_1 and k_2 from Table 3-1, arbitrarily select R_1 and C (try 10K and 0.01μF), and then calculate R and R_f from the equations in the figure. The parameter, f_c, is the cutoff frequency of the filter, in hertz.

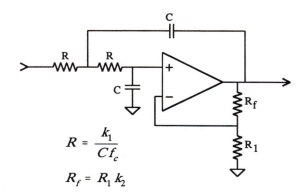

$$R = \frac{k_1}{Cf_c}$$

$$R_f = R_1\,k_2$$

TABLE 3-1
Parameters for designing Bessel, Butterworth, and Chebyshev (6% ripple) filters.

# poles		Bessel k_1	Bessel k_2	Butterworth k_1	Butterworth k_2	Chebyshev k_1	Chebyshev k_2
2	stage 1	0.1251	0.268	0.1592	0.586	0.1293	0.842
4	stage 1	0.1111	0.084	0.1592	0.152	0.2666	0.582
	stage 2	0.0991	0.759	0.1592	1.235	0.1544	1.660
6	stage 1	0.0990	0.040	0.1592	0.068	0.4019	0.537
	stage 2	0.0941	0.364	0.1592	0.586	0.2072	1.448
	stage 3	0.0834	1.023	0.1592	1.483	0.1574	1.846
8	stage 1	0.0894	0.024	0.1592	0.038	0.5359	0.522
	stage 2	0.0867	0.213	0.1592	0.337	0.2657	1.379
	stage 3	0.0814	0.593	0.1592	0.889	0.1848	1.711
	stage 4	0.0726	1.184	0.1592	1.610	0.1582	1.913

better. The particular op amp use isn't critical, as long as the unity gain frequency is more than 30 to 100 times higher than the filter's cutoff frequency. This is an easy requirement as long as the filter's cutoff frequency is below about 100 kHz.

Four, six, and eight pole filters are formed by cascading 2,3, and 4 of these circuits, respectively. For example, Fig. 3-9 shows the schematic of a 6 pole

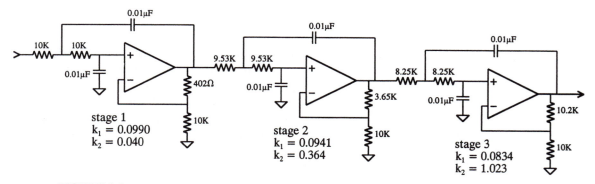

FIGURE 3-9
A six pole Bessel filter formed by cascading three Sallen-Key circuits. This is a low-pass filter with a cutoff frequency of 1 kHz.

Bessel filter created by cascading three stages. Each stage has different values for k_1 and k_2 as provided by Table 3-1, resulting in different resistors and capacitors being used. Need a high-pass filter? Simply swap the R and C components in the circuits (leaving R_f and R_1 alone).

This type of circuit is very common for small quantity manufacturing and R&D applications; however, serious production requires the filter to be made as an *integrated circuit*. The problem is, it is difficult to make resistors directly in silicon. The answer is the **switched capacitor filter**. Figure 3-10 illustrates its operation by comparing it to a simple RC network. If a step function is fed into an RC low-pass filter, the output rises exponentially until it matches the input. The voltage on the capacitor doesn't change instantaneously, because the resistor restricts the flow of electrical charge.

The switched capacitor filter operates by replacing the basic resistor-capacitor network with two capacitors and an electronic switch. The newly added capacitor is much smaller in value than the already existing capacitor, say, 1% of its value. The switch alternately connects the small capacitor between the input and the output at a very high frequency, typically 100 times faster than the cutoff frequency of the filter. When the switch is connected to the input, the small capacitor rapidly charges to whatever voltage is presently on the input. When the switch is connected to the output, the charge on the small capacitor is transferred to the large capacitor. In a resistor, the rate of charge transfer is determined by its resistance. In a switched capacitor circuit, the rate of charge transfer is determined by the value of the small capacitor *and* by the switching frequency. This results in a very useful feature of switched capacitor

Resistor-Capacitor

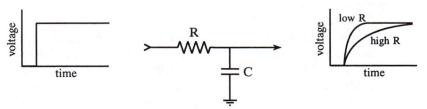

Switched Capacitor

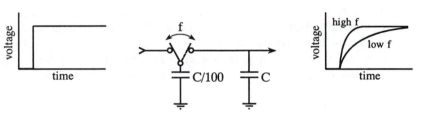

FIGURE 3-10
Switched capacitor filter operation. Switched capacitor filters use switches and capacitors to mimic resistors. As shown by the equivalent step responses, two capacitors and one switch can perform the same function as a resistor-capacitor network.

filters: *the cutoff frequency of the filter is directly proportional to the clock frequency used to drive the switches*. This makes the switched capacitor filter ideal for data acquisition systems that operate with more than one sampling rate. These are easy-to-use devices; pay ten bucks and have the performance of an eight pole filter inside a single 8 pin IC.

Now for the important part: the characteristics of the three classic filter types. The first performance parameter we want to explore is **cutoff frequency sharpness**. A low-pass filter is designed to block all frequencies above the cutoff frequency (the **stopband**), while passing all frequencies below (the **passband**). Figure 3-11 shows the frequency response of these three filters on a logarithmic (dB) scale. These graphs are shown for filters with a one hertz cutoff frequency, but they can be directly scaled to whatever cutoff frequency you need to use. How do these filters rate? The Chebyshev is clearly the best, the Butterworth is worse, and the Bessel is absolutely ghastly! As you probably surmised, this is what the Chebyshev is designed to do, **roll-off** (drop in amplitude) as rapidly as possible.

Unfortunately, even an 8 pole Chebyshev isn't as good as you would like for an antialias filter. For example, imagine a 12 bit system sampling at 10,000 samples per second. The sampling theorem dictates that any frequency above 5 kHz will be aliased, something you want to avoid. With a little guess work, you decide that all frequencies above 5 kHz must be reduced in amplitude by a factor of 100, insuring that any aliased frequencies will have an amplitude of less than one percent. Looking at Fig. 3-11c, you find that an 8 pole Chebyshev filter, with a cutoff frequency of 1 hertz, doesn't reach an attenuation (signal reduction) of 100 until about 1.35 hertz. Scaling this to the example, the filter's cutoff frequency must be set to 3.7 kHz so that everything above 5 kHz will have the required attenuation. This results in the frequency band between 3.7 kHz and 5 kHz being wasted on the inadequate roll-off of the analog filter.

A subtle point: the attenuation factor of 100 in this example is probably sufficient even though there 4096 steps in 12 bit. From Fig. 3-4, 5100 hertz will alias to 4900 hertz, 6000 hertz will alias to 4000 hertz, etc. You don't care what the amplitudes of the signals between 5000 and 6300 hertz are, because they alias into the unusable region between 3700 hertz and 5000 hertz. In order for a frequency to alias into the filter's passband (0 to 3.7 kHz), it must be greater than 6300 hertz, or 1.7 times the filter's cutoff frequency of 3700 hertz. As shown in Fig. 3-11c, the attenuation provided by an 8 pole Chebyshev filter at 1.7 times the cutoff frequency is about 1300, much more adequate than the 100 we started the analysis with. The moral to this story: *In most systems, the frequency band between about 0.4 and 0.5 of the sampling frequency is an unusable wasteland of filter roll-off and aliased signals*. This is a direct result of the limitations of analog filters.

The frequency response of the perfect low-pass filter is *flat* across the entire passband. All of the filters look great in this respect in Fig. 3-11, but only because the vertical axis is displayed on a *logarithmic* scale. Another story is told when the graphs are converted to a *linear* vertical scale, as is shown

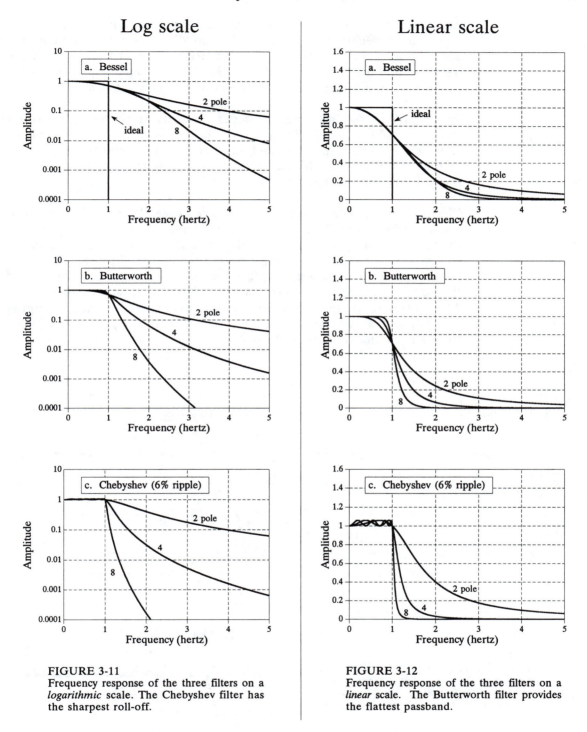

Log scale

a. Bessel

ideal
2 pole
4
8

Amplitude
Frequency (hertz)

b. Butterworth

2 pole
4
8

Amplitude
Frequency (hertz)

c. Chebyshev (6% ripple)

2 pole
4
8

Amplitude
Frequency (hertz)

FIGURE 3-11
Frequency response of the three filters on a *logarithmic* scale. The Chebyshev filter has the sharpest roll-off.

Linear scale

a. Bessel

ideal
2 pole
4
8

Amplitude
Frequency (hertz)

b. Butterworth

2 pole
4
8

Amplitude
Frequency (hertz)

c. Chebyshev (6% ripple)

2 pole
4
8

Amplitude
Frequency (hertz)

FIGURE 3-12
Frequency response of the three filters on a *linear* scale. The Butterworth filter provides the flattest passband.

in Fig. 3-12. **Passband ripple** can now be seen in the Chebyshev filter (wavy variations in the amplitude of the passed frequencies). In fact, the Chebyshev filter obtains its excellent roll-off by *allowing* this passband ripple. When more passband ripple is allowed in a filter, a faster roll-off can be achieved. All the Chebyshev filters designed by using Table 3-1 have a passband ripple of about 6% (0.5 dB), a good compromise, and a common choice. A similar design, the **elliptic filter**, allows ripple in both the passband *and* the stopband. Although harder to design, elliptic filters can achieve an even better tradeoff between roll-off and passband ripple.

In comparison, the Butterworth filter is optimized to provide the sharpest roll-off possible *without* allowing ripple in the passband. It is commonly called the *maximally flat filter*, and is identical to a Chebyshev designed for zero passband ripple. The Bessel filter has no ripple in the passband, but the roll-off far worse than the Butterworth.

The last parameter to evaluate is the **step response,** how the filter responds when the input rapidly changes from one value to another. Figure 3-13 shows the step response of each of the three filters. The horizontal axis is shown for filters with a 1 hertz cutoff frequency, but can be scaled (inversely) for higher cutoff frequencies. For example, a 1000 hertz cutoff frequency would show a step response in *milliseconds*, rather than *seconds*. The Butterworth and Chebyshev filters **overshoot** and show **ringing** (oscillations that slowly decreasing in amplitude). In comparison, the Bessel filter has neither of these nasty problems.

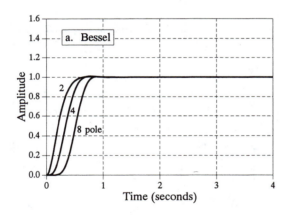

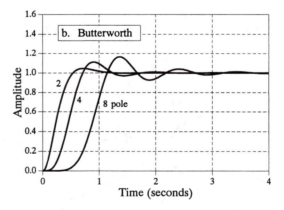

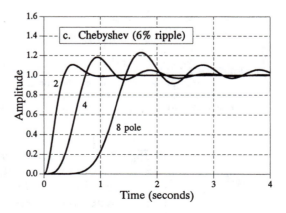

FIGURE 3-13
Step response of the three filters. The times shown on the horizontal axis correspond to a one hertz cutoff frequency. The Bessel is the optimum filter when overshoot and ringing must be minimized.

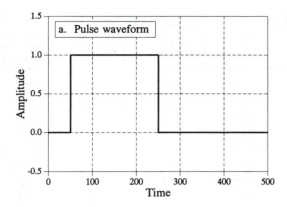

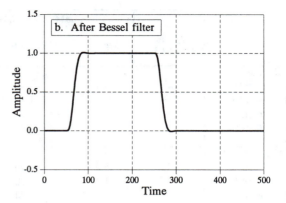

FIGURE 3-14
Pulse response of the Bessel and Chebyshev filters. A key property of the Bessel filter is that the rising and falling edges in the filter's output looking similar. In the jargon of the field, this is called *linear phase*. Figure (b) shows the result of passing the pulse waveform in (a) through a 4 pole Bessel filter. Both edges are smoothed in a similar manner. Figure (c) shows the result of passing (a) through a 4 pole Chebyshev filter. The left edge overshoots on the *top*, while the right edge overshoots on the *bottom*. Many applications cannot tolerate this distortion.

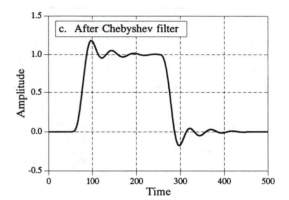

Figure 3-14 further illustrates this very favorable characteristic of the Bessel filter. Figure (a) shows a pulse waveform, which can be viewed as a rising step followed by a falling step. Figures (b) and (c) show how this waveform would appear after Bessel and Chebyshev filters, respectively. If this were a video signal, for instance, the distortion introduced by the Chebyshev filter would be devastating! The overshoot would change the brightness of the *edges* of objects compared to their *centers*. Worse yet, the left side of objects would look bright, while the right side of objects would look dark. Many applications cannot tolerate poor performance in the step response. This is where the Bessel filter shines; no overshoot and symmetrical edges.

Selecting The Antialias Filter

Table 3-2 summarizes the characteristics of these three filters, showing how each optimizes a particular parameter at the expense of everything else. The Chebyshev optimizes the *roll-off*, the Butterworth optimizes the *passband flatness*, and the Bessel optimizes the *step response*.

The selection of the antialias filter depends almost entirely on one issue: *how information is represented in the signals you intend to process*. While

		Step Response			Frequency Response		
	Voltage gain at DC	Overshoot	Time to settle to 1%	Time to settle to 0.1%	Ripple in passband	Frequency for x100 attenuation	Frequency for x1000 attenuation
Bessel							
2 pole	1.27	0.4%	0.60	1.12	0%	12.74	40.4
4 pole	1.91	0.9%	0.66	1.20	0%	4.74	8.45
6 pole	2.87	0.7%	0.74	1.18	0%	3.65	5.43
8 pole	4.32	0.4%	0.80	1.16	0%	3.35	4.53
Butterworth							
2 pole	1.59	4.3%	1.06	1.66	0%	10.0	31.6
4 pole	2.58	10.9%	1.68	2.74	0%	3.17	5.62
6 pole	4.21	14.3%	2.74	3.92	0%	2.16	3.17
8 pole	6.84	16.4%	3.50	5.12	0%	1.78	2.38
Chebyshev							
2 pole	1.84	10.8%	1.10	1.62	6%	12.33	38.9
4 pole	4.21	18.2%	3.04	5.42	6%	2.59	4.47
6 pole	10.71	21.3%	5.86	10.4	6%	1.63	2.26
8 pole	28.58	23.0%	8.34	16.4	6%	1.34	1.66

TABLE 3-2
Characteristics of the three classic filters. The Bessel filter provides the best step response, making it the choice for time domain encoded signals. The Chebyshev and Butterworth filters are used to eliminate frequencies in the stopband, making them ideal for frequency domain encoded signals. Values in this table are in the units of *seconds* and *hertz*, for a one hertz cutoff frequency.

there are many ways for information to be encoded in an analog waveform, only two methods are common, **time domain encoding**, and **frequency domain encoding**. The difference between these two is critical in DSP, and will be a reoccurring theme throughout this book.

In *frequency domain encoding*, the information is contained in *sinusoidal waves* that combine to form the signal. Audio signals are an excellent example of this. When a person hears speech or music, the perceived sound depends on the frequencies present, and not on the particular *shape* of the waveform. This can be shown by passing an audio signal through a circuit that changes the phase of the various sinusoids, but retains their frequency and amplitude. The resulting signal *looks* completely different on an oscilloscope, but *sounds* identical. The pertinent information has been left intact, even though the waveform has been significantly altered. Since aliasing misplaces and overlaps frequency components, it directly destroys information encoded in the frequency domain. Consequently, digitization of these signals usually involves an antialias filter with a sharp cutoff, such as a Chebyshev, Elliptic, or Butterworth. What about the nasty step response of these filters? It doesn't matter; the encoded information isn't affected by this type of distortion.

In contrast, *time domain encoding* uses the *shape of the waveform* to store information. For example, physicians can monitor the electrical activity of a person's heart by attaching electrodes to their chest and arms (an electrocardiogram or EKG). The *shape* of the EKG waveform provides the information being sought, such as when the various chambers contract during a heartbeat. Images are another example of this type of signal. Rather than a waveform that varies over *time*, images encode information in the shape of a waveform that varies over *distance*. Pictures are formed from regions of brightness and color, and how they relate to other regions of brightness and color. You don't look at the *Mona Lisa* and say, "*My, what an interesting collection of sinusoids.*"

Here's the problem: The sampling theorem is an analysis of what happens in the frequency domain during digitization. This makes it ideal to understand the analog-to-digital conversion of signals having their information encoded in the frequency domain. However, the sampling theorem is little help in understanding how time domain encoded signals should be digitized. Let's take a closer look.

Figure 3-15 illustrates the choices for digitizing a time domain encoded signal. Figure (a) is an example analog signal to be digitized. In this case, the information we want to capture is the *shape* of the rectangular pulses. A short burst of a high frequency sine wave is also included in this example signal. This represents wideband noise, interference, and similar junk that always appears on analog signals. The other figures show how the digitized signal would appear with different antialias filter options: a Chebyshev filter, a Bessel filter, and no filter.

It is important to understand that *none* of these options will allow the original signal to be reconstructed from the sampled data. This is because the original signal inherently contains frequency components greater than one-half of the sampling rate. Since these frequencies cannot exist in the digitized signal, the reconstructed signal cannot contain them either. These high frequencies result from two sources: (1) noise and interference, which you would like to eliminate, and (2) sharp edges in the waveform, which probably contain information you want to retain.

The Chebyshev filter, shown in (b), attacks the problem by aggressively removing all high frequency components. This results in a filtered analog signal that *can* be sampled and later perfectly reconstructed. However, the reconstructed analog signal is identical to the *filtered signal*, not the *original signal*. Although nothing is lost in sampling, the waveform has been severely distorted by the antialias filter. As shown in (b), the cure is worse than the disease! Don't do it!

The Bessel filter, (c), is designed for just this problem. Its output closely resembles the original waveform, with only a gentle rounding of the edges. By adjusting the filter's cutoff frequency, the smoothness of the edges can be traded for elimination of high frequency components in the signal. Using more poles in the filter allows a *better* tradeoff between these two parameters. A common guideline is to set the cutoff frequency at about one-quarter of the sampling frequency. This results in about two samples along the rising portion of each edge.   Notice that both the Bessel and the Chebyshev filter have removed the burst of high frequency noise present in the original signal.

The last choice is to use no antialias filter at all, as is shown in (d).   This has the strong advantage that the value of each sample is *identical* to the value of the original analog signal.   In other words, it has perfect edge sharpness; a change in the original signal is immediately mirrored in the digital data.   The disadvantage is that aliasing can distort the signal.   This takes two different forms.   First, high frequency interference and noise, such as the example sinusoidal burst, will turn into meaningless samples, as shown in (d).   That is, any high frequency noise present in the analog signal will appear as aliased noise in the digital signal.   In a more general sense, this is not a problem of the sampling, but a problem of the upstream analog electronics.   It is not the ADC's purpose to reduce noise and interference; this is the responsibility of the analog electronics before the digitization takes place.   It may turn out that a Bessel filter should be placed before the digitizer to control this problem.   However, this means the filter should be viewed as part of the analog processing, not something that is being done for the sake of the digitizer.

The second manifestation of aliasing is more subtle.   When an event occurs in the analog signal (such as an edge), the digital signal in (d) detects the change on the *next* sample.   There is no information in the digital data to indicate what happens *between* samples.   Now, compare using *no filter* with using a *Bessel filter* for this problem.   For example, imagine drawing straight lines between the samples in (c).   The time when this constructed line crosses one-half the amplitude of the step provides a *subsample* estimate of when the edge occurred in the analog signal.   When no filter is used, this subsample information is completely lost.   You don't need a fancy theorem to evaluate how this will affect your particular situation, just a good understanding of what you plan to do with the data once is it acquired.

Multirate Data Conversion

There is a strong trend in electronics to replace *analog circuitry* with *digital algorithms*.   Data conversion is an excellent example of this.   Consider the design of a digital voice recorder, a system that will digitize a voice signal, store the data in digital form, and later reconstruct the signal for playback.   To recreate intelligible speech, the system must capture the frequencies between about 100 and 3000 hertz.   However, the analog signal produced by the microphone also contains much higher frequencies, say to 40 kHz.   The brute force approach is to pass the analog signal through an eight pole low-pass Chebyshev filter at 3 kHz, and then sample at 8 kHz.   On the other end, the DAC reconstructs the analog signal at 8 kHz with a zeroth order hold.   Another Chebyshev filter at 3 kHz is used to produce the final voice signal.

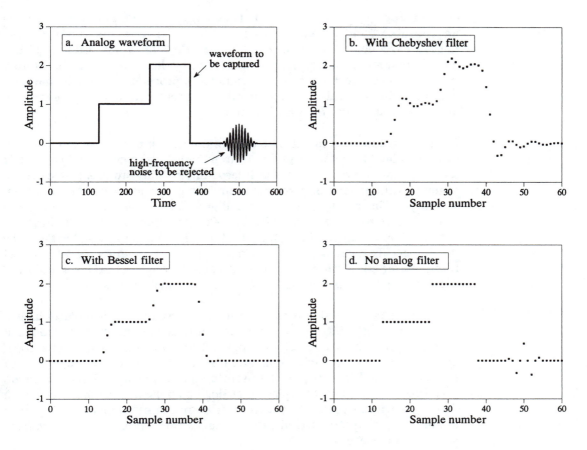

FIGURE 3-15
Three antialias filter options for time domain encoded signals. The goal is to eliminate high frequencies (that will alias during sampling), while simultaneously retaining edge sharpness (that carries information). Figure (a) shows an example analog signal containing both sharp edges and a high frequency noise burst. Figure (b) shows the digitized signal using a *Chebyshev filter*. While the high frequencies have been effectively removed, the edges have been grossly distorted. This is usually a terrible solution. The *Bessel filter*, shown in (c), provides a gentle edge smoothing while removing the high frequencies. Figure (d) shows the digitized signal using *no antialias filter*. In this case, the edges have retained perfect sharpness; however, the high frequency burst has aliased into several meaningless samples.

There are many useful benefits in sampling *faster* than this direct analysis. For example, imagine redesigning the digital voice recorder using a 64 kHz sampling rate. The antialias filter now has an easier task: pass all frequencies below 3 kHz, while rejecting all frequencies above 32 kHz. A similar simplification occurs for the reconstruction filter. In short, the higher sampling rate allows the eight pole filters to be replaced with simple RC networks. The problem is, the digital system is now swamped with data from the higher sampling rate.

The next level of sophistication involves **multirate** techniques, using more than one sampling rate in the same system. It works like this for the digital voice recorder example. First, pass the voice signal through a simple RC low-pass filter and sample the data at 64 kHz. The resulting digital data contains the desired voice band between 100 and 3000 hertz, but also has an unusable band between 3 kHz and 32 kHz. Second, remove these unusable frequencies in *software*, by using a *digital* low-pass filter at 3 kHz. Third, resample the digital signal from 64 kHz to 8 kHz by simply discarding every seven out of eight samples, a procedure called **decimation.** The resulting digital data is equivalent to that produced by aggressive analog filtering and direct 8 kHz sampling.

Multirate techniques can also be used in the output portion of our example system. The 8 kHz data is pulled from memory and converted to a 64 kHz sampling rate, a procedure called **interpolation.** This involves placing seven samples, with a value of zero, between each of the samples obtained from memory. The resulting signal is a digital *impulse train*, containing the desired voice band between 100 and 3000 hertz, plus spectral duplications between 3 kHz and 32 kHz. Refer back to Figs. 3-6 a&b to understand why this it true. Everything above 3 kHz is then removed with a *digital* low-pass filter. After conversion to an analog signal through a DAC, a simple RC network is all that is required to produce the final voice signal.

Multirate data conversion is valuable for two reasons: (1) it replaces analog components with software, a clear economic advantage in mass-produced products, and (2) it can achieve higher levels of performance in critical applications. For example, compact disc audio systems use techniques of this type to achieve the best possible sound quality. This increased performance is a result of replacing analog components (1% precision), with digital algorithms (0.0001% precision from round-off error). As discussed in upcoming chapters, digital filters outperform analog filters by *hundreds of times* in key areas.

Single Bit Data Conversion

A popular technique in telecommunications and high fidelity music reproduction is **single bit ADC and DAC**. These are multirate techniques where a higher sampling rate is traded for a lower number of bits. In the extreme, only a single bit is needed for each sample. While there are many different circuit configurations, most are based on the use of **delta modulation**. Three example circuits will be presented to give you a flavor of the field. All of these circuits are implemented in IC's, so don't worry where all of the individual transistors and op amps should go. No one is going to ask you to build one of these circuits from basic components.

Figure 3-16 shows the block diagram of a typical delta modulator. The analog input is a voice signal with an amplitude of a few volts, while the output signal is a stream of digital ones and zeros. A comparator decides which has the greater voltage, the incoming analog signal, or the voltage stored on the capacitor. This decision, in the form of a digital one or zero, is applied to the input of the latch. At each clock pulse, typically at a few hundred kilohertz, the latch transfers whatever digital state appears on its

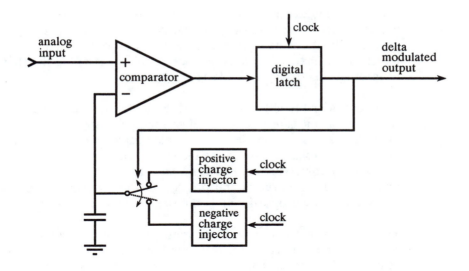

FIGURE 3-16
Block diagram of a delta modulation circuit. The input voltage is compared with the voltage stored on the capacitor, resulting in a digital zero or one being applied to the input of the latch. The output of the latch is updated in synchronization with the clock, and used in a feedback loop to cause the capacitor voltage to track the input voltage.

input, to its output. This latch insures that the output is synchronized with the clock, thereby defining the sampling rate, i.e., the rate at which the 1 bit output can update itself.

A feedback loop is formed by taking the digital output and using it to drive an electronic switch. If the output is a digital *one*, the switch connects the capacitor to a *positive charge injector*. This is a very loose term for a circuit that increases the voltage on the capacitor by a fixed amount, say 1 millivolt per clock cycle. This may be nothing more than a resistor connected to a large positive voltage. If the output is a digital *zero*, the switch is connected to a *negative charge injector*. This *decreases* the voltage on the capacitor by the same fixed amount.

Figure 3-17 illustrates the signals produced by this circuit. At time equal zero, the analog input and the voltage on the capacitor both start with a voltage of zero. As shown in (a), the input signal suddenly increases to 9.5 volts on the eighth clock cycle. Since the input signal is now more positive than the voltage on the capacitor, the digital output changes to a *one*, as shown in (b). This results in the switch being connected to the positive charge injector, and the voltage on the capacitor increasing by a small amount on each clock cycle. Although an increment of 1 volt per clock cycle is shown in (a), this is only for illustration, and a value of 1 millivolt is more typical. This staircase increase in the capacitor voltage continues until it exceeds the voltage of the input signal. Here the system reached an equilibrium with the output oscillating between a digital one and zero, causing the voltage on the capacitor to oscillate between 9 volts and 10 volts. In this manner, the feedback of the circuit forces the capacitor voltage to track the voltage of the input signal. If the input signal changes very rapidly, the voltage on the capacitor changes at a constant rate until a match is obtained. This constant rate of change is called the **slew rate**, just as in other electronic devices such as op amps.

Now, consider the characteristics of the delta modulated output signal. If the analog input is *increasing* in value, the output signal will consist of more ones than zeros. Likewise, if the analog input is *decreasing* in value, the output will consist of more zeros than ones. If the analog input is constant, the digital output will alternate between zero and one with an equal number of each. Put in more general terms, the relative number of ones versus zeros is directly proportional to the *slope* (derivative) of the analog input.

This circuit is a cheap method of transforming an analog signal into a serial stream of ones and zeros for transmission or digital storage. An especially attractive feature is that all the bits have the same meaning, unlike the conventional serial format: *start bit, LSB,* ··· *,MSB, stop bit*. The circuit at the receiver is identical to the feedback portion of the transmitting circuit. Just as the voltage on the capacitor in the transmitting circuit follows the analog input, so does the voltage on the capacitor in the receiving circuit. That is, the capacitor voltage shown in (a) also represents how the reconstructed signal would appear.

A critical limitation of this circuit is the unavoidable tradeoff between (1) maximum slew rate, (2) quantization size, and (3) data rate. In particular, if the maximum slew rate and quantization size are adjusted to acceptable values for voice communication, the data rate ends up in the MHz range. This is too high to be of commercial value. For instance, conventional sampling of a voice signal requires only about 64,000 bits per second.

A solution to this problem is shown in Fig. 3-18, the Continuously Variable Slope Delta (**CVSD**) modulator, a technique implemented in the Motorola MC3518 family. In this approach, the clock rate and the quantization size are set to something acceptable, say 30 kHz, and 2000 levels. This results in a terrible slew rate, which you correct with additional circuitry. In operation, a shift resister continually looks at the last four bits that the system has produced. If the circuit is in a slew rate limited condition, the last four bits will be all ones (positive slope) or all zeros (negative slope). A logic circuit detects this situation and produces an analog signal that increase the level of charge produced by the charge injectors. This boosts the slew rate by increasing the size of the voltage steps being applied to the capacitor.

An analog filter is usually placed between the logic circuitry and the charge injectors. This allows the step size to depend on how long the circuit has been in a slew limited condition. As long as the circuit is slew limited, the step size keeps getting larger and larger. This is often called a *syllabic filter*, since its characteristics depend on the average length of the syllables making up speech. With proper optimization (from the chip manufacturer's

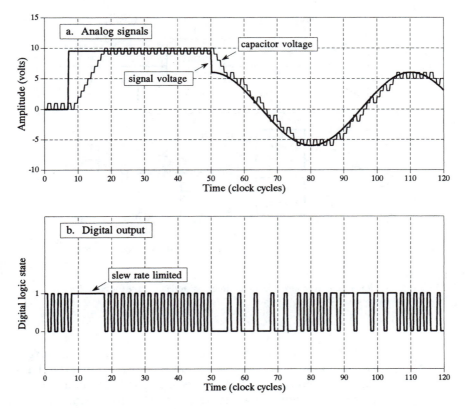

FIGURE 3-17
Example of signals produced by the delta modulator in Fig. 3-16. Figure (a) shows the analog input signal, and the corresponding voltage on the capacitor. Figure (b) shows the delta modulated output, a digital stream of ones and zeros.

spec sheet, not your own work), data rates of 16 to 32 kHz produce acceptable quality speech. The continually changing step size makes the digital data difficult to understand, but fortunately, you don't need to. At the receiver, the analog signal is reconstructed by incorporating a syllabic filter that is identical to the one in the transmission circuit. If the two filters are matched, little distortion results from the CVSD modulation. CVSD is probably the easiest way to digitally transmit a voice signal.

While CVSD modulation is great for encoding voice signals, it cannot be used for general purpose analog-to-digital conversion. Even if you get around the fact that the digital data is related to the *derivative* of the input signal, the *changing step size* will confuse things beyond repair. In addition, the DC level of the analog signal is usually not captured in the digital data.

The **delta-sigma** converter, shown in Fig. 3-19, eliminates these problems by cleverly combining analog electronics with DSP algorithms. Notice that the voltage on the capacitor is now being compared with ground potential. The feedback loop has also been modified so that the voltage on the

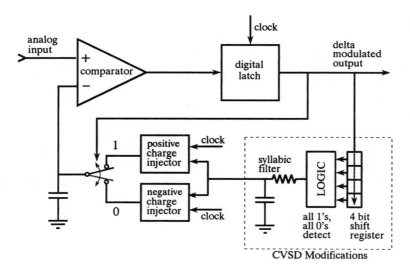

FIGURE 3-18
CVSD modulation block diagram. A logic circuit is added to the basic delta modulator
to improve the slew rate.

capacitor is *decreased* when the circuit's output is a digital *one*, and
increased when it is a digital *zero*. As the input signal increases and
decreases in voltage, it tries to raise and lower the voltage on the capacitor.
This change in voltage is detected by the comparator, resulting in the
charge injectors producing a *counteracting* charge to keep the capacitor at
zero volts.

If the input voltage is positive, the digital output will be composed of more
ones than zeros. The excess number of ones being needed to generate the
negative charge that cancels with the *positive* input signal. Likewise, if the
input voltage is negative, the digital output will be composed of more zeros
than ones, providing a net positive charge injection. If the input signal is
equal to zero volts, an equal number of ones and zeros will be generated in
the output, providing an overall charge injection of zero.

The relative number of ones and zeros in the output is now related to the
level of the input voltage, not the *slope* as in the previous circuit. This is
much simpler. For instance, you could form a 12 bit ADC by feeding the
digital output into a counter, and counting the number of *ones* over 4096
clock cycles. A digital number of 4095 would correspond to the maximum
positive input voltage. Likewise, digital number 0 would correspond to the
maximum negative input voltage, and 2048 would correspond to an input
voltage of zero. This also shows the origin of the name, *delta-sigma*: delta
modulation followed by summation (sigma).

The ones and zeros produced by this type of delta modulator are very easy
to transform back into an analog signal. All that is required is an analog
low-pass filter, which might be as simple as a single RC network. The high

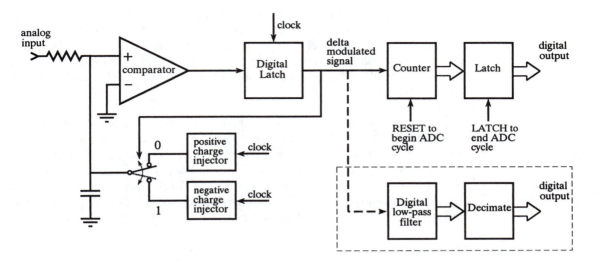

FIGURE 3-19
Block diagram of a delta-sigma analog-to-digital converter. In the simplest case, the pulses from a delta
modulator are counted for a predetermined number of clock cycles. The output of the counter is then
latched to complete the conversion. In a more sophisticated circuit, the pulses are passed through a digital
low-pass filter and then resampled (decimated) to a lower sampling rate.

and low voltages corresponding to the digital ones and zeros average out to
form the correct analog voltage. For example, suppose that the ones and
zeros are represented by 5 volts and 0 volts, respectively. If 80% of the bits
in the data stream are *ones*, and 20% are *zeros*, the output of the low-pass
filter will be 4 volts.

This method of transforming the single bit data stream back into the
original waveform is important for several reasons. First, it describes a
slick way to replace the counter in the delta-sigma ADC circuit. Instead
of simply counting the pulses from the delta modulator, the binary signal
is passed through a *digital* low-pass filter, and then *decimated* to reduce the
sampling rate. For example, this procedure might start by changing each
of the ones and zeros in the digital stream into a 12 bit sample; ones
become a value of 4095, while zeros become a value of 0. Using a digital
low-pass filter on this signal produces a digitized version of the original
waveform, just as an analog low-pass filter would form an analog recreation.
Decimation then reduces the sampling rate by discarding most of the
samples. This results in a digital signal that is equivalent to direct sampling
of the original waveform.

This approach is used in many commercial ADC's for digitizing voice and
other audio signals. An example is the National Semiconductor ADC16071,
which provides 16 bit analog-to-digital conversion at sampling rates up to
192 kHz. At a sampling rate of 100 kHz, the delta modulator operates with
a clock frequency of 6.4 MHz. The low-pass digital filter is a 246 point
FIR, such as described in Chapter 16. This removes all frequencies in the

digital data above 50 kHz, ½ of the eventual sampling rate. Conceptually, this can be viewed as forming a digital signal at 6.4 MHz, with each sample represented by 16 bits. The signal is then decimated from 6.4 MHz to 100 kHz, accomplished by deleting every 63 out of 64 samples. In actual operation, much more goes on inside of this device than described by this simple discussion.

Delta-sigma converters can also be used for digital-to-analog conversion of voice and audio signals. The digital signal is retrieved from memory, and converted into a delta modulated stream of ones and zeros. As mentioned above, this single bit signal can easily be changed into the reconstructed analog signal with a simple low-pass analog filter. As with the antialias filter, usually only a single RC network is required. This is because the majority of the filtration is handled by the high-performance digital filters.

Delta-sigma ADC's have several quirks that limit their use to specific applications. For example, it is difficult to multiplex their inputs. When the input is switched from one signal to another, proper operation is not established until the digital filter can clear itself of data from the previous signal. Delta-sigma converters are also limited in another respect: you don't know exactly *when* each sample was taken. Each acquired sample is a composite of the one bit information taken over a segment of the input signal. This is not a problem for signals encoded in the frequency domain, such as audio, but it is a significant limitation for time domain encoded signals. To understand the shape of a signal's waveform, you often need to know the precise instant each sample was taken. Lastly, most of these devices are specifically designed for audio applications, and their performance specifications are quoted accordingly. For example, a 16 bit ADC used for voice signals does not necessarily mean that each sample has 16 bits of precision. Much more likely, the manufacturer is stating that *voice signals* can be digitized to 16 bits of *dynamic range*. Don't expect to get a full 16 bits of useful information from this device for general purpose data acquisition.

While these explanations and examples provide an introduction to single bit ADC and DAC, it must be emphasized that they are simplified descriptions of sophisticated DSP and integrated circuit technology. You wouldn't expect the manufacturer to tell their *competitors* all the internal workings of their chips, so don't expect them to tell *you*.

DSP Software

DSP applications are usually programmed in the same languages as other science and engineering tasks, such as: C, BASIC and assembly. The power and versatility of C makes it the language of choice for computer scientists and other professional programmers. On the other hand, the simplicity of BASIC makes it ideal for scientists and engineers who only occasionally visit the programming world. Regardless of the language you use, most of the important DSP software issues are buried far below in the realm of whirling ones and zeros. This includes such topics as: how numbers are represented by bit patterns, round-off error in computer arithmetic, the computational speed of different types of processors, etc. This chapter is about the things you can do at the *high level* to avoid being trampled by the *low level* internal workings of your computer.

Computer Numbers

Digital computers are very proficient at storing and recalling numbers; unfortunately, this process isn't without error. For example, you instruct your computer to store the number: 1.41421356. The computer does its best, storing the closest number it *can* represent: 1.41421354. In some cases this error is quite insignificant, while in other cases it is disastrous. As another illustration, a classic computational error results from the addition of two numbers with very different values, for example, 1 and 0.00000001. We would like the answer to be 1.00000001, but the computer replies with 1. An understanding of how computers store and manipulate numbers allows you to anticipate and correct these problems *before* your program spits out meaningless data.

These problems arise because a fixed number of bits are allocated to store each number, usually 8, 16, 32 or 64. For example, consider the case where eight bits are used to store the value of a variable. Since there are $2^8 = 256$ possible bit patterns, the variable can only take on 256 different values. This is a fundamental limitation of the situation, and there is nothing we can do about it. The part we *can* control is what value we declare each bit pattern to represent. In the simplest cases, the 256 bit patterns might represent the integers from 0 to 255, 1 to 256, -127 to 128, etc. In a more unusual scheme, the 256 bit patterns might represent 256 exponentially related numbers: $1, 10, 100, 1000, \cdots, 10^{254}, 10^{255}$. Everyone accessing the data must understand what value each bit pattern represents. This is usually provided by an algorithm or formula for converting between the represented value and the corresponding bit pattern, and back again.

While many encoding schemes are possible, only two general formats have become common, *fixed point* (also called integer numbers) and *floating point* (also called real numbers). In this book's BASIC programs, fixed point variables are indicated by the % symbol as the last character in the name, such as: I%, N%, SUM%, etc. All other variables are floating point, for example: X, Y, MEAN, etc. When you evaluate the formats presented in the next few pages, try to understand them in terms of their **range** (the largest and smallest numbers they can represent) and their **precision** (the size of the gaps between numbers).

Fixed Point (Integers)

Fixed point representation is used to store *integers*, the positive and negative whole numbers: $\cdots -3, -2, -1, 0, 1, 2, 3, \cdots$. High level programs, such as C and BASIC, usually allocate 16 bits to store each integer. In the simplest case, the $2^{16} = 65,536$ possible bit patterns are assigned to the numbers 0 through 65,535. This is called **unsigned integer** format, and a simplified example is shown in Fig. 4-1 (using only 4 bits per number). Conversion between the bit pattern and the number being represented is nothing more than changing between base 2 (binary) and base 10 (decimal). The disadvantage of unsigned integer is that negative numbers cannot be represented.

Offset binary is similar to unsigned integer, except the decimal values are *shifted* to allow for negative numbers. In the 4 bit example of Fig. 4-1, the decimal numbers are offset by *seven*, resulting in the 16 bit patterns corresponding to the integer numbers -7 through 8. In this same manner, a 16 bit representation would use 32,767 as an offset, resulting in a range between -32,767 and 32,768. Offset binary is not a standardized format, and you will find other offsets used, such 32,768. The most important use of offset binary is in ADC and DAC. For example, the input voltage range of -5v to 5v might be mapped to the digital numbers 0 to 4095, for a 12 bit conversion.

Sign and magnitude is another simple way of representing negative integers. The far left bit is called the **sign bit**, and is made a *zero* for positive numbers, and a *one* for negative numbers. The other bits are a standard binary representation of the absolute value of the number. This results in one wasted bit pattern, since there are two representations for zero, 0000 (positive zero) and 1000 (negative zero). This encoding scheme results in 16 bit numbers having a range of -32,767 to 32,767.

UNSIGNED INTEGER		OFFSET BINARY		SIGN AND MAGNITUDE		TWO'S COMPLEMENT	
Decimal	Bit Pattern	Decimal	Bit Pattern	Decimal	Bit Pattern	Decimal	Bit Pattern
15	1111	8	1111	7	0111	7	0111
14	1110	7	1110	6	0110	6	0110
13	1101	6	1101	5	0101	5	0101
12	1100	5	1100	4	0100	4	0100
11	1011	4	1011	3	0011	3	0011
10	1010	3	1010	2	0010	2	0010
9	1001	2	1001	1	0001	1	0001
8	1000	1	1000	0	0000	0	0000
7	0111	0	0111	0	1000	-1	1111
6	0110	-1	0110	-1	1001	-2	1110
5	0101	-2	0101	-2	1010	-3	1101
4	0100	-3	0100	-3	1011	-4	1100
3	0011	-4	0011	-4	1100	-5	1011
2	0010	-5	0010	-5	1101	-6	1010
1	0001	-6	0001	-6	1110	-7	1001
0	0000	-7	0000	-7	1111	-8	1000

16 bit range: 0 to 65,535	16 bit range -32,767 to 32,768	16 bit range -32,767 to 32,767	16 bit range -32,768 to 32,767

FIGURE 4-1

Common formats for fixed point (integer) representation. Unsigned integer is a simple binary format, but cannot represent negative numbers. Offset binary and sign & magnitude allow negative numbers, but they are difficult to implement in hardware. Two's complement is the easiest to design hardware for, and is the most common format for general purpose computing.

These first three representations are conceptually simple, but difficult to implement in hardware. Remember, when A=B+C is entered into a computer program, some hardware engineer had to figure out how to make the bit pattern representing B, combine with the bit pattern representing C, to form the bit pattern representing A.

Two's complement is the format loved by hardware engineers, and is how integers are usually represented in computers. To understand the encoding pattern, look first at decimal number zero in Fig. 4-1, which corresponds to a binary zero, 0000. As we count upward, the decimal number is simply the binary equivalent (0 = 0000, 1 = 0001, 2 = 0010, 3 = 0011, etc.). Now, remember that these four bits are stored in a register consisting of 4 flip-flops. If we again start at 0000 and begin subtracting, the digital hardware automatically counts in two's complement: 0 = 0000, -1 = 1111, -2 = 1110, -3 = 1101, etc. This is analogous to the odometer in a new automobile. If driven forward, it changes: 00000, 00001, 00002, 00003, and so on. When driven backwards, the odometer changes: 00000, 99999, 99998, 99997, etc.

Using 16 bits, two's complement can represent numbers from -32,768 to 32,767. The left most bit is a 0 if the number is positive or zero, and a 1 if the number is negative. Consequently, the left most bit is called the **sign bit**, just as in sign & magnitude representation. Converting between decimal and two's complement is straightforward for positive numbers, a simple decimal to binary conversion. For negative numbers, the following algorithm is often used: (1) take the absolute value of the decimal number, (2) convert it to binary, (3) complement all of the bits (ones become zeros and zeros become ones), (4) add 1 to the binary number. For example: -5 → 5 → 0101 → 1010 → 1011. Two's complement is hard for humans, but easy for digital electronics.

Floating Point (Real Numbers)

The encoding scheme for floating point numbers is more complicated than for fixed point. The basic idea is the same as used in scientific notation, where a **mantissa** is multiplied by ten raised to some **exponent**. For instance, 5.4321×10^6, where 5.4321 is the *mantissa* and 6 is the *exponent*. Scientific notation is exceptional at representing very large and very small numbers. For example: 1.2×10^{50}, the number of atoms in the earth, or 2.6×10^{-23}, the distance a turtle crawls in one second, compared to the diameter of our galaxy. Notice that numbers represented in scientific notation are *normalized* so that there is only a single nonzero digit left of the decimal point. This is achieved by adjusting the exponent as needed.

Floating point representation is similar to scientific notation, except everything is carried out in base two, rather than base ten. While several similar formats are in use, the most common is ANSI/IEEE Std. 754-1985. This standard defines the format for 32 bit numbers called **single precision**, as well as 64 bit numbers called **double precision**. As shown in Fig. 4-2, the 32 bits used in single precision are divided into three separate groups: bits 0 through 22 form the mantissa, bits 23 through 30 form the exponent, and bit 31 is the sign bit. These bits form the floating point number, v, by the following relation:

EQUATION 4-1
Equation for converting a bit pattern into a floating point number. The number is represented by v, S is the value of the sign bit, M is the value of the mantissa, and E is the value of the exponent.

$$v = (-1)^S \times M \times 2^{E-127}$$

The term: $(-1)^S$, simply means that the sign bit, S, is 0 for a positive number and 1 for a negative number. The variable, E, is the number between 0 and 255 represented by the eight exponent bits. Subtracting 127 from this number allows the exponent term to run from 2^{-127} to 2^{128}. In other words, the exponent is stored in *offset binary* with an offset of 127.

The mantissa, M, is formed from the 23 bits as a *binary fraction*. For example, the decimal fraction: 2.783, is interpreted: $2 + 7/10 + 8/100 + 3/1000$. The binary fraction: 1.0101, means: $1 + 0/2 + 1/4 + 0/8 + 1/16$. Floating point numbers are *normalized* in the same way as scientific notation, that is, there is only one nonzero digit left of the decimal point (called a *binary point* in

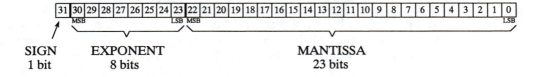

SIGN 1 bit	EXPONENT 8 bits

MANTISSA
23 bits

Example 1

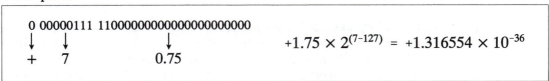

$$+1.75 \times 2^{(7-127)} = +1.316554 \times 10^{-36}$$

Example 2

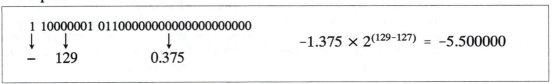

$$-1.375 \times 2^{(129-127)} = -5.500000$$

FIGURE 4-2
Single precision floating point storage format. The 32 bits are broken into three separate parts, the sign bit, the exponent and the mantissa. Equations 4-1 and 4-2 shows how the represented number is found from these three parts.

base 2). Since the only nonzero number that exists in base two is 1, the leading digit in the mantissa will always be a 1, and therefore does not need to be stored. Removing this redundancy allows the number to have an additional one bit of precision. The 23 stored bits, referred to by the notation: $m_{22}, m_{21}, m_{21}, \cdots, m_0$, form the mantissa according to:

EQUATION 4-2
Algorithm for converting the bit pattern into the mantissa, M, used in Eq. 4-1.

$$M = 1.m_{22}m_{21}m_{20}m_{19} \cdots m_2 m_1 m_0$$

In other words, $M = 1 + m_{22}2^{-1} + m_{21}2^{-2} + m_{20}2^{-3} \cdots$. If bits 0 through 22 are all *zeros*, M takes on the value of one. If bits 0 through 22 are all *ones*, M is just a hair under two, i.e., $2 - 2^{-23}$.

Using this encoding scheme, the largest number that can be represented is: $\pm(2-2^{-23}) \times 2^{128} = \pm 6.8 \times 10^{38}$. Likewise, the smallest number that can be represented is: $\pm 1.0 \times 2^{-127} = \pm 5.9 \times 10^{-39}$. The IEEE standard reduces this range slightly to free bit patterns that are assigned special meanings. In particular, the largest and smallest numbers allowed in the standard are

$\pm 3.4 \times 10^{38}$ and $\pm 1.2 \times 10^{-38}$, respectively. The freed bit patterns allow three special classes of numbers: (1) ± 0 is defined as all of the mantissa and exponent bits being zero. (2) $\pm \infty$ is defined as all of the mantissa bits being zero, and all of the exponent bits being one. (3) A group of very small *unnormalized* numbers between $\pm 1.2 \times 10^{-38}$ and $\pm 1.4 \times 10^{-45}$. These are lower precision numbers obtained by removing the requirement that the leading digit in the mantissa be a one. Besides these three special classes, there are bit patterns that are not assigned a meaning, commonly referred to as NANs (Not A Number).

The IEEE standard for double precision simply adds more bits to the single precision format. Of the 64 bits used to store a double precision number, bits 0 through 51 are the mantissa, bits 52 through 62 are the exponent, and bit 63 is the sign bit. As before, the mantissa is between one and just under two, i.e., $M = 1 + m_{51}2^{-1} + m_{50}2^{-2} + m_{49}2^{-3}\cdots$. The 11 exponent bits form a number between 0 and 2047, with an offset of 1023, allowing exponents from 2^{-1023} to 2^{1024}. The largest and smallest numbers allowed are $\pm 1.8 \times 10^{308}$ and $\pm 2.2 \times 10^{-308}$, respectively. These are incredibly large and small numbers! It is quite uncommon to find an application where single precision is not adequate. You will probably never find a case where double precision limits what you want to accomplish.

Number Precision

The errors associated with number representation are very similar to quantization errors during ADC. You *want* to store a continuous range of values; however, you *can* represent only a finite number of quantized levels. Every time a new number is generated, after a math calculation for example, it must be rounded to the nearest value that can be stored in the format you are using.

As an example, imagine that you allocate 32 bits to store a number. Since there are exactly $2^{32} = 4,294,967,296$ different bit patterns possible, you can represent exactly 4,294,967,296 different numbers. Some programming languages allow a variable called a **long integer**, stored as 32 bits, fixed point, two's complement. This means that the 4,294,967,296 possible bit patterns represent the integers between -2,147,483,648 and 2,147,483,647. In comparison, single precision floating point spreads these 4,294,967,296 bit patterns over the much larger range: -3.4×10^{38} to 3.4×10^{38}.

With fixed point variables, the gaps between adjacent numbers are always *exactly one*. In floating point notation, the gaps between adjacent numbers vary over the represented number range. If we randomly pick a floating point number, the gap next to that number is approximately *ten million times smaller* than the number itself (to be exact, 2^{-24} to 2^{-23} times the number). This is a key concept of floating point notation: large numbers have large gaps between them, while small numbers have small gaps. Figure 4-3 illustrates this by showing consecutive floating point numbers, and the gaps that separate them.

0.00001233862713	spacing = 0.00000000000091
0.00001233862804	*(1 part in 13 million)*
0.00001233862895	
0.00001233862986	

$\vdots$

1.000000000	spacing = 0.000000119
1.000000119	*(1 part in 8 million)*
1.000000238	
1.000000358	

$\vdots$

1.996093750	spacing = 0.000000119
1.996093869	*(1 part in 17 million)*
1.996093988	
1.996094108	

$\vdots$

636.0312500	spacing = 0.0000610
636.0313110	*(1 part in 10 million)*
636.0313720	
636.0314331	

$\vdots$

217063424.0	spacing = 16.0
217063440.0	*(1 part in 14 million)*
217063456.0	
217063472.0	

FIGURE 4-3
Examples of the spacing between single precision floating point numbers. The spacing between adjacent numbers is always between about 1 part in 8 million and 1 part in 17 million of the value of the number.

The program in Table 4-1 illustrates how **round-off error** (quantization error in math calculations) causes problems in DSP. Within the program loop, two random numbers are added to the floating point variable X, and then subtracted back out again. Ideally, this should do *nothing*. In reality, the round-off error from each of the arithmetic operations causes the value of X to gradually drift away from its initial value. This drift can take one of two forms depending on how the errors add together. If the round-off errors are randomly positive and negative, the value of the variable will randomly increase and decrease. If the errors are predominately of the same sign, the value of the variable will drift away much more rapidly and uniformly.

TABLE 4-1
Program for demonstrating floating point error accumulation. This program initially sets the value of X to 1.000000, and then runs through a loop that should ideally do *nothing*. During each loop, two random numbers, A and B, are added to X, and then subtracted back out. The accumulated error from these additions and subtraction causes X to wander from its initial value. As Fig. 4-4 shows, the error may be random or additive.

```
100  X = 1                'initialize X
110  '
120  FOR I% = 0 TO 2000
130     A = RND           'load random numbers
140     B = RND           'into A and B
150  '
160     X = X + A         'add A and B to X
170     X = X + B
180     X = X - A         'undo the additions
190     X = X - B
200  '
210     PRINT X           'ideally, X should be 1
220  NEXT I%
230  END
```

FIGURE 4-4
Accumulation of round-off error in floating point variables. These curves are generated by the program shown in Table 4-1. When a floating point variable is repeatedly used in arithmetic operations, accumulated round-off error causes the variable's value to drift. If the errors are both positive and negative, the value will increase and decrease in a random fashion. If the round-off errors are predominately of the same sign, the value will change in a much more rapid and uniform manner.

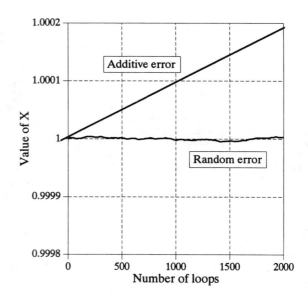

Figure 4-4 shows how the variable, X, in this example program drifts in value. An obvious concern is that *additive error* is much worse than *random error*. This is because random errors tend to cancel with each other, while the additive errors simply accumulate. The additive error is roughly equal to the round-off error from a single operation, multiplied by the total number of operations. In comparison, the random error only increases in proportion to the *square root* of the number of operations. As shown by this example, additive error can be hundreds of times worse than random error for common DSP algorithms.

Unfortunately, it is nearly impossible to control or predict which of these two behaviors a particular algorithm will experience. For example, the program in Table 4-1 generates an additive error. This can be changed to a random error by merely making a slight modification to the numbers being added and subtracted. In particular, the random error curve in Fig. 4-4 was generated by defining: $A = EXP(RND)$ and $B = EXP(RND)$, rather than: $A = RND$ and $B = RND$. Instead of A and B being randomly distributed numbers between 0 and 1, they become exponentially distributed values between 1 and 2.718. Even this small change is sufficient to toggle the mode of error accumulation.

Since we can't control which way the round-off errors accumulate, keep in mind the worse case scenario. Expect that every single precision number will have an error of about *one part in forty million, multiplied by the number of operations it has been through*. This is based on the assumption of additive error, and the average error from a single operation being one-quarter of a quantization level. Through the same analysis, every double precision number has an error of about *one part in forty quadrillion, multiplied by the number of operations*.

```
100 'Floating Point Loop Control
110 FOR X = 0 TO 10 STEP 0.01
120   PRINT X
130 NEXT X
```

```
100 'Integer Loop Control
110 FOR I% = 0 TO 1000
120   X = I%/100
130   PRINT X
140 NEXT I%
```

Program Output:

```
  0.00
  0.01
  0.02
  0.03
   ⋮
9.960132
9.970133
9.980133
9.990133
```

Program Output:

```
 0.00
 0.01
 0.02
 0.03
  ⋮
 9.96
 9.97
 9.98
 9.99
10.00
```

TABLE 4-2

Comparison of floating point and integer variables for loop control. The left hand program controls the FOR-NEXT loop with a floating point variable, X. This results in an accumulated round-off error of 0.000133 by the end of the program, causing the last loop value, $X = 10.0$, to be omitted. In comparison, the right hand program uses an integer, $I\%$, for the loop index. This provides *perfect* precision, and guarantees that the proper number of loop cycles will be completed.

Table 4-2 illustrates a particularly annoying problem of round-off error. Each of the two programs in this table perform the same task: printing 1001 numbers equally spaced between 0 and 10. The left-hand program uses the floating point variable, X, as the loop index. When instructed to execute a loop, the computer begins by setting the index variable to the starting value of the loop (0 in this example). At the end of each loop cycle, the step size (0.01 in the case) is *added* to the index. A decision is then made: are more loops cycles required, or is the loop completed? The loop ends when the computer finds that the value of the index is *greater than* the termination value (in this example, 10.0). As shown by the generated output, round-off error in the additions cause the value of X to accumulate a significant discrepancy over the course of the loop. In fact, the accumulated error *prevents* the execution of the last loop cycle. Instead of X having a value of 10.0 on the last cycle, the error make the last value of X equal to 10.000133. Since X is greater than the termination value, the computer thinks its work is done, and the loop prematurely ends. This missing last value is a common bug in many computer programs.

In comparison, the program on the right uses an integer variable, $I\%$, to control the loop. The addition, subtraction, or multiplication of two integers always produces another integer. This means that fixed point notation has absolutely no round-off error with these operations. Integers are ideal for controlling loops, as well as other variables that undergo multiple mathematical operations. The last loop cycle is guaranteed to execute! Unless you have some strong motivation to do otherwise, always use integers for loop indexes and counters.

If you must use a floating point variable as a loop index, try to use fractions that are a power of *two* (such as: 1/2, 1/4, 3/8, 27/16), instead of a power of *ten* (such as: 0.1, 0.6, 1.4, 2.3, etc.). For instance, it would be better to use: FOR X = 1 TO 10 STEP 0.125, rather than: FOR X = 1 to 10 STEP 0.1. This allows the index to always have an *exact* binary representation, thereby reducing round-off error. For example, the decimal number: 1.125, can be represented exactly in binary notation: $1.001000000000000000000000 \times 2^0$. In comparison, the decimal number: 1.1, falls *between* two floating point numbers: 1.0999999046 and 1.1000000238 (in binary these numbers are: $1.0001100110011001100 \times 2^0$ and $1.0001100110011001101 \times 2^0$). This results in an inherent error each time 1.1 is encountered in a program.

A useful fact to remember: single precision floating point has an *exact* binary representation for every whole number between ±16.8 million (to be exact, $\pm 2^{24}$). Above this value, the gaps between the levels are larger than one, causing some whole number values to be missed. This allows floating point whole numbers (between ±16.8 million) to be added, subtracted and multiplied, with no round-off error.

Execution Speed: Program Language

DSP programming can be loosely divided into three levels of sophistication: *Assembly*, *Compiled*, and *Application Specific*. To understand the difference between these three, we need to start with the very basics of digital electronics. All microprocessors are based around a set of internal binary registers, that is, a group of flip-flops that can store a series of ones and zeros. For example, the 8088 microprocessor, the core of the original IBM PC, has *four* general purpose registers, each consisting of 16 bits. These are identified by the names: AX, BX, CX, and DX. There are also *nine* additional registers with special purposes, called: SI, DI, SP, BP, CS, DS, SS, ES, and IP. For example, IP, the Instruction Pointer, keeps track of where in memory the next instruction resides.

Suppose you write a program to add the numbers: 1234 and 4321. When the program begins, IP contains the address of a section of memory that contains a pattern of ones and zeros, as shown in Table 4-3. Although it looks meaningless to most humans, this pattern of ones and zeros contains all of the commands and data required to complete the task. For example, when the microprocessor encounters the bit pattern: 00000011 11000011, it interpreters it as a command to take the 16 bits stored in the BX register, add them in binary to the 16 bits stored in the AX register, and store the result in the AX register. This level of programming is called **machine code**, and is only a hair above working with the actual electronic circuits.

Since working in binary will eventually drive even the most patient engineer crazy, these patterns of ones and zeros are assigned names according to the function they perform. This level of programming is called **assembly**, and an example is shown in Table 4-4. Although an assembly program is much easier to understand, it is fundamentally the same as programming in

10111001	00000000
11010010	10100001
00000100	00000000
10001001	00000000
00001110	10001011
00000000	00011110
00000000	00000010
10111001	00000000
11100001	00000011
00010000	11000011
10001001	10100011
00001110	00000100
00000010	00000000

TABLE 4-3
A *machine code* program for adding
1234 and 4321. This is the lowest level
of programming: direct manipulation
of the digital electronics.

machine code, since there is a one-to-one correspondence between the
program commands and the action taken in the microprocessor. For
example: ADD AX, BX translates to: 00000011 11000011. A program
called an **assembler** is used to convert the assembly code in Table 4-4
(called the **source code**) into the patterns of ones and zeros shown in Table
4-3 (called the **object code** or **executable code**). This executable code can
be directly run on the microprocessor. Obviously, assembly programming
requires an extensive understanding of the internal construction of the
particular microprocessor you intend to use.

TABLE 4-4
An *assembly* program for adding 1234
and 4321. An *assembler* is a program
that converts an assembly program into
machine code.

MOV CX,1234	;store 1234 in register CX, and then
MOV DS:[0],CX	;transfer it to memory location DS:[0]
MOV CX,4321	;store 4321 in register CX, and then
MOV DS:[2],CX	;transfer it to memory location DS:[0]
MOV AX,DS:[0]	;move variables stored in memory at
MOV BX,DS:[2]	;DS:[0] and DS:[2] into AX & BX
ADD AX,BX	;add AX and BX, store sum in AX
MOV DS:[4],AX	;move the sum into memory at DS:[4]

Assembly programming involves the direct manipulation of the digital
electronics: registers, memory locations, status bits, etc. The next level of
sophistication can manipulate *abstract* variables without any reference to the
particular hardware. These are called **compiled** or **high-level** languages. A
dozen or so are in common use, such as: C, BASIC, FORTRAN, PASCAL,
APL, COBOL, LISP, etc. Table 4-5 shows a BASIC program for adding
1234 and 4321. The programmer only knows about the variables A, B, and
C, and nothing about the hardware.

TABLE 4-5
A *BASIC* program for adding 1234 and
4321. A *compiler* is a program that
converts this type of high-level source
code into machine code.

```
100  A = 1234
110  B = 4321
120  C = A+B
130  END
```

A program called a **compiler** is used to transform the high-level source code directly into machine code. This requires the compiler to *assign* hardware memory locations to each of the abstract variables being referenced. For example, the first time the compiler encounters the variable A in Table 4-5 (line 100), it understands that the programmer is using this symbol to mean a single precision floating point variable. Correspondingly, the compiler designates four bytes of memory that will be used for nothing but to hold the value of this variable. Each subsequent time that an A appears in the program, the computer knows to update the value of the four bytes as needed. The compiler also breaks complicated mathematical expressions, such as: $Y = LOG(X^{COS(Z)})$, into more basic arithmetic. Microprocessors only know how to add, subtract, multiply and divide. Anything more complicated must be done as a series of these elementary operations.

High-level languages isolate the programmer from the hardware. This makes the programming much easier and allows the source code to be transported between different types of microprocessors. Most important, the programmer who uses a compiled language needs to know *nothing* about the internal workings of the computer. Another programmer has assumed this responsibility, the one who wrote the compiler.

Most compilers operate by converting the *entire* program into machine code *before* it is executed. An exception to this is a type of compiler called an **interpreter**, of which **interpreter BASIC** is the most common example. An interpreter converts a *single line* of source code into machine code, executes that machine code, and then goes on to the next line of source code. This provides an interactive environment for simple programs, although the execution speed is extremely slow (think a factor of 100).

The highest level of programming sophistication is found in **applications** packages for DSP. These come in a variety of forms, and are often provided to support specific hardware. Suppose you buy a newly developed DSP microprocessor to embed in your current project. These devices often have lots of built-in features for DSP: analog inputs, analog outputs, digital I/O, antialias and reconstruction filters, etc. The question is: how do you program it? In the worst case, the manufacturer will give you an *assembler*, and expect you to learn the internal architecture of the device. In a more typical scenario, a C *compiler* will be provided, allowing you to program without being bothered by how the microprocessor actually operates.

In the best case, the manufacturer will provide a sophisticated software package to help in the programming: libraries of algorithms, prewritten routines for I/O, debugging tools, etc. You might simply connect icons to form the desired system in an easy-to-use graphical display. The *things* you manipulate are signal pathways, algorithms for processing signals, analog I/O parameters, etc. When you are satisfied with the design, it is transformed into suitable machine code for execution in the hardware. Other types of applications packages are used with image processing, spectral analysis, instrumentation and control, digital filter design, etc. This is the shape of the future.

The distinction between these three levels can be very fuzzy. For example, most complied languages allow you to directly manipulate the hardware. Likewise, a high-level language with a well stocked library of DSP functions is very close to being an applications package. The point of these three catagories is understand what you are manipulating: (1) hardware, (2) abstract variables, or (3) entire procedures and algorithms.

There is also another important concept behind these classifications. When you use a high-level language, you are relying on the programmer who wrote the compiler to understand the best techniques for hardware manipulation. Similarly, when you use an applications package, you are relying on the programmer who wrote the package to understand the best DSP techniques. Here's the rub: these programmers have never seen the particular problem you are dealing with. Therefore, they cannot always provide you with an optimal solution. As you operate on a *higher* level, expect that the final machine code will be *less* efficient in terms of memory usage, speed, and precision.

Which programming language should you use? That depends on *who* you are and *what* you plan to do. Most computer scientists and programmers use C (or the more advanced C++). Power, flexibility, modularity; C has it all. C is so popular, the question becomes: Why would anyone program their DSP application in something other than C? Three answers come to mind. First, DSP has grown so rapidly that some organizations and individuals are stuck in the mode of other languages, such as FORTRAN and PASCAL. This is especially true of military and government agencies that are notoriously slow to change. Second, some applications require the utmost efficiency, only achievable by assembly programming. This falls into the category of "a little more speed for a lot more work." Third, C is not an especially easy language to master, especially for part time programmers. This includes a wide range of engineers and scientists who occasionally need DSP techniques to assist in their research or design activities. This group often turns to BASIC because of its simplicity.

Why was BASIC chosen for this book? This book is about *algorithms*, not programming style. You should be concentrating on DSP techniques, and not be distracted by the quirks of a particular language. For instance, all the programs in this book have *line numbers*. This makes it easy to describe how the program operates: "line 100 does such-and-such, line 110 does this-and-that," etc. Of course, you will probably never use line numbers in your actual programs. The point is, *learning DSP* has different requirements than *using DSP*. There are many books on the market that provide exquisite source code for DSP algorithms. If you are simply looking for prewritten code to copy into your program, you are in the wrong place.

Comparing the execution speed of hardware or software is a thankless task; no matter what the result, the loser will cry that the match was unfair! Programmers who like high-level languages (such as traditional computer scientists), will argue that assembly is only 50% faster than compiled code, but five times more trouble. Those who like assembly (typically, scientists and hardware engineers) will claim the reverse: assembly is five times faster, but only 50% more difficult to use. As in most controversies, both sides can provide selective data to support their claims.

As a rule-of-thumb, expect that a subroutine written in assembly will be between 1.5 and 3.0 times faster than the comparable high-level program. The only way to know the exact value is to write the code and conduct speed tests. Since personal computers are increasing in speed about 40% every year, writing a routine in assembly is equivalent to about a two year jump in hardware technology.

Most professional programmers are rather offended at the idea of using assembly, and gag if you suggest BASIC. Their rational is quite simple: assembly and BASIC discourage the use of good software practices. Good code should be *portable* (able to move from one type of computer to another), *modular* (broken into a well defined subroutine structure), and *easy to understand* (lots of comments and descriptive variable names). The weak structure of assembly and BASIC makes it difficult to achieve these standards. This is compounded by the fact that the people who are attracted to assembly and BASIC often have little formal training in proper software structure and documentation.

Assembly lovers respond to this attack with a zinger of their own. Suppose you write a program in C, and your competitor writes the same program in assembly. The end user's first impression will be that your program is junk because it is twice as slow. No one would suggest that you write large programs in assembly, only those portions of the program that need rapid execution. For example, many functions in DSP software libraries are written in assembly, and then accessed from larger programs written in C. Even the staunchest software purist will *use* assembly code, as long as they don't have to *write* it.

Execution Speed: Hardware

Computing power is increasing so rapidly, any *book* on the subject will be obsolete before it is published. It's an author's nightmare! The original IBM PC was introduced in 1981, based around the 8088 microprocessor with a 4.77 MHz clock and an 8 bit data bus. This was followed by a new generation of personal computers being introduced every 3-4 years: 8088 → 80286 → 80386 → 80486 → 80586 (Pentium). Each of these new systems boosted the computing speed by a factor of about *five* over the previous technology. By 1996, the clock speed had increased to 200 MHz, and the data bus to 32 bits. With other improvements, this resulted in an increase in computing power of nearly *one thousand* in only 15 years! You should expect *another* factor of one thousand in the *next* 15 years.

The only way to obtain up-to-date information in this rapidly changing field is directly from the manufacturers: advertisements, specification sheets, price lists, etc. Forget books for performance data, look in magazines and your daily newspaper. Expect that raw computational speed will more than double each two years. Learning about the current state of computer power is simply not enough; you need to understand and track how it is evolving.

Keeping this in mind, we can jump into an overview of how execution speed is limited by computer hardware. Since computers are composed of many subsystems, the time required to execute a particular task will depend on two primary factors: (1) the speed of the individual subsystems, and (2) the time it takes to transfer data between these blocks. Figure 4-5 shows a simplified diagram of the most important speed limiting components in a typical personnel computer. The **Central Processing Unit (CPU)** is the heart of the system. As previously described, it consists of a dozen or so registers, each capable of holding 32 bits (in present generation personnel computers). Also included in the CPU is the digital electronics needed for rudimentary operations, such as moving bits around and fixed point arithmetic.

More involved mathematics is handled by transferring the data to a special hardware circuit called a **math coprocessor** (also called an **arithmetic logic unit**, or **ALU**). The math coprocessor may be contained in the same chip as the CPU, or it may be a separate electronic device. For example, the addition of two floating point numbers would require the CPU to transfer 8 bytes (4 for each number) to the math coprocessor, and several bytes that describe what to do with the data. After a short computational time, the math coprocessor would pass four bytes back to the CPU, containing the floating point number that is the sum. The most inexpensive computer systems don't have a math coprocessor, or provide it only as an option. For example, the 80486DX microprocessor has an internal math coprocessor, while the 80486SX does not. These lower performance systems replace *hardware* with *software*. Each of the mathematical functions is broken into

FIGURE 4-5
Architecture of a typical computer system. The computational speed is limited by: (1) the speed of the individual subsystems, and (2) the rate at which data can be transferred between these subsystems.

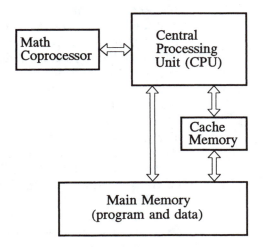

elementary binary operations that can be handled directly within the CPU. While this provides the same result, the execution time is much slower, say, a factor of 10 to 20.

Most personal computer software can be used with or without a math coprocessor. This is accomplished by having the compiler generate machine code to handle both cases, all stored in the final executable program. If a math coprocessor is present on the particular computer being used, one section of the code will be run. If a math coprocessor is not present, the other section of the code will be used. The compiler can also be directed to generate code for only one of these situations. For example, you will occasionally find a program that requires that a math coprocessor be present, and will crash if run on a computer that does not have one. Applications such as word processing usually do not benefit from a math coprocessor. This is because they involve moving data around in memory, not the calculation of mathematical expressions. Likewise, calculations involving fixed point variables (integers) are unaffected by the presence of a math coprocessor, since they are handled within the CPU. On the other hand, the execution speed of DSP and other computational programs using floating point calculations can be an order of magnitude different with and without a math coprocessor.

The CPU and main memory are contained in separate chips in most computer systems. For obvious reasons, you would like the main memory to be very large and very fast. Unfortunately, this makes the memory very expensive. The transfer of data between the main memory and the CPU is a very common bottleneck for speed. The CPU *asks* the main memory for the binary information at a particular memory address, and then must *wait* to receive the information. A common technique to get around this problem is to use a **memory cache**. This is a small amount of very fast memory used as a buffer between the CPU and the main memory. A few hundred kilobytes is typical. When the CPU requests the main memory to provide the binary data at a particular address, high speed digital electronics copies a *section* of the main memory around this address into the memory cache. The next time that the CPU requests memory information, it is very likely that it will already be contained in the memory cache, making the retrieval very rapid. This is based on the fact that programs tend to access memory locations that are nearby neighbors of previously accessed data. In typical personnel computer applications, the addition of a memory cache can improve the overall speed by several times. The memory cache may be in the same chip as the CPU, or it may be an external electronic device.

The rate at which data can be transferred between subsystems depends on the number of parallel data lines provided, and the maximum rate that digital signals that can be passed along each line. Digital data can generally be transferred at a much higher rate within a single chip as compared to transferring data between chips. Likewise, data paths that must pass through electrical connectors to other printed circuit boards (i.e., a bus structure) will be slower still. This is a strong motivation for stuffing as much electronics as possible inside the CPU.

A particularly nasty problem for computer speed is *backward compatibility*. When a computer company introduces a new product, say a data acquisition card or a software program, they want to sell it into the largest possible market. This means that it must be compatible with most of the computers currently in use, which could span several generations of technology. This frequently limits the performance of the hardware or software to that of a much older system. For example, suppose you buy an I/O card that plugs into the bus of your 200 MHz Pentium personal computer, providing you with eight digital lines that can transmit and receive data one byte at a time. You then write an assembly program to rapidly transfer data between your computer and some external device, such as a scientific experiment or another computer. Much to your surprise, the maximum data transfer rate is only about 100,000 bytes per second, more than *one thousand* times slower than the microprocessor clock rate! The villain is the ISA bus, a technology that is *backward compatible* to the computers of the early 1980s.

Table 4-6 provides execution times for several generations of computers. Obviously, you should treat these as very rough approximations. If you want to understand *your* system, take measurements on *your* system. It's quite easy; write a loop that executes a *million* of some operation, and use your watch to time how long it takes. The first three systems, the 80286, 80486, and Pentium, are the standard desk-top personal computers of 1986, 1993 and 1996, respectively. The forth is a 1994 microprocessor designed especially for DSP tasks, the Texas Instruments TMS320C40.

	80286 (12 MHz)	80486 (33 MHz)	PENTIUM (100 MHz)	TMS320C40 (40 MHz)
INTEGER				
A% = B%+C%	1.6	0.12	0.04	
A% = B%−C%	1.6	0.12	0.04	
A% = B%×C%	2.7	0.59	0.13	
A% = B%÷C%	64	9.2	1.5	
FLOATING POINT				
A = B+C	33	2.5	0.50	0.10
A = B−C	35	2.5	0.50	0.10
A = B×C	35	2.5	0.50	0.10
A = B÷C	49	4.5	0.87	0.80
A = SQR(B)	45	5.3	1.3	0.90
A = LOG(B)	186	19	3.4	1.7
A = EXP(B)	246	25	5.5	1.7
A = B^C	311	31	5.3	2.4
A = SIN(B)	262	30	6.6	1.1
A = ARCTAN(B)	168	21	4.4	2.2

TABLE 4-6
Measured execution times for various computers. Times are in microseconds. The 80286, 80486, and Pentium are three generations of personal computers, while the TMS320C40 is a micro-processor specifically designed for DSP tasks. All of the personal computers include a math coprocessor. Use these times only as a general estimate; times on your computer will vary according to the particular hardware and software used.

The Pentium is faster than the 80286 system for four reasons, (1) the greater clock speed, (2) more lines in the data bus, (3) the addition of a memory cache, and (4) a more efficient internal design, requiring fewer clock cycles per instruction.

If the Pentium was a Cadillac, the TMS320C40 would be a Ferrari: less comfort, but blinding speed. This chip is representative of several microprocessors specifically designed to decrease the execution time of DSP algorithms. Others in this category are the Intel i860, AT&T DSP3210, Motorola DSP96002, and the Analog Devices ADSP-2171. These often go by the name: **DSP microprocessor**, or **RISC** (Reduced Instruction Set Computer). This last name reflects that the increased speed results from fewer assembly level instructions being made available to the programmer. In comparison, more traditional microprocessors, such as the Pentium, are called **CISC** (Complex Instruction Set Computer).

DSP microprocessors are used in two ways: as slave modules under the control of a more conventional computer, or as an imbedded processor in a dedicated application, such as a cellular telephone. Some models only handle fixed point numbers, while others can work with floating point. The internal architecture used to obtain the increased speed includes: (1) lots of very fast cache memory contained within the chip, (2) separate buses for the program and data, allowing the two to be accessed simultaneously (called a **Harvard Architecture**), (3) fast hardware for math calculations contained directly in the microprocessor, and (4) a *pipeline* design.

A **pipeline** architecture breaks the *hardware* required for a certain task into several successive stages. For example, the addition of two numbers may be done in three pipeline stages. The first stage of the pipeline does nothing but fetch the numbers to be added from memory. The only task of the second stage is to add the two numbers together. The third stage does nothing but store the result in memory. If each stage can complete its task in a single clock cycle, the entire procedure will take three clock cycles to execute. The key feature of the pipeline structure is that another task can be started before the previous task is completed. In this example, we could begin the addition of *another* two numbers as soon as the first stage is idle, at the end of the first clock cycle. For a large number of operations, the speed of the system will be quoted as one addition per clock cycle, even though the addition of any two numbers requires three clock cycles to complete. Pipelines are great for speed, but they can be difficult to program. The algorithm must allow a new calculation to begin, even though the results of previous calculations are unavailable (because they are still in the pipeline).

Execution Speed: Programming Tips

While computer hardware and programming languages are important for maximizing execution speed, they are not something you change on a day-to-day basis. In comparison, *how you program* can be changed at any time, and will drastically affect how long the program will require to execute. Here are three suggestions.

First, use integers instead of floating point variables whenever possible. Conventional microprocessors, such as used in personal computers, process integers 10 to 20 times faster than floating point numbers. On systems without a math coprocessor, the difference can be 200 to 1. An exception to this is integer division, which is often accomplished by converting the values into floating point. This makes the operation ghastly slow compared to other integer calculations. See Table 4-6 for details.

Second, avoid using functions such as: sin(x), log(x), y^x, etc. These transcendental functions are calculated as a series of additions, subtractions and multiplications. For example, the Maclaurin power series provides:

EQUATION 4-3
Maclaurin power series expansion for three transcendental functions. This is how computers calculate functions of this type, and why they execute so slowly.

$$\sin(x) = x - \frac{x^3}{3!} + \frac{x^5}{5!} - \frac{x^7}{7!} + \frac{x^9}{9!} - \frac{x^{11}}{11!} + \cdots$$

$$\cos(x) = 1 - \frac{x^2}{2!} + \frac{x^4}{4!} - \frac{x^6}{6!} + \frac{x^8}{8!} - \frac{x^{10}}{10!} + \cdots$$

$$e^x = 1 + x + \frac{x^2}{2!} + \frac{x^3}{3!} + \frac{x^4}{4!} + \frac{x^5}{5!} + \cdots$$

While these relations are infinite in length, the terms rapidly become small enough to be ignored. For example:

$$\sin(1) = 1 - 0.166666 + 0.008333 - 0.000198 + 0.000002 - \cdots$$

These functions require about ten times longer to calculate than a single addition or multiplication (see Table 4-6). Several tricks can be used to bypass these calculations, such as: $x^3 = x \cdot x \cdot x$; $\sin(x) \approx x$, when x is very small; $\sin(-x) = \sin(x)$, where you already know one of the values and need to find the other, etc. Most languages only provide a few transcendental functions, and expect you to derive the others by means of the relations in Table 4-7. Not surprisingly, these derived calculations are even slower.

Another option is to precalculate these slow functions, and store the values in a **look-up table (LUT)**. For example, imagine an 8 bit data acquisition system used to continually monitor the *voltage* across a resistor. If the parameter of interest is the *power* being dissipated in the resistor, the measured voltage can be used to calculate: $P = V^2/R$. As a faster alternative, the power corresponding to each of the possible 256 voltage

FUNCTION	EQUATION FOR CALCULATING
Secant (X) =	1/COS(X)
Cosecant (X) =	1/SIN(X)
Cotangent (X) =	1/TAN(X)
Arc Sine (X) =	ATN(X/SQR(1-X*X))
Arc Cosine (X) =	-ATN(X/SQR(1-X*X)) + PI/2
Arc Secant (X) =	ATN(SQR(X*X-1)) + (SGN(X)-1) * PI/2
Arc Cosecant (X) =	ATN(1/SQR(X*X-1)) + (SGN(X)-1) *
Arc Cotangent (X) =	-ATN(X) + PI/2
Hyperbolic Sine (X) =	(EXP(X)-EXP(-X))/2
Hyperbolic Cosine (X) =	(EXP(X)+EXP(-X))/2
Hyperbolic Tangent (X) =	(EXP(X)-EXP(-X))/(EXP(X)+EXP(-X))
Hyperbolic Secant (X) =	1/HYPERBOLIC COSINE
Hyperbolic Cosecant (X) =	1/HYPERBOLIC SINE
Hyperbolic Cotangent (X) =	1/HYPERBOLIC TANGENT
Arc Hyperbolic Sine (X) =	LOG(X+SQR(X*X+1))
Arc Hyperbolic Cosine (X) =	LOG(X+SQR(X*X-1))
Arc Hyperbolic Tangent (X) =	LOG((1+X) /(1-X))/2
Arc Hyperbolic Secant (X) =	LOG((SQR(1-X*X)+1)/X)
Arc Hyperbolic Cosecant (X) =	LOG(1+SGN(X)*SQR(1+X*X))/X
Arc Hyperbolic Cotangent (X) =	LOG((X+1)/(X-1))/2
$LOG_{10}(X)$ =	LOG(X)/LOG(10) = 0.4342945 LOG(X)
PI =	4*ATN(1) = 3.141592653589794

TABLE 4-7
Calculating rarely used functions from more common ones. All angles are in radians,
ATN(X) is the arctangent, LOG(X) is the natural logarithm, SGN(X) is the sign of X
(i.e., -1 for X≤0, 1 for X>0), EXP(X) is e^X.

measurements can be calculated beforehand, and stored in a LUT. When
the system is running, the measured voltage, a digital number between 0
and 255, becomes an index in the LUT to find the corresponding power.
Look-up tables can be hundreds of times faster than direct calculation.

Third, learn what is *fast* and what is *slow* on your particular system. This
comes with experience and testing, and there will always be surprises. Pay
particular attention to *graphics* commands and *I/O*. There are usually several
ways to handle these requirements, and the speeds can be tremendously
different. For example, the BASIC command: *BLOAD*, transfers a data file
directly into a section of memory. Reading the same file into memory byte-
by-byte (in a loop) can be 100 times as slow. As another example, the
BASIC command: *LINE,* can be used to draw a colored box on the video
screen. Drawing the same box pixel-by-pixel can also take 100 times as
long. Even putting a *print* statement within a loop (to keep track of what
it is doing) can slow the operation by *thousands*!

CHAPTER 5

Linear Systems

Most DSP techniques are based on a divide-and-conquer strategy called *superposition*. The signal being processed is broken into simple components, each component is processed individually, and the results reunited. This approach has the tremendous power of breaking a single complicated problem into many easy ones. Superposition can only be used with *linear systems*, a term meaning that certain mathematical rules apply. Fortunately, most of the applications encountered in science and engineering fall into this category. This chapter presents the foundation of DSP: what it means for a system to be linear, various ways for breaking signals into simpler components, and how superposition provides a variety of signal processing techniques.

Signals and Systems

A **signal** is a description of how one parameter varies with another parameter. For instance, voltage changing over time in an electronic circuit, or brightness varying with distance in an image. A **system** is any process that produces an *output signal* in response to an *input signal*. This is illustrated by the block diagram in Fig. 5-1. Continuous systems input and output continuous signals, such as in analog electronics. Discrete systems input and output discrete signals, such as computer programs that manipulate the values stored in arrays.

Several rules are used for naming signals. These aren't always followed in DSP, but they are very common and you should memorize them. The mathematics is difficult enough without a clear notation. First, *continuous* signals use parentheses, such as: $x(t)$ and $y(t)$, while *discrete* signals use brackets, as in: $x[n]$ and $y[n]$. Second, signals use lower case letters. Upper case letters are reserved for the frequency domain, discussed in later chapters. Third, the name given to a signal is usually descriptive of the parameters it represents. For example, a *voltage* depending on *time* might be called: $v(t)$, or a stock market *price* measured each *day* could be: $p[d]$.

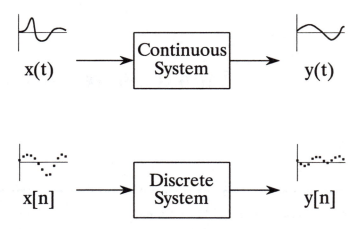

FIGURE 5-1
Terminology for signals and systems. A system is any process that generates an output signal in response to an input signal. Continuous signals are usually represented with parentheses, while discrete signals use brackets. All signals use lower case letters, reserving the upper case for the frequency domain (presented in later chapters). Unless there is a better name available, the input signal is called: $x(t)$ or $x[n]$, while the output is called: $y(t)$ or $y[n]$.

Signals and systems are frequently discussed without knowing the exact parameters being represented. This is the same as using x and y in algebra, without assigning a physical meaning to the variables. This brings in a fourth rule for naming signals. If a more descriptive name is not available, the input signal to a discrete system is usually called: $x[n]$, and the output signal: $y[n]$. For continuous systems, the signals: $x(t)$ and $y(t)$ are used.

There are many reasons for wanting to understand a *system*. For example, you may want to *design* a system to remove noise in an electrocardiogram, sharpen an out-of-focus image, or remove echoes in an audio recording. In other cases, the system might have a distortion or interfering effect that you need to characterize or measure. For instance, when you speak into a telephone, you expect the other person to hear something that resembles your voice. Unfortunately, the input signal to a transmission line is seldom identical to the output signal. If you understand how the transmission line (the system) is changing the signal, maybe you can compensate for its effect. In still other cases, the system may represent some physical process that you want to study or analyze. Radar and sonar are good examples of this. These methods operate by comparing the transmitted and reflected signals to find the characteristics of a remote object. In terms of system theory, the problem is to find the system that changes the transmitted signal into the received signal.

At first glance, it may seem an overwhelming task to understand all of the possible systems in the world. Fortunately, most useful systems fall into a category called **linear systems**. This fact is extremely important. *Without* the linear system concept, we would be forced to examine the individual characteristics of many unrelated systems. *With* this approach, we can focus on the traits of the linear system category as a whole. Our first task is to identify what properties make a system linear, and how they fit into the everyday notion of electronics, software, and other signal processing systems.

Requirements for Linearity

A system is called *linear* if it has two mathematical properties: **homogeneity** (hōma-gen-ā-ity) and **additivity**. If you can show that a system has both properties, then you have proven that the system is linear. Likewise, if you can show that a system doesn't have one or both properties, you have proven that it isn't linear. A third property, **shift invariance**, is not a strict requirement for linearity, but it is a mandatory property for most DSP techniques. When you see the term *linear system* used in DSP, you should assume it includes *shift invariance* unless you have reason to believe otherwise. These three properties form the mathematics of how linear system theory is defined and used. Later in this chapter we will look at more intuitive ways of understanding linearity. For now, let's go through these formal mathematical properties.

As illustrated in Fig. 5-2, homogeneity means that a change in the input signal's amplitude results in a corresponding change in the output signal's amplitude. In mathematical terms, if an input signal of $x[n]$ results in an output signal of $y[n]$, an input of $kx[n]$ results in an output of $ky[n]$, for any input signal and constant, k.

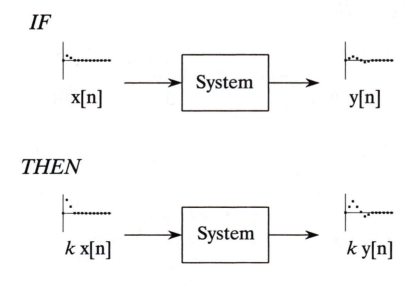

FIGURE 5-2
Definition of homogeneity. A system is said to be *homogeneous* if an amplitude change in the input results in an identical amplitude change in the output. That is, if $x[n]$ results in $y[n]$, then $kx[n]$ results in $ky[n]$, for any signal, $x[n]$, and any constant, k.

A simple resistor provides a good example of both homogenous and non-homogeneous systems.   If the input to the system is the voltage across the resistor, $v(t)$, and the output from the system is the current through the resistor, $i(t)$, the system is homogeneous. Ohm's law guarantees this; if the voltage is increased or decreased, there will be a corresponding increase or decrease in the current. Now, consider another system where the input signal is the voltage across the resistor, $v(t)$, but the output signal is the power being dissipated in the resistor, $p(t)$. Since power is proportional to the square of the voltage, if the input signal is increased by a factor of *two*, the output signal is increase by a factor of *four*. This system is not homogeneous and therefore cannot be linear.

The property of additivity is illustrated in Fig. 5-3. Consider a system where an input of $x_1[n]$ produces an output of $y_1[n]$. Further suppose that a different input, $x_2[n]$, produces another output, $y_2[n]$. The system is said to be *additive*, if an input of $x_1[n]+x_2[n]$ results in an output of $y_1[n]+y_2[n]$, for all possible input signals. In words, signals added at the input produce signals that are added at the output.

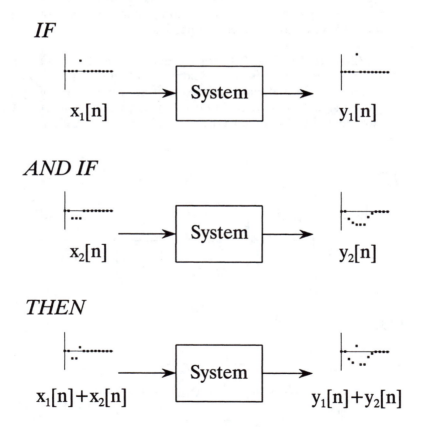

IF

$x_1[n]$ System $y_1[n]$

AND IF

$x_2[n]$ System $y_2[n]$

THEN

$x_1[n]+x_2[n]$ System $y_1[n]+y_2[n]$

FIGURE 5-3
Definition of additivity. A system is said to be *additive* if added signals pass through it without interacting. Formally, if $x_1[n]$ results in $y_1[n]$, and if $x_2[n]$ results in $y_2[n]$, then $x_1[n]+x_2[n]$ results in $y_2[n]+y_2[n]$.

The important point is that *added* signals pass through the system without interacting. As an example, think about a telephone conversation with your Aunt Edna and Uncle Bernie. Aunt Edna begins a rather lengthy story about how well her radishes are doing this year. In the background, Uncle Bernie is yelling at the dog for having an accident in his favorite chair. The two voice signals are added and electronically transmitted through the telephone network. Since this system is additive, the sound you hear is the sum of the two voices as they would sound if transmitted individually. You hear *Edna* and *Bernie*, not the creature, *Ednabernie*.

A good example of a *nonadditive* circuit is the mixer stage in a radio transmitter. Two signals are present: an audio signal that contains the voice or music, and a carrier wave that can propagate through space when applied to an antenna. The two signals are added and applied to a nonlinearity, such as a pn junction diode. This results in the signals *merging* to form a third signal, a modulated radio wave capable of carrying the information over great distances.

As shown in Fig. 5-4, shift invariance means that a shift in the input signal will result in nothing more than an identical shift in the output signal. In more formal terms, if an input signal of $x[n]$ results in an output of $y[n]$, an input signal of $x[n+s]$ results in an output of $y[n+s]$, for any input signal and any constant, s. Pay particular notice to how the mathematics of this shift is written, it will be used in upcoming chapters. By adding a constant, s, to the independent variable, n, the waveform can be advanced or retarded in the horizontal direction. For example, when $s=2$, the signal is shifted *left* by two samples; when $s=-2$, the signal is shifted *right* by two samples.

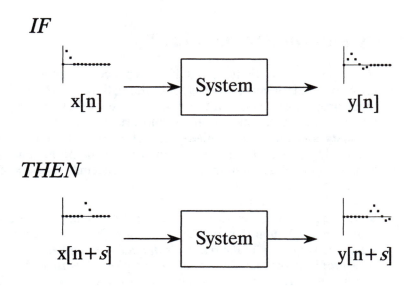

FIGURE 5-4
Definition of shift invariance. A system is said to be *shift invariant* if a shift in the input signal causes an identical shift in the output signal. In mathematical terms, if $x[n]$ produces $y[n]$, then $x[n+s]$ produces $y[n+s]$, for any signal, $x[n]$, and any constant, s.

Shift invariance is important because it means the characteristics of the system do not change with time (or whatever the independent variable happens to be). If a *blip* in the input causes a *blop* in the output, you can be assured that another *blip* will cause an identical *blop*. Most of the systems you encounter will be shift invariant. This is fortunate, because it is difficult to deal with systems that change their characteristics while in operation. For example, imagine that you have designed a digital filter to compensate for the degrading effects of a telephone transmission line. Your filter makes the voices sound more natural and easier to understand. Much to your surprise, along comes winter and you find the characteristics of the telephone line have changed with temperature. Your compensation filter is now mismatched and doesn't work especially well. This situation may require a more sophisticated algorithm that can *adapt* to changing conditions.

Why do homogeneity and additivity play a critical role in linearity, while shift invariance is something on the side? This is because linearity is a very broad concept, encompassing much more than just signals and systems. For example, consider a farmer selling oranges for $2 per crate and apples for $5 per crate. If the farmer sells only oranges, he will receive $20 for 10 crates, and $40 for 20 crates, making the exchange *homogenous*. If he sells 20 crates of oranges and 10 crates of apples, the farmer will receive: $20 \times \$2 + 10 \times \$5 = \$90$. This is the same amount as if the two had been sold individually, making the transaction *additive*. Being both homogenous and additive, this sale of goods is a linear process. However, since there are no signals involved, this is not a *system*, and *shift invariance* has no meaning. Shift invariance can be thought of as an additional aspect of linearity needed when signals and systems are involved.

Static Linearity and Sinusoidal Fidelity

Homogeneity, additivity, and shift invariance are important because they provide the mathematical basis for defining linear systems. Unfortunately, these properties alone don't provide most scientists and engineers with an intuitive feeling of what linear systems are about. The properties of **static linearity** and **sinusoidal fidelity** are often of help here. These are not especially important from a mathematical standpoint, but relate to how humans think about and understand linear systems. You should pay special attention to this section.

Static linearity defines how a linear system reacts when the signals aren't changing, i.e., when they are *DC* or *static*. The static response of a linear system is very simple: *the output is the input multiplied by a constant*. That is, a graph of the possible input values plotted against the corresponding output values is a straight line that passes through the origin. This is shown in Fig. 5-5 for two common linear systems: Ohm's law for resistors, and Hooke's law for springs. For comparison, Fig. 5-6 shows the static relationship for two nonlinear systems: a pn junction diode, and the magnetic properties of iron.

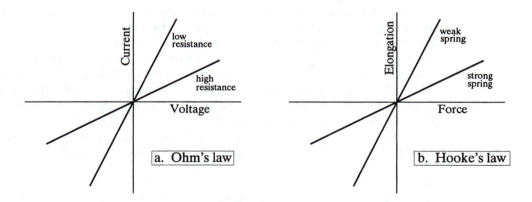

FIGURE 5-5
Two examples of static linearity. In (a), Ohm's law: the current through a resistor is equal to the voltage across the resistor divided by the resistance. In (b), Hooke's law: The elongation of a spring is equal to the applied force multiplied by the spring stiffness coefficient.

All linear systems have the property of *static linearity*. The opposite is usually true, but not always. There are systems that show static linearity, but are not linear with respect to changing signals. However, a very common class of systems can be completely understood with static linearity alone. In these systems it doesn't matter if the input signal is static or changing. These are called **memoryless** systems, because the output depends only on the present state of the input, and not on its history. For example, the instantaneous current in a resistor depends only on the instantaneous voltage across it, and not on how the signals came to be the value they are. If a system has static linearity, and is memoryless, then the system must be linear. This provides an important way to understand (and prove) the linearity of these simple systems.

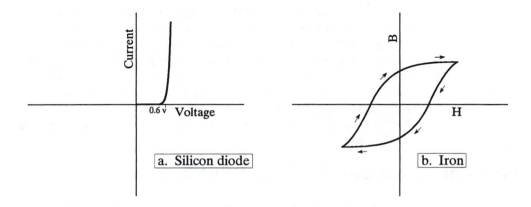

FIGURE 5-6
Two examples of DC nonlinearity. In (a), a silicon diode has an exponential relationship between voltage and current. In (b), the relationship between magnetic intensity, H, and flux density, B, in iron depends on the history of the sample, a behavior called *hysteresis*.

An important characteristic of linear systems is how they behave with sinusoids, a property we will call **sinusoidal fidelity**: *If the input to a linear system is a sinusoidal wave, the output will also be a sinusoidal wave, and at exactly the same frequency as the input.* Sinusoids are the only waveform that have this property. For instance, there is no reason to expect that a square wave entering a linear system will produce a square wave on the output. Although a sinusoid on the input guarantees a sinusoid on the output, the two may be different in *amplitude* and *phase*. This should be familiar from your knowledge of electronics: a circuit can be described by its *frequency response*, graphs of how the circuit's gain and phase vary with frequency.

Now for the reverse question: If a system always produces a sinusoidal output in response to a sinusoidal input, is the system guaranteed to be linear? The answer is no, but the exceptions are rare and usually obvious. For example, imagine an evil demon hiding inside a system, with the goal of trying to mislead you. The demon has an oscilloscope to observe the input signal, and a sine wave generator to produce an output signal. When you feed a sine wave into the input, the demon quickly measures the frequency and adjusts his signal generator to produce a corresponding output. Of course, this system is not linear, because it is not additive. To show this, place the sum of two sine waves into the system. The demon can only respond with a single sine wave for the output. This example is not as contrived as you might think; *phase lock loops* operate in much this way.

To get a better feeling for linearity, think about a technician trying to determine if an electronic device is linear. The technician would attach a sine wave generator to the input of the device, and an oscilloscope to the output. With a sine wave input, the technician would look to see if the output is also a sine wave. For example, the output cannot be clipped on the top or bottom, the top half cannot look different from the bottom half, there must be no distortion where the signal crosses zero, etc. Next, the technician would vary the amplitude of the input and observe the effect on the output signal. If the system is linear, the amplitude of the output must track the amplitude of the input. Lastly, the technician would vary the input signal's frequency, and verify that the output signal's frequency changes accordingly. As the frequency is changed, there will likely be amplitude and phase changes seen in the output, but these are perfectly permissible in a linear system. At some frequencies, the output may even be *zero*, that is, a sinusoid with zero amplitude. If the technician sees all these things, he will conclude that the system is linear. While this conclusion is not a rigorous mathematical proof, the level of confidence is justifiably high.

Examples of Linear and Nonlinear Systems

Table 5-1 provides examples of common linear and nonlinear systems. As you go through the lists, keep in mind the mathematician's view of linearity (*homogeneity*, *additivity*, and *shift invariance*), as well as the informal way most scientists and engineers use (*static linearity* and *sinusoidal fidelity*).

Examples of *Linear* Systems

Wave propagation such as sound and electromagnetic waves

Electrical circuits composed of resistors, capacitors, and inductors

Electronic circuits, such as amplifiers and filters

Mechanical motion from the interaction of masses, springs, and dashpots (dampeners)

Systems described by differential equations such as resistor-capacitor-inductor networks

Multiplication by a constant, that is, amplification or attenuation of the signal

Signal changes, such as echoes, resonances, and image blurring

The unity system where the output is always equal to the input

The null system where the output is always equal to the zero, regardless of the input

Differentiation and integration, and the analogous operations of *first difference* and *running sum* for discrete signals

Small perturbations in an otherwise nonlinear system, for instance, a small signal being amplified by a properly biased transistor

Convolution, a mathematical operation where each value in the output is expressed as the sum of values in the input multiplied by a set of weighing coefficients.

Recursion, a technique similar to convolution, except previously calculated values in the output are used in addition to values from the input

Examples of *Nonlinear* Systems

Systems that do not have static linearity, for instance, the voltage and power in a resistor: $P = V^2 R$, the radiant energy emission of a hot object depending on its temperature: $R = kT^4$, the intensity of light transmitted through a thickness of translucent material: $I = e^{-\alpha T}$, etc.

Systems that do not have sinusoidal fidelity, such as electronics circuits for: peak detection, squaring, sine wave to square wave conversion, frequency doubling, etc.

Common electronic distortion, such as clipping, crossover distortion and slewing

Multiplication of one signal by another signal, such as in amplitude modulation and automatic gain controls

Hysteresis phenomena, such as magnetic flux density versus magnetic intensity in iron, or mechanical stress versus strain in vulcanized rubber

Saturation, such as electronic amplifiers and transformers driven too hard

Systems with a threshold, for example, digital logic gates, or seismic vibrations that are strong enough to pulverize the intervening rock

Table 5-1
Examples of linear and nonlinear systems. Formally, linear systems are defined by the properties of *homogeneity*, *additivity*, and *shift invariance*. Informally, most scientists and engineers think of linear systems in terms of *static linearity* and *sinusoidal fidelity*.

Special Properties of Linearity

Linearity is **commutative**, a property involving the combination of two or more systems. Figure 5-10 shows the general idea. Imagine two systems combined in a **cascade**, that is, the output of one system is the input to the next. If each system is linear, then the overall combination will also be linear. The commutative property states that the order of the systems in the cascade can be rearranged without affecting the characteristics of the overall combination. You probably have used this principle in electronic circuits. For example, imagine a circuit composed of two stages, one for amplification, and one for filtering. Which is best, amplify and then filter, or filter and then amplify? If both stages are linear, the order doesn't make any difference and the overall result is the same. Keep in mind that actual electronics has *nonlinear* effects that may make the order important, for instance: interference, DC offsets, internal noise, slew rate distortion, etc.

FIGURE 5-7
The commutative property for linear systems. When two or more linear systems are arranged in a cascade, the order of the systems does not affect the characteristics of the overall combination.

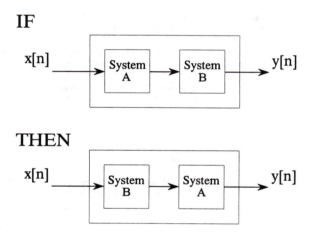

Figure 5-11 shows the next step in linear system theory: multiple inputs and outputs. A system with multiple inputs and/or outputs will be linear if it is composed of *linear subsystems* and *additions of signals*. The complexity does not matter, only that nothing *nonlinear* is allowed inside of the system.

To understand what linearity means for systems with multiple inputs and/or outputs, consider the following thought experiment. Start by placing a signal on one input while the other inputs are held at zero. This will cause the multiple outputs to respond with some pattern of signals. Next, repeat the procedure by placing another signal on a different input. Just as before, keep all of the other inputs at zero. This second input signal will result in another pattern of signals appearing on the multiple outputs. To finish the experiment, place both signals on their respective inputs simultaneously. The signals appearing on the outputs will simply be the superposition (sum) of the output signals produced when the input signals were applied separately.

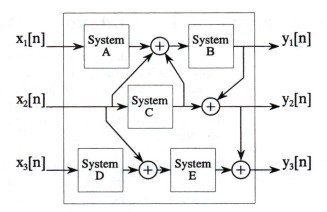

FIGURE 5-8
Any system with multiple inputs and/or
outputs will be linear if it is composed
of linear systems and signal additions.

The use of multiplication in linear systems is frequently misunderstood.
This is because multiplication can be either linear or nonlinear, depending
on what the signal is multiplied by. Figure 5-9 illustrates the two cases. A
system that multiplies the input signal by a *constant*, is linear. This system
is an amplifier or an attenuator, depending if the constant is greater or less
than one, respectively. In contrast, multiplying a signal by *another signal* is
nonlinear. Imagine a sinusoid multiplied by another sinusoid; the resulting
waveform is clearly not sinusoidal.

Another commonly misunderstood situation relates to parasitic signals
added in electronics, such as DC offsets and thermal noise. Is the addition
of these extraneous signals linear or nonlinear? The answer depends on
where the contaminating signals are viewed as originating. If they are
viewed as coming from *within* the system, the process is nonlinear. This is
because a sinusoidal input does not produce a pure sinusoidal output.
Conversely, the extraneous signal can be viewed as *externally* entering the
system on a separate input of a multiple input system. This makes the
process linear, since only a signal addition is required within the system.

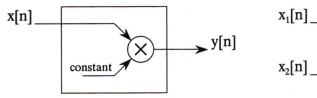

Linear
a. Multiplication by a constant

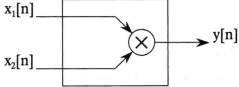

Nonlinear
b. Multiplication of two signals

FIGURE 5-9
Linearity of multiplication. Multiplying a signal by a constant is a linear operation. In
contrast, the multiplication of two signals is nonlinear.

Superposition: the Foundation of DSP

When we are dealing with linear systems, the only way signals can be combined is by *scaling* (multiplication of the signals by constants) followed by *addition*. For instance, a signal cannot be multiplied by another signal. Figure 5-10 shows an example: three signals: $x_0[n]$, $x_1[n]$, and $x_2[n]$ are added to form a fourth signal, $x[n]$. This process of combining signals through scaling and addition is called **synthesis**.

Decomposition is the inverse operation of synthesis, where a single signal is broken into two or more additive components. This is more involved than synthesis, because there are infinite possible decompositions for any given signal. For example, the numbers 15 and 25 can only be synthesized (added) into the number 40. In comparison, the number 40 can be decomposed into: 1+39 or 2+38 or –30.5+60+10.5, etc.

Now we come to the heart of DSP: **superposition**, the overall strategy for understanding how signals and systems can be analyzed. Consider an input

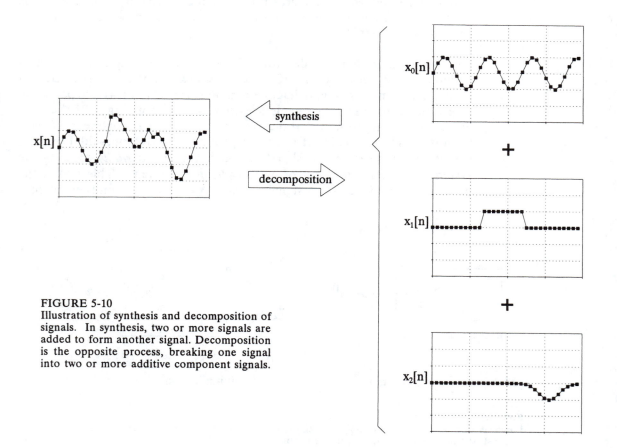

FIGURE 5-10
Illustration of synthesis and decomposition of signals. In synthesis, two or more signals are added to form another signal. Decomposition is the opposite process, breaking one signal into two or more additive component signals.

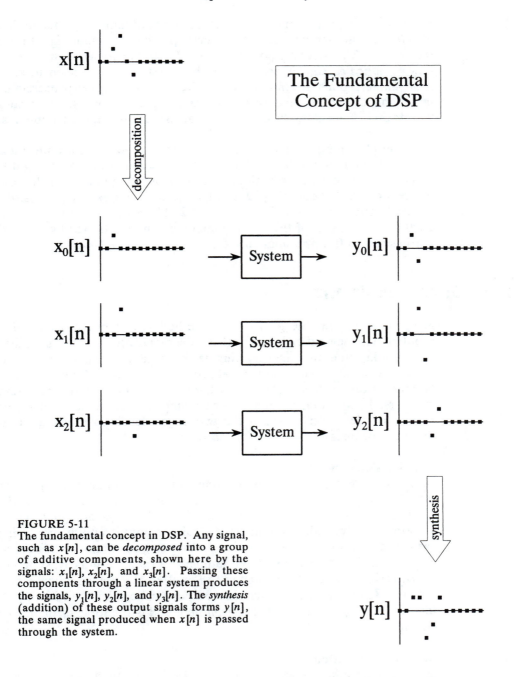

The Fundamental
Concept of DSP

FIGURE 5-11
The fundamental concept in DSP. Any signal,
such as $x[n]$, can be *decomposed* into a group
of additive components, shown here by the
signals: $x_1[n]$, $x_2[n]$, and $x_3[n]$. Passing these
components through a linear system produces
the signals, $y_1[n]$, $y_2[n]$, and $y_3[n]$. The *synthesis*
(addition) of these output signals forms $y[n]$,
the same signal produced when $x[n]$ is passed
through the system.

signal, called $x[n]$, passing through a linear system, resulting in an output
signal, $y[n]$. As illustrated in Fig. 5-11, the input signal can be decomposed
into a group of simpler signals: $x_1[n]$, $x_2[n]$, $x_3[n]$, etc. We will call these
the **input signal components.** Next, each input signal component is
individually passed through the system, resulting in a set of **output signal
components:** $y_1[n]$, $y_2[n]$, $y_3[n]$, etc. These output signal components are
then synthesized into the output signal, $y[n]$.

Here is the important part: the output signal obtained by this method is *identical* to the one produced by directly passing the input signal through the system. This is a very powerful idea. Instead of trying to understanding how *complicated* signals are changed by a system, all we need to know is how *simple* signals are modified. In the jargon of signal processing, the input and output signals are viewed as a *superposition* (sum) of simpler waveforms. This is the basis of nearly all signal processing techniques.

As a simple example of how superposition is used, multiply the number 2041 by the number 4, in your head. How did you do it? You might have imagined 2041 match sticks floating in your mind, quadrupled the mental image, and started counting. Much more likely, you used superposition to simplify the problem. The number 2041 can be decomposed into: $2000 + 40 + 1$. Each of these components can be multiplied by 4 and then synthesized to find the final answer, i.e., $8000 + 160 + 4 = 8164$.

Common Decompositions

Keep in mind that the goal of this method is to replace a complicated problem with several easy ones. If the decomposition doesn't simplify the situation in some way, then nothing has been gained. There are two main ways to decompose signals in signal processing: *impulse decomposition* and *Fourier decomposition*. There are described in detail in the next several chapters. In addition, several minor decompositions are occasionally used. Here are brief descriptions of the two major decompositions, along with three minor ones.

Impulse Decomposition
As shown in Fig. 5-12, impulse decomposition breaks an N samples signal into N component signals, each containing N samples. Each of the component signals contains one point from the original signal, with the remainder of the values being zero. A single nonzero point in a string of zeros is called an *impulse*. Impulse decomposition is important because it allows signals to be examined one sample at a time. Similarly, systems are characterized by how they respond to impulses. By knowing how a system responds to an impulse, the system's output can be calculated for any given input. This approach is called *convolution*, and is the topic of the next two chapters.

Step Decomposition
Step decomposition, shown in Fig. 5-13, also breaks an N sample signal into N component signals, each composed of N samples. Each component signal is a *step*, that is, the first samples have a value of zero, while the last samples are some constant value. Consider the decomposition of an N point signal, $x[n]$, into the components: $x_0[n], x_1[n], x_2[n], \cdots x_{N-1}[n]$. The kth component signal, $x_k[n]$, is composed of zeros for points 0 through $k-1$, while the remaining points have a value of: $x[k] - x[k-1]$. For example, the 5th component signal, $x_5[n]$, is composed of zeros for points 0 through 4, while the remaining samples have a value of: $x[5] - x[4]$ (the difference between

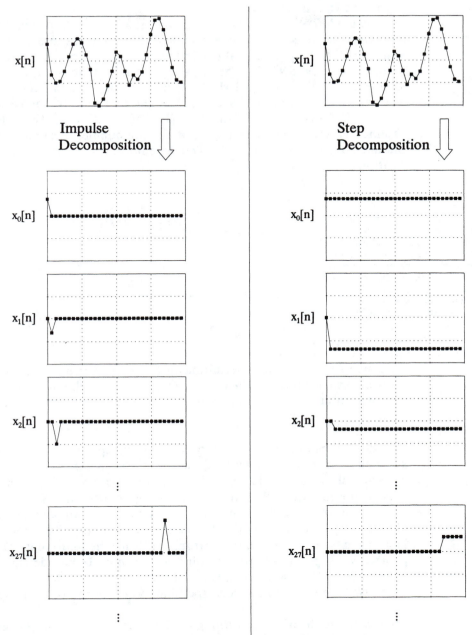

FIGURE 5-12
Example of impulse decomposition. An *N* point signal is broken into *N* components, each consisting of a single nonzero point.

FIGURE 5-13
Example of step decomposition. An *N* point signal is broken into *N* signals, each consisting of a step function

sample 4 and 5 of the original signal). As a special case, $x_0[n]$ has all of its samples equal to $x[0]$. Just as impulse decomposition looks at signals one point at a time, step decomposition characterizes signals by the *difference* between adjacent samples. Likewise, systems are characterized by how they respond to a *change* in the input signal.

Even/Odd Decomposition

The even/odd decomposition, shown in Fig. 5-14, breaks a signal into two component signals, one having **even symmetry** and the other having **odd symmetry**. An N point signal is said to have even symmetry if it is a mirror image around point $N/2$. That is, sample $x[N/2+1]$ must equal $x[N/2-1]$, sample $x[N/2+2]$ must equal $x[N/2-2]$, etc. Similarly, odd symmetry occurs when the matching points have equal magnitudes but are opposite in sign, such as: $x[N/2+1] = -x[N/2-1]$, $x[N/2+2] = -x[N/2-2]$, etc. These definitions assume that the signal is composed of an even number of samples, and that the indexes run from 0 to $N-1$. The decomposition is calculated form the relations:

EQUATION 5-1
Equations for even/odd decomposition. These equations separate a signal, $x[n]$, into its even part, $x_E[n]$, and its odd part, $x_O[n]$. Since this decomposition is based on circularly symmetry, the zeroth samples in the even and odd signals are calculated: $x_E[0] = x[0]$, and $x_O[0] = 0$. All of the signals are N samples long, with indexes running from 0 to $N-1$

$$x_E[n] = \frac{x[n] + x[N-n]}{2}$$

$$x_O[n] = \frac{x[n] - x[N-n]}{2}$$

This may seem a strange definition of left-right symmetry, since $N/2-\frac{1}{2}$ (between two samples) is the exact center of the signal, not $N/2$. Likewise, this off-center symmetry means that sample zero needs special handling. What's this all about?

This decomposition is part of an important concept in DSP called **circular symmetry**. It is based on viewing the *end* of the signal as connected to the *beginning* of the signal. Just as point $x[4]$ is next to point $x[5]$, point $x[N-1]$ is next to point $x[0]$. Picture a snake biting its own tail. When even and odd signals are viewed in this circular manner, there are actually *two* lines of symmetry, one at point $x[N/2]$ and another at point $x[0]$. For example, in an even signal, this symmetry around $x[0]$ means that point $x[1]$ equals point $x[N-1]$, point $x[2]$ equals point $x[N-2]$, etc. In an odd signal, point 0 and point $N/2$ always have a value of zero. In an even signal, point 0 and point $N/2$ are equal to the corresponding points in the original signal.

What is the motivation for viewing the last sample in a signal as being next to the first sample? There is nothing in conventional data acquisition to support this circular notion. In fact, the first and last samples generally have less in common than any other two points in the sequence. It's common sense! The missing piece to this puzzle is a DSP technique called *Fourier analysis*. The mathematics of Fourier analysis inherently views the signal as being circular, although it usually has no physical meaning in terms of where the data came from. We will look at this in more detail in Chapter 10. For now, the important thing to understand is that Eq. 5-1 provides a valid decomposition, simply because the even and odd parts can be added together to reconstruct the original signal.

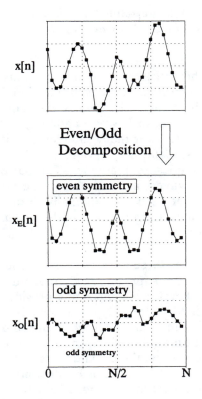

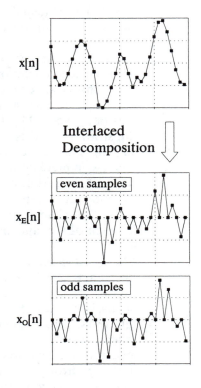

FIGURE 5-14
Example of even/odd decomposition. A N point signal is broken into two N point signals, one with even symmetry, and the other with odd symmetry.

FIGURE 5-15
Example of interlaced decomposition. An N point signal is broken into two N point signals, one with the odd samples set to zero, the other with the even samples set to zero.

Interlaced Decomposition

As shown in Fig. 5-15, the interlaced decomposition breaks the signal into two component signals, the *even sample* signal and the *odd sample* signal (not to be confused with even and odd symmetry signals). To find the even sample signal, start with the original signal and set all of the odd numbered samples to zero. To find the odd sample signal, start with the original signal and set all of the even numbered samples to zero. It's that simple.

At first glance, this decomposition might seem trivial and uninteresting. This is ironic, because the interlaced decomposition is the basis for an extremely important algorithm in DSP, the Fast Fourier Transform (FFT). The procedure for calculating the Fourier decomposition has been know for several hundred years. Unfortunately, it is frustratingly slow, often requiring minutes or hours to execute on present day computers. The FFT is a family of algorithms developed in the 1960s to reduce this computation time. The strategy is an exquisite example of DSP: reduce the signal to elementary components by repeated use of the interlace transform; calculate the Fourier decomposition of the individual components; synthesized the results into the final answer. The results are dramatic; it is common for the speed to be improved by a factor of *hundreds* or *thousands*.

Fourier Decomposition

Fourier decomposition is very mathematical and not at all obvious. Figure 5-16 shows an example of the technique. Any N point signal can be decomposed into $N+2$ signals, half of them sine waves and half of them cosine waves. The lowest frequency cosine wave (called $x_{C0}[n]$ in this illustration), makes *zero* complete cycles over the N samples, i.e., it is a DC signal. The next cosine components: $x_{C1}[n]$, $x_{C2}[n]$, and $x_{C3}[n]$, make 1, 2, and 3 complete cycles over the N samples, respectively. This pattern holds for the remainder of the cosine waves, as well as for the sine wave components. Since the frequency of each component is fixed, the only thing that changes for different signals being decomposed is the *amplitude* of each of the sine and cosine waves.

Fourier decomposition is important for three reasons. First, a wide variety of signals are inherently created from superimposed sinusoids. Audio signals are a good example of this. Fourier decomposition provides a direct analysis of the information contained in these types of signals. Second, linear systems respond to sinusoids in a unique way: a sinusoidal input always results in a sinusoidal output. In this approach, systems are characterized by how they change the amplitude and phase of sinusoids passing through them. Since an input signal can be decomposed into sinusoids, knowing how a system will react to sinusoids allows the output of the system to be found. Third, the Fourier decomposition is the basis for a broad and powerful area of mathematics called *Fourier analysis*, and the even more advanced *Laplace* and *z-transforms*. Most cutting-edge DSP algorithms are based on some aspect of these techniques.

Why is it even possible to decompose an arbitrary signal into sine and cosine waves? How are the amplitudes of these sinusoids determined for a particular signal? What kinds of systems can be designed with this technique? These are the questions to be answered in later chapters. The details of the Fourier decomposition are too involved to be presented in this brief overview. For now, the important idea to understand is that when all of the component sinusoids are added together, the original signal is exactly reconstructed. Much more on this in Chapter 8.

Alternatives to Linearity

To appreciate the importance of linear systems, consider that there is only *one* major strategy for analyzing systems that are nonlinear. That strategy is to make the nonlinear system *resemble* a linear system. There are three common ways of doing this:

First, ignore the nonlinearity. If the nonlinearity is small enough, the system can be approximated as linear. Errors resulting from the original assumption are tolerated as noise or simply ignored.

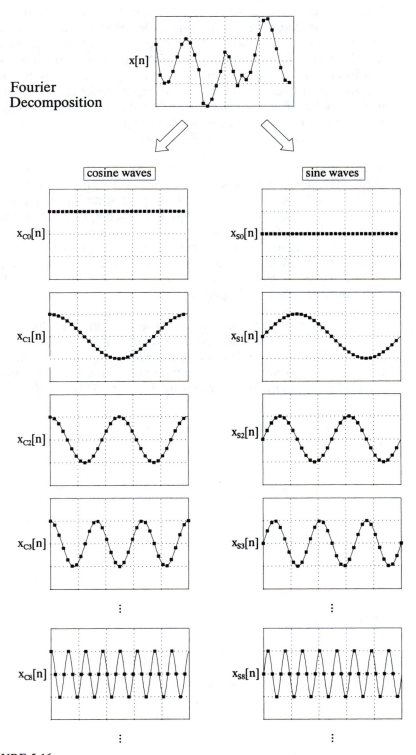

FIGURE 5-16
Illustration of Fourier decomposition. An N point signal is decomposed into $N+2$ signals, each having N points. Half of these signals are cosine waves, and half are sine waves. The frequencies of the sinusoids are fixed; only the amplitudes can change.

Second, keep the signals very small. Many nonlinear systems appear linear if the signals have a very small amplitude. For instance, transistors are very nonlinear over their full range of operation, but provide accurate linear amplification when the signals are kept under a few millivolts. Operational amplifiers take this idea to the extreme. By using very high open-loop gain together with negative feedback, the input signal to the op amp (i.e., the difference between the inverting and noninverting inputs) is kept to only a few microvolts. This minuscule input signal results in excellent linearity from an otherwise ghastly nonlinear circuit.

Third, apply a linearizing transform. For example, consider two signals being multiplied to make a third: $a[n] = b[n] \times c[n]$. Taking the logarithm of the signals changes the nonlinear process of multiplication into the linear process of addition: $\log(a[n]) = \log(b[n]) + \log(c[n])$. The fancy name for this approach is *homomorphic* signal processing. For example, a visual image can be modeled as the reflectivity of the scene (a two-dimensional signal) being multiplied by the ambient illumination (another two-dimensional signal). Homomorphic techniques enable the illumination signal to be made more uniform, thereby improving the image.

In the next chapters we examine the two main techniques of signal processing: *convolution* and *Fourier analysis*. Both are based on the strategy presented in this chapter: (1) decompose signals into simple additive components, (2) process the components in some useful manner, and (3) synthesize the components into a final result. This is DSP.

CHAPTER 6

Convolution

Convolution is a mathematical way of combining two signals to form a third signal. It is the single most important technique in Digital Signal Processing. Using the strategy of impulse decomposition, systems are described by a signal called the *impulse response*. Convolution is important because it relates the three signals of interest: the input signal, the output signal, and the impulse response. This chapter presents convolution from two different viewpoints, called the input side algorithm and the output side algorithm. Convolution provides the mathematical framework for DSP; there is nothing more important in this book.

The Delta Function and Impulse Response

The previous chapter describes how a signal can be decomposed into a group of components called **impulses**. An impulse is a signal composed of all zeros, except a single nonzero point. In effect, impulse decomposition provides a way to analyze signals one sample at a time. The previous chapter also presented the fundamental concept of DSP: the input signal is decomposed into simple additive components, each of these components is passed through a linear system, and the resulting output components are synthesized (added). The signal resulting from this divide-and-conquer procedure is identical to that obtained by directly passing the original signal through the system. While many different decompositions are possible, two form the backbone of signal processing: impulse decomposition and Fourier decomposition. When impulse decomposition is used, the procedure can be described by a mathematical operation called **convolution**. In this chapter (and most of the following ones) we will only be dealing with *discrete* signals. Convolution also applies to *continuous* signals, but the mathematics is more complicated. We will look at how continious signals are processed in Chapter 13.

Figure 6-1 defines two important terms used in DSP. The first is the **delta function**, symbolized by the Greek letter delta, $\delta[n]$. The delta function is a *normalized* impulse, that is, sample number zero has a value of one, while all other samples have a value of zero. For this reason, the delta function is frequently called the **unit impulse**.

The second term defined in Fig. 6-1 is the **impulse response**. As the name suggests, the impulse response is the signal that exits a system when a delta function (unit impulse) is the input. If two systems are different in any way, they will have different impulse responses. Just as the input and output signals are often called $x[n]$ and $y[n]$, the impulse response is usually given the symbol, $h[n]$. Of course, this can be changed if a more descriptive name is available, for instance, $f[n]$ might be used to identify the impulse response of a *filter*.

Any impulse can be represented as a *shifted* and *scaled* delta function. Consider a signal, $a[n]$, composed of all zeros except sample number 8, which has a value of -3. This is the same as a delta function shifted to the right by 8 samples, and multiplied by -3. In equation form: $a[n] = -3\delta[n-8]$. Make sure you understand this notation, it is used in nearly all DSP equations.

If the input to a system is an impulse, such as $-3\delta[n-8]$, what is the system's output? This is where the properties of homogeneity and shift invariance are used. Scaling and shifting the input results in an identical scaling and shifting of the output. If $\delta[n]$ results in $h[n]$, it follows that $-3\delta[n-8]$ results in $-3h[n-8]$. In words, the output is a version of the impulse response that has been *shifted* and *scaled* by the same amount as the delta function on the input. If you know a system's impulse response, you immediately know how it will react to *any* impulse.

Convolution

Let's summarize this way of understanding how a system changes an input signal into an output signal. First, the input signal can be decomposed into a set of impulses, each of which can be viewed as a scaled and shifted delta function. Second, the output resulting from each impulse is a scaled and shifted version of the impulse response. Third, the overall output signal can be found by adding these scaled and shifted impulse responses. In other words, if we know a system's impulse response, then we can calculate what the output will be for any possible input signal. This means we know *everything* about the system. There is nothing more that can be learned about a linear system's characteristics. (However, in later chapters we will show that this information can be represented in different forms).

The impulse response goes by a different name in some applications. If the system being considered is a *filter*, the impulse response is called the **filter kernel**, the **convolution kernel**, or simply, the **kernel**. In image processing, the impulse response is called the **point spread function**. While these terms are used in slightly different ways, they all mean the same thing, the signal produced by a system when the input is a delta function.

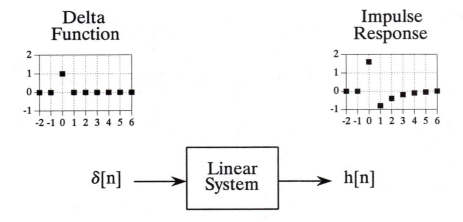

FIGURE 6-1
Definition of *delta function* and *impulse response*. The delta function is a normalized impulse. All of its samples have a value of zero, except for sample number zero, which has a value of one. The Greek letter delta, $\delta[n]$, is used to identify the delta function. The *impulse response* of a linear system, usually denoted by $h[n]$, is the output of the system when the input is a delta function.

Convolution is a formal mathematical operation, just as multiplication, addition, and integration. Addition takes two *numbers* and produces a third *number*, while convolution takes two *signals* and produces a third *signal*. Convolution is used in the mathematics of many fields, such as probability and statistics. In linear systems, convolution is used to describe the relationship between three signals of interest: the input signal, the impulse response, and the output signal.

Figure 6-2 shows the notation when convolution is used with linear systems. An input signal, $x[n]$, enters a linear system with an impulse response, $h[n]$, resulting in an output signal, $y[n]$. In equation form: $x[n] * h[n] = y[n]$. Expressed in words, the input signal convolved with the impulse response is equal to the output signal. Just as addition is represented by the plus, +, and multiplication by the cross, ×, convolution is represented by the star, ∗. It is unfortunate that most programming languages also use the star to indicate multiplication. A star in a computer program means multiplication, while a star in an equation means convolution.

FIGURE 6-2
How convolution is used in DSP. The output signal from a linear system is equal to the input signal *convolved* with the system's impulse response. Convolution is denoted by a star when writing equations.

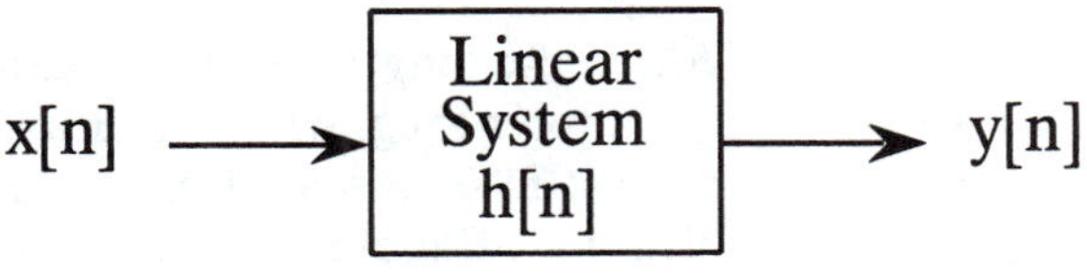

$$x[n] \quad \longrightarrow \quad \boxed{\begin{array}{c} \text{Linear} \\ \text{System} \\ h[n] \end{array}} \quad \longrightarrow \quad y[n]$$

$$x[n] * h[n] = y[n]$$

a. Low-pass Filter

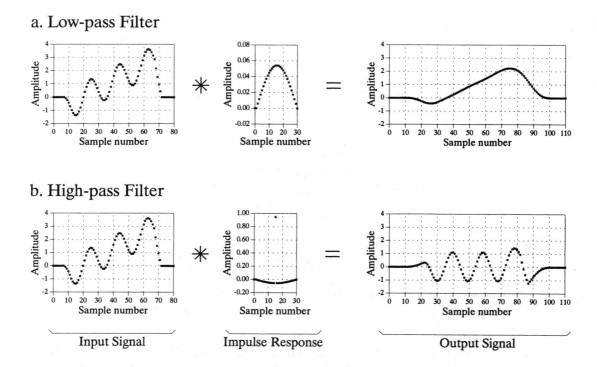

b. High-pass Filter

| Input Signal | Impulse Response | Output Signal |

FIGURE 6-3
Examples of low-pass and high-pass filtering using convolution. In this example, the input signal is a few cycles of a sine wave plus a slowly rising ramp. These two components are separated by using properly selected impulse responses.

Figure 6-3 shows convolution being used for low-pass and high-pass filtering.   The example input signal is the sum of two components: three cycles of a sine wave (representing a high frequency), plus a slowly rising ramp (composed of low frequencies). In (a), the impulse response for the low-pass filter is a smooth arch, resulting in only the slowly changing ramp waveform being passed to the output. Similarly, the high-pass filter, (b), allows only the more rapidly changing sinusoid to pass.

Figure 6-4 illustrates two additional examples of how convolution is used to process signals. The inverting attenuator, (a), flips the signal top-for-bottom, and reduces its amplitude. The discrete derivative (also called the first difference), shown in (b), results in an output signal related to the *slope* of the input signal.

Notice the lengths of the signals in Figs. 6-3 and 6-4. The input signals are 81 samples long, while each impulse response is composed of 31 samples. In most DSP applications, the input signal is hundreds, thousands, or even millions of samples in length. The impulse response is usually much shorter, say, a few points to a few hundred points. The mathematics behind convolution doesn't restrict how long these signals are. It does, however, specify the length of the output signal. The length of the output signal is

a. Inverting Attenuator

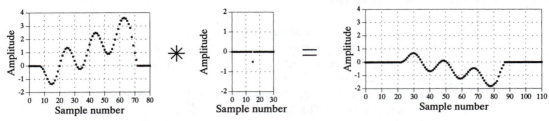

b. Discrete Derivative

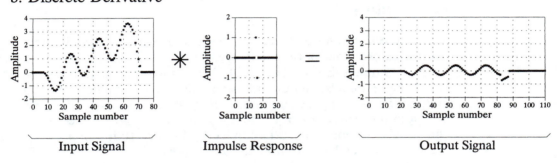

| Input Signal | Impulse Response | Output Signal |

FIGURE 6-4
Examples of signals being processed using convolution. Many signal processing tasks use very simple impulse responses. As shown in these examples, dramatic changes can be achieved with only a few nonzero points.

equal to the length of the input signal, plus the length of the impulse response, minus one. For the signals in Figs. 6-3 and 6-4, each output signal is: $81 + 31 - 1 = 111$ samples long. The input signal runs from sample 0 to 80, the impulse response from sample 0 to 30, and the output signal from sample 0 to 110.

Now we come to the detailed mathematics of convolution. As used in Digital Signal Processing, convolution can be understood in two separate ways. The first looks at convolution from the **viewpoint of the input signal**. This involves analyzing how each sample in the input signal *contributes* to many points in the output signal. The second way looks at convolution from the **viewpoint of the output signal**. This examines how each sample in the output signal has *received* information from many points in the input signal.

Keep in mind that these two perspectives are different ways of thinking about the same mathematical operation. The first viewpoint is important because it provides a *conceptual* understanding of how convolution pertains to DSP. The second viewpoint describes the *mathematics* of convolution. This typifies one of the most difficult tasks you will encounter in DSP: making your conceptual understanding fit with the jumble of mathematics used to communicate the ideas.

The Input Side Algorithm

Figure 6-5 shows a simple convolution problem: a 9 point input signal, $x[n]$, is passed through a system with a 4 point impulse response, $h[n]$, resulting in a $9+4-1 = 12$ point output signal, $y[n]$. In mathematical terms, $x[n]$ is convolved with $h[n]$ to produce $y[n]$. This first viewpoint of convolution is based on the fundamental concept of DSP: decompose the input, pass the components through the system, and synthesize the output. In this example, each of the nine samples in the input signal will contribute a scaled and shifted version of the impulse response to the output signal. These nine signals are shown in Fig. 6-6. Adding these nine signals produces the output signal, $y[n]$.

Let's look at several of these nine signals in detail. We will start with sample number four in the input signal, i.e., $x[4]$. This sample is at index number four, and has a value of 1.4. When the signal is decomposed, this turns into an impulse represented as: $1.4\delta[n-4]$. After passing through the system, the resulting output component will be: $1.4h[n-4]$. This signal is shown in the center box of the nine signals in Fig. 6-6. Notice that this is the impulse response, $h[n]$, multiplied by 1.4, and shifted four samples to the right. Zeros have been added at samples 0-3 and at samples 8-11 to serve as place holders. To make this more clear, Fig. 6-6 uses *squares* to represent the data points that come from the shifted and scaled impulse response, and *diamonds* for the added zeros.

Now examine sample $x[8]$, the last point in the input signal. This sample is at index number eight, and has a value of -0.5. As shown in the lower-right graph of Fig. 6-6, $x[8]$ results in an impulse response that has been shifted to the right by eight points and multiplied by -0.5. Place holding zeros have been added at points 0-7. Lastly, examine the effect of points $x[0]$ and $x[7]$. Both these samples have a value of zero, and therefore produce output components consisting of all zeros.

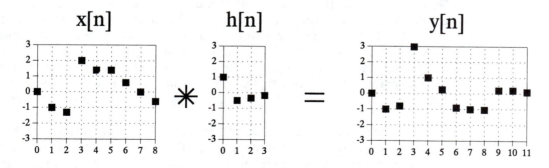

FIGURE 6-5

Example convolution problem. A nine point input signal, convolved with a four point impulse response, results in a twelve point output signal. Each point in the input signal contributes a scaled and shifted impulse response to the output signal. These nine scaled and shifted impulse responses are shown in Fig. 6-6.

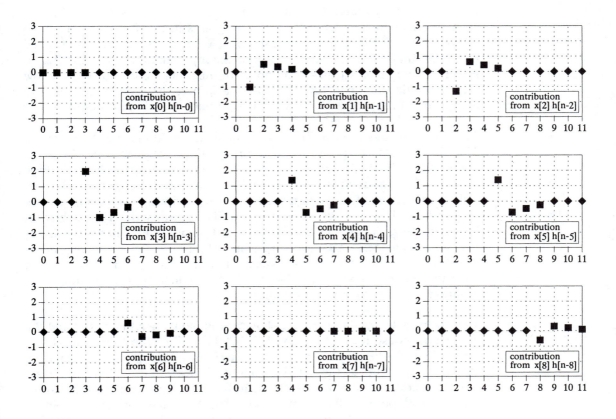

FIGURE 6-6
Output signal components for the convolution in Fig. 6-5. In these signals, each point that results from a scaled and shifted impulse response is represented by a square marker. The remaining data points, represented by diamonds, are zeros that have been added as place holders.

In this example, $x[n]$ is a nine point signal and $h[n]$ is a four point signal. In our next example, shown in Fig. 6-7, we will reverse the situation by making $x[n]$ a four point signal, and $h[n]$ a nine point signal. The same two waveforms are used, they are just swapped. As shown by the output signal components, the four samples in $x[n]$ result in four shifted and scaled versions of the nine point impulse response. Just as before, leading and trailing zeros are added as place holders.

But wait just one moment! The output signal in Fig. 6-7 is *identical* to the output signal in Fig. 6-5. This isn't a mistake, but an important property. Convolution is *commutative*: $a[n]*b[n] = b[n]*a[n]$. The mathematics does not care which is the input signal and which is the impulse response, only that *two signals are convolved with each other*. Although the mathematics may allow it, exchanging the two signals has no physical meaning in system theory. The input signal and impulse response are two totally different things and exchanging them doesn't make sense. What the commutative property provides is a *mathematical tool* for manipulating equations to achieve various results.

A program for calculating convolutions using the input side algorithm is shown in Table 6-1. Remember, the programs in this book are meant to convey *algorithms* in the simplest form, even at the expense of good programming style. For instance, all of the input and output is handled in **mythical** subroutines (lines 160 and 280), meaning we do not define how these operations are conducted. Do not skip over these programs; they are a key part of the material and you need to understand them in detail.

The program convolves an 81 point input signal, held in array X[ ], with a 31 point impulse response, held in array H[ ], resulting in a 111 point output signal, held in array Y[ ]. These are the same lengths shown in Figs. 6-3 and 6-4. Notice that the names of these arrays use upper case letters. This is a violation of the naming conventions previously discussed, because upper case letters are reserved for frequency domain signals. Unfortunately, the simple BASIC used in this book does not allow lower case variable names. Also notice that line 240 uses a star for *multiplication*. Remember, a star in a program means multiplication, while a star in an equation means convolution. A star in text (such as documentation or program comments) can mean either.

The mythical subroutine in line 160 places the input signal into X[ ] and the impulse response into H[ ]. Lines 180-200 set all of the values in Y[ ] to zero. This is necessary because Y[ ] is used as an accumulator to sum the output components as they are calculated. Lines 220 to 260 are the heart of the program. The FOR statement in line 220 controls a loop that steps through each point in the input signal, X[ ]. For each sample in the input signal, an inner loop (lines 230-250) calculates a scaled and shifted version of the impulse response, and adds it to the array accumulating the output signal, Y[ ]. This nested loop structure (one loop within another loop) is a key characteristic of convolution programs; become familiar with it.

```
100  'CONVOLUTION USING THE INPUT SIDE ALGORITHM
110                          '
120 DIM X[80]                'The input signal, 81 points
130 DIM H[30]                'The impulse response, 31 points
140 DIM Y[110]               'The output signal, 111 points
150                          '
160 GOSUB XXXX               'Mythical subroutine to load X[ ] and H[ ]
170                          '
180 FOR I% = 0 TO 110        'Zero the output array
190    Y(I%) = 0
200 NEXT I%
210                          '
220 FOR I% = 0 TO 80         'Loop for each point in X[ ]
230    FOR J% = 0 TO 30      'Loop for each point in H[ ]
240       Y[I%+J%] = Y[I%+J%] + X[I%]*H[J%]
250    NEXT J%
260 NEXT I%
270                          '
280 GOSUB XXXX               'Mythical subroutine to store Y[ ]
290                          '
300 END
```

TABLE 6-1

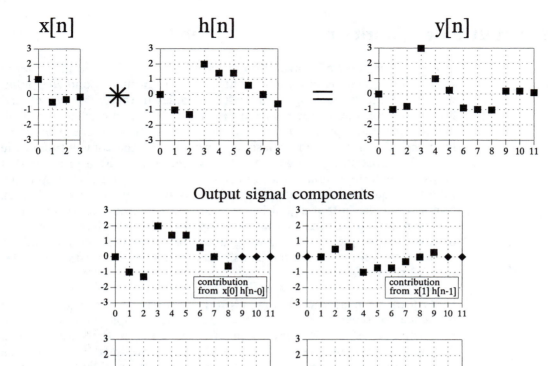

FIGURE 6-7
A second example of convolution. The waveforms for the input signal and impulse response are exchanged from the example of Fig. 6-5. Since convolution is commutative, the output signals for the two examples are identical.

Keeping the indexing straight in line 240 can drive you crazy! Let's say we are halfway through the execution of this program, so that we have just begun action on sample X[40], i.e., I% = 40. The inner loop runs through each point in the impulse response doing three things. First, the impulse response is *scaled* by multiplying it by the value of the input sample. If this were the only action taken by the inner loop, line 240 could be written, Y[J%] = X[40]*H[J%]. Second, the scaled impulse is *shifted* 40 samples to the right by adding this number to the index used in the output signal. This second action would change line 240 to: Y[40+J%] = X[40]*H[J%]. Third, Y[ ] must accumulate (*synthesize*) all the signals resulting from each sample in the input signal. Therefore, the new information must be added to the information that is already in the array. This results in the final command: Y[40+J%] = Y[40+J%] + X[40]*H[J%]. Study this carefully; it is *very* confusing, but *very* important.

The Output Side Algorithm

The first viewpoint of convolution analyzes how each sample in the *input signal* affects many samples in the output signal. In this second viewpoint, we reverse this by looking at individual samples in the *output signal*, and finding the contributing points from the input. This is important from both mathematical and practical standpoints. Suppose that we are given some input signal and impulse response, and want to find the convolution of the two. The most straightforward method would be to write a program that loops through the *output signal*, calculating one sample on each loop cycle. Likewise, equations are written in the form: $y[n] = $ *blah blah blah*. That is, sample n in the output signal is equal to some combination of the many values in the input signal and impulse response. This requires a knowledge of how each sample in the output signal can be calculated independently of all other samples in the output signal. The output side algorithm provides this information.

Let's look an example of how a single point in the output signal is influenced by several points from the input. The example point we will use is $y[6]$ in Fig. 6-5. This point is equal to the sum of all the sixth points in the nine output components, shown in Fig. 6-6. Now, look closely at these nine output components and identify which can affect $y[6]$. That is, find which of these nine signals contains a nonzero sample at the sixth position. Five of the output components only have *added* zeros (the diamond markers) at the sixth sample, and can therefore be ignored. Only four of the output components are capable of having a nonzero value in the sixth position. These are the output components generated from the input samples: $x[3]$, $x[4]$, $x[5]$, and $x[6]$. By adding the sixth sample from each of these output components, it can be found that $y[6]$ is determined from the relation: $y[6] = x[3]h[3] + x[4]h[2] + x[5]h[1] + x[6]h[0]$. That is, four samples from the input signal are multiplied by the four samples in the impulse response, and the products added.

Figure 6-8 illustrates the output side algorithm as a **convolution machine**, a flow diagram of how convolution occurs. Think of the input signal, $x[n]$, and the output signal, $y[n]$, as fixed on the page. The convolution machine, everything inside the dashed box, is free to move left and right as needed. The convolution machine is positioned so that its output is aligned with the output sample being calculated. Four samples from the input signal fall into the inputs of the convolution machine. These values are multiplied by the indicated samples in the impulse response, and the products are added. This produces the value for the output signal, which drops into its proper place. For example, $y[6]$ is shown being calculated from the four input samples: $x[3]$, $x[4]$, $x[5]$, and $x[6]$.

To calculate $y[7]$, the convolution machine moves one sample to the right. This results in another four samples entering the machine, $x[4]$ through $x[7]$, and the value for $y[7]$ dropping into the proper place. This process is repeated for all points in the output signal needing to be calculated.

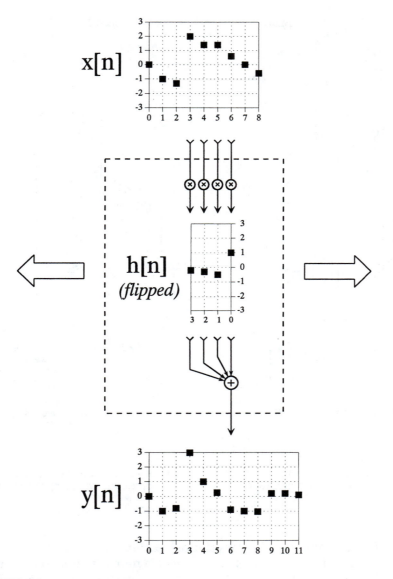

FIGURE 6-8
The convolution machine. This is a flow diagram showing how each sample in the output signal is influenced by the input signal and impulse response. See the text for details.

The arrangement of the impulse response *inside* the convolution machine is very important. The impulse response is *flipped left-for-right*. This places sample number zero on the right, and increasingly positive sample numbers running to the left. Compare this to the normal impulse response in Fig. 6-5 to understand the geometry of this flip. Why is this flip needed? It simply falls out of the mathematics. The impulse response describes how each point in the input signal affects the output signal. This results in each point in the output signal being affected by points in the input signal weighted by a *flipped* impulse response.

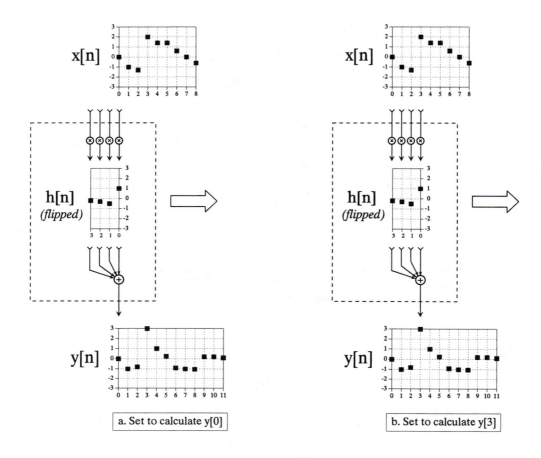

FIGURE 6-9
The convolution machine in action. Figures (a) through (d) show the convolution machine set to calculate four different output signal samples, y[0], y[3], y[8], and y[11].

Figure 6-9 shows the convolution machine being used to calculate several samples in the output signal. This diagram also illustrates a real nuisance in convolution. In (a), the convolution machine is located fully to the left with its output aimed at $y[0]$. In this position, it is trying to receive input from samples: $x[-3]$, $x[-2]$, $x[-1]$, and $x[0]$. The problem is, three of these samples: $x[-3]$, $x[-2]$, and $x[-1]$, do not exist! This same dilemma arises in (d), where the convolution machine tries to accept samples to the right of the defined input signal, points $x[9]$, $x[10]$, and $x[11]$.

One way to handle this problem is by *inventing* the nonexistent samples. This involves adding samples to the ends of the input signal, with each of the added samples having a value of *zero*. This is called **padding** the signal with zeros. Instead of trying to access a nonexistent value, the convolution machine receives a sample that has a value of zero. Since this zero is eliminated during the multiplication, the result is mathematically the same as *ignoring* the nonexistent inputs.

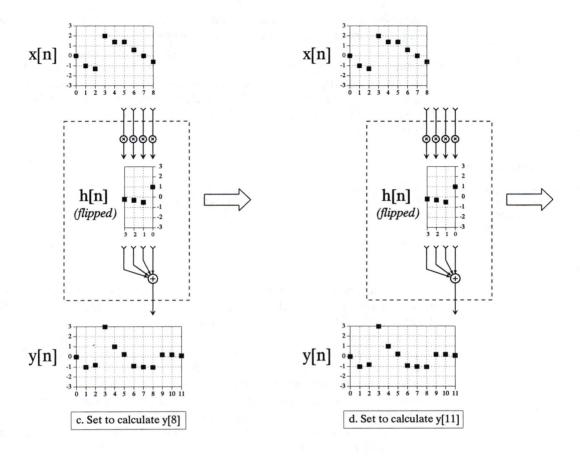

c. Set to calculate y[8]

d. Set to calculate y[11]

FIGURE 6-9 (continued)

The important part is that the far left and far right samples in the output signal are based on *incomplete* information. In DSP jargon, **the impulse response is not fully immersed in the input signal.** If the impulse response is M points in length, the first *and* last M-1 samples in the output signal are based on less information than the samples between. This is analogous to an electronic circuit requiring a certain amount of time to stabilize after the power is applied. The difference is that this transient is easy to ignore in electronics, but very prominent in DSP.

Figure 6-10 shows an example of the trouble these end effects can cause. The input signal is a sine wave plus a DC component. The desire is to remove the DC part of the signal, while leaving the sine wave intact. This calls for a high-pass filter, such as the impulse response shown in the figure. The problem is, the first and last 30 points are a mess! The shape of these end regions can be understood by imagining the input signal padded with 30 zeros on the left side, samples $x[-1]$ through $x[-30]$, and 30 zeros on the right, samples $x[81]$ through $x[110]$. The output signal can then be viewed as a filtered version of this longer waveform. These "end effect" problems are widespread in DSP. As a general rule, expect that the beginning and ending samples in processed signals will be quite useless.

Now the math. Using the convolution machine as a guideline, we can write the standard equation for convolution. If $x[n]$ is an N point signal running from 0 to N-1, and $h[n]$ is an M point signal running from 0 to M-1, the convolution of the two: $y[n] = x[n] * h[n]$, is an $N+M$-1 point signal running from 0 to $N+M$-2, given by:

EQUATION 6-1
The convolution summation. This is the formal definition of convolution, written in the shorthand: $y[n] = x[n] * h[n]$. In this equation, $h[n]$ is an M point signal with indexes running from 0 to M-1.

$$y[i] = \sum_{j=0}^{M-1} h[j]\, x[i-j]$$

This equation is called the **convolution sum**. It allows each point in the output signal to be calculated independently of all other points in the output signal. The index, i, determines which sample in the output signal is being calculated, and therefore corresponds to the left-right position of the convolution machine. In computer programs performing convolution, a loop makes this index run through each sample in the output signal. To calculate one of the output samples, the index, j, is used *inside* of the convolution machine. As j runs through 0 to M-1, each sample in the impulse response, $h[j]$, is multiplied by the proper sample from the input signal, $x[i-j]$. All these products are added to produce the output sample being calculated. Study Eq. 6-1 until you fully understand how it is implemented by the convolution machine. Much of DSP is based on this equation. (Don't be confused by the n in $y[n] = x[n] * h[n]$. This is merely a place holder to indicate that *some* variable is the index into the array. Sometimes the equations are written: $y[\ ] = x[\ ] * h[\ ]$, just to avoid having to bring in a meaningless symbol).

Table 6-2 shows a program for performing convolutions using the *output side algorithm*, a direct use of Eq. 6-1. This program produces the same output signal as the program for the *input side algorithm*, shown previously in Table 6-1. Notice the main difference between these two programs: the input side algorithm loops through each sample in the *input signal* (line 220 of Table 6-1), while the output side algorithm loops through each sample in the *output signal* (line 180 of Table 6-2).

Here is a detailed operation of this program. The FOR-NEXT loop in lines 180 to 250 steps through each sample in the output signal, using I% as the index. For each of these values, an inner loop, composed of lines 200 to 230, calculates the value of the output sample, Y[I%]. The value of Y[I%] is set to zero in line 190, allowing it to accumulate the products inside of the convolution machine. The FOR-NEXT loop in lines 200 to 240 provide a direct implementation of Eq. 6-2. The index, J%, steps through each

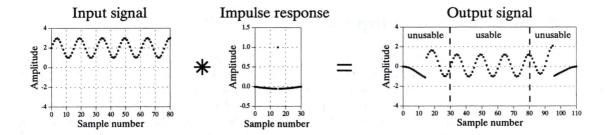

FIGURE 6-10
End effects in convolution. When an input signal is convolved with an *M* point impulse response, the first and last *M*-1 points in the output signal may not be usable. In this example, the impulse response is a high-pass filter used to remove the DC component from the input signal.

sample in the impulse response. Line 230 provides the multiplication of each sample in the impulse response, H[J%], with the appropriate sample from the input signal, X[I%-J%], and adds the result to the accumulator.

In line 230, the sample taken from the input signal is: X[I%-J%]. Lines 210 and 220 prevent this from being outside the defined array, X[0] to X[80]. In other words, this program handles undefined samples in the input signal by *ignoring* them. Another alternative would be to define the input signal's array from X[-30] to X[110], allowing 30 zeros to be padded on each side of the true data. As a third alternative, the FOR-NEXT loop in line 180 could be changed to run from 30 to 80, rather than 0 to 110. That is, the program would only calculate the samples in the output signal where the impulse response is *fully immersed* in the input signal. The important thing is that you must use one of these three techniques. If you don't, the program will crash when it tries to read the out-of-bounds data.

```
100 'CONVOLUTION USING THE OUTPUT SIDE ALGORITHM
110                         '
120 DIM X[80]               'The input signal, 81 points
130 DIM H[30]               'The impulse response, 31 points
140 DIM Y[110]              'The output signal, 111 points
150                         '
160 GOSUB XXXX              'Mythical subroutine to load X[ ] and H[ ]
170                         '
180 FOR I% = 0 TO 110       'Loop for each point in Y[ ]
190   Y[I%] = 0             'Zero the sample in the output array
200   FOR J% = 0 TO 30      'Loop for each point in H[ ]
210     IF (I%-J% < 0)   THEN GOTO 240
220     IF (I%-J% > 80)  THEN GOTO 240
230     Y(I%) = Y(I%) + H(J%) * X(I%-J%)
240   NEXT J%
250 NEXT I%
260                         '
270 GOSUB XXXX              'Mythical subroutine to store Y[ ]
280                         '
290 END
```

TABLE 6-2

The Sum of Weighted Inputs

The characteristics of a linear system are completely described by its impulse response. This is the basis of the input side algorithm: each point in the input signal contributes a scaled and shifted version of the impulse response to the output signal. The mathematical consequences of this lead to the output side algorithm: each point in the output signal receives a contribution from many points in the input signal, multiplied by a *flipped* impulse response. While this is all true, it doesn't provide the full story on why convolution is important in signal processing.

Look back at the convolution machine in Fig. 6-8, and ignore that the signal inside the dotted box is an *impulse response*. Think of it as a set of **weighing coefficients** that happen to be embedded in the flow diagram. In this view, each sample in the output signal is equal to a *sum of weighted inputs*. Each sample in the output is influenced by a region of samples in the input signal, as determined by what the weighing coefficients are chosen to be. For example, imagine there are ten weighing coefficients, each with a value of one-tenth. This makes each sample in the output signal the *average* of ten samples from the input.

Taking this further, the weighing coefficients do not need to be restricted to the *left side* of the output sample being calculated. For instance, Fig. 6-8 shows $y[6]$ being calculated from: $x[3]$, $x[4]$, $x[5]$, and $x[6]$. Viewing the convolution machine as a sum of weighted inputs, the weighing coefficients could be chosen *symmetrically* around the output sample. For example, $y[6]$ might receive contributions from: $x[4]$, $x[5]$, $x[6]$, $x[7]$, and $x[8]$. Using the same indexing notation as in Fig. 6-8, the weighing coefficients for these five inputs would be held in: $h[2]$, $h[1]$, $h[0]$, $h[-1]$, and $h[-2]$. In other words, the impulse response that corresponds to our selection of symmetrical weighing coefficients requires the use of *negative indexes*. We will return to this in the next chapter.

Mathematically, there is only one concept here: convolution as defined by Eq. 6-1. However, science and engineering problems approach this single concept from two distinct directions. Sometimes you will want to think of a system in terms of what its impulse response looks like. Other times you will understand the system as a set of weighing coefficients. You need to become familiar with both views, and how to toggle between them.

CHAPTER 7

Properties of Convolution

A linear system's characteristics are completely specified by the system's impulse response, as governed by the mathematics of convolution. This is the basis of many signal processing techniques. For example: Digital filters are created by *designing* an appropriate impulse response. Enemy aircraft are detected with radar by *analyzing* a measured impulse response. Echo suppression in long distance telephone calls is accomplished by creating an impulse response that *counteracts* the impulse response of the reverberation. The list goes on and on. This chapter expands on the properties and usage of convolution in several areas. First, several common impulse responses are discussed. Second, methods are presented for dealing with cascade and parallel combinations of linear systems. Third, the technique of *correlation* is introduced. Forth, a nasty problem with convolution is examined, the computation time can be unacceptably long using conventional algorithms and computers.

Common Impulse Responses

Delta Function
The simplest impulse response is nothing more that a delta function, as shown in Fig. 7-1a. That is, an impulse on the input produces an identical impulse on the output. This means that *all* signals are passed through the system *without change*. Convolving any signal with a delta function results in exactly the same signal. Mathematically, this is written:

EQUATION 7-1
The delta function is the identity for convolution. Any signal convolved with a delta function is left unchanged.

$$x[n] * \delta[n] = x[n]$$

This property makes the delta function the **identity** for convolution. This is analogous to *zero* being the identity for addition ($a+0=a$), and *one* being the identity for multiplication ($a \times 1 = a$). At first glance, this type of system may seem trivial and uninteresting. Not so! Such systems are the ideal for data storage, communication and measurement. Much of DSP is concerned with passing information through systems without change or degradation.

Figure 7-1b shows a slight modification to the delta function impulse response. If the delta function is made larger or smaller in amplitude, the resulting system is an **amplifier** or **attenuator**, respectively. In equation form, amplification results if *k* is *greater than one*, and attenuation results if *k* is *less than one*:

EQUATION 7-2
A system that amplifies or attenuates has a scaled delta function for an impulse response. In this equation, *k* determines the amplification or attenuation.

$$x[n] * k\delta[n] = kx[n]$$

The impulse response in Fig. 7-1c is a delta function with a **shift**. This results in a system that introduces an identical shift between the input and output signals. This could be described as a signal **delay**, or a signal **advance**, depending on the direction of the shift. Letting the shift be represented by the parameter, *s*, this can be written as the equation:

EQUATION 7-3
A relative shift between the input and output signals corresponds to an impulse response that is a shifted delta function. The variable, s, determines the amount of shift in this equation.

$$x[n] * \delta[n+s] = x[n+s]$$

Science and engineering are filled with cases where one signal is a shifted version of another. For example, consider a radio signal transmitted from a remote space probe, and the corresponding signal received on the earth. The time it takes the radio wave to propagate over the distance causes a delay between the two signals. In biology, the electrical signals in adjacent nerve cells are shifted versions of each other, as determined by the time it takes an action potential to cross the synaptic junction that connects the two.

Figure 7-1d shows an impulse response composed of a delta function plus a shifted and scaled delta function. By superposition, the output of this system is the input signal plus a delayed version of the input signal, i.e., an *echo*. Echoes are important in many DSP applications. The addition of echoes is a key part in making audio recordings sound natural and pleasant. Radar and sonar analyze echoes to detect aircraft and submarines. Geophysicists use echoes to find oil. Echoes are also very important in telephone networks, because you want to *avoid* them.

a. Identity
The delta function is the *identity* for convolution. Convolving a signal with the delta function leaves the signal unchanged. This is the goal of systems that transmit or store signals.

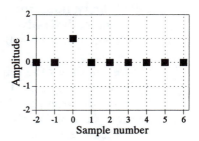

b. Amplification & Attenuation
Increasing or decreasing the amplitude of the delta function forms an impulse response that *amplifies* or *attenuates*, respectively. This impulse response will amplify the signal by 1.6.

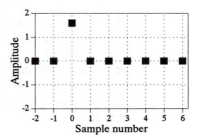

c. Shift
Shifting the delta function produces a corresponding shift between the input and output signals. Depending on the direction, this can be called a *delay* or an *advance*. This impulse response delays the signal by four samples.

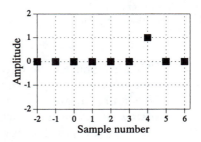

d. Echo
A delta function plus a shifted and scaled delta function results in an *echo* being added to the original signal. In this example, the echo is delayed by four samples and has an amplitude of 60% of the original signal.

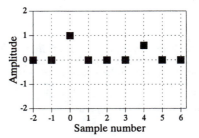

FIGURE 7-1
Simple impulse responses using shifted and scaled delta functions.

Calculus-like Operations
Convolution can change discrete signals in ways that resemble integration and differentiation. Since the terms "derivative" and "integral" specifically refer to operations on *continuous* signals, other names are given to their discrete counterparts. The discrete operation that mimics the *first derivative* is called the **first difference**. Likewise, the discrete form of the *integral* is

called the **running sum**. It is also common to hear these operations called the **discrete derivative** and the **discrete integral**, although mathematicians frown when they hear these informal terms used.

Figure 7-2 shows the impulse responses that implement the first difference and the running sum. Figure 7-3 shows an example using these operations. In 7-3a, the original signal is composed of several sections with varying slopes. Convolving this signal with the first difference impulse response produces the signal in Fig. 7-3b. Just as with the first derivative, the amplitude of each point in the first difference signal is equal to the *slope* at the corresponding location in the original signal. The running sum is the inverse operation of the first difference. That is, convolving the signal in (b), with the running sum's impulse response, produces the signal in (a).

These impulse responses are simple enough that a full convolution program is usually not needed to implement them. Rather, think of them in the alternative mode: each sample in the output signal is a *sum of weighted samples from the input*. For instance, the first difference can be calculated:

EQUATION 7-4
Calculation of the first difference. In this relation, $x[n]$ is the original signal, and $y[n]$ is the first difference.

$$y[n] = x[n] - x[n-1]$$

That is, each sample in the output signal is equal to the difference between two adjacent samples in the input signal. For instance, $y[40] = x[40] - x[39]$. It should be mentioned that this is not the only way to define a *discrete derivative*. Another common method is to define the slope symmetrically around the point being examined, such as: $y[n] = (x[n+1] - x[n-1])/2$.

a. First Difference
This is the discrete version of the *first derivative*. Each sample in the output signal is equal to the *difference* between adjacent samples in the input signal. In other words, the output signal is the *slope* of the input signal.

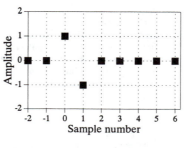

b. Running Sum
The running sum is the discrete version of the *integral*. Each sample in the output signal is equal to the sum of all samples in the input signal to the *left*. Note that the impulse response extends to infinity, a rather nasty feature.

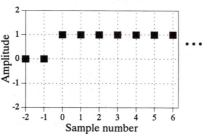

FIGURE 7-2
Impulse responses that mimic calculus operations.

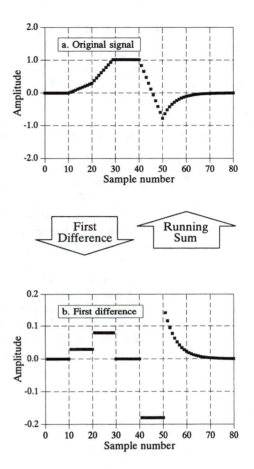

FIGURE 7-3
Example of calculus-like operations. The signal in (b) is the *first difference* of the signal in (a). Correspondingly, the signal is (a) is the *running sum* of the signal in (b). These processing methods are used with discrete signals the same as differentiation and integration are used with continuous signals.

Using this same approach, each sample in the running sum can be calculated by summing all points in the original signal to the *left* of the sample's location. For instance, if $y[n]$ is the running sum of $x[n]$, then sample $y[40]$ is found by adding samples $x[0]$ through $x[40]$. Likewise, sample $y[41]$ is found by adding samples $x[0]$ through $x[41]$. Of course, it would be very inefficient to calculate the running sum in this manner. For example, if $y[40]$ has already been calculated, $y[41]$ can be calculated with only a single addition: $y[41] = x[41]+y[40]$. In equation form:

EQUATION 7-5
Calculation of the running sum. In this relation, $x[n]$ is the original signal, and $y[n]$ is the running sum.

$$y[n] = x[n] + y[n-1]$$

Relations of this type are called **recursion equations** or **difference equations**. We will revisit them in Chapter 19. For now, the important idea to understand is that these relations are *identical* to convolution using the impulse responses of Fig. 7-2. Table 7-1 provides computer programs that implement these calculus-like operations.

```
100 'Calculation of the First Difference        100 'Calculation of the running sum
110 Y[0] = 0                                    110 Y[0] = X[0]
110 FOR I% = 1 TO N%-1                          120 FOR I% = 1 TO N%-1
120   Y[I%] = X[I%] - Y[I%-1]                   120   Y[I%] = Y[I%-1] + X[I%]
130 NEXT I%                                     130 NEXT I%
```

Table 7-1
Programs for calculating the first difference and running sum. The original signal is held in X[ ], and the processed signal (the first difference or running sum) is held in Y[ ]. Both arrays run from 0 to N%-1.

Low-pass and High-pass Filters

The design of digital filters is covered in detail in later chapters. For now, be satisfied to understand the general shape of low-pass and high-pass *filter kernels* (another name for a filter's impulse response). Figure 7-4 shows several common low-pass filter kernels. In general, low-pass filter kernels are composed of a group of *adjacent positive points*. This results in each sample in the output signal being a weighted average of many adjacent points from the input signal. This averaging *smoothes* the signal, thereby removing high-frequency components. As shown by the sinc function in (c), some low-pass filter kernels include a few negative valued samples in the tails. Just as in analog electronics, digital low-pass filters are used for noise reduction, signal separation, wave shaping, etc.

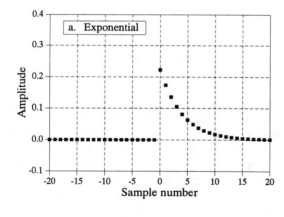

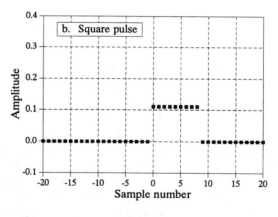

FIGURE 7-4
Typical low-pass filter kernels. Low-pass filter kernels are formed from a group of adjacent positive points that provide an averaging (smoothing) of the signal. As discussed in later chapters, each of these filter kernels is best for a particular purpose. The exponential, (a), is the simplest recursive filter. The rectangular pulse, (b), is best at reducing noise while maintaining edge sharpness. The sinc function in (c), a curve of the form: $\sin(x)/(x)$, is used to separate one band of frequencies from another.

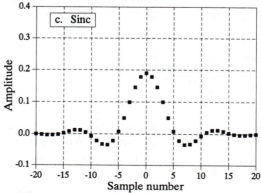

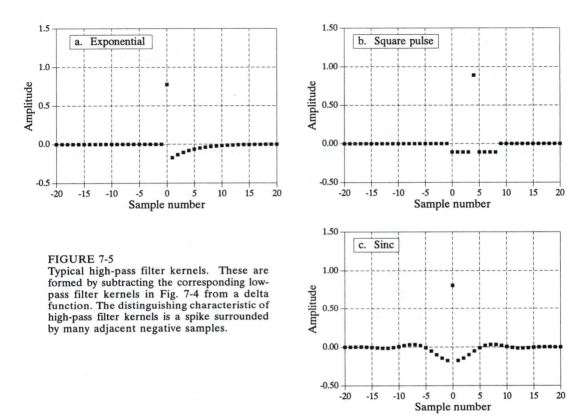

FIGURE 7-5
Typical high-pass filter kernels. These are formed by subtracting the corresponding low-pass filter kernels in Fig. 7-4 from a delta function. The distinguishing characteristic of high-pass filter kernels is a spike surrounded by many adjacent negative samples.

The cutoff frequency of the filter is changed by making filter kernel wider or narrower. If a low-pass filter has a gain of *one* at DC (zero frequency), then the sum of all of the points in the impulse response must be equal to *one*. As illustrated in (a) and (c), some filter kernels *theoretically* extend to infinity without dropping to a value of zero. In actual practice, the tails are truncated after a certain number of samples, allowing it to be represented by a finite number of points. How else could it be stored in a computer?

Figure 7-5 shows three common high-pass filter kernels, derived from the corresponding low-pass filter kernels in Fig. 7-4. This is a common strategy in filter design: first devise a low-pass filter and then transform it to what you need, high-pass, band-pass, band-reject, etc. To understand the low-pass to high-pass transform, remember that a delta function impulse response passes the entire signal, while a low-pass impulse response passes only the low-frequency components. By superposition, a filter kernel consisting of a delta function minus the low-pass filter kernel will pass the entire signal minus the low-frequency components. A high-pass filter is born! As shown in Fig. 7-5, the delta function is usually added at the center of symmetry, or sample zero if the filter kernel is not symmetrical. High-pass filters have zero gain at DC (zero frequency), achieved by making the sum of all the points in the filter kernel equal to *zero*.

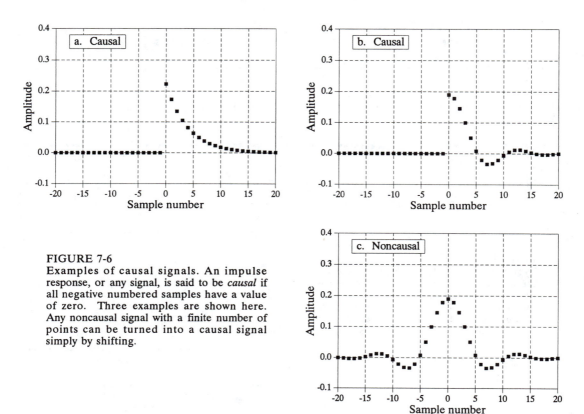

FIGURE 7-6
Examples of causal signals. An impulse
response, or any signal, is said to be *causal* if
all negative numbered samples have a value
of zero. Three examples are shown here.
Any noncausal signal with a finite number of
points can be turned into a causal signal
simply by shifting.

Causal and Noncausal Signals

Imagine a simple analog electronic circuit. If you apply a short pulse to the
input, you will see a response on the output. This is the kind of cause and
effect that our universe is based on. One thing we definitely know: *any
effect must happen after the cause.* This is a basic characteristic of what we
call *time.* Now compare this to a DSP system that changes an input signal
into an output signal, both stored in arrays in a computer. If this mimics
a real world system, it must follow the same principle of *causality* as the real
world does. For example, the value at sample number eight in the input
signal can only affect sample number eight or greater in the output signal.
Systems that operate in this manner are said to be **causal.** Of course,
digital processing doesn't necessarily have to function this way. Since both
the input and output signals are arrays of numbers stored in a computer,
any of the input signal values can affect any of the output signal values.

As shown by the examples in Fig. 7-6, the impulse response of a causal
system must have a value of zero for all *negative numbered* samples. Think
of this from the input side view of convolution. To be causal, an impulse
in the input signal at sample number *n* must only affect those points in the
output signal with a sample number of *n* or greater. In common usage, the
term *causal* is applied to *any* signal where all the negative numbered
samples have a value of zero, whether it is an impulse response or not.

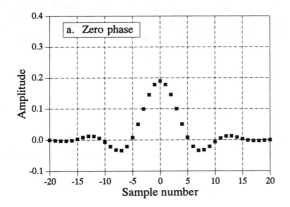

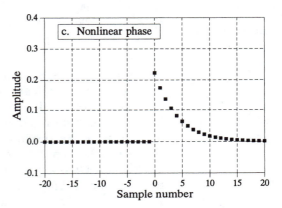

FIGURE 7-7
Examples of phase linearity. Signals that have a left-right symmetry are said to be *linear phase*. If the axis of symmetry occurs at sample number zero, they are additionally said to be *zero phase*. Any linear phase signal can be transformed into a zero phase signal simply by shifting. Signals that do not have a left-right symmetry are said to be *nonlinear phase*. Do not confuse these terms with the *linear* in linear systems. They are completely different concepts.

Zero Phase, Linear Phase, and Nonlinear Phase
As shown in Fig. 7-7, a signal is said to be **zero phase** if it has left-right symmetry around sample number zero. A signal is said to be **linear phase** if it has left-right symmetry, but around some point other than zero. This means that any linear phase signal can be changed into a zero phase signal simply by shifting left or right. Lastly, a signal is said to be **nonlinear phase** if it does not have left-right symmetry.

You are probably thinking that these names don't seem to follow from their definitions. What does *phase* have to do with *symmetry*? The answer lies in the frequency spectrum, and will be discussed in more detail in later chapters. Briefly, the frequency spectrum of any signal is composed of two parts, the magnitude and the phase. The frequency spectrum of a signal that is symmetrical around zero has a phase that is zero. Likewise, the frequency spectrum of a signal that is symmetrical around some nonzero point has a phase that is a straight line, i.e., a linear phase. Lastly, the frequency spectrum of a signal that is not symmetrical has a phase that is not a straight line, i.e., it has a nonlinear phase.

A special note about the potentially confusing terms: *linear* and *nonlinear phase*. What does this have to do the concept of system linearity discussed in previous chapters? Absolutely nothing! System linearity is the broad concept that nearly all of DSP is based on (superposition, homogeneity, additivity, etc). *Linear* and *nonlinear phase* mean that the phase is, or is not, a straight line. In fact, a system must be *linear* even to say that the phase is zero, linear, or nonlinear.

Mathematical Properties

Commutative Property
The commutative property for convolution is expressed in mathematical form:

EQUATION 7-6
The commutative property of convolution. This states that the order in which signals are convolved can be exchanged.

$$a[n] * b[n] = b[n] * a[n]$$

In words, the order in which two signals are convolved makes no difference; the results are identical. As shown in Fig. 7-8, this has a strange meaning for system theory. In any linear system, the input signal and the system's impulse response can be *exchanged* without changing the output signal. This is interesting, but usually doesn't have any physical meaning. The input signal and the impulse response are very different things. Just because the mathematics *allows* you to do something, doesn't mean that it makes sense to do it. For example, suppose you make: \$10/hour × 2,000 hours/year = \$20,000/year. The commutative property for multiplication provides that you can make the same annual salary by only working 10 hours/year at \$2000/hour. Let's see you convince your boss that this is meaningful! In spite of this, the commutative property sees great use in DSP for manipulating equations, just as in ordinary algebra.

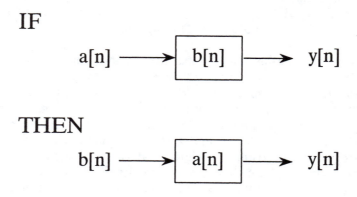

FIGURE 7-8
The commutative property in system theory. The commutative property of convolution allows the input signal and the impulse response of a system to be exchanged without changing the output. While interesting, this usually has no physical significance.

Associative Property

Is it possible to convolve three or more signals? The answer is yes, and the associative property describes how: convolve two of the signals to produce an intermediate signal, then convolve the intermediate signal with the third signal. The associative property provides that the order of the convolutions doesn't matter. As an equation:

EQUATION 7-7
The associative property of convolution describes how three or more signals are convolved.

$$\Big[a[n] * b[n] \Big] * c[n] = a[n] * \Big[b[n] * c[n] \Big]$$

The associative property is used in system theory to describe how **cascaded systems** behave. As shown in Fig. 7-9, two or more systems are said to be in a *cascade* if the output of one system is used as the input for the next system. From the associative property, the order of the systems can be rearranged without changing the overall response of the cascade. Further, any number of cascaded systems can be replaced with a *single* system. The impulse response of the replacement system is found by convolving the impulse responses of all of the original systems.

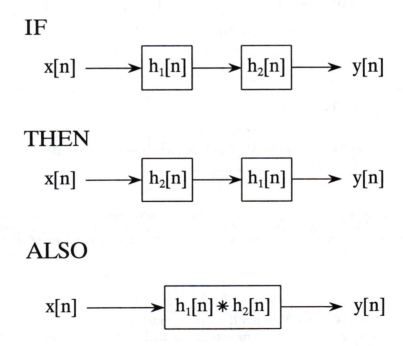

FIGURE 7-9
The associative property in system theory. The associative property provides two important characteristics of *cascaded* linear systems. First, the order of the systems can be rearranged without changing the overall operation of the cascade. Second, two or more systems in a cascade can be replaced by a single system. The impulse response of the replacement system is found by convolving the impulse responses of the stages being replaced.

Distributive Property

In equation form, the distributive property is written:

EQUATION 7-8
The distributive property of convolution describes how parallel
systems are analyzed.

$$a[n]*b[n] + a[n]*c[n] = a[n] * \Big[b[n]+c[n]\Big]$$

The distributive property describes the operation of **parallel systems with added outputs**. As shown in Fig. 7-10, two or more systems can share the same input, $x[n]$, and have their outputs added to produce $y[n]$. The distributive property allows this combination of systems to be replaced with a single system, having an impulse response equal to the *sum* of the impulse responses of the original systems.

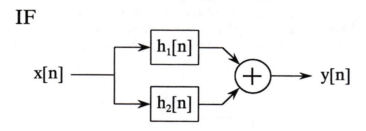

FIGURE 7-10
The distributive property in system theory. The distributive property shows that parallel systems with added outputs can be replaced with a single system. The impulse response of the replacement system is equal to the sum of the impulse responses of all the original systems.

Transference between the Input and Output

Rather than being a formal mathematical property, this is a way of thinking about a common situation in signal processing. As illustrated in Fig. 7-11, imagine a linear system receiving an input signal, $x[n]$, and generating an output signal, $y[n]$. Now suppose that the input signal is changed in some linear way, resulting in a new input signal, which we will call $x'[n]$. This results in a new output signal, $y'[n]$. The question is, how does the change in the input signal relate to the change in the output signal? The answer is: *the output signal is changed in exactly the same linear way that the input signal was changed*. For example, if the input signal is amplified by a factor of two, the output signal will also be amplified by a factor of two. If the derivative is taken of the input signal, the derivative will also be taken of the output signal. If the input is filtered in some way, the output will be filtered in an identical manner. This can easily be proven by using the associative property.

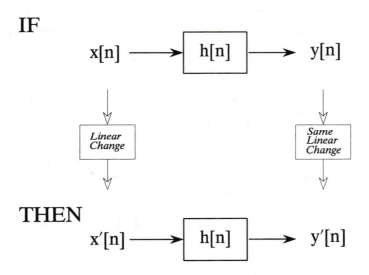

FIGURE 7-11
Tranference between the input and output. This is a way of thinking about a common situation in signal processing. A linear change made to the input signal results in the same linear change being made to the output signal.

The Central Limit Theorem

The Central Limit Theorem is an important tool in probability theory because it mathematically explains why the Gaussian probability distribution is observed so commonly in nature. For example: the amplitude of thermal noise in electronic circuits follows a Gaussian distribution; the cross-sectional intensity of a laser beam is Gaussian; even the pattern of holes around a dart board bull's eye is Gaussian. In its simplest form, the Central Limit Theorem states that a Gaussian distribution results when the observed variable is the sum of many random processes. Even if the component processes do not have a Gaussian distribution, the sum of them will.

The Central Limit Theorem has an interesting implication for convolution. If a pulse-like signal is convolved with *itself* many times, a Gaussian is produced. Figure 7-12 shows an example of this. The signal in (a) is an

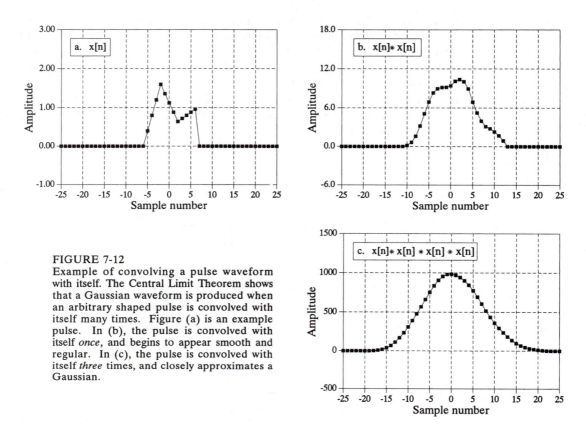

FIGURE 7-12
Example of convolving a pulse waveform with itself. The Central Limit Theorem shows that a Gaussian waveform is produced when an arbitrary shaped pulse is convolved with itself many times. Figure (a) is an example pulse. In (b), the pulse is convolved with itself *once*, and begins to appear smooth and regular. In (c), the pulse is convolved with itself *three* times, and closely approximates a Gaussian.

irregular pulse, purposely chosen to be very unlike a Gaussian. Figure (b) shows the result of convolving this signal with itself *one* time. Figure (c) shows the result of convolving this signal with itself *three* times. Even with only three convolutions, the waveform looks very much like a Gaussian. In mathematics jargon, the procedure *converges* to a Gaussian *very quickly*. The width of the resulting Gaussian (i.e., σ in Eq. 2-7 or 2-8) is equal to the width of the original pulse (expressed as σ in Eq. 2-7) multiplied by the square root of the number of convolutions.

Correlation

The concept of correlation can best be presented with an example. Figure 7-13 shows the key elements of a radar system. A specially designed antenna transmits a short burst of radio wave energy in a selected direction. If the propagating wave strikes an object, such as the helicopter in this illustration, a small fraction of the energy is reflected back toward a radio receiver located near the transmitter. The transmitted pulse is a specific shape that we have selected, such as the triangle shown in this example. The received signal will consist of two parts: (1) a shifted and scaled version of the transmitted pulse, and (2) random noise, resulting from interfering radio waves, thermal noise in the electronics, etc. Since radio signals travel at a known rate, the speed of light, the shift between the transmitted and received pulse is a direct measure of the distance to the object being detected. This is the problem: given a signal of some known shape, what is the best way to determine where (or if) the signal occurs in *another* signal. Correlation is the answer.

Correlation is a mathematical operation that is very similar to convolution. Just as with convolution, correlation uses two signals to produce a third signal. This third signal is called the **cross-correlation** of the two input signals. If a signal is correlated with *itself*, the resulting signal is instead called the **autocorrelation**. The convolution machine was presented in the last chapter to show how convolution is performed. Figure 7-14 is a similar

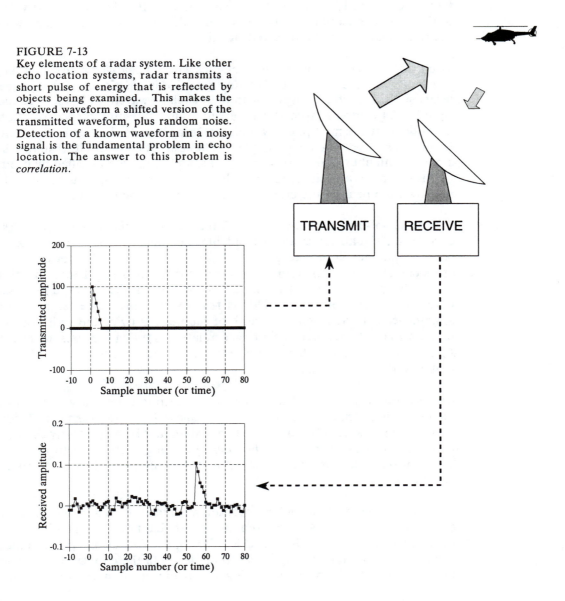

FIGURE 7-13
Key elements of a radar system. Like other echo location systems, radar transmits a short pulse of energy that is reflected by objects being examined. This makes the received waveform a shifted version of the transmitted waveform, plus random noise. Detection of a known waveform in a noisy signal is the fundamental problem in echo location. The answer to this problem is *correlation*.

illustration of a **correlation machine**. The received signal, $x[n]$, and the cross-correlation signal, $y[n]$, are fixed on the page. The waveform we are looking for, $t[n]$, commonly called the **target** signal, is contained *within* the correlation machine. Each sample in $y[n]$ is calculated by moving the correlation machine left or right until it points to the sample being worked on. Next, the indicated samples from the received signal fall into the correlation machine, and are multiplied by the corresponding points in the target signal. The sum of these products then moves into the proper sample in the cross-correlation signal.

The amplitude of each sample in the cross-correlation signal is a measure of how much the received signal *resembles* the target signal, *at that location*. This means that a peak will occur in the cross-correlation signal for every target signal that is present in the received signal. In other words, the value of the cross-correlation is maximized when the target signal is *aligned* with the same features in the received signal.

What if the target signal contains samples with a negative value? Nothing changes. Imagine that the correlation machine is positioned such that the target signal is perfectly aligned with the matching waveform in the received signal. As samples from the received signal fall into the correlation machine, they are multiplied by their matching samples in the target signal. Neglecting noise, a positive sample will be multiplied by itself, resulting in a positive number. Likewise, a negative sample will be multiplied by itself, also resulting in a positive number. Even if the target signal is completely negative, the peak in the cross-correlation will still be positive.

If there is noise on the received signal, there will also be noise on the cross-correlation signal. It is an unavoidable fact that random noise looks a certain amount like any target signal you can choose. The noise on the cross-correlation signal is simply measuring this similarity. Except for this noise, the peak generated in the cross-correlation signal is symmetrical between its left and right. This is true even if the target signal isn't symmetrical. In addition, the width of the peak is twice the width of the target signal. Remember, the cross-correlation is trying to *detect* the target signal, not *recreate* it. There is no reason to expect that the peak will even look like the target signal.

Correlation is the *optimal* technique for detecting a known waveform in random noise. That is, the peak is higher above the noise using correlation than can be produced by any other linear system. (To be perfectly correct, it is only optimal for *random white noise*). Using correlation to detect a known waveform is frequently called **matched filtering**. More on this in Chapter 17.

The correlation machine and convolution machine are identical, except for one small difference. As discussed in the last chapter, the signal inside of the convolution machine is *flipped* left-for-right. This means that samples numbers: 1, 2, 3 ⋯ run from the right to the left. In the correlation machine this flip doesn't take place, and the samples run in the normal direction.

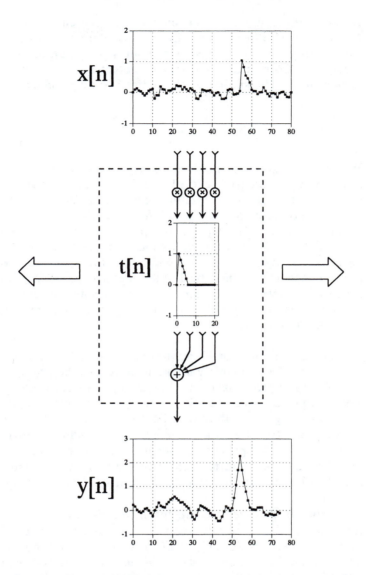

FIGURE 7-14
The correlation machine. This is a flowchart showing how the cross-correlation of two signals is calculated. In this example, $y[n]$ is the cross-correlation of $x[n]$ and $t[n]$. The dashed box is moved left or right so that its output points at the sample being calculated in $y[n]$. The indicated samples from $x[n]$ are multiplied by the corresponding samples in $t[n]$, and the products added. The correlation machine is identical to the convolution machine (Figs. 6-8 and 6-9), except that the signal inside of the dashed box is *not* reversed. In this illustration, the only samples calculated in $y[n]$ are where $t[n]$ is fully *immersed* in $x[n]$.

Since this signal reversal is the only difference between the two operations, it is possible to represent *correlation* using the same mathematics as *convolution*. This requires *preflipping* one of the two signals being correlated, so that the left-for-right flip inherent in convolution is canceled. For instance, when $a[n]$ and $b[n]$, are convolved to produce $c[n]$, the equation is written: $a[n]*b[n] = c[n]$. In comparison, the cross-correlation of $a[n]$ and $b[n]$ can be written: $a[n]*b[-n] = c[n]$. That is, flipping $b[n]$ left-for-right is accomplished by reversing the sign of the index, i.e., $b[-n]$.

Don't let the mathematical similarity between convolution and correlation fool you; they represent very different DSP procedures. Convolution is the relationship between a system's input signal, output signal, and impulse response. Correlation is a way to detect a known waveform in a noisy background. The similar mathematics is only a convenient coincidence.

Speed

Writing a program to convolve one signal by another is a simple task, only requiring a few lines of code. Executing the program may be more painful. The problem is the large number of additions and multiplications required by the algorithm, resulting in long execution times. As shown by the programs in the last chapter, the time-consuming operation is composed of multiplying two numbers and adding the result to an accumulator. Other parts of the algorithm, such as indexing the arrays, are very quick. The multiply-accumulate is a basic building block in DSP, and we will see it repeated in several other important algorithms. In fact, the speed of DSP computers is often *specified* by how long it takes to preform this multiply-accumulate operation.

If a signal composed of N samples is convolved with a signal composed of M samples, $N \times M$ multiply-accumulations must be preformed. This can be seen from the programs of the last chapter. Personal computers of the mid 1990's requires about one microsecond per multiply-accumulation (100 MHz Pentium using single precision floating point, see Table 4-6). Therefore, convolving a 10,000 sample signal with a 100 sample signal requires about one second. To process a one million point signal with a 3000 point impulse response requires nearly an hour. A decade earlier (80286 at 12 MHz), this calculation would have required three days!

The problem of excessive execution time is commonly handled in one of three ways. First, simply keep the signals as short as possible and use integers instead of floating point. If you only need to run the convolution a few times, this will probably be the best trade-off between execution time and programming effort. Second, use a computer designed for DSP. DSP microprocessors are available with multiply-accumulate times of only a few tens of nanoseconds. This is the route to go if you plan to perform the convolution many times, such as in the design of commercial products.

The third solution is to use a better algorithm for implementing the convolution. Chapter 17 describes a very sophisticated algorithm called *FFT convolution*. FFT convolution produces exactly the same result as the convolution algorithms presented in the last chapter; however, the execution time is dramatically reduced. For signals with thousands of samples, FFT convolution can be hundreds of times faster. The disadvantage is program complexity. Even if you are familiar with the technique, expect to spend several hours getting the program to run.

CHAPTER 8

The Discrete Fourier Transform

Fourier analysis is a family of mathematical techniques, all based on decomposing signals into sinusoids. The discrete Fourier transform (DFT) is the family member used with *digitized* signals. This is the first of four chapters on the **real DFT**, a version of the discrete Fourier transform that uses real numbers to represent the input and output signals. The **complex DFT**, a more advanced technique that uses complex numbers, will be discussed in Chapter 29. In this chapter we look at the mathematics and algorithms of the Fourier decomposition, the heart of the DFT.

The Family of Fourier Transform

Fourier analysis is named after **Jean Baptiste Joseph Fourier** (1768-1830), a French mathematician and physicist. (Fourier is pronounced: for·ē·ā, and is always capitalized). While many contributed to the field, Fourier is honored for his mathematical discoveries and insight into the practical usefulness of the techniques. Fourier was interested in heat propagation, and presented a paper in 1807 to the Institut de France on the use of sinusoids to represent temperature distributions. The paper contained the controversial claim that any continuous periodic signal could be represented as the sum of properly chosen sinusoidal waves. Among the reviewers were two of history's most famous mathematicians, Joseph Louis Lagrange (1736-1813), and Pierre Simon de Laplace (1749-1827).

While Laplace and the other reviewers voted to publish the paper, Lagrange adamantly protested. For nearly 50 years, Lagrange had insisted that such an approach could not be used to represent signals with *corners*, i.e., discontinuous slopes, such as in square waves. The Institut de France bowed to the prestige of Lagrange, and rejected Fourier's work. It was only after Lagrange died that the paper was finally published, some 15 years later. Luckily, Fourier had other things to keep him busy, political activities, expeditions to Egypt with Napoleon, and trying to avoid the guillotine after the French Revolution (literally!).

Who was right? It's a split decision. Lagrange was correct in his assertion that a summation of sinusoids cannot form a signal with a corner. However, you can get *very* close. So close that the difference between the two has *zero energy*. In this sense, Fourier was right, although 18th century science knew little about the concept of energy. This phenomenon now goes by the name: *Gibbs Effect*, and will be discussed in Chapter 11.

Figure 8-1 illustrates how a signal can be decomposed into sine and cosine waves. Figure (a) shows an example signal, 16 points long, running from sample number 0 to 15. Figure (b) shows the Fourier decomposition of this signal, nine cosine waves and nine sine waves, each with a different frequency and amplitude. Although far from obvious, these 18 sinusoids

FIGURE 8-1a
(see facing page)

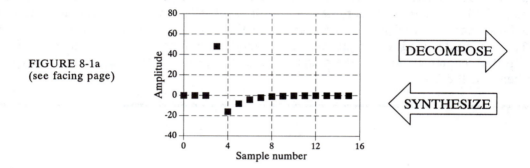

add to produce the waveform in (a). It should be noted that the objection made by Lagrange only applies to *continuous* signals. For *discrete* signals, this decomposition is mathematically exact. There is no difference between the signal in (a) and the *sum* of the signals in (b), just as there is no difference between 7 and 3+4.

Why are sinusoids used instead of, for instance, square or triangular waves? Remember, there are an infinite number of ways that a signal can be decomposed. The goal of decomposition is to end up with something *easier* to deal with than the original signal. For example, impulse decomposition allows signals to be examined one point at a time, leading to the powerful technique of convolution. The component sine and cosine waves are simpler than the original signal because they have a property that the original signal does not have: *sinusoidal fidelity*. As discussed in Chapter 5, a sinusoid input to a system is guaranteed to produce a sinusoidal output. Only the amplitude and phase of the signal can change; the frequency and wave shape must remain the same. Sinusoids are the only waveform that have this useful property. While square and triangular decompositions are *possible*, there is no general reason for them to be *useful*.

The general term: *Fourier transform*, can be broken into four categories, resulting from the four basic types of signals that can be encountered.

Cosine Waves

Sine Waves

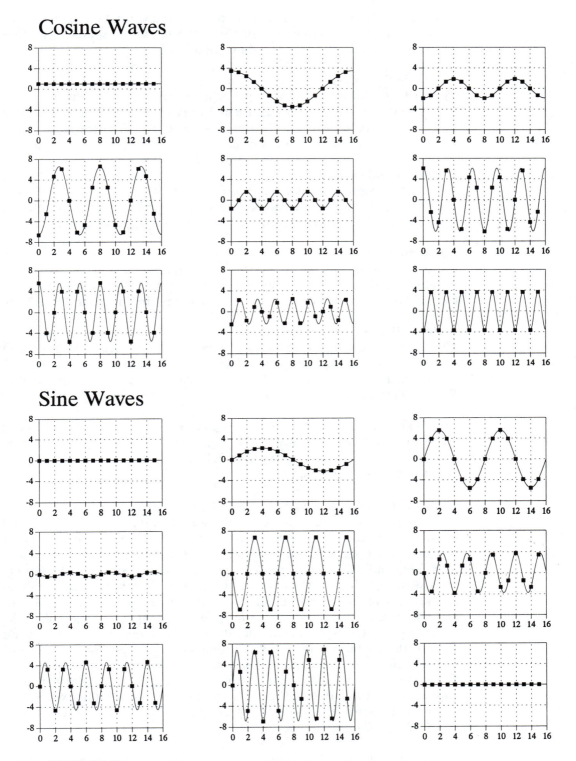

FIGURE 8-1b
Example of Fourier decomposition. A 16 point signal (opposite page) is decomposed into 9 cosine waves and 9 sine waves. The frequency of each sinusoid is fixed; only the amplitude is changed depending on the shape of the waveform being decomposed.

A signal can be either *continuous* or *discrete*, and it can be either *periodic* or *aperiodic*. The combination of these two features generates the four categories, described below and illustrated in Fig. 8-2.

Aperiodic-Continuous

This includes, for example, decaying exponentials and the Gaussian curve. These signals extend to both positive and negative infinity *without* repeating in a periodic pattern. The Fourier Transform for this type of signal is simply called the **Fourier Transform**.

Periodic-Continuous

Here the examples include: sine waves, square waves, and any waveform that repeats itself in a regular pattern from negative to positive infinity. This version of the Fourier transform is called the **Fourier Series**.

Aperiodic-Discrete

These signals are only defined at discrete points between positive and negative infinity, and do not repeat themselves in a periodic fashion. This type of Fourier transform is called the **Discrete Time Fourier Transform**.

Periodic-Discrete

These are discrete signals that repeat themselves in a periodic fashion from negative to positive infinity. This class of Fourier Transform is sometimes called the Discrete Fourier Series, but is most often called the **Discrete Fourier Transform**.

You might be thinking that the names given to these four types of Fourier transforms are confusing and poorly organized. You're right, the names have evolved rather haphazardly over 200 years. There is nothing you can do but memorize them and move on.

These four classes of signals all extend to positive and negative *infinity*. Hold on, you say! What if you only have a finite number of samples stored in your computer, say a signal formed from 1024 points. Isn't there a version of the Fourier Transform that uses finite length signals? No, there isn't. Sine and cosine waves are *defined* as extending from negative infinity to positive infinity. You cannot use a group of infinitely long signals to synthesize something finite in length. The way around this dilemma is to make the finite data *look like* an infinite length signal. This is done by imagining that the signal has an infinite number of samples on the left and right of the actual points. If all these imaginary samples have a value of zero, the signal looks *discrete* and *aperiodic*, and the Discrete Time Fourier Transform applies. As an alternative, the imaginary samples can be a duplication of the actual 1024 points. In this case, the signal looks discrete and periodic, with a period of 1024 samples. This calls for the Discrete Fourier Transform to be used.

As it turns out, an *infinite* number of sinusoids are required to synthesize a signal that is *aperiodic*. This makes it impossible to calculate the Discrete Time Fourier Transform in a computer algorithm. By elimination, the only

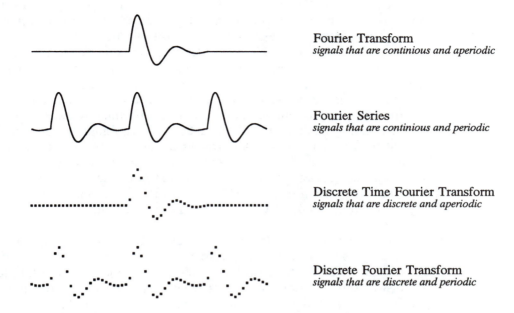

Fourier Transform
signals that are continious and aperiodic

Fourier Series
signals that are continious and periodic

Discrete Time Fourier Transform
signals that are discrete and aperiodic

Discrete Fourier Transform
signals that are discrete and periodic

FIGURE 8-2
Illustration of the four Fourier transforms. A signal may be continuous or discrete, and it may be periodic or aperiodic. Together these define four possible combinations, each having its own version of the Fourier transform. The names are not well organized; simply memorize them.

type of Fourier transform that can be used in DSP is the DFT. In other words, digital computers can only work with information that is *discrete* and *finite* in length. When you struggle with theoretical issues, grapple with homework problems, and ponder mathematical mysteries, you may find yourself using the first three members of the Fourier transform family. When you sit down to your computer, you will only use the DFT. We will briefly look at these other Fourier transforms in future chapters. For now, concentrate on understanding the Discrete Fourier Transform.

Look back at the example DFT decomposition in Fig. 8-1. On the face of it, it appears to be a 16 point signal being decomposed into 18 sinusoids, each consisting of 16 points. In more formal terms, the 16 point signal, shown in (a), must be viewed as a single period of an infinitely long periodic signal. Likewise, each of the 18 sinusoids, shown in (b), represents a 16 point segment from an infinitely long sinusoid. Does it really matter if we view this as a 16 point signal being synthesized from 16 point sinusoids, or as an infinitely long periodic signal being synthesized from infinitely long sinusoids? The answer is: *usually no, but sometimes, yes.* In upcoming chapters we will encounter properties of the DFT that seem baffling if the signals are viewed as finite, but become obvious when the periodic nature is considered. The key point to understand is that this periodicity is invoked in order to use a *mathematical tool*, i.e., the DFT. It is usually meaningless in terms of where the signal originated or how it was acquired.

Each of the four Fourier Transforms can be subdivided into **real** and **complex** versions. The real version is the simplest, using ordinary numbers and algebra for the synthesis and decomposition. For instance, Fig. 8-1 is an example of the **real DFT**. The complex versions of the four Fourier transforms are immensely more complicated, requiring the use of *complex numbers*. These are numbers such as: $3+4j$, where j is equal to $\sqrt{-1}$ (electrical engineers use the variable j, while mathematicians use the variable, i). Complex mathematics can quickly become overwhelming, even to those that specialize in DSP. In fact, a primary goal of this book is to present the fundamentals of DSP *without* the use of complex math, allowing the material to be understood by a wider range of scientists and engineers. The complex Fourier transforms are the realm of those that specialize in DSP, and are willing to sink to their necks in the swamp of mathematics. If you are so inclined, Chapters 28-31 will take you there.

The mathematical term: **transform**, is extensively used in Digital Signal Processing, such as: Fourier transform, Laplace transform, Z transform, Hilbert transform, Discrete Cosine transform, etc. Just what is a transform? To answer this question, remember what a *function* is. A function is an algorithm or procedure that changes one value into another value. For example, $y = 2x+1$ is a function. You pick some value for x, plug it into the equation, and out pops a value for y. Functions can also change *several* values into a single value, such as: $y = 2a + 3b + 4c$, where a, b, and c are changed into y.

Transforms are a direct extension of this, allowing both the input and output to have *multiple* values. Suppose you have a signal composed of 100 samples. If you devise some equation, algorithm, or procedure for changing these 100 samples into another 100 samples, you have yourself a transform. If you think it is useful enough, you have the perfect right to attach your last name to it and expound its merits to your colleagues. (This works best if you are an eminent 18th century French mathematician). Transforms are not limited to any specific type or number of data. For example, you might have 100 samples of discrete data for the input and 200 samples of discrete data for the output. Likewise, you might have a continuous signal for the input and a continuous signal for the output. Mixed signals are also allowed, discrete in and continuous out, and vice versa. In short, a transform is any fixed procedure that changes one chunk of data into another chunk of data. Let's see how this applies to the topic at hand: the Discrete Fourier transform.

Notation and Format of the Real DFT

As shown in Fig. 8-3, the discrete Fourier transform changes an N point input signal into two $N/2+1$ point output signals. The input signal contains the signal being decomposed, while the two output signals contain the *amplitudes* of the component sine and cosine waves (scaled in a way we will discuss shortly). The input signal is said to be in the **time domain**. This is because the most common type of signal entering the DFT is composed of

Time Domain

Frequency Domain

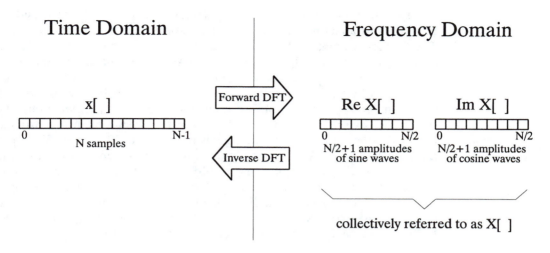

FIGURE 8-3
DFT terminology. In the time domain, $x[\ ]$ consists of N points running from 0 to N-1. In the frequency domain, the DFT produces two signals, the real part, written: $Re\ X[\ ]$, and the imaginary part, written: $Im\ X[\ ]$. Each of these frequency domain signals are $N/2$+1 points long, and run from 0 to $N/2$. The Forward DFT transforms from the time domain to the frequency domain, while the Inverse DFT transforms from the frequency domain to the time domain. (Take note: this figure describes the **real DFT**. The **complex DFT**, discussed in Chapter 29, changes N complex points into another set of N complex points).

samples taken at regular intervals of *time*. Of course, any kind of sampled data can be fed into the DFT, regardless of how it was acquired. When you see the term "time domain" in Fourier analysis, it may actually refer to samples taken over time, or it might be a general reference to any discrete signal that is being decomposed. The term **frequency domain** is used to describe the amplitudes of the sine and cosine waves (including the special scaling we promised to explain).

The frequency domain contains exactly the same information as the time domain, just in a different form. If you know one domain, you can calculate the other. Given the time domain signal, the process of calculating the frequency domain is called **decomposition**, **analysis,** the **forward DFT**, or simply, **the DFT**. If you know the frequency domain, calculation of the time domain is called **synthesis**, or the **inverse DFT**. Both synthesis and analysis can be represented in equation form and computer algorithms.

The number of samples in the time domain is usually represented by the **variable N**. While N can be any positive integer, a power of two is usually chosen, i.e., 128, 256, 512, 1024, etc. There are two reasons for this. First, digital data storage uses binary addressing, making powers of two a natural signal length. Second, the most efficient algorithm for calculating the DFT, the Fast Fourier Transform (FFT), usually operates with N that is a power of two. Typically, N is selected between 32 and 4096. In most cases, the samples run from 0 to N-1, rather than 1 to N.

Standard DSP notation uses **lower case letters** to represent time domain signals, such as $x[\ ]$, $y[\ ]$, and $z[\ ]$. The corresponding **upper case letters** are used to represent their frequency domains, that is, $X[\ ]$, $Y[\ ]$, and $Z[\ ]$. For illustration, assume an N point time domain signal is contained in $x[\ ]$. The frequency domain of this signal is called $X[\ ]$, and consists of two parts, each an array of $N/2+1$ samples. These are called the **Real part of $X[\ ]$**, written as: *$Re\ X[\ ]$*, and the **Imaginary part of $X[\ ]$**, written as: *$Im\ X[\ ]$*. The values in $Re\ X[\ ]$ are the amplitudes of the cosine waves, while the values in $Im\ X[\ ]$ are the amplitudes of the sine waves (not worrying about the scaling factors for the moment). Just as the time domain runs from $x[0]$ to $x[N-1]$, the frequency domain signals run from $Re\ X[0]$ to $Re\ X[N/2]$, and from $Im\ X[0]$ to $Im\ X[N/2]$. Study these notations carefully; they are critical to understanding the equations in DSP. Unfortunately, some computer languages don't distinguish between lower and upper case, making the variable names up to the individual programmer. The programs in this book use the array XX[ ] to hold the time domain signal, and the arrays REX[ ] and IMX[ ] to hold the frequency domain signals.

The names *real part* and *imaginary part* originate from the complex DFT, where they are used to distinguish between *real* and *imaginary* numbers. Nothing so complicated is required for the real DFT. Until you get to Chapter 29, simply think that "real part" means the *cosine wave amplitudes*, while "imaginary part" means the *sine wave amplitudes*. Don't let these suggestive names mislead you; everything here uses ordinary numbers.

Likewise, don't be misled by the *lengths* of the frequency domain signals. It is common in the DSP literature to see statements such as: "The DFT changes an N point time domain signal into an N point frequency domain signal." This is referring to the *complex DFT*, where each "point" is a complex number (consisting of real and imaginary parts). For now, focus on learning the real DFT, the difficult math will come soon enough.

The Frequency Domain's Independent Variable

Figure 8-4 shows an example DFT with $N = 128$. The time domain signal is contained in the array: $x[0]$ to $x[127]$. The frequency domain signals are contained in the two arrays: $Re\ X[0]$ to $Re\ X[64]$, and $Im\ X[0]$ to $Im\ X[64]$. Notice that 128 points in the time domain corresponds to <u>65</u> points in each of the frequency domain signals, with the frequency indexes running from 0 to 64. That is, N points in the time domain corresponds to $N/2+1$ points in the frequency domain (not $N/2$ points). Forgetting about this extra point is a common bug in DFT programs.

The horizontal axis of the frequency domain can be referred to in **four different ways**, all of which are common in DSP. In the first method, the horizontal axis is labeled from 0 to 64, corresponding to the 0 to $N/2$ samples in the arrays. When this labeling is used, the index for the frequency domain is an integer, for example, $Re\ X[k]$ and $Im\ X[k]$, where k runs from 0 to $N/2$ in steps of one. Programmers like this method because it is how they write code, using an index to access array locations. This notation is used in Fig. 8-4b.

Time Domain

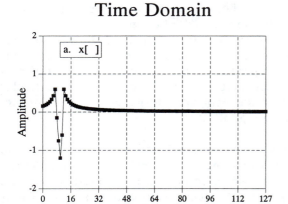

Frequency Domain

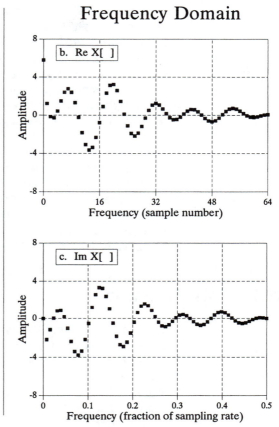

FIGURE 8-4
Example of the DFT. The DFT converts the time domain signal, $x[\ ]$, into the frequency domain signals, $Re\,X[\ ]$ and $Im\,X[\ ]$. The horizontal axis of the frequency domain can be labeled in one of three ways: (1) as an array index that runs between 0 and $N/2$, (2) as a fraction of the sampling frequency, running between 0 and 0.5, (3) as a natural frequency, running between 0 and π. In the example shown here, (b) uses the first method, while (c) use the second method.

In the second method, used in (c), the horizontal axis is labeled as a *fraction of the sampling rate*. This means that the values along the horizonal axis always run between 0 and 0.5, since discrete data can only contain frequencies between DC and one-half the sampling rate. The index used with this notation is f, for frequency. The real and imaginary parts are written: $Re\,X[f]$ and $Im\,X[f]$, where f takes on $N/2+1$ equally spaced values between 0 and 0.5. To convert from the first notation, k, to the second notation, f, divide the horizontal axis by N. That is, $f = k/N$. Most of the graphs in this book use this second method, reinforcing that discrete signals only contain frequencies between 0 and 0.5 of the sampling rate.

The third style is similar to the second, except the horizontal axis is multiplied by 2π. The index used with this labeling is ω, a lower case Greek *omega*. In this notation, the real and imaginary parts are written: $Re\,X[\omega]$ and $Im\,X[\omega]$, where ω takes on $N/2+1$ equally spaced values between 0 and π. The parameter, ω, is called the **natural frequency**, and has the units of **radians**. This is based on the idea that there are 2π radians in a circle. Mathematicians like this method because it makes the equations shorter. For instance, consider how a cosine wave is written in each of these first three notations: using k: $c[n] = \cos(2\pi k n/N)$, using f: $c[n] = \cos(2\pi f n)$, and using ω: $c[n] = \cos(\omega n)$.

The fourth method is to label the horizontal axis in terms of the analog frequencies used in a *particular* application. For instance, if the system being examined has a sampling rate of 10 kHz (i.e., 10,000 samples per second), graphs of the frequency domain would run from 0 to 5 kHz. This method has the advantage of presenting the frequency data in terms of a *real world* meaning. The disadvantage is that it is tied to a particular sampling rate, and is therefore not applicable to general DSP algorithm development, such as designing digital filters.

All of these four notations are used in DSP, and you need to become comfortable with converting between them. This includes both graphs and mathematical equations. To find which notation is being used, look at the independent variable and its range of values. You should find one of four notations: k (or some other integer index), running from 0 to $N/2$; f, running from 0 to 0.5; ω, running from 0 to π; or a frequency expressed in hertz, running from DC to one-half of an actual sampling rate.

DFT Basis Functions

The sine and cosine waves used in the DFT are commonly called the DFT **basis functions**. In other words, the output of the DFT is a set of numbers that represent amplitudes. The basis functions are a set of sine and cosine waves with *unity* amplitude. If you assign each amplitude (the frequency domain) to the proper sine or cosine wave (the basis functions), the result is a set of *scaled* sine and cosine waves that can be added to form the time domain signal.

The DFT basis functions are generated from the equations:

EQUATION 8-1
Equations for the DFT basis functions. In these equations, $c_k[i]$ and $s_k[i]$ are the cosine and sine waves, each N points in length, running from $i = 0$ to $N-1$. The parameter, k, determines the frequency of the wave. In an N point DFT, k takes on values between 0 and $N/2$.

$$c_k[i] = \cos(2\pi k i / N)$$

$$s_k[i] = \sin(2\pi k i / N)$$

where: $c_k[\ ]$ is the cosine wave for the amplitude held in *Re* $X[k]$, and $s_k[\ ]$ is the sine wave for the amplitude held in *Im* $X[k]$. For example, Fig. 8-5 shows some of the 17 sine and 17 cosine waves used in an $N = 32$ point DFT. Since these sinusoids add to form the input signal, they must be the same *length* as the input signal. In this case, each has 32 points running from $i = 0$ to 31. The parameter, k, sets the frequency of each sinusoid. In particular, $c_1[\ ]$ is the cosine wave that makes *one* complete cycle in N points, $c_5[\ ]$ is the cosine wave that makes *five* complete cycles in N points, etc. This is an important concept in understanding the basis functions; the frequency parameter, k, is equal to the number of complete cycles that occur over the N points of the signal.

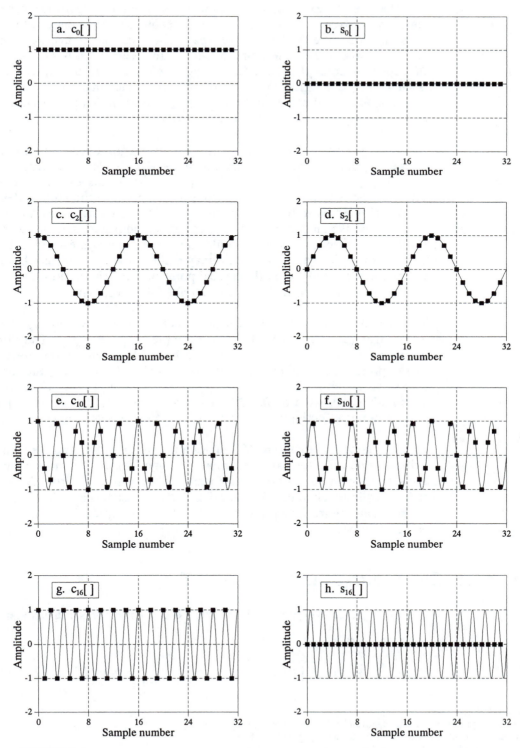

FIGURE 8-5
DFT basis functions. A 32 point DFT has 17 discrete cosine waves and 17 discrete sine waves for its basis functions. Eight of these are shown in this figure. These are discrete signals; the continious lines are shown in these graphs only to help the reader's eye follow the waveforms.

Let's look at several of these basis functions in detail. Figure (a) shows the cosine wave $c_0[\ ]$. This is a cosine wave of zero frequency, which is a constant value of one. This means that $Re\,X[0]$ holds the average value of all the points in the time domain signal. In electronics, it would be said that $Re\,X[0]$ holds the **DC offset**. The sine wave of zero frequency, $s_0[\ ]$, is shown in (b), a signal composed of all *zeros*. Since this can not affect the time domain signal being synthesized, the value of $Im\,X[0]$ is *irrelevant*, and always set to zero. More about this shortly.

Figures (c) & (d) show $c_2[\ ]$ & $s_2[\ ]$, the sinusoids that complete *two* cycles in the N points. These correspond to $Re\,X[2]$ & $Im\,X[2]$, respectively. Likewise, (e) & (f) show $c_{10}[\ ]$ & $s_{10}[\ ]$, the sinusoids that complete *ten* cycles in the N points. These sinusoids correspond to the amplitudes held in $Re\,X[10]$ & $Im\,X[10]$. The problem is, the samples in (e) and (f) no longer *look* like sine and cosine waves. If the continuous curves were not present in these graphs, you would have a difficult time even detecting thepattern of the waveforms. This may make you a little uneasy, but don't worry about it. From a mathematical point of view, these samples do form discrete sinusoids, even if your eye cannot follow the pattern.

The highest frequencies in the basis functions are shown in (g) and (h). These are $c_{N/2}[\ ]$ & $s_{N/2}[\ ]$, or in this example, $c_{16}[\ ]$ & $s_{16}[\ ]$. The discrete cosine wave alternates in value between 1 and -1, which can be interpreted as sampling a continuous sinusoid at the *peaks*. In contrast, the discrete sine wave contains all zeros, resulting from sampling at the *zero crossings*. This makes the value of $Im\,X[N/2]$ the same as $Im\,X[0]$, always equal to zero, and not affecting the synthesis of the time domain signal.

Here's a puzzle: If there are N samples entering the DFT, and $N+2$ samples exiting, where did the extra information come from? The answer: two of the output samples contain *no* information, allowing the other N samples to be fully independent. As you might have guessed, the points that carry no information are $Im\,X[0]$ and $Im\,X[N/2]$, the samples that always have a value of zero.

Synthesis, Calculating the Inverse DFT

Pulling together everything said so far, we can write the **synthesis equation**:

$$x[i] = \sum_{k=0}^{N/2} Re\bar{X}[k] \cos(2\pi ki/N) + \sum_{k=0}^{N/2} Im\bar{X}[k] \sin(2\pi ki/N)$$

EQUATION 8-2
The synthesis equation. In this relation, $x[i]$ is the signal being synthesized, with i running from 0 to $N-1$. $Re\,\bar{X}[k]$ and $Im\,\bar{X}[k]$ hold the amplitudes of the cosine and sine waves, respectively, with k running from 0 to $N/2$. Equation 8-3 provides the normalization to change this equation into the inverse DFT.

In words, any N point signal, $x[i]$, can be created by adding $N/2+1$ cosine waves and $N/2+1$ sine waves. The amplitudes of the cosine and sine waves are held in the arrays $Im\bar{X}[k]$ and $Re\bar{X}[k]$, respectively. The synthesis equation multiplies these amplitudes by the basis functions to create a set of scaled sine and cosine waves. Adding the scaled sine and cosine waves produces the time domain signal, $x[i]$.

In Eq. 8-2, the arrays are called $Im\bar{X}[k]$ and $Re\bar{X}[k]$, rather than $Im\,X[k]$ and $Re\,X[\underline{k}]$. This is because the *amplitudes needed for synthesis* (called here: $Im\bar{X}[k]$ and $Re\bar{X}[k]$), are slightly different from the *frequency domain of a signal* (denoted by: $Im\,X[k]$ and $Re\,X[k]$). This is the scaling factor issue we referred to earlier. Although the conversion is only a simple normalization, it is a common bug in computer programs. Look out for it! In equation form, the conversion between the two is given by:

EQUATIONS 8-3
Conversion between the sinusoidal amplitudes and the frequency domain values. In these equations, $Re\bar{X}[k]$ and $Im\bar{X}[k]$ hold the amplitudes of the cosine and sine waves needed for synthesis, while $Re\,X[k]$ and $Im\,X[k]$ hold the real and imaginary parts of the frequency domain. As usual, N is the number of points in the time domain signal, and k is an index that runs from 0 to $N/2$.

$$Re\bar{X}[k] \;=\; \frac{Re\,X[k]}{N/2}$$

$$Im\bar{X}[k] \;=\; -\frac{Im\,X[k]}{N/2}$$

except for two special cases:

$$Re\bar{X}[0] \;=\; \frac{Re\,X[0]}{N}$$

$$Re\bar{X}[N/2] \;=\; \frac{Re\,X[N/2]}{N}$$

Suppose you are given a frequency domain representation, and asked to synthesize the corresponding time domain signal. To start, you must find the amplitudes of the sine and cosine waves. In other words, given $Im\,X[k]$ and $Re\,X[k]$, you must find $Im\bar{X}[k]$ and $Re\bar{X}[k]$. Equation 8-3 shows this in a mathematical form. To do this in a computer program, three actions must be taken. First, divide all the values in the frequency domain by $N/2$. Second, change the sign of all the imaginary values. Third, divide the first and last samples in the real part, $Re\,X[0]$ and $Re\,X[N/2]$, by two. This provides the amplitudes needed for the synthesis described by Eq. 8-2. Taken together, Eqs. 8-2 and 8-3 *define* the inverse DFT.

The entire Inverse DFT is shown in the computer program listed in Table 8-1. There are two ways that the synthesis (Eq. 8-2) can be programmed, and both are shown. In the first method, each of the scaled sinusoids are generated one at a time and added to an accumulation array, which ends up becoming the time domain signal. In the second method, each sample in the time domain signal is calculated one at a time, as the sum of all the

```
100 'THE INVERSE DISCRETE FOURIER TRANSFORM
110 'The time domain signal, held in XX[ ], is calculated from the frequency domain signals,
120 'held in REX[ ] and IMX[ ].
130 '
140 DIM XX[511]                'XX[ ] holds the time domain signal
150 DIM REX[256]               'REX[ ] holds the real part of the frequency domain
160 DIM IMX[256]               'IMX[ ] holds the imaginary part of the frequency domain
170 '
180 PI = 3.14159265            'Set the constant, PI
190 N% = 512                   'N% is the number of points in XX[ ]
200 '
210 GOSUB XXXX                 'Mythical subroutine to load data into REX[ ] and IMX[ ]
220 '
230
240 '                          'Find the cosine and sine wave amplitudes using Eq. 8-3
250 FOR K% = 0 TO 256
260   REX[K%] =  REX[K%] / (N%/2)
270   IMX[K%] = -IMX[K%] / (N%/2)
280 NEXT K%
290 '
300 REX[0] = REX[0] / 2
310 REX[256] = REX[256] / 2
320 '
330 '
340 FOR I% = 0 TO 511          'Zero XX[ ] so it can be used as an accumulator
350   XX[I%] = 0
360 NEXT I%
370 '
380 '                          Eq. 8-2 SYNTHESIS METHOD #1.  Loop through each
390 '                          frequency generating the entire length of the sine and cosine
400 '                          waves, and add them to the accumulator signal, XX[ ]
410 '
420 FOR K% = 0 TO 256          'K% loops through each sample in REX[ ] and IMX[ ]
430   FOR I% = 0 TO 511        'I% loops through each sample in XX[ ]
440     '
450     XX[I%] = XX[I%] + REX[K%] * COS(2*PI*K%*I%/N%)
460     XX[I%] = XX[I%] + IMX[K%] * SIN(2*PI*K%*I%/N%)
470     '
480   NEXT I%
490 NEXT K%
500 '
510 END
```

Alternate code for lines 380 to 510

```
380 '                          Eq. 8-2 SYNTHESIS METHOD #2.  Loop through each
390 '                          sample in the time domain, and sum the corresponding
400 '                          samples from each cosine and sine wave
410 '
420 FOR I% = 0 TO 511          'I% loops through each sample in XX[ ]
430   FOR K% = 0 TO 256        'K% loops through each sample in REX[ ] and IMX[ ]
440     '
450     XX[I%] = XX[I%] + REX[K%] * COS(2*PI*K%*I%/N%)
460     XX[I%] = XX[I%] + IMX[K%] * SIN(2*PI*K%*I%/N%)
470     '
480   NEXT K%
490 NEXT I%
500 '
510 END
```

TABLE 8-1

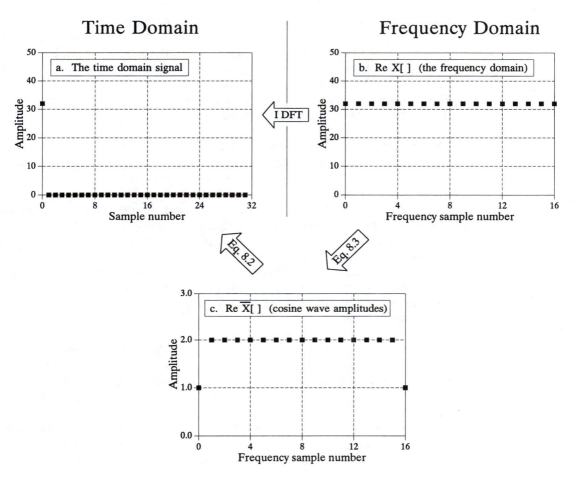

FIGURE 8-6

Example of the Inverse DFT. Figure (a) shows an example time domain signal, an impulse at sample zero with an amplitude of 32. Figure (b) shows the real part of the frequency domain of this signal, a constant value of 32. The imaginary part of the frequency domain (not shown) is composed of all zeros. Figure(c) shows the amplitudes of the cosine waves needed to reconstruct (a) using Eq. 8-2. The values in (c) are found from (b) by using Eq. 8-3.

corresponding samples in the cosine and sine waves. Both methods produce the same result. The difference between these two programs is very minor; the inner and outer loops are swapped during the synthesis.

Figure 8-6 illustrates the operation of the Inverse DFT, and the slight differences between the frequency domain and the amplitudes needed for synthesis. Figure 8-6a is an example signal we wish to synthesize, an impulse at sample zero with an amplitude of 32. Figure 8-6b shows the frequency domain representation of this signal. The real part of the frequency domain is a constant value of 32. The imaginary part (not shown) is composed of all zeros. As discussed in the next chapter, this is an important DFT *pair*: an impulse in the time domain corresponds to a constant value in the frequency domain. For now, the important point is that (b) is the *DFT* of (a), and (a) is the *Inverse DFT* of (b).

Equation 8-3 is used to convert the frequency domain signal, (b), into the amplitudes of the cosine waves, (c). As shown, all of the cosine waves have an amplitude of *two*, except for samples 0 and 16, which have a value of *one*. The amplitudes of the sine waves are not shown in this example because they have a value of zero, and therefore provide no contribution. The synthesis equation, Eq. 8-2, is then used to convert the amplitudes of the cosine waves, (b), into the time domain signal, (a).

This describes *how* the frequency domain is different from the sinusoidal amplitudes, but it doesn't explain *why* it is different. The difference occurs because the frequency domain is defined as a **spectral density**. Figure 8-7 shows how this works. The example in this figure is the real part of the frequency domain of a 32 point signal. As you should expect, the samples run from 0 to 16, representing 17 frequencies equally spaced between 0 and 1/2 of the sampling rate. *Spectral density* describes how much signal (amplitude) is present *per unit of bandwidth*. To convert the sinusoidal amplitudes into a spectral density, divide each amplitude by the bandwidth represented by each amplitude. This brings up the next issue: how do we determine the bandwidth of each of the discrete frequencies in the frequency domain?

As shown in the figure, the bandwidth can be defined by drawing dividing lines between the samples. For instance, sample number 5 occurs in the band between 4.5 and 5.5; sample number 6 occurs in the band between 5.5 and 6.5, etc. Expressed as a fraction of the total bandwidth (i.e., $N/2$), the bandwidth of each sample is $2/N$. An exception to this is the samples on each end, which have one-half of this bandwidth, $1/N$. This accounts for the $2/N$ scaling factor between the sinusoidal amplitudes and frequency domain, as well as the additional factor of two needed for the first and last samples.

Why the negation of the imaginary part? This is done solely to make the *real DFT* consistent with its big brother, the *complex DFT*. In Chapter 29 we will show that it is necessary to make the mathematics of the complex DFT work. When dealing only with the real DFT, many authors do not include this negation. For that matter, many authors do not even include

FIGURE 8-7
The bandwidth of frequency domain samples. Each sample in the frequency domain can be thought of as being contained in a frequency band of width $2/N$, expressed as a fraction of the total bandwidth. An exception to this is the first and last samples, which have a bandwidth only one-half this wide, $1/N$.

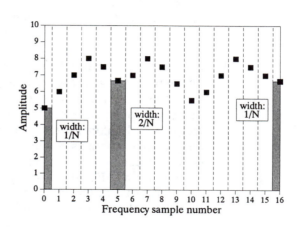

the 2/*N* scaling factor. Be prepared to find both of these missing in some discussions. They are included here for a tremendously important reason: The most efficient way to calculate the DFT is through the Fast Fourier Transform (FFT) algorithm, presented in Chapter 12. The FFT generates a frequency domain defined according to Eq. 8-2 and 8-3. If you start messing with these normalization factors, your programs containing the FFT are not going to work as expected.

Analysis, Calculating the DFT

The DFT can be calculated in three completely different ways. First, the problem can be approached as a set of *simultaneous equations*. This method is useful for understanding the DFT, but it is too inefficient to be of practical use. The second method brings in an idea from the last chapter: *correlation*. This is based on detecting a known waveform in another signal. The third method, called the Fast Fourier Transform (FFT), is an ingenious algorithm that decomposes a DFT with *N* points, into *N* DFTs each with a single point. The FFT is typically hundreds of times faster than the other methods. The first two methods are discussed here, while the FFT is the topic of Chapter 12. It is important to remember that all three of these methods produce an identical output. Which should you use? In actual practice, *correlation* is the preferred technique if the DFT has less than about 32 points, otherwise the *FFT* is used.

DFT by Simultaneous Equations
Think about the DFT calculation in the following way. You are given *N* values from the time domain, and asked to calculate the *N* values of the frequency domain (ignoring the two frequency domain values that you know must be zero). Basic algebra provides the answer: to solve for *N* unknowns, you must be able to write *N* linearly independent equations. To do this, take the first sample from each sinusoid and add them together. The sum must be equal to the first sample in the time domain signal, thus providing the first equation. Likewise, an equation can be written for each of the remaining points in the time domain signal, resulting in the required *N* equations. The solution can then be found by using established methods for solving simultaneous equations, such as Gauss Elimination. Unfortunately, this method requires a tremendous number of calculations, and is virtually never used in DSP. However, it is important for another reason, it shows *why* it is possible to decompose a signal into sinusoids, how *many* sinusoids are needed, and that the basis functions must be linearly independent (more about this shortly).

DFT by Correlation
Let's move on to a better way, the *standard way* of calculating the DFT. An example will show how this method works. Suppose we are trying to calculate the DFT of a 64 point signal. This means we need to calculate the 33 points in the real part, and the 33 points in the imaginary part of the frequency domain. In this example we will only show how to calculate a single sample, *Im X*[3], i.e., the amplitude of the sine wave that makes three

complete cycles between point 0 and point 63. All of the other frequency domain values are calculated in a similar manner.

Figure 8-8 illustrates using correlation to calculate *Im X*[3]. Figures (a) and (b) show two example time domain signals, called: *x*1[ ] and *x*2[ ], respectively. The first signal, *x*1[ ], is composed of nothing but a sine wave that makes three cycles between points 0 and 63. In contrast, *x*2[ ] is composed of several sine and cosine waves, *none* of which make three cycles between points 0 and 63. These two signals illustrate what the algorithm for calculating *Im X*[3] must do. When fed *x*1[ ], the algorithm must produce a value of 32, the amplitude of the sine wave present in the signal (modified by the scaling factors of Eq. 8-3). In comparison, when the algorithm is fed the other signal, *x*2[ ], a value of zero must be produced, indicating that this particular sine wave is not present in this signal.

The concept of correlation was introduced in Chapter 7. Briefly, to detect a known waveform contained in another signal, multiply the two and add the points in the resulting product. The single number that results from this procedure is a measure of how similar the two signals are. Figure 8-8 illustrates this approach. Figures (c) and (d) both display the signal we are looking for, a sine wave that makes 3 cycles between samples 0 and 63. Figure (e) shows the result of multiplying (a) and (c). Likewise, (f) shows the result of multiplying (b) and (d). The sum of all the points in (e) is 32, while the sum of all the points in (f) is zero, showing we have found the desired algorithm.

The other samples in the frequency domain are calculated in the same way. This procedure is formalized in the *analysis equation*, the mathematical way to calculate the frequency domain from the time domain:

EQUATION 8-4
The analysis equations for calculating the DFT. In these equations, *x*[*i*] is the time domain signal being analyzed, and *Re X*[*k*] & *Im X*[*k*] are the frequency domain signals being calculated. The index *i* runs from 0 to *N*-1, while the index *k* runs from 0 to*N*/2.

$$Re\,X[k] \;=\; \sum_{i=0}^{N-1} x[i]\,\cos\left(2\pi k i/N\right)$$

$$Im\,X[k] \;=\; -\sum_{i=0}^{N-1} x[i]\,\sin\left(2\pi k i/N\right)$$

In words, each sample in the frequency domain is found by multiplying the time domain signal by the sine or cosine wave being looked for, and adding the resulting points. If someone asks you what you are doing, say with confidence: "I am correlating the input signal with each basis function." Table 8-2 shows a computer program for calculating the DFT in this way.

The analysis equation does *not* require special handling of the first and last points, as did the synthesis equation. There is, however, a negative sign in the imaginary part in Eq. 8-4. Just as before, this negative sign makes the *real DFT* consistent with the *complex DFT*, and is not always included.

Example 1

Example 2

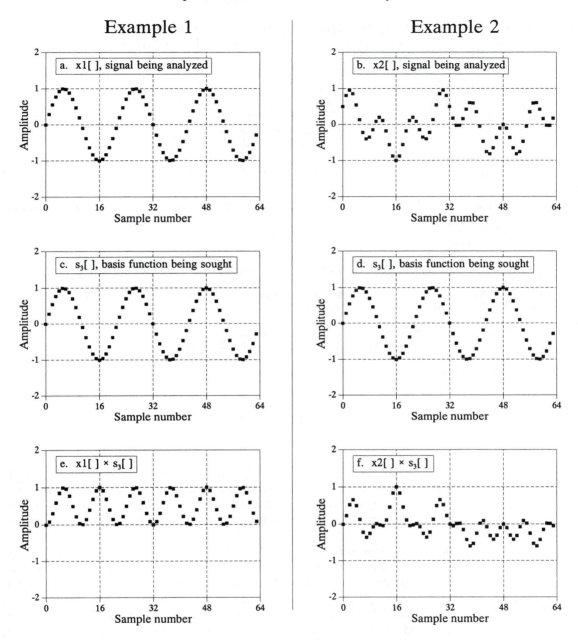

FIGURE 8-8
Two example signals, (a) and (b), are analyzed for containing the specific basis function shown in (c) and (d). Figures (e) and (f) show the result of multiplying each example signal by the basis function. Figure (e) has an average of 0.5, indicating that $x1[\ ]$ contains the basis function with an amplitude of 1.0. Conversely, (f) has a zero average, indicating that $x2[\ ]$ does not contain the basis function.

In order for this correlation algorithm to work, the basis functions must have an interesting property: each of them must be completely *uncorrelated* with all of the others. This means that if you multiply any two of the basis functions, the sum of the resulting points will be equal to zero. Basis functions that have this property are called **orthognal**. Many other

```
100 'THE DISCRETE FOURIER TRANSFORM
110 'The frequency domain signals, held in REX[ ] and IMX[ ], are calculated from
120 'the time domain signal, held in XX[ ].
130 '
140 DIM XX[511]                'XX[ ] holds the time domain signal
150 DIM REX[256]               'REX[ ] holds the real part of the frequency domain
160 DIM IMX[256]               'IMX[ ] holds the imaginary part of the frequency domain
170 '
180 PI = 3.14159265            'Set the constant, PI
190 N% = 512                   'N% is the number of points in XX[ ]
200 '
210 GOSUB XXXX                 'Mythical subroutine to load data into XX[ ]
220 '
230 '
240 FOR K% = 0 TO 256          'Zero REX[ ] & IMX[ ] so they can be used as accumulators
250   REX[K%] = 0
260   IMX[K%] = 0
270 NEXT K%
280 '
290 '                          'Correlate XX[ ] with the cosine and sine waves, Eq. 8-4
300 '
310 FOR K% = 0 TO 256          'K% loops through each sample in REX[ ] and IMX[ ]
320   FOR I% = 0 TO 511        'I% loops through each sample in XX[ ]
330     '
340     REX[K%] = REX[K%] + X[I%] * COS(2*PI*K%*I%/N%)
350     IMX[K%] = IMX[K%] - X[I%] * SIN(2*PI*K%*I%/N%)
360     '
370   NEXT I%
380 NEXT K%
390 '
400 END
```

TABLE 8-2

orthognal basis functions exist, including: square waves, triangle waves, impulses, etc. Signals can be decomposed into these other orthognal basis functions using correlation, just as done here with sinusoids. This is not to suggest that this is *useful*, only that it is *possible*.

As previously shown in Table 8-1, the *Inverse DFT* has two ways to be implemented in a computer program. This difference involves *swapping* the inner and outer loops during the synthesis. While this does not change the output of the program, it makes a difference in how you *view* what is being done. The *DFT* program in Table 8-2 can also be changed in this fashion, by swapping the inner and outer loops in lines 310 to 380. Just as before, the output of the program is the same, but the way you *think* about the calculation is different. (These two different ways of viewing the DFT and inverse DFT could be described as "input side" and "output side" algorithms, just as for convolution).

As the program in Table 8-2 is written, it describes how an individual sample in the frequency domain is affected by all of the samples in the time domain. That is, the program calculates each of the values in the frequency domain in succession, not as a group. When the inner and outer loops are exchanged, the program loops through each sample in the time domain, calculating the contribution of that point to the frequency domain. The overall frequency domain is found by adding the contributions from the individual time domain points. This brings up our next question: what kind of contribution does an individual sample in the time domain provide to the frequency domain? The answer is contained in an interesting aspect of the Fourier domain called *duality*.

Duality

The synthesis and analysis equations (Eqs. 8-2 and 8-4) are strikingly similar.   To move from one domain to the other, the known values are multiplied by the basis functions, and the resulting products added.   The fact that the *DFT* and the *Inverse DFT* use this same mathematical approach is really quite remarkable, considering the totally different way we arrived at the two procedures.   In fact, the only significant difference between the two equations is a result of the time domain being *one* signal of N points, while the frequency domain is *two* signals of $N/2+1$ points. As discussed in later chapters, the *complex DFT* expresses both the time and the frequency domains as complex signals of N points each. This makes the two domains completely symmetrical, and the equations for moving between them virtually *identical*.

This symmetry between the time and frequency domains is called **duality**, and gives rise to many interesting properties. For example, a single point in the frequency domain corresponds to a sinusoid in the time domain. By duality, the inverse is also true, a single point in the time domain corresponds to a sinusoid in the frequency domain. As another example, convolution in the time domain corresponds to multiplication in the frequency domain. By duality, the reverse is also true: convolution in the frequency domain corresponds to multiplication in the time domain. These and other duality relationships are discussed in more detail in Chapters 10 and 11.

Polar Notation

As it has been described so far, the frequency domain is a group of amplitudes of cosine and sine waves (with slight scaling modifications). This is called **rectangular** notation. Alternatively, the frequency domain can be expressed in **polar** form. In this notation, *Re X*[ ] & *Im X*[ ] are replaced with two other arrays, called the **Magnitude of X**[ ], written in equations as: *Mag X*[ ], and the **Phase of X**[ ], written as: *Phase X*[ ]. The magnitude and phase are a pair-for-pair replacement for the real and imaginary parts. For example, *Mag X*[0] and *Phase X*[0] are calculated using only *Re X*[0] and *Im X*[0].   Likewise, *Mag X*[14] and *Phase X*[14] are calculated using only *Re X*[14] and *Im X*[14], and so forth. To understand the conversion, consider what happens when you add a cosine wave and a sine wave of the same frequency. The result is a cosine wave of the same frequency, but with a new amplitude and a phase shift. In equation form, the two representations are related:

EQUATION 8-5
The addition of a cosine and sine wave results in a cosine wave with a different amplitude and phase shift. The information contained in *A* & *B* is transferred to two other variables, *M* and θ.

$$A\cos(x) + B\sin(x) = M\cos(x + \theta)$$

The important point is that no information is lost in this process; given one representation you can calculate the other. In other words, the information contained in the amplitudes *A* and *B*, is also contained in the variables *M* and θ. Although this equation involves sine and cosine waves, it follows the same conversion equations as do simple vectors. Figure 8-9 shows the analogous vector representation of how the two variables, *A* and *B*, can be viewed in a rectangular coordinate system, while *M* and θ are parameters in polar coordinates.

FIGURE 8-9
Rectangular-to-polar conversion. The addition of a cosine wave and a sine wave (of the same frequency) follows the same mathematics as the addition of simple vectors.

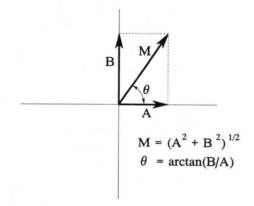

$$M = (A^2 + B^2)^{1/2}$$
$$\theta = \arctan(B/A)$$

In polar notation, *Mag X*[ ] holds the amplitude of the cosine wave (*M* in Eq. 8-4 and Fig. 8-9), while *Phase X*[ ] holds the phase angle of the cosine wave (θ in Eq. 8-4 and Fig. 8-9). The following equations convert the frequency domain from rectangular to polar notation, and vice versa:

EQUATION 8-6
Rectangular-to-polar conversion. The rectangular representation of the frequency domain, *Re X*[k] and *Im X*[k], is changed into the polar form, *Mag X*[k] and *Phase X*[k].

$$Mag\,X[k] = \left(Re\,X[k]^2 + Im\,X[k]^2\right)^{1/2}$$

$$Phase\,X[k] = \arctan\left(\frac{Im\,X[k]}{Re\,X[k]}\right)$$

EQUATION 8-7
Polar-to-rectangular conversion. The arrays, *Mag X*[k] and *Phase X*[k], are converted into *Re X*[k] and *Im X*[k].

$$Re\,X[k] = Mag\,X[k]\,\cos\big(Phase\,X[k]\big)$$

$$Im\,X[k] = Mag\,X[k]\,\sin\big(Phase\,X[k]\big)$$

Rectangular and polar notation allow you to think of the DFT in two different ways. With rectangular notation, the DFT decomposes an N point signal into $N/2+1$ cosine waves and $N/2+1$ sine waves, each with a specified *amplitude*. In polar notation, the DFT decomposes an N point signal into $N/2+1$ cosine waves, each with a specified *amplitude* (called the *magnitude*) and *phase shift*. Why does polar notation use cosine waves instead of sine waves? Sine waves cannot represent the DC component of a signal, since a sine wave of zero frequency is composed of all zeros (see Figs. 8-5 a&b).

Even though the polar and rectangular representations contain exactly the same information, there are many instances where one is easier to use that the other. For example, Fig. 8-10 shows a frequency domain signal in both rectangular and polar form. Warning: Don't try to understand the shape of the real and imaginary parts; your head will explode! In comparison, the polar curves are straightforward: only frequencies below about 0.25 are present, and the phase shift is approximately proportional to the frequency. This is the frequency response of a low-pass filter.

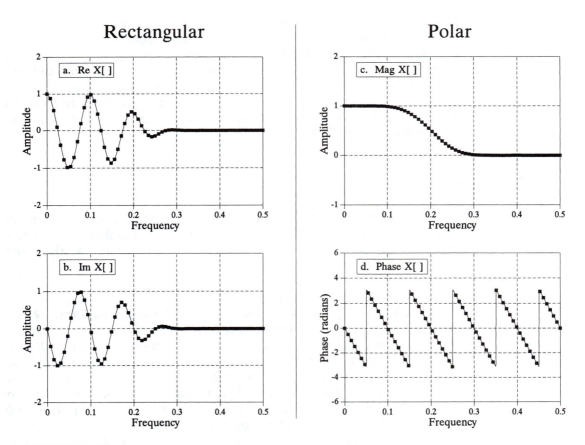

FIGURE 8-10
Example of rectangular and polar frequency domains. This example shows a frequency domain expressed in both rectangular and polar notation. As in this case, polar notation usually provides human observers with a better understanding of the characteristics of the signal. In comparison, the rectangular form is almost always used when math computations are required. Pay special notice to the fact that the first and last samples in the phase must be zero, just as they are in the imaginary part.

When should you use rectangular notation and when should you use polar? Rectangular notation is usually the best choice for calculations, such as in equations and computer programs. In comparison, graphs are almost always in polar form. As shown by the previous example, it is nearly impossible for *humans* to understand the characteristics of a frequency domain signal by looking at the real and imaginary parts. In a typical program, the frequency domain signals are kept in rectangular notation until an observer needs to look at them, at which time a rectangular-to-polar conversion is done.

Why is it easier to understand the frequency domain in polar notation? This question goes to the heart of why decomposing a signal into sinusoids is *useful*. Recall the property of *sinusoidal fidelity* from Chapter 5: if a sinusoid enters a linear system, the output will also be a sinusoid, and at exactly the same frequency as the input. Only the amplitude and phase can change. Polar notation directly represents signals in terms of the amplitude and phase of the component cosine waves. In turn, systems can be represented by how they modify the amplitude and phase of each of these cosine waves.

Now consider what happens if rectangular notation is used with this scenario. A mixture of cosine and sine waves enter the linear system, resulting in a mixture of cosine and sine waves leaving the system. The problem is, a cosine wave on the input may result in both cosine and sine waves on the output. Likewise, a sine wave on the input can result in both cosine and sine waves on the output. While these cross-terms can be straightened out, the overall method doesn't match with why we wanted to use sinusoids in the first place.

Polar Nuisances

There are many nuisances associated with using polar notation. None of these are overwhelming, just really annoying! Table 8-3 shows a computer program for converting between rectangular and polar notation, and provides solutions for some of these pests.

Nuisance 1: Radians vs. Degrees
It is possible to express the phase in either *degrees* or *radians*. When expressed in degrees, the values in the phase signal are between -180 and 180. Using radians, each of the values will be between $-\pi$ and π, that is, -3.141592 to 3.141592. Most computer languages require the use radians for their trigonometric functions, such as cosine, sine, arctangent, etc. It can be irritating to work with these long decimal numbers, and difficult to interpret the data you receive. For example, if you want to introduce a 90 degree phase shift into a signal, you need to add 1.570796 to the phase. While it isn't going to kill you to type this into your program, it does become tiresome. The best way to handle this problem is to define the constant, $PI = 3.141592$, at the beginning of your program. A 90 degree phase shift can then be written as $PI/2$. Degrees and radians are both widely used in DSP and you need to become comfortable with both.

```
100 'RECTANGULAR-TO-POLAR & POLAR-TO-RECTANGULAR CONVERSION
110 '
120 DIM REX[256]                'REX[ ]    holds the real part
130 DIM IMX[256]                'IMX[ ]    holds the imaginary part
140 DIM MAG[256]                'MAG[ ]   holds the magnitude
150 DIM PHASE[256]              'PHASE[ ] holds the phase
160 '
170 PI = 3.14159265
180 '
190 GOSUB XXXX                  'Mythical subroutine to load data into REX[ ] and IMX[ ]
200 '
210 '
220 '                          'Rectangular-to-polar conversion, Eq. 8-6
230 FOR K% = 0 TO 256
240   MAG[K%] = SQR( REX[K%]^2 + IMX[K%]^2 )   'from Eq. 8-6
250   IF REX[K%] = 0 THEN REX[K%] = 1E-20       'prevent divide by 0 (nuisance 2)
260   PHASE[K%] = ATN( IMX[K%] / REX[K%] )      'from Eq. 8-6
270   '                                         'correct the arctan (nuisance 3)
280   IF REX[K%] < 0 AND IMX[K%] <   0 THEN PHASE[K%] = PHASE[K%] - PI
290   IF REX[K%] < 0 AND IMX[K%] >= 0 THEN PHASE[K%] = PHASE[K%] + PI
300 NEXT K%
310 '
320 '
330 '                          'Polar-to-rectangular conversion, Eq. 8-7
340 FOR K% = 0 TO 256
350   REX[K%] = MAG[K%] * COS( PHASE[K%] )
360   IMX[K%] = MAG[K%] * SIN( PHASE[K%] )
370 NEXT K%
380 '
390 END
```

TABLE 8-3

Nuisance 2: Divide by zero error
When converting from rectangular to polar notation, it is very common to
find frequencies where the real part is zero and the imaginary part is some
nonzero value. This simply means that the phase is exactly 90 or -90
degrees. Try to tell your computer this! When your program tries to
calculate the phase from: *Phase X[k]* = arctan$(Im\,X[k]\,/\,Re\,X[k])$, a *divide by
zero error* occurs. Even if the program execution doesn't halt, the phase you
obtain for this frequency won't be correct. To avoid this problem, the real
part must be tested for being zero before the division. If it is zero, the
imaginary part must be tested for being positive or negative, to determine
whether to set the phase to $\pi/2$ or $-\pi/2$, respectively. Lastly, the division
needs to be bypassed. Nothing difficult in all these steps, just the potential
for aggravation.   An alternative way to handle this problem is shown in line
250 of Table 8-3. If the real part is zero, change it to a negligibly small
number to keep the math processor happy during the division.

Nuisance 3: Incorrect arctan
Consider a frequency domain sample where *Re X[k]* = 1 and *Im X[k]* = 1.
Equation 8-6 provides the corresponding polar values of *Mag X[k]* = 1.414

and *Phase X[k]* = 45°. Now consider another sample where *Re X[k]* = -1 and *Im X[k]* = -1. Again, Eq. 8-6 provides the values of *Mag X[k]* = 1.414 and *Phase X[k]* = 45°. The problem is, the phase is wrong! It should be -135°. This error occurs whenever the real part is negative. This problem can be corrected by testing the real and imaginary parts after the phase has been calculated. If both the real and imaginary parts are negative, subtract 180° (or π radians) from the calculated phase. If the real part is negative and the imaginary part is positive, add 180° (or π radians). Lines 340 and 350 of the program in Table 8-3 show how this is done. If you fail to catch this problem, the calculated value of the phase will only run between $-\pi/2$ and $\pi/2$, rather than between $-\pi$ and π. Drill this into your mind. If you see the phase only extending to ±1.5708, you have forgotten to correct the ambiguity in the arctangent calculation.

Nuisance 4: Phase of very small magnitudes

Imagine the following scenario. You are grinding away at some DSP task, and suddenly notice that part of the phase doesn't look right. It might be noisy, jumping all over, or just plain *wrong*. After spending the next hour looking through hundreds of lines of computer code, you find the answer. The corresponding values in the magnitude are so small that they are buried in round-off noise. If the magnitude is negligibly small, the phase doesn't have any meaning, and can assume unusual values. An example of this is shown in Fig. 8-11. It is usually obvious when an *amplitude* signal is lost in noise; the values are so small that you are forced to suspect that the values are meaningless. The phase is different. When a polar signal is contaminated with noise, the values in the phase are random numbers between $-\pi$ and π. Unfortunately, this often *looks* like a real signal, rather than the nonsense it really is.

Nuisance 5: 2π ambiguity of the phase

Look again at Fig. 8-10d, and notice the several discontinuities in the data. Every time a point looks as if it is going to dip below -3.14592, it snaps back to 3.141592. This is a result of the periodic nature of sinusoids. For

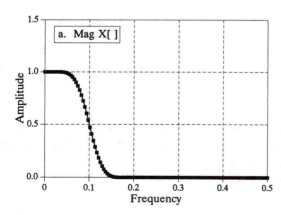

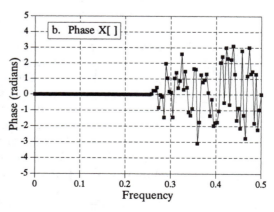

FIGURE 8-11
The phase of small magnitude signals. At frequencies where the magnitude drops to a very low value, round-off noise can cause wild excursions of the phase. Don't make the mistake of thinking this is a meaningful signal.

FIGURE 8-12
Example of phase unwrapping. The top curve shows a typical phase signal obtained from a rectangular-to-polar conversion routine. Each value in the signal must be between -π and π (-3.14159 and 3.14159). As shown in the lower curve, the phase can be *unwrapped* by adding or subtracting integer multiplies of 2π from each sample, where the integer is chosen to minimize the discontinuities between points.

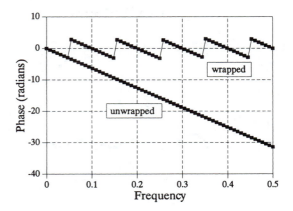

example, a phase shift of θ, is exactly the same as a phase shift of θ+2π, θ+4π, θ+6π, etc. Any sinusoid is unchanged when you add an integer multiple of 2π to the phase. The apparent discontinuities in the signal are a result of the computer algorithm picking its favorite choice from an infinite number of equivalent possibilities. The smallest possible value is always chosen, keeping the phase between -π and π.

It is often easier to understand the phase if it does not have these discontinuities, even if it means that the phase extends above π, or below -π. This is called **unwrapping the phase**, and an example is shown in Fig. 8-12. As shown by the program in Table 8-4, a multiple of 2π is added or subtracted from each value of the phase. The exact value is determined by an algorithm that minimizes the difference between adjacent samples.

Nuisance 6: The magnitude is always positive (π ambiguity of the phase)
Figure 8-13 shows a frequency domain signal in rectangular and polar form. The real part is smooth and quite easy to understand, while the imaginary part is entirely zero. In comparison, the polar signals contain abrupt

```
100 ' PHASE UNWRAPPING
110 '
120 DIM PHASE[256]              'PHASE[ ]   holds the original phase
130 DIM UWPHASE[256]            'UWPHASE[ ] holds the unwrapped phase
140 '
150 PI = 3.14159265
160 '
170 GOSUB XXXX                  'Mythical subroutine to load data into PHASE[ ]
180 '
190 UWPHASE[0] = 0              'The first point of all phase signals is zero
200 '
210 '                           'Go through the unwrapping algorithm
220 FOR K% = 1 TO 256
230   C% = CINT( (UWPHASE[K%-1] - PHASE[K%]) / (2 * PI) )
240   UWPHASE[K%] = PHASE[K%] + C%*2*PI
250 NEXT K%
260 '
270 END
```

TABLE 8-4

Rectangular   ## Polar

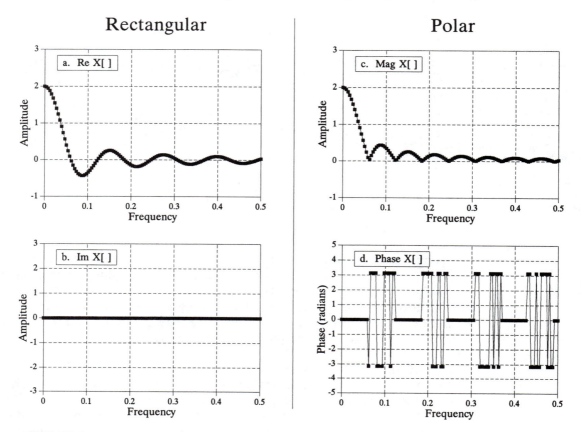

FIGURE 8-13
Example signals in rectangular and polar form. Since the magnitude must always be positive (by definition), the magnitude and phase may contain abrupt discontinuities and sharp corners. Figure (d) also shows another nuisance: random noise can cause the phase to rapidly oscillate between π or $-\pi$.

discontinuities and sharp corners. This is because the magnitude must always be positive, *by definition*. Whenever the real part dips below zero, the magnitude remains positive by changing the phase by π (or $-\pi$, which is the same thing). While this is not a problem for the mathematics, the irregular curves can be difficult to interpret.

One solution is to allow the magnitude to have *negative* values. In the example of Fig. 8-13, this would make the magnitude appear the same as the real part, while the phase would be entirely zero. There is nothing wrong with this if it helps your understanding. Just be careful not to call a signal with negative values the "magnitude" since this violates its formal definition. In this book we use the weasel words: *unwrapped magnitude* to indicate a "magnitude" that is allowed to have negative values.

Nuisance 7: Spikes between π and $-\pi$
Since π and $-\pi$ represent exactly the same phase shift, round-off noise can cause adjacent points in the phase to rapidly switch between the two values. As shown in (d), this can produce sharp breaks and spikes in an otherwise smooth curve. Don't be fooled, the phase isn't really this discontinuous.

CHAPTER 9

Applications of the DFT

The Discrete Fourier Transform (DFT) is one of the most important tools in Digital Signal Processing. This chapter discusses three common ways it is used. First, the DFT can calculate a signal's *frequency spectrum*. This is a direct examination of information encoded in the frequency, phase, and amplitude of the component sinusoids. For example, human speech and hearing use signals with this type of encoding. Second, the DFT can find a system's frequency response from the system's impulse response, and vice versa. This allows systems to be analyzed in the *frequency domain*, just as convolution allows systems to be analyzed in the *time domain*. Third, the DFT can be used as an intermediate step in more elaborate signal processing techniques. The classic example of this is *FFT convolution*, an algorithm for convolving signals that is hundreds of times faster than conventional methods.

Spectral Analysis of Signals

It is very common for information to be encoded in the sinusoids that form a signal. This is true of naturally occurring signals, as well as those that have been created by humans. Many things oscillate in our universe. For example, speech is a result of vibration of the human vocal cords; stars and planets change their brightness as they rotate on their axes and revolve around each other; ship's propellers generate periodic displacement of the water, and so on. The *shape* of the time domain waveform is not important in these signals; the key information is in the *frequency*, *phase* and *amplitude* of the component sinusoids. The DFT is used to extract this information.

An example will show how this works. Suppose we want to investigate the sounds that travel through the ocean. To begin, a microphone is placed in the water and the resulting electronic signal amplified to a reasonable level, say a few volts. An analog low-pass filter is then used to remove all frequencies above 80 hertz, so that the signal can be digitized at 160 samples per second. After acquiring and storing several thousand samples, what next?

The first thing is to simply *look* at the data. Figure 9-1a shows 256 samples from our imaginary experiment. All that can be seen is a noisy waveform that conveys little information to the human eye. For reasons explained shortly, the next step is to multiply this signal by a smooth curve called a **Hamming window**, shown in (b). (Chapter 16 provides the equations for the Hamming and other windows; see Eqs. 16-1 and 16-2, and Fig. 16-2a). This results in a 256 point signal where the samples near the ends have been reduced in amplitude, as shown in (c).

Taking the DFT, and converting to polar notation, results in the 129 point frequency spectrum in (d). Unfortunately, this also looks like a noisy mess. This is because there is not enough information in the original 256 points to obtain a well behaved curve. Using a longer DFT does nothing to help this problem. For example, if a 2048 point DFT is used, the frequency spectrum becomes 1025 samples long. Even though the original 2048 points contain more information, the greater number of samples in the spectrum dilutes the information by the same factor. Longer DFTs provide better frequency resolution, but the same noise level.

The answer is to use more of the original signal in a way that doesn't increase the number of points in the frequency spectrum. This can be done by breaking the input signal into many 256 point *segments*. Each of these segments is multiplied by the Hamming window, run through a 256 point DFT, and converted to polar notation. The resulting frequency spectra are then *averaged* to form a single 129 point frequency spectrum. Figure (e) shows an example of averaging 100 of the frequency spectra typified by (d). The improvement is obvious; the noise has been reduced to a level that allows interesting features of the signal to be observed. Only the *magnitude* of the frequency domain is averaged in this manner; the *phase* is usually discarded because it doesn't contain useful information. The random noise reduces in proportion to the *square-root* of the number of segments. While 100 segments is typical, some applications might average *millions* of segments to bring out weak features.

There is also a second method for reducing spectral noise. Start by taking a very long DFT, say 16,384 points. The resulting frequency spectrum is high resolution (8193 samples), but very noisy. A low-pass digital filter is then used to *smooth* the spectrum, reducing the noise at the expense of the resolution. For example, the simplest digital filter might average 64 adjacent samples in the original spectrum to produce each sample in the filtered spectrum. Going through the calculations, this provides about the same noise and resolution as the first method, where the 16,384 points would be broken into 64 segments of 256 points each.

Which method should you use? The first method is easier, because the digital filter isn't needed. The second method has the *potential* of better performance, because the digital filter can be tailored to optimize the trade-off between noise and resolution. However, this improved performance is seldom worth the trouble. This is because both noise and resolution can be improved by using *more data* from the input signal. For example,

Time Domain

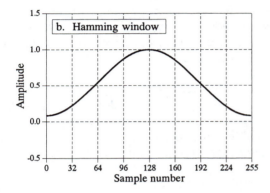

Frequency Domain

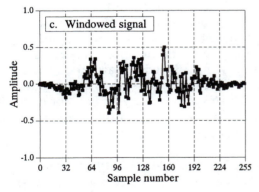

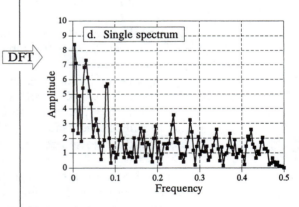

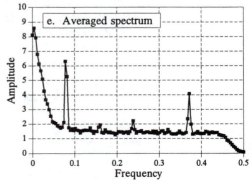

FIGURE 9-1
An example of spectral analysis. Figure (a) shows 256 samples taken from a (simulated) undersea microphone at a rate of 160 samples per second. This signal is multiplied by the Hamming window shown in (b), resulting in the windowed signal in (c). The frequency spectrum of the windowed signal is found using the DFT, and is displayed in (d). Averaging 100 of these spectra reduces the random noise, resulting in the averaged frequency spectrum shown in (e).

imagine breaking the acquired data into 10,000 segments of 16,384 samples each. This resulting frequency spectrum is high resolution (8193 points) *and* low noise (10,000 averages). Problem solved! For this reason, we will only look at the averaged segment method in this discussion.

Figure 9-2 shows an example spectrum from our undersea microphone, illustrating the features that commonly appear in the frequency spectra of acquired signals. Ignore the sharp peaks for a moment. Between 10 and 70 hertz, the signal consists of a relatively flat region. This is called **white noise** because it contains an equal amount of all frequencies, the same as white light. It results from the noise on the time domain waveform being *uncorrelated* from sample-to-sample. That is, knowing the noise value present on any one sample provides no information on the noise value present on any other sample. For example, the random motion of electrons in electronic circuits produces white noise. As a more familiar example, the sound of the water spray hitting the shower floor is white noise. The white noise shown in Fig. 9-2 could be originating from any of several sources, including the analog electronics, or the ocean itself.

Above 70 hertz, the white noise rapidly decreases in amplitude. This is a result of the roll-off of the antialias filter. An ideal filter would pass all frequencies below 80 hertz, and block all frequencies above. In practice, a perfectly sharp cutoff isn't possible, and you should expect to see this gradual drop. If you don't, suspect that an aliasing problem is present.

Below about 10 hertz, the noise rapidly increases due to a curiosity called **1/f noise** (one-over-f noise). 1/f noise is a mystery. It has been measured in very diverse systems, such as traffic density on freeways and electronic noise in transistors. It probably could be measured in all systems, if you look low enough in frequency. In spite of its wide occurrence, a general theory and understanding of 1/f noise has eluded researchers. The cause of this noise can be identified in some specific systems; however, this doesn't answer the question of why 1/f noise is everywhere. For common analog electronics and most physical systems, the transition between white noise and 1/f noise occurs between about 1 and 100 hertz.

Now we come to the sharp peaks in Fig. 9-2. The easiest to explain is at 60 hertz, a result of electromagnetic interference from commercial electrical power. Also expect to see smaller peaks at multiples of this frequency (120, 180, 240 hertz, etc.) since the power line waveform is not a *perfect* sinusoid. It is also common to find interfering peaks between 25-40 kHz, a favorite for designers of switching power supplies. Nearby radio and television stations produce interfering peaks in the megahertz range. Low frequency peaks can be caused by components in the system vibrating when shaken. This is called *microphonics*, and typically creates peaks at 10 to 100 hertz.

Now we come to the actual signals. There is a strong peak at 13 hertz, with weaker peaks at 26 and 39 hertz. As discussed in the next chapter, this is the frequency spectrum of a nonsinusoidal periodic waveform. The peak at 13 hertz is called the fundamental frequency, while the peaks at 26 and 39

FIGURE 9-2
Example frequency spectrum. Three types of features appear in the spectra of acquired signals: (1) random noise, such as white noise and 1/f noise, (2) interfering signals from power lines, switching power supplies, radio and TV stations, microphonics, etc., and (3) real signals, usually appearing as a fundamental plus harmonics. This spectrum shows several of these features.

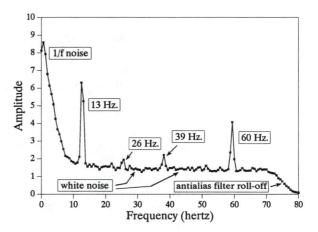

hertz are referred to as the second and third harmonic respectively. You would also expect to find peaks at other multiples of 13 hertz, such as 52, 65, 78 hertz, etc. You don't see these in Fig. 9-2 because they are buried in the white noise. This 13 hertz signal might be generated, for example, by a submarines's three bladed propeller turning at 4.33 revolutions per second. This is the basis of *passive* sonar, identifying undersea sounds by their frequency and harmonic content.

Suppose there are peaks very close together, such as shown in Fig. 9-3. There are two factors that limit the frequency resolution that can be obtained, that is, how close the peaks can be without merging into a single entity. The first factor is the length of the DFT. The frequency spectrum produced by an N point DFT consists of $N/2+1$ samples equally spaced between zero and one-half of the sampling frequency. To separate two closely spaced frequencies, the sample spacing must be *smaller* than the distance between the two peaks. For example, a 512 point DFT is sufficient to separate the peaks in Fig. 9-3, while a 128 point DFT is not.

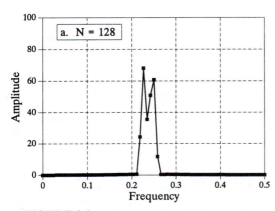
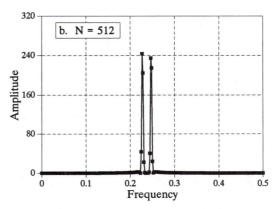

FIGURE 9-3
Frequency spectrum resolution. The longer the DFT, the better the ability to separate closely spaced features. In this example, a 128 point DFT cannot resolve the two peaks, while a 512 point DFT can.

The second factor limiting resolution is more subtle. Imagine a signal created by adding two sine waves with only a slight difference in their frequencies. Over a short segment of this signal, say a few periods, the waveform will look like a *single* sine wave. The closer the frequencies, the longer the segment must be to conclude that more than one frequency is present. In other words, the *length* of the signal limits the frequency resolution. This is distinct from the first factor, because the *length of the input signal* does not have to be the same as the *length of the DFT*. For example, a 256 point signal could be padded with zeros to make it 2048 points long. Taking a 2048 point DFT produces a frequency spectrum with 1025 samples. The added zeros don't change the shape of the spectrum, they only provide more samples in the frequency domain. In spite of this very close sampling, the ability to separate closely spaced peaks would be only slightly better than using a 256 point DFT. When the DFT is the same length as the input signal, the resolution is limited about equally by these two factors. We will come back to this issue shortly.

Next question: What happens if the input signal contains a sinusoid with a frequency *between* two of the basis functions? Figure 9-4a shows the answer. This is the frequency spectrum of a signal composed of two sine waves, one having a frequency *matching* a basis function, and the other with a frequency *between* two of the basis functions. As you should expect, the first sine wave is represented as a single point. The other peak is more difficult to understand. Since it cannot be represented by a single sample, it becomes a peak with **tails** that extend a significant distance away.

The solution? Multiply the signal by a Hamming window before taking the DFT, as was previously discussed. Figure (b) shows that the spectrum is changed in three ways by using the window. First, the two peaks are made to look more alike. This is good. Second, the tails are greatly reduced.

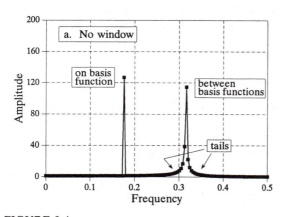
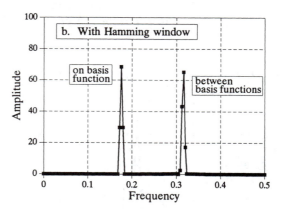

FIGURE 9-4

Example of using a window in spectral analysis. Figure (a) shows the frequency spectrum of a signal consisting of two sine waves. One sine wave has a frequency exactly equal to a basis function, allowing it to be represented by a single sample. The other sine wave has a frequency *between* two of the basis functions, resulting in *tails* on the peak. Figure (b) shows the frequency spectrum of the same signal, but with a Blackman window applied before taking the DFT. The window makes the peaks look the same and reduces the tails, but broadens the peaks.

This is also good. Third, the window reduces the resolution in the spectrum by making the peaks wider. This is bad. In DSP jargon, windows provide a trade-off between *resolution* (the width of the peak) and *spectral leakage* (the amplitude of the tails).

To explore the theoretical aspects of this in more detail, imagine an infinitely long discrete sine wave at a frequency of 0.1 the sampling rate. The frequency spectrum of this signal is an infinitesimally narrow peak, with all other frequencies being zero. Of course, neither this signal nor its frequency spectrum can be brought into a digital computer, because of their infinite and infinitesimal nature. To get around this, we change the signal in two ways, both of which distort the true frequency spectrum.

First, we *truncate* the information in the signal, by multiplying it by a window. For example, a 256 point *rectangular window* would allow 256 points to retain their correct value, while all the other samples in the infinitely long signal would be set to a value of zero. Likewise, the Hamming window would *shape* the retained samples, besides setting all points outside the window to zero. The signal is still infinitely long, but only a finite number of the samples have a nonzero value.

How does this windowing affect the frequency domain? When two time domain signals are *multiplied*, the corresponding frequency domains are *convolved*. Since the original spectrum is an infinitesimally narrow peak (i.e., a delta function), the spectrum of the windowed signal is the spectrum of the window shifted to the location of the peak. Figure 9-5 shows how the spectral peak would appear using three different window options. Figure 9-5a results from a rectangular window. Figures (b) and (c) result from using two popular windows, the Hamming and the Blackman (as previously mentioned, see Eqs. 16-1 and 16-2, and Fig. 16-2a for information on these windows).

As shown in Fig. 9-5, all these windows have degraded the original spectrum by broadening the peak and adding tails composed of numerous side lobes. This is an unavoidable result of using only a portion of the original time domain signal. Here we can see the tradeoff between the three windows. The Blackman has the widest main lobe (bad), but the lowest amplitude tails (good). The rectangular window has the narrowest main lobe (good) but the largest tails (bad). The Hamming window sits between these two.

Notice in Fig. 9-5 that the frequency spectra are continuous curves, not discrete samples. After windowing, the time domain signal is still infinitely long, even though most of the samples are zero. This means that the frequency spectrum consists of $\infty/2+1$ samples between 0 and 0.5, which is the same as a continuous line.

This brings in the second way we need to modify the time domain signal to allow it to be represented in a computer: *select N points from the signal*. These N points must contain all the nonzero points identified by the window, but may also include any number of the zeros. This has the effect

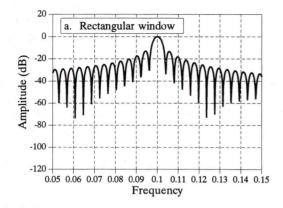

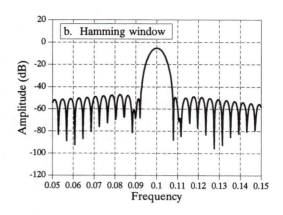

FIGURE 9-5
Detailed view of a spectral peak using various windows. Each peak in the frequency spectrum is a central lobe surrounded by tails formed from side lobes. By changing the window shape, the amplitude of the side lobes can be reduced at the expense of making the main lobe wider. The rectangular window, (a), has the narrowest main lobe but the largest amplitude side lobes. The Hamming window, (b), and the Blackman window, (c), have lower amplitude side lobes at the expense of a wider main lobe. Unless the time domain signal is padded with zeros to improve the frequency domain resolution, these peaks and valleys will not be seen; the spectra will appear as in Fig. 9-4. These curves are for 256 point DFTs.

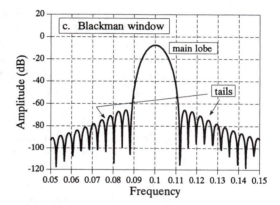

of *sampling* the frequency spectrum's continuous curve. For example, if N is chosen to be 1024, the spectrum's continuous curve will be sampled 513 times between 0 and 0.5. If N is chosen to be much larger than the window length, the samples in the frequency domain will be close enough that the peaks and valleys of the continuous curve will be preserved in the new spectrum. If N is made the same as the window length, the fewer number of samples in the spectrum results in the regular pattern of peaks and valleys turning into irregular tails, depending on where the samples happen to fall. This explains why the two peaks in Fig. 9-4a do not look alike. Each peak in Fig 9-4a is a *sampling* of the underlying curve in Fig. 9-5a. The presence or absence of the tails depends on where the samples are taken in relation to the peaks and valleys. If the sine wave exactly matches a basis function, the samples occur exactly at the valleys, eliminating the tails. If the sine wave is between two basis functions, the samples occur somewhere along the peaks and valleys, resulting in various patterns of tails.

Frequency Response of Systems

Systems are analyzed in the *time domain* by using convolution. A similar analysis can be done in the *frequency domain*. Using the Fourier transform,

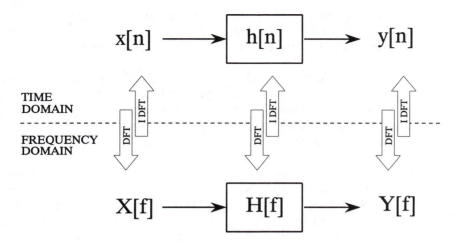

FIGURE 9-6
Comparing system operation in the time and frequency domains. In the time domain, an input signal is *convolved* with an impulse response, resulting in the output signal, that is, $x[n] * h[n] = y[n]$. In the frequency domain, an input spectrum is *multiplied* by a frequency response, resulting in the output spectrum, that is, $X[f] \times H[f] = Y[f]$. The DFT and the Inverse DFT relate the signals in the two domain.

every input signal can be represented as a group of cosine waves, each with a specified amplitude and phase shift. Likewise, the DFT can be used to represent every output signal in a similar form. This means that any linear system can be *completely* described by how it changes the amplitude and phase of cosine waves passing through it. This information is called the system's **frequency response.** Since both the impulse response and the frequency response contain complete information about the system, there must be a one-to-one correspondence between the two. Given one, you can calculate the other. The relationship between the impulse response and the frequency response is one of the foundations of signal processing: *A system's frequency response is the Fourier Transform of its impulse response.* Figure 9-6 illustrates these relationships.

Keeping with standard DSP notation, impulse responses use lower case variables, while the corresponding frequency responses are upper case. Since $h[\ ]$ is the common symbol for the impulse response, $H[\ ]$ is used for the frequency response. Systems are described in the time domain by convolution, that is: $x[n] * h[n] = y[n]$. In the frequency domain, the input spectrum is *multiplied* by the frequency response, resulting in the output spectrum. As an equation: $X[f] \times H[f] = Y[f]$. In other words, *convolution* in the time domain corresponds to *multiplication* in the frequency domain.

Figure 9-7 shows an example of using the DFT to convert a system's impulse response into its frequency response. Figure (a) is the impulse response of the system. Looking at this curve isn't going to give you the slightest idea what the system does. Taking a 64 point DFT of this impulse response produces the frequency response of the system, shown in (b). Now the function of this system becomes obvious, it passes frequencies between 0.2 and 0.3, and rejects all others. It is a band-pass filter. The *phase* of the frequency response could also be examined; however, it is *more* difficult to interpret and *less* interesting. It will be discussed in upcoming chapters.

Figure (b) is very jagged due to the low number of samples defining the curve. This situation can be improved by padding the impulse response with zeros before taking the DFT. For example, adding zeros to make the impulse response 512 samples long, as shown in (c), results in the higher resolution frequency response shown in (d).

How much resolution can you obtain in the frequency response? The answer is: *infinitely* high, if you are willing to pad the impulse response with an *infinite* number of zeros. In other words, there is nothing limiting the frequency resolution except the length of the DFT. This leads to a very important concept. Even though the impulse response is a *discrete* signal, the corresponding frequency response is *continuous*. An N point DFT of the impulse response provides $N/2+1$ *samples* of this continuous curve. If you make the DFT longer, the resolution improves, and you obtain a better idea of what the continuous curve looks like. Remember what the frequency response represents: amplitude and phase changes experienced by cosine waves as they pass through the system. Since the input signal can contain *any* frequency between 0 and 0.5, the system's frequency response *must* be a continuous curve over this range.

This can be better understood by bringing in another member of the Fourier transform family, the **Discrete Time Fourier Transform (DTFT)**. Consider an N sample signal being run through an N point DFT, producing an $N/2+1$ sample frequency domain. Remember from the last chapter that the DFT considers the time domain signal to be *infinitely long* and *periodic*. That is, the N points are repeated over and over from negative to positive infinity. Now consider what happens when we start to pad the time domain signal with an ever increasing number of zeros, to obtain a finer and finer sampling in the frequency domain. Adding zeros makes the period of the time domain *longer*, while simultaneously making the frequency domain samples *closer together*.

Now we will take this to the extreme, by adding an *infinite* number of zeros to the time domain signal. This produces a different situation in two respects. First, the time domain signal now has an infinitely long period. In other words, it has turned into an *aperiodic* signal. Second, the frequency domain has achieved an infinitesimally small spacing between samples. That is, it has become a *continuous signal*. This is the DTFT, the procedure that changes a discrete aperiodic signal into a frequency domain that is a continuous curve. In mathematical terms, a system's frequency response is found by taking the DTFT of its impulse response. Since this cannot be done in a computer, the DFT is used to calculate a *sampling* of the true frequency response. This is the difference between what you do in a computer (the DFT) and what you do with mathematical equations (the DTFT).

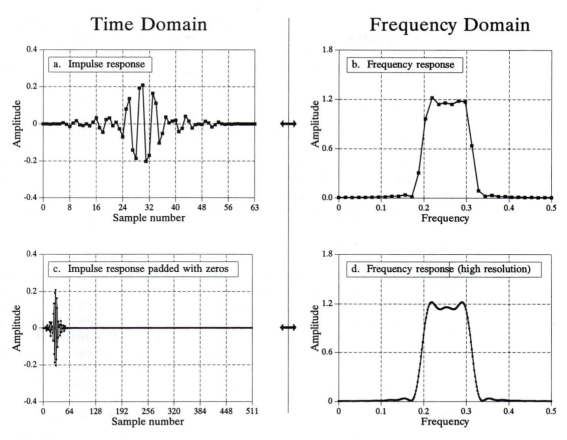

FIGURE 9-7
Finding the frequency response from the impulse response. By using the DFT, a system's impulse response, (a), can be transformed into the system's frequency response, (b). By padding the impulse response with zeros (c), higher resolution can be obtained in the frequency response, (d). Only the magnitude of the frequency response is shown in this example; discussion of the phase is postponed until the next chapter.

Convolution via the Frequency Domain

Suppose that you despise convolution. What are you going to do if given an input signal and impulse response, and need to find the resulting output signal? Figure 9-8 provides an answer: transform the two signals into the frequency domain, multiply them, and then transform the result back into the time domain. This replaces one convolution with two DFTs, a multiplication, and an Inverse DFT. Even though the intermediate steps are very different, the output is *identical* to the standard convolution algorithm.

Does anyone hate convolution enough to go to this trouble? The answer is yes. Convolution is avoided for two reasons. First, convolution is *mathematically* difficult to deal with. For instance, suppose you are given a system's impulse response, and its output signal. How do you calculate what the input signal is? This is called **deconvolution**, and is virtually impossible to understand in the time domain. However, deconvolution can be carried out in the frequency domain as a simple *division*, the inverse operation of multiplication. The frequency domain becomes attractive whenever the complexity of the Fourier Transform is less than the complexity of the convolution. This isn't a matter of which you like better; it is a matter of which you hate less.

The second reason for avoiding convolution is *computation speed*. For example, suppose you design a digital filter with a kernel (impulse response) containing 512 samples. Using a 200 MHz personal computer with floating point numbers, each sample in the output signal requires about one millisecond to calculate, using the standard convolution algorithm. In other words, the throughput of the system is only about 1,000 samples per second. This is 40 times too slow for high-fidelity audio, and 10,000 times too slow for television quality video!

The standard convolution algorithm is slow because of the large number of multiplications and additions that must be calculated. Unfortunately, simply bringing the problem into the frequency domain via the DFT doesn't help at all. Just as many calculations are required to calculate the DFTs, as are required to directly calculate the convolution. A breakthrough was made in the problem in the early 1960s when the *Fast Fourier Transform* (FFT) was developed.   The FFT is a clever algorithm for rapidly calculating the DFT.   Using the FFT, convolution by multiplication in the frequency domain can be hundreds of times faster than conventional convolution. Problems that take hours of calculation time are reduced to only minutes. This is why people get excited about the FFT, and processing signals in the frequency domain.   The FFT will be presented in Chapter 12, and the method of FFT convolution in Chapter 18. For now, focus on how signals are convolved by frequency domain multiplication.

To start, we need to define how to multiply one frequency domain signal by another, i.e., what it means to write: $X[f] \times H[f] = Y[f]$. In polar form, the magnitudes are multiplied: $Mag\, Y[f] = Mag\, X[f] \times Mag\, H[f]$, and the phases are added: $Phase\, Y[f] = Phase\, X[f] + Phase\, H[f]$. To understand this, imagine a cosine wave entering a system with some amplitude and phase. Likewise, the output signal is also a cosine wave with some amplitude and phase. The polar form of the frequency response directly describes how the two amplitudes are related and how the two phases are related.

When frequency domain multiplication is carried out in *rectangular form* there are cross terms between the real and imaginary parts. For example, a sine wave entering the system can produce both cosine and sine waves in the output. To multiply frequency domain signals in rectangular notation:

EQUATION 9-1
Multiplication of frequency domain signals in rectangular form: $Y[f] = X[f] \times H[f]$.

$$Re\, Y[f] \;=\; Re\, X[f]\, Re\, H[f] \;-\; Im\, X[f]\, Im\, H[f]$$

$$Im\, Y[f] \;=\; Im\, X[f]\, Re\, H[f] \;+\; Re\, X[f]\, Im\, H[f]$$

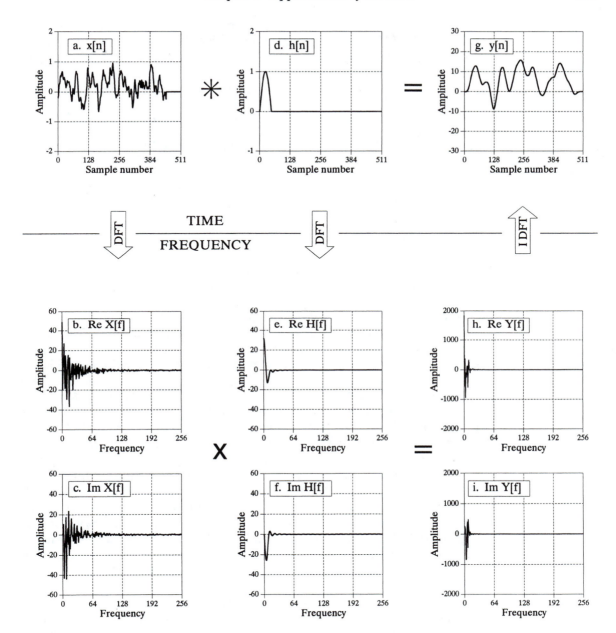

FIGURE 9-8
Frequency domain convolution. In the *time domain*, $x[n]$ is convolved with $h[n]$ resulting in $y[n]$, as is shown in Figs. (a), (d), and (g). This same procedure to be accomplished in the *frequency domain*. The DFT is used to find the frequency spectrum of the input signal, (b) & (c), and the system's frequency response, (e) & (f). Multiplying these two frequency domain signals results in the frequency spectrum of the output signal, (h) & (i). The Inverse DFT is then used to find the output signal, (g).

Focus on understanding multiplication using *polar notation*, and the idea of cosine waves passing through the system. Then simply accept that these more elaborate equations result when the same operations are carried out in rectangular form. For instance, let's look at the *division* of one frequency domain signal by another. In polar form, the division of frequency domain signals is achieved by the inverse operations we used for multiplication. To calculate: $H[f] = Y[f]/X[f]$, divide the magnitudes and subtract the phases, i.e., *Mag H[f] = Mag Y[f] / Mag X[f]*, *Phase H[f] = Phase Y[f] - Phase X[f]*. In rectangular form this becomes:

EQUATION 9-2
Division of frequency domain signals in rectangular form, where: $H[f] = Y[f]/X[f]$.

$$Re\,H[f] = \frac{Re\,Y[f]\ Re\,X[f] + Im\,Y[f]\ Im\,X[f]}{Re\,X[f]^2 + Im\,X[f]^2}$$

$$Im\,H[f] = \frac{Im\,Y[f]\ Re\,X[f] - Re\,Y[f]\ Im\,X[f]}{Re\,X[f]^2 + Im\,X[f]^2}$$

Now back to frequency domain convolution. You may have noticed that we cheated slightly in Fig. 9-8. Remember, the convolution of an *N* point signal with an *M* point impulse response results in an *N+M*-1 point output signal. We cheated by making the last part of the input signal all *zeros* to allow this expansion to occur. Specifically, (a) contains 453 nonzero samples, and (b) contains 60 nonzero samples. This means the convolution of the two, shown in (c), can fit comfortably in the 512 points provided.

Now consider the more general case in Fig. 9-9. The input signal, (a), is 256 points long, while the impulse response, (b), contains 51 nonzero points. This makes the convolution of the two signals 306 samples long, as shown in (c). The problem is, if we use frequency domain multiplication to perform the convolution, there are only 256 samples *allowed* in the output signal. In other words, 256 point DFTs are used to move (a) and (b) into the frequency domain. After the multiplication, a 256 point Inverse DFT is used to find the output signal. How do you squeeze 306 values of the correct signal into the 256 points provided by the frequency domain algorithm? The answer is, you can't! The 256 points end up being a distorted version of the correct signal. This process is called **circular convolution**. It is important because you want to *avoid* it.

To understand circular convolution, remember that an *N* point DFT views the time domain as being an infinitely long periodic signal, with *N* samples per period. Figure (d) shows three periods of how the DFT views the output signal in this example. Since $N = 256$, each period consists of 256 points: 0-255, 256-511, and 512-767. Frequency domain convolution tries to place the 306 point *correct output signal*, shown in (c), into each of these 256 point periods. This results in 49 of the samples being pushed into the neighboring period to the right, where they overlap with the samples that are legitimately there. These overlapping sections add, resulting in each of the periods appearing as shown in (e), the *circular convolution*.

Once the nature of circular convolution is understood, it is quite easy to avoid. Simply pad each of the signals being convolved with enough zeros

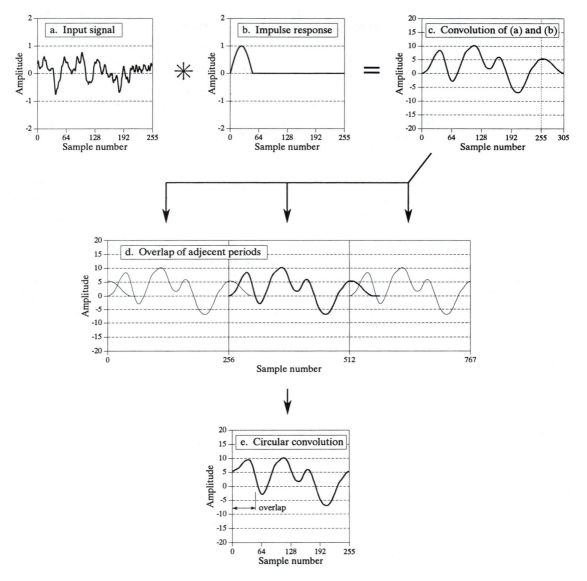

FIGURE 9-9
Circular convolution. A 256 sample signal, (a), convolved with a 51 sample impulse response, (b), results in a 306 sample signal, (c). If this convolution is performed in the frequency domain using 256 point DFTs, the 306 points in the correct convolution cannot fit into the 256 samples provided. As shown in (d), samples 256 through 305 of the output signal are pushed into the next period to the right, where they *add* to the beginning of the next period's signal. Figure (e) is a single period of the resulting signal.

to allow the output signal room to handle the $N+M-1$ points in the correct convolution. For example, the signals in (a) and (b) could be padded with zeros to make them 512 points long, allowing the use of 512 point DFTs. After the frequency domain convolution, the output signal would consist of 306 nonzero samples, plus 206 samples with a value of zero. Chapter 18 explains this procedure in detail.

Why is it called *circular* convolution? Look back at Fig. 9-9d and examine the center period, samples 256 to 511. Since all of the periods are the same, the portion of the signal that flows out of this period to the *right*, is the same that flows into this period from the *left*. If you only consider a single period, such as in (e), it *appears* that the right side of the signal is somehow *connected* to the left side. Imagine a snake biting its own tail; sample 255 is located next to sample 0, just as sample 100 is located next to sample 101. When a portion of the signal exits to the right, it magically reappears on the left. In other words, the N point time domain behaves as if it were *circular*.

In the last chapter we posed the question: does it really matter if the DFT's time domain is viewed as being N points, rather than an infinitely long periodic signal of period N? Circular convolution is an example where it *does* matter. If the time domain signal is understood to be *periodic*, the distortion encountered in circular convolution can be simply explained as the signal expanding from one period to the next. In comparison, a rather bizarre conclusion is reached if only N points of the time domain are considered. That is, frequency domain convolution acts as if the time domain is somehow wrapping into a circular ring with sample 0 being positioned next to sample N-1.

CHAPTER 10

Fourier Transform Properties

The time and frequency domains are alternative ways of representing signals. The Fourier transform is the mathematical relationship between these two representations. If a signal is modified in one domain, it will also be changed in the other domain, although usually not in the same way. For example, it was shown in the last chapter that *convolving* time domain signals results in their frequency spectra being *multiplied*. Other mathematical operations, such as addition, scaling and shifting, also have a matching operation in the opposite domain. These relationships are called *properties* of the Fourier Transform, how a mathematical change in one domain results in a mathematical change in the other domain.

Linearity of the Fourier Transform

The Fourier Transform is *linear,* that is, it possesses the properties of *homogeneity* and *additivity*. This is true for all four members of the Fourier transform family (Fourier transform, Fourier Series, DFT, and DTFT).

Figure 10-1 provides an example of how homogeneity is a property of the Fourier transform. Figure (a) shows an arbitrary time domain signal, with the corresponding frequency spectrum shown in (b). We will call these two signals: $x[\ ]$ and $X[\ ]$, respectively. *Homogeneity* means that a change in amplitude in one domain produces an identical change in amplitude in the other domain. This should make intuitive sense: when the amplitude of a time domain waveform is changed, the amplitude of the sine and cosine waves making up that waveform must also change by an equal amount.

In mathematical form, if $x[\ ]$ and $X[\ ]$ are a Fourier Transform pair, then $kx[\ ]$ and $kX[\ ]$ are also a Fourier Transform pair, for any constant k. If the frequency domain is represented in *rectangular* notation, $kX[\ ]$ means that both the real part and the imaginary part are multiplied by k. If the frequency domain is represented in *polar* notation, $kX[\ ]$ means that the magnitude is multiplied by k, while the phase remains unchanged.

Time Domain   Frequency Domain

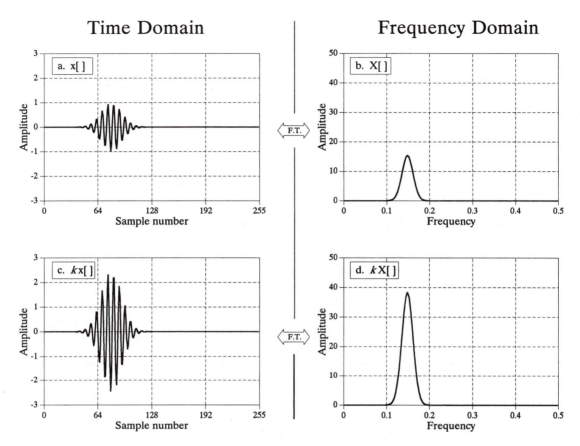

FIGURE 10-1
Homogeneity of the Fourier transform. If the amplitude is changed in one domain, it is changed by the same amount in the other domain. In other words, *scaling* in one domain corresponds to *scaling* in the other domain.

Additivity of the Fourier transform means that *addition* in one domain corresponds to *addition* in the other domain. An example of this is shown in Fig. 10-2. In this illustration, (a) and (b) are signals in the time domain called $x_1[\ ]$ and $x_2[\ ]$, respectively. Adding these signals produces a third time domain signal called $x_3[\ ]$, shown in (c). Each of these three signals has a frequency spectrum consisting of a real and an imaginary part, shown in (d) through (i). Since the two time domain signals *add* to produce the third time domain signal, the two corresponding spectra *add* to produce the third spectrum. Frequency spectra are added in rectangular notation by adding the real parts to the real parts and the imaginary parts to the imaginary parts. If: $x_1[n] + x_2[n] = x_3[n]$, then: $Re\, X_1[f] + Re\, X_2[f] = Re\, X_3[f]$ and $Im\, X_1[f] + Im\, X_2[f] = Im\, X_3[f]$. Think of this in terms of cosine and sine waves. All the cosine waves add (the real parts) and all the sine waves add (the imaginary parts) with no interaction between the two.

Frequency spectra in polar form cannot be directly added; they must be converted into rectangular notation, added, and then reconverted back to

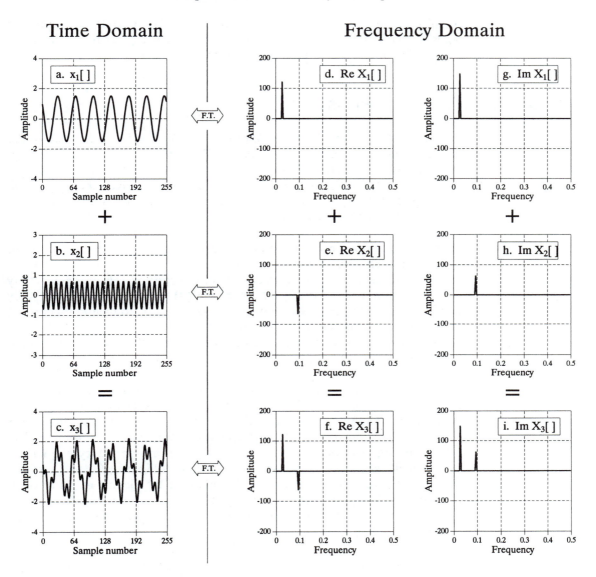

FIGURE 10-2
Additivity of the Fourier transform.   Adding two or more signals in one domain results in the
corresponding signals being added in the other domain.   In this illustration, the time domain
signals in (a) and (b) are added to produce the signal in (c).   This results in the corresponding real
and imaginary parts of the frequency spectra being added.

polar form. This can also be understood in terms of how sinusoids behave.
Imagine adding two sinusoids having the same frequency, but with different
amplitudes (A_1 and A_2) and phases (ϕ_1 and ϕ_2). If the two phases happen
to be same ($\phi_1 = \phi_2$), the amplitudes will add ($A_1 + A_2$) when the sinusoids
are added.   However, if the two phases happen to be exactly opposite
($\phi_1 = -\phi_2$), the amplitudes will *subtract* ($A_1 - A_2$) when the sinusoids are
added. The point is, when sinusoids (or spectra) are in polar form, they
cannot be added by simply adding the magnitudes and phases.

In spite of being linear, the Fourier transform is *not* shift invariant. In other words, a shift in the time domain *does not* correspond to a shift in the frequency domain. This is the topic of the next section.

Characteristics of the Phase

In mathematical form: if $x[n] \leftrightarrow Mag\ X[f]$ & $Phase\ X[f]$, then a shift in the time domain results in: $x[n+s] \leftrightarrow Mag\ X[f]$ & $Phase\ X[f] + 2\pi sf$, (where f is expressed as a fraction of the sampling rate, running between 0 and 0.5). In words, a shift of s samples in the time domain leaves the magnitude unchanged, but adds a linear term to the phase, $2\pi sf$. Let's look at an example of how this works.

Figure 10-3 shows how the phase is affected when the time domain waveform is shifted to the left or right. The magnitude has not been included in this illustration because it isn't interesting; it is not changed by the time domain shift. In Figs. (a) through (d), the waveform is gradually shifted from having the peak centered on sample 128, to having it centered on sample 0. This sequence of graphs takes into account that the DFT views the time domain as *circular*; when portions of the waveform exit to the right, they reappear on the left.

The time domain waveform in Fig. 10-3 is symmetrical around a vertical axis, that is, the left and right sides are mirror images of each other. As mentioned in Chapter 7, signals with this type of symmetry are called *linear phase*, because the phase of their frequency spectrum is a *straight line*. Likewise, signals that don't have this left-right symmetry are called *nonlinear phase*, and have phases that are something other than a straight line. Figures (e) through (h) show the phase of the signals in (a) through (d). As described in Chapter 7, these phase signals are *unwrapped*, allowing them to appear without the discontinuities associated with keeping the value between π and -π.

When the time domain waveform is shifted to the right, the phase remains a straight line, but experiences a *decrease* in slope. When the time domain is shifted to the left, there is an *increase* in the slope. This is the main property you need to remember from this section; a shift in the time domain corresponds to changing the slope of the phase.

Figures (b) and (f) display a unique case where the phase is entirely zero. This occurs when the time domain signal is *symmetrical* around sample *zero*. At first glance, this symmetry may not be obvious in (b); it may appear that the signal is symmetrical around sample 256 (i.e., $N/2$) instead. Remember that the DFT views the time domain as circular, with sample zero inherently connected to sample N-1. Any signal that is symmetrical around sample zero will also be symmetrical around sample $N/2$, and vice versa. When using members of the Fourier Transform family that do not view the time domain as periodic (such as the DTFT), the symmetry must be around sample zero to produces a zero phase.

Time Domain

Frequency Domain

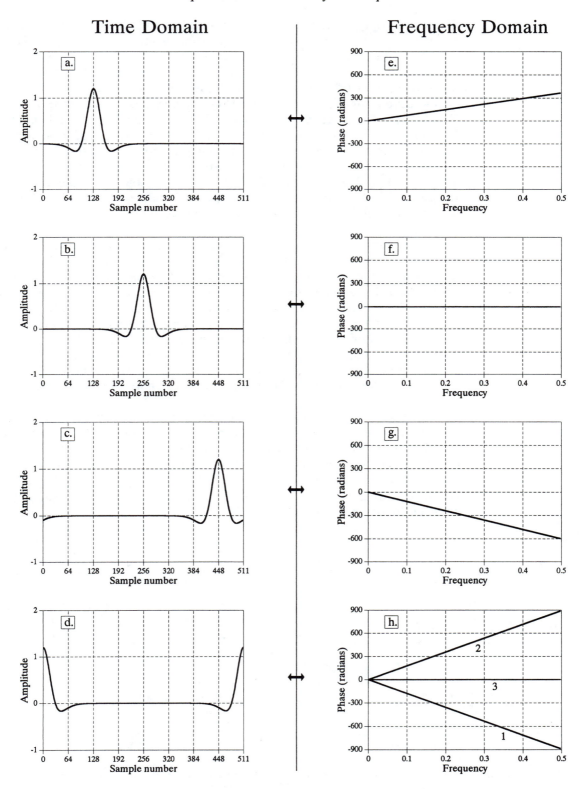

FIGURE 10-3
Phase changes resulting from a time domain shift.

Figures (d) and (h) shows something of a riddle. First imagine that (d) was formed by shifting the waveform in (c) slightly more to the right. This means that the phase in (h) would have a slightly more negative slope than in (g). This phase is shown as line 1. Next, imagine that (d) was formed by starting with (a) and shifting it to the left. In this case, the phase should have a slightly more positive slope than (e), as is illustrated by line 2. Lastly, notice that (d) is symmetrical around sample $N/2$, and should therefore have a zero phase, as illustrated by line 3. Which of these three phases is correct? They all are, depending on how the π and 2π phase ambiguities (discussed in Chapter 8) are arranged. For instance, every sample in line 2 differs from the corresponding sample in line 1 by an integer multiple of 2π, making them equal. To relate line 3 to lines 1 and 2, the π ambiguities must also be taken into account.

To understand why the phase behaves as it does, imagine shifting a waveform by *one* sample to the right. This means that all of the sinusoids that compose the waveform must also be shifted by *one* sample to the right. Figure 10-4 shows two sinusoids that might be a part of the waveform. In (a), the sine wave has a very low frequency, and a one sample shift is only a small fraction of a full cycle. In (b), the sinusoid has a frequency of one-half of the sampling rate, the highest frequency that can exist in sampled data. A one sample shift at this frequency is equal to an entire 1/2 cycle, or π radians. That is, when a shift is expressed in terms of a phase change, it becomes *proportional* to the frequency of the sinusoid being shifted.

For example, consider a waveform that is symmetrical around sample zero, and therefore has a zero phase. Figure 10-5a shows how the phase of this signal changes when it is shifted left or right. At the highest frequency, one-half of the sampling rate, the phase increases by π for each one sample shift to the left, and decreases by π for each one sample shift to the right. At zero frequency there is no phase shift, and all of the frequencies between follow in a straight line.

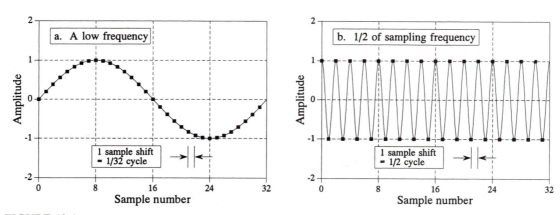

FIGURE 10-4
The relationship between samples and phase. Figures (a) and (b) show low and high frequency sinusoids, respectively. In (a), a one sample shift is equal to 1/32 of a cycle. In (b), a one sample shift is equal to 1/2 of a cycle. This is why a shift in the waveform changes the phase more at high frequencies than at low frequencies.

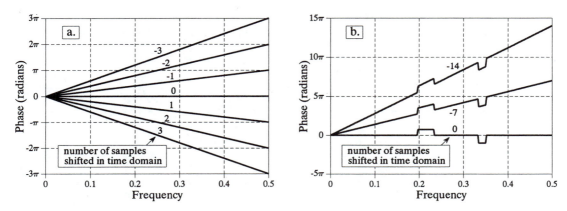

FIGURE 10-5
Phases resulting from time domain shifting. For each sample that a time domain signal is shifted in the positive direction (i.e., to the right), the phase at frequency 0.5 will decrease by π radians. For each sample shifted in the negative direction (i.e., to the left), the phase at frequency 0.5 will increase by π radians. Figure (a) shows this for a linear phase (a straight line), while (b) is an example using a nonlinear phase.

All of the examples we have used so far are *linear* phase. Figure 10-5b shows that *nonlinear* phase signals react to shifting in the same way. In this example the nonlinear phase is a straight line with two rectangular pulses. When the time domain is shifted, these nonlinear features are simply superimposed on the changing slope.

What happens in the *real* and *imaginary parts* when the time domain waveform is shifted? Recall that frequency domain signals in rectangular notation are nearly impossible for humans to understand. The real and imaginary parts typically look like random oscillations with no apparent pattern. When the time domain signal is shifted, the wiggly patterns of the real and imaginary parts become even more oscillatory and difficult to interpret. Don't waste your time trying to understand these signals, or how they are changed by time domain shifting.

Figure 10-6 is an interesting demonstration of what information is contained in the *phase*, and what information is contained in the *magnitude*. The waveform in (a) has two very distinct features: a rising edge at sample number 55, and a falling edge at sample number 110. Edges are very important when information is encoded in the *shape* of a waveform. An edge indicates *when* something happens, dividing whatever is on the left from whatever is on the right. It is time domain encoded information in its purest form. To begin the demonstration, the DFT is taken of the signal in (a), and the frequency spectrum converted into polar notation. To find the signal in (b), the phase is replaced with random numbers between -π and π, and the inverse DFT used to reconstruct the time domain waveform. In other words, (b) is based only on the information contained in the *magnitude*. In a similar manner, (c) is found by replacing the magnitude with small random numbers before using the inverse DFT. This makes the reconstruction of (c) based solely on the information contained in the *phase*.

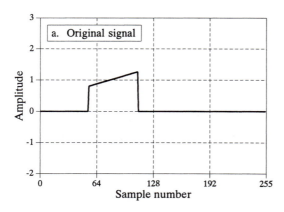

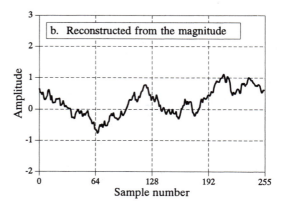

FIGURE 10-6
Information contained in the phase. Figure (a) shows a pulse-like waveform. The signal in (b) is created by taking the DFT of (a), replacing the *phase* with random numbers, and taking the Inverse DFT. The signal in (c) is found by taking the DFT of (a), replacing the *magnitude* with random numbers, and taking the Inverse DFT. The location of the *edges* is retained in (c) but not in (b). This shows that the phase contains information on the location of events in the time domain signal.

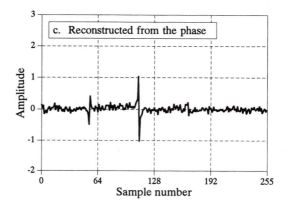

The result? The locations of the edges are clearly present in (c), but totally absent in (b). This is because an edge is formed when many sinusoids *rise* at the same location, possible only when their *phases* are coordinated. In short, much of the information about the shape of the time domain waveform is contained in the *phase*, rather than the *magnitude*. This can be contrasted with signals that have their information encoded in the frequency domain, such as audio signals. The magnitude is most important for these signals, with the phase playing only a minor role. In later chapters we will see that this type of understanding provides strategies for designing filters and other methods of processing signals. Understanding how information is represented in signals is always the first step in successful DSP.

Why does left-right symmetry correspond to a zero (or linear) phase? Figure 10-7 provides the answer. Such a signal can be decomposed into a left half and a right half, as shown in (a), (b) and (c). The sample at the center of symmetry (zero in this case) is divided equally between the left and right halves, allowing the two sides to be perfect mirror images of each other. The magnitudes of these two halves will be *identical*, as shown in (e) and (f), while the phases will be opposite in sign, as in (h) and (i). Two important concepts fall out of this. First, every signal that is symmetrical between the left and right will have a linear phase *because* the nonlinear phase of the left half exactly cancels the nonlinear phase of the right half.

Time Domain

Frequency Domain

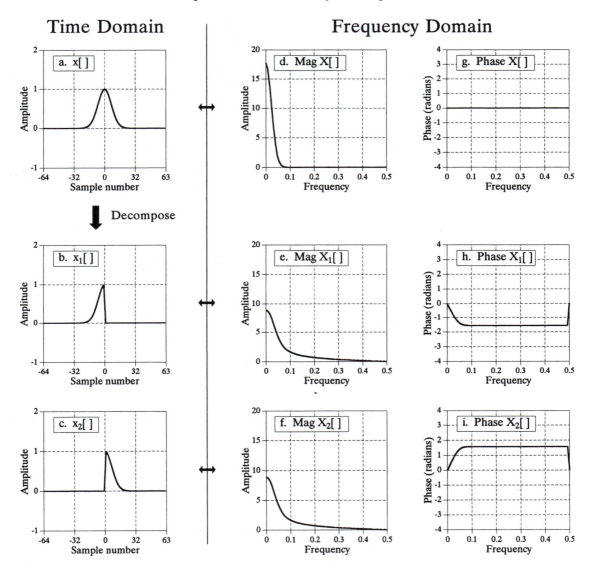

FIGURE 10-7
Phase characteristics of left-right symmetry. A signal with left-right symmetry, shown in (a), can be decomposed into a right half, (b), and a left half, (c). The magnitudes of the two halves are identical, (e) and (f), while the phases are the negative of each other, (h) and (i).

Second, imagine flipping (b) such that it becomes (c). This left-right flip in the time domain does nothing to the magnitude, but changes the sign of every point in the phase. Likewise, changing the sign of the phase flips the time domain signal left-for-right. If the signals are continuous, the flip is around zero. If the signals are discrete, the flip is around sample zero *and* sample *N/2*, simultaneously.

Changing the sign of the phase is a common enough operation that it is given its own name and symbol. The name is **complex conjugation**, and it is represented by placing a star to the upper-right of the variable. For example, if $X[f]$ consists of *Mag X[f]* and *Phase X[f]*, then $X^*[f]$ is called the *complex conjugate* and is composed of *Mag X[f]* and *-Phase X[f]*. In rectangular notation, the complex conjugate is found by leaving the real part alone, and changing the sign of the imaginary part. In mathematical terms, if $X[f]$ is composed of *Re X[f]* and *Im X[f]*, then $X^*[f]$ is made up of *Re X[f]* and *-Im X[f]*.

Here are several examples of how the complex conjugate is used in DSP. If $x[n]$ has a Fourier transform of $X[f]$, then $x[-n]$ has a Fourier transform of $X^*[f]$. In words, flipping the time domain left-for-right corresponds to changing the sign of the phase. As another example, recall from Chapter 7 that correlation can be performed as a convolution. This is done by flipping one of the signals left-for-right. In mathematical form, $a[n] * b[n]$ is convolution, while $a[n] * b[-n]$ is correlation. In the frequency domain these operations correspond to $A[f] \times B[f]$ and $A[f] \times B^*[f]$, respectively. As the last example, consider an arbitrary signal, $x[n]$, and its frequency spectrum, $X[f]$. The frequency spectrum can be changed to *zero phase* by multiplying it by its complex conjugate, that is, $X[f] \times X^*[f]$.   In words, whatever phase $X[f]$ happens to have will be canceled by adding its opposite (remember, when frequency spectra are multiplied, their phases are added).   In the time domain, this means that $x[n] * x[-n]$ (a signal convolved with a left-right flipped version of itself) will have left-right symmetry around sample zero, regardless of what $x[n]$ is.

To many engineers and mathematicians, this kind of manipulation *is* DSP. If you want to be able to communicate with this group, get used to using their language.

Periodic Nature of the DFT

Unlike the other three Fourier Transforms, the DFT views *both* the time domain and the frequency domain as *periodic*. This can be confusing and inconvenient since most of the signals used in DSP are *not* periodic. Nevertheless, if you want to use the DFT, you must conform with the DFT's view of the world.

Figure 10-8 shows two different interpretations of the time domain signal. First, look at the upper signal, the time domain viewed as N points. This represents how digital signals are typically acquired in scientific experiments and engineering applications. For instance, these 128 samples might have been acquired by sampling some parameter at regular intervals of *time*. Sample 0 is distinct and separate from sample 127 because they were acquired at *different* times. From the way this signal was formed, there is no reason to think that the samples on the left of the signal are even related to the samples on the right.

Unfortunately, the DFT doesn't see things this way. As shown in the lower figure, the DFT views these 128 points to be a single period of an infinitely long periodic signal. This means that the left side of the acquired signal is connected to the right side of a duplicate signal. Likewise, the right side of the acquired signal is connected to the left side of an identical period. This can also be thought of as the right side of the acquired signal wrapping around and connecting to its left side. In this view, sample 127 occurs next to sample 0, just as sample 43 occurs next to sample 44. This is referred to as being **circular**, and is identical to viewing the signal as being *periodic*.

The most serious consequence of this periodicity is **time domain aliasing**. To illustrate this, suppose we take a time domain signal and pass it through the DFT to find its frequency spectrum. We could immediately pass this frequency spectrum through an Inverse DFT to reconstruct the original time domain signal, but the entire procedure wouldn't be very interesting. Instead, we will modify the frequency spectrum in some manner before using the Inverse DFT. For instance, selected frequencies might be deleted, changed in amplitude or phase, shifted around, etc. These are the kinds of things routinely done in DSP. Unfortunately, these changes in the frequency domain can create a time domain signal that is too long to fit into

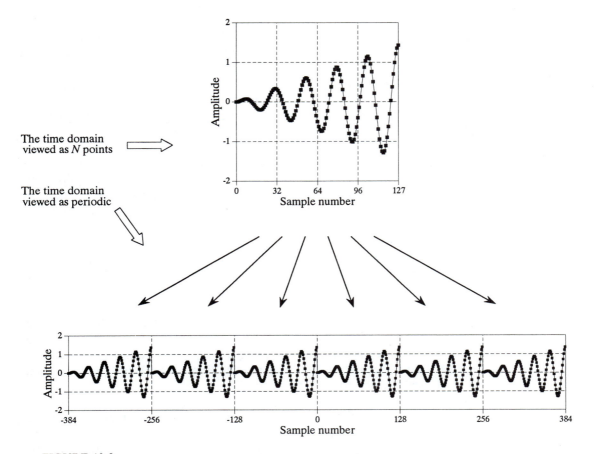

The time domain viewed as *N* points

The time domain viewed as periodic

FIGURE 10-8

Periodicity of the DFT's time domain signal. The time domain can be viewed as *N* samples in length, shown in the upper figure, or as an infinitely long periodic signal, shown in the lower figure.

a single period. This forces the signal to spill over from one period into the adjacent periods. When the time domain is view as *circular*, portions of the signal that overflow on the right suddenly seem to reappear on the left side of the signal, and vice versa. That is, the overflowing portions of the signal *alias* themselves to a new location in the time domain. If this new location happens to already contain an existing signal, the whole mess adds, resulting in a loss of information. Circular convolution resulting from frequency domain multiplication (discussed in Chapter 9), is an excellent example of this type of aliasing.

Periodicity in the frequency domain behaves in much the same way, but is more complicated. Figure 10-9 shows an example. The upper figures show the magnitude and phase of the frequency spectrum, viewed as being composed of $N/2+1$ samples spread between 0 and 0.5 of the sampling rate. This is the simplest way of viewing the frequency spectrum, but it doesn't explain many of the DFT's properties.

The lower two figures show how the DFT views this frequency spectrum as being periodic. The key feature is that the frequency spectrum between 0 and 0.5 appears to have a *mirror image* of frequencies that run between 0 and -0.5. This mirror image of **negative frequencies** is slightly different for the magnitude and the phase signals. In the magnitude, the signal is flipped left-for-right. In the phase, the signal is flipped left-for-right, *and* changed in sign. As you recall, these two types of symmetry are given names: the magnitude is said to be an **even** signal (it has *even* symmetry), while the phase is said to be an **odd** signal (it has *odd* symmetry). If the frequency spectrum is converted into the real and imaginary parts, the *real part* will always be *even*, while the *imaginary part* will always be *odd*.

Taking these negative frequencies into account, the DFT views the frequency domain as periodic, with a period of 1.0 times the sampling rate, such as -0.5 to 0.5, or 0 to 1.0. In terms of sample numbers, this makes the length of the frequency domain period equal to N, the same as in the time domain.

The periodicity of the frequency domain makes it susceptible to **frequency domain aliasing**, completely analogous to the previously described time domain aliasing. Imagine a time domain signal that corresponds to some frequency spectrum. If the time domain signal is modified, it is obvious that the frequency spectrum will also be changed. If the modified frequency spectrum cannot fit in the space provided, it will push into the adjacent periods. Just as before, this aliasing causes two problems: frequencies aren't where they should be, and overlapping frequencies from different periods add, destroying information.

Frequency domain aliasing is more difficult to understand than time domain aliasing, since the periodic pattern is more complicated in the frequency domain. Consider a single frequency that is being forced to move from 0.01 to 0.49 in the frequency domain. The corresponding negative frequency is therefore moving from -0.01 to -0.49. When the positive frequency moves

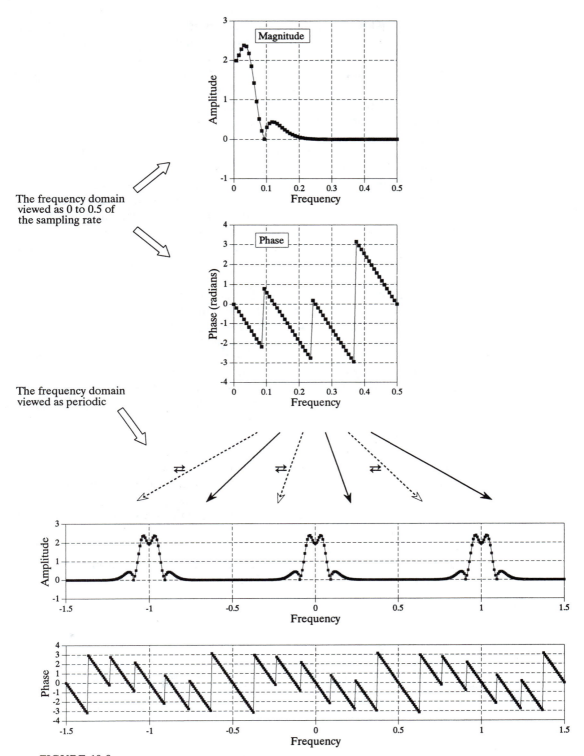

The frequency domain viewed as 0 to 0.5 of the sampling rate

The frequency domain viewed as periodic

FIGURE 10-9
Periodicity of the DFT's frequency domain. The frequency domain can be viewed as running from 0 to 0.5 of the sampling rate (upper two figures), or an infinity long periodic signal with every other 0 to 0.5 segment flipped left-for-right (lower two figures).

across the 0.5 barrier, the negative frequency is pushed across the -0.5 barrier. Since the frequency domain is periodic, these same events are occurring in the other periods, such as between 0.5 and 1.5. A clone of the positive frequency is crossing frequency 1.5 from left to right, while a clone of the negative frequency is crossing 0.5 from right to left. Now imagine what this looks like if you can only see the frequency band of 0 to 0.5. It appears that a frequency leaving to the *right*, reappears on the *right*, but moving in the opposite direction.

Figure 10-10 illustrates how aliasing appears in the time and frequency domains when only a single period is viewed. As shown in (a), if one end of a time domain signal is too long to fit inside a single period, the protruding end will be *cut off* and *pasted* onto the other side. In comparison, (b) shows that when a frequency domain signal overflows the period, the protruding end is *folded over*. Regardless of where the aliased segment ends up, it adds to whatever signal is already there, destroying information.

Hold on, you say! What are these strange things called *negative frequencies?* Are they just some bizarre artifact of the mathematics, or do they have a real world meaning? Figure 10-11 shows what they are about. Figure (a) is a discrete signal composed of 32 samples. Imagine that you are given the task of finding the frequency spectrum that corresponds to these 32 points. To make your job easier, you are told that these points represent a discrete cosine wave. In other words, you must find the frequency and phase shift (f and θ) such that $x[n] = \cos(2\pi n f/N + \theta)$ matches the given samples. It isn't long before you come up with the solution shown in (b), that is, $f = 3$ and $\theta = -\pi/4$.

If you stopped your analysis at this point, you only get 1/3 credit for the problem. This is because there are two other solutions that you have missed. As shown in (c), the second solution is $f = -3$ and $\theta = \pi/4$. Even if the idea of a *negative frequency* offends your sensibilities, it doesn't

a. Time domain aliasing

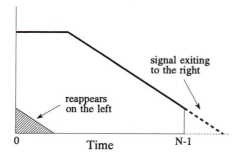

b. Frequency domain aliasing

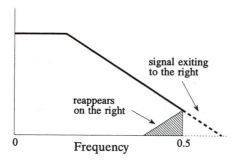

FIGURE 10-10
Examples of aliasing in the time and frequency domains. In the time domain, shown in (a), portions of the signal that exits to the right, reappear on the left. In the frequency domain, (b), portions of the signal that exit to the right, reappear on the right as if they had been folded over.

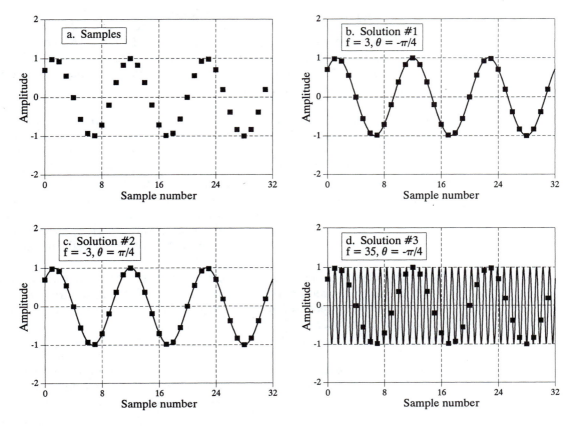

FIGURE 10-11
The meaning of negative frequencies. The problem is to find the frequency spectrum of the discrete signal shown in (a). That is, we want to find the frequency and phase of the sinusoid that passed through all of the samples. Figure (b) is a solution using a *positive* frequency, while (c) is a solution using a *negative* frequency. Figure (d) represents a family of solutions to the problem.

change the fact that it is a mathematically valid solution to the defined problem. Every *positive* frequency sinusoid can alternately be expressed as a *negative* frequency sinusoid. This applies to continuous as well as discrete signals.

The third solution is not a single answer, but an infinite family of solutions. As shown in (d), the sinusoid with $f = 35$ and $\theta = -\pi/4$ passes through all of the discrete points, and is therefore a correct solution. The fact that it shows oscillation between the samples may be confusing, but it doesn't disqualify it from being an authentic answer. Likewise, $f = \pm 29$, $f = \pm 35$, $f = \pm 61$, and $f = \pm 67$ are all solutions with multiple oscillations between the points. This third group of solutions requires the original signal to be discrete, rather than continuous. With continuous signals, you can't have oscillations between the samples, because you don't have samples.

Each of these three solutions corresponds to a different section of the frequency spectrum. For discrete signals, the first solution corresponds to frequencies between 0 and 0.5 of the sampling rate. The second solution results in frequencies between 0 and -0.5. Lastly, the third solution makes up the infinite number of duplicated frequencies below -0.5 and above 0.5. With continuous signals, the first solution results in frequencies from zero to positive infinity, while the second solution results in frequencies from zero to negative infinity.

Many DSP techniques do not require the use of negative frequencies, or an understanding of the DFT's periodicity. For example, two common ones were described in the last chapter, *spectral analysis*, and the *frequency response* of systems. For these applications, it is completely sufficient to view the time domain as extending from sample 0 to N-1, and the frequency domain from zero to one-half of the sampling frequency. These techniques can use a simpler view of the world because they never result in portions of one period moving into another period. With this restriction, looking at a single period is no different from looking at the entire periodic signal.

However, certain procedures can *only* be analyzed by considering how signals overflow between periods. Two examples of this have already been presented, *circular convolution* and *analog-to-digital conversion*. In circular convolution, multiplication of the frequency spectra results in the time domain signals being convolved. If the resulting time domain signal is too long to fit inside a single period, it overflows into the adjacent periods, resulting in *time domain aliasing*. In contrast, analog-to-digital conversion is an example of *frequency domain aliasing*. A nonlinear action is taken in the time domain, that is, changing a continuous signal into a discrete signal by sampling. The problem is, the spectrum of the original analog signal may be too long to fit inside the discrete signal's spectrum. When we force the situation, the ends of the spectrum protrude into adjacent periods. Let's look at two more examples where the periodic nature of the DFT is important, *compression & expansion* of signals, and *amplitude modulation*.

Compression and Expansion, Multirate methods

As shown in Fig. 10-12, a *compression* of the signal in one domain results in an *expansion* in the other, and vice versa. For continuous signals, if $X(f)$ is the Fourier Transform of $x(t)$, then $1/k \times X(f/k)$ is the Fourier Transform of $x(kt)$, where k is the parameter controlling the expansion or contraction. If an event happens *faster* (it is compressed in time), it must be composed of *higher* frequencies. If an event happens *slower* (it is expanded in time), it must be composed of *lower* frequencies. This pattern holds if taken to either of the two extremes. That is, if the time domain signal is compressed so far that it becomes an *impulse*, the corresponding frequency spectrum is expanded so far that it becomes a *constant value*. Likewise, if the time domain is expanded until it becomes a constant value, the frequency domain becomes an impulse.

Discrete signals behave in a similar fashion, but there are a few more details. The first issue with discrete signals is *aliasing*. Imagine that the

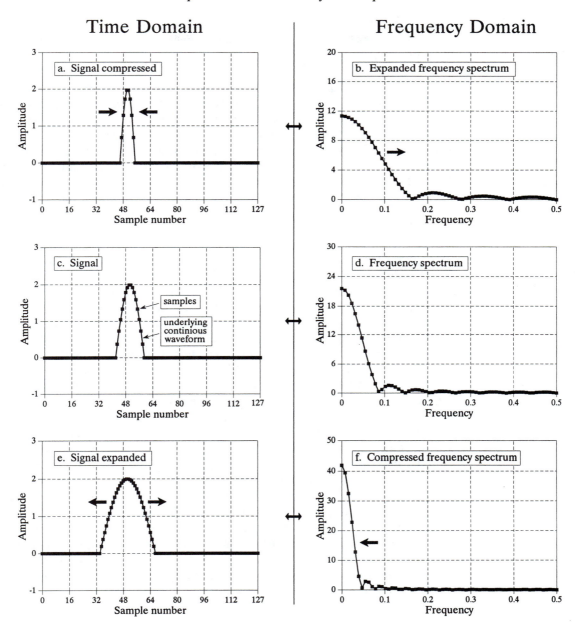

FIGURE 10-12
Compression and expansion. Compressing a signal in one domain results in the signal being expanded in the other domain, and vice versa. Figures (c) and (d) show a discrete signal and its spectrum, respectively. In (a) and (b), the time domain signal has been compressed, resulting in the frequency spectrum being expanded. Figures (e) and (f) show the opposite process. As shown in these figures, discrete signals are expanded or contracted by expanding or contracting the underlying continuous waveform. This underlying waveform is then resampled to find the new discrete signal.

pulse in (a) is compressed several times more than is shown. The frequency spectrum is expanded by an equal factor, and several of the humps in (b) are pushed to frequencies beyond 0.5. The resulting aliasing breaks the simple expansion/contraction relationship. This type of aliasing can also happen in the time domain. Imagine that the frequency spectrum in (f) is compressed much harder, resulting in the time domain signal in (e) expanding into neighboring periods.

A second issue is to define exactly what it means to compress or expand a discrete signal. As shown in Fig. 10-12a, a discrete signal is compressed by compressing the underlying *continuous* curve that the samples lie on, and then resampling the new continuous curve to find the new discrete signal. Likewise, this same process for the expansion of discrete signals is shown in (e). When a discrete signal is compressed, events in the signal (such as the width of the pulse) happen over a *fewer* number of samples. Likewise, events in an expanded signal happen over a *greater* number of samples.

An equivalent way of looking at this procedure is to keep the underlying continuous waveform the same, but resample it at a different sampling rate. For instance, look at Fig. 10-13a, a discrete Gaussian waveform composed of 50 samples. In (b), the same underlying curve is represented by 400 samples. The change between (a) and (b) can be viewed in two ways: (1) the sampling rate has been kept constant, but the underlying waveform has been expanded to be eight times wider, or (2) the underlying waveform has been kept constant, but the sampling rate has increased by a factor of eight. Methods for changing the sampling rate in this way are called **multirate** techniques. If more samples are added, it is called **interpolation**. If fewer samples are used to represent the signal, it is called **decimation**. Chapter 3 describes how multirate techniques are used in ADC and DAC.

Here is the problem: if we are given an arbitrary discrete signal, how do we know what the underlying continuous curve is? It depends on if the signal's information is encoded in the *time domain* or in the *frequency domain*. For time domain encoded signals, we want the underlying continious waveform to be a smooth curve that passes through all the samples. In the simplest case, we might drawing straight lines between the points and then round the rough corners. The next level of sophistication is to use a curve fitting algorithm, such as a spline function or polynomial fit. There is not a single "correct" answer to this problem. This approach is based on minimize irregularities in the *time domain* waveform, and completely ignores the freqeuncy domain.

When a signal has information encoded in the frequency domain, we ignore the time domain waveform and concentrate on the frequency spectrum. As discussed in the last chapter, a finer sampling of a frequency spectrum (more samples between frequency 0 and 0.5) can be obtained by padding the time domain signal with zeros before taking the DFT. Duality allows this to work in the opposite direction. If we want a finer sampling in the time domain (interpolation), pad the frequency spectrum with zeros before taking the Inverse DFT. Say we want to interpolate a 50 sample signal into a 400 sample signal. It's done like this: (1) Take the 50 samples and add zeros to make the signal 64 samples long. (2) Use a 64 point DFT to find the frequency spectrum, which will consist of a 33 point real part and a 33 point imaginary part. (3) Pad the right side of the frequency spectrum

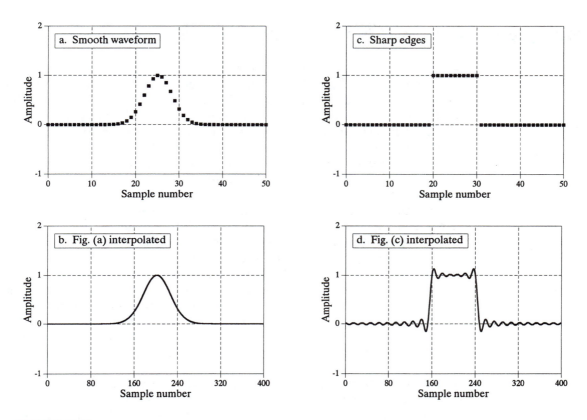

FIGURE 10-13
Interpolation by padding the frequency domain. Figures (a) and (c) each consist of 50 samples. These are interpolated to 400 samples by padding the frequency domain with zeros, resulting in (b) and (d), respectively. (Figures (b) and (d) are discrete signals, but are drawn as continuous lines because of the large number of samples).

(both the real and imaginary parts) with 224 zeros to make the frequency spectrum 257 points long. (4) Use a 512 point Inverse DFT to transform the data back into the time domain. This will result in a 512 sample signal that is a high resolution version of the 64 sample signal. The first 400 samples of this signal are an interpolated version of the original 50 samples.

The key feature of this technique is that the interpolated signal is composed of *exactly* the same frequencies as the original signal. This may or may not provide a well-behaved fit in the time domain. For example, Figs. 10-13 (a) and (b) show a 50 sample signal being interpolated into a 400 sample signal by this method. The interpolation is a smooth fit between the original points, much as if a curve fitting routine had been used. In comparison, (c) and (d) show another example where the time domain is a mess! The oscillatory behavior shown in (d) arises at edges or other discontinuities in the signal. This also includes any discontinuity between sample zero and *N*-1, since the time domain is viewed as being circular. This overshoot at discontinuities is called the *Gibbs effect*, and is discussed in Chapter 11. Another frequency domain interpolation technique is presented in Chapter 3, adding zeros between the time domain samples and low-pass filtering.

Multiplying Signals (Amplitude Modulation)

An important Fourier transform property is that *convolution* in one domain corresponds to *multiplication* in the other domain. One side of this was discussed in the last chapter: time domain signals can be convolved by multiplying their frequency spectra. Amplitude modulation is an example of the reverse situation, multiplication in the time domain corresponds to convolution in the frequency domain. In addition, amplitude modulation provides an excellent example of how the elusive *negative* frequencies enter into everyday science and engineering problems.

Audio signals are great for short distance communication; when you speak, someone across the room hears you. On the other hand, radio frequencies are very good at propagating long distances. For instance, if a 100 volt, 1 MHz sine wave is fed into an antenna, the resulting radio wave can be detected in the next *room*, the next *country*, and even on the next *planet*. **Modulation** is the process of merging two signals to form a third signal with desirable characteristics of both. This always involves nonlinear processes such as multiplication; you can't just add the two signals together. In radio communication, modulation results in radio signals that can propagate long distances *and* carry along audio or other information.

Radio communication is an extremely well developed discipline, and many modulation schemes have been developed. One of the simplest is called **amplitude modulation**. Figure 10-14 shows an example of how amplitude modulation appears in both the time and frequency domains. Continuous signals will be used in this example, since modulation is usually carried out in analog electronics. However, the whole procedure could be carried out in discrete form if needed (the shape of the future!).

Figure (a) shows an audio signal with a DC bias such that the signal always has a positive value. Figure (b) shows that its frequency spectrum is composed of frequencies from 300 Hz to 3 kHz, the range needed for voice communication, plus a spike for the DC component. All other frequencies have been removed by analog filtering. Figures (c) and (d) show the **carrier wave**, a pure sinusoid of much higher frequency than the audio signal. In the time domain, amplitude modulation consists of *multiplying* the audio signal by the carrier wave. As shown in (e), this results in an oscillatory waveform that has an instantaneous amplitude proportional to the original audio signal. In the jargon of the field, the *envelope* of the carrier wave is equal to the modulating signal. This signal can be routed to an antenna, converted into a radio wave, and then detected by a receiving antenna. This results in a signal identical to (e) being generated in the radio receiver's electronics. A *detector* or *demodulator* circuit is then used to convert the waveform in (e) back into the waveform in (a).

Since the time domain signals are multiplied, the corresponding frequency spectra are convolved. That is, (f) is found by convolving (b) & (d). Since the spectrum of the carrier is a shifted delta function, the spectrum of the

Time Domain

Frequency Domain

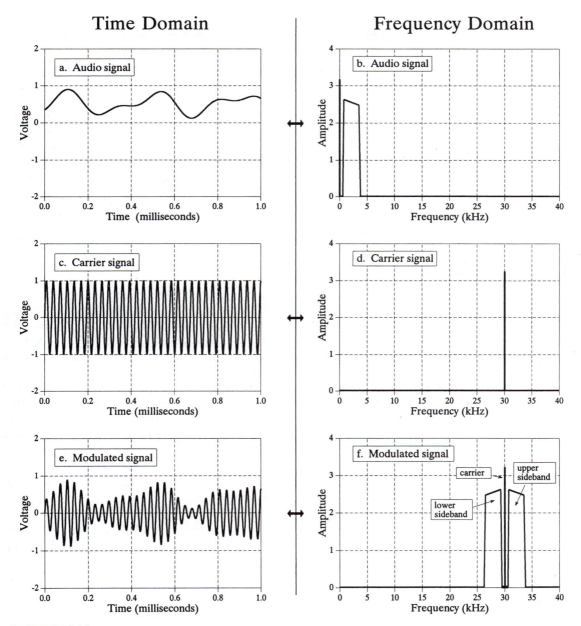

FIGURE 10-14
Amplitude modulation. In the time domain, amplitude modulation is achieved by multiplying the audio signal, (a), by the carrier signal, (c), to produce the modulated signal, (e). Since multiplication in the time domain corresponds to convolution in the frequency domain, the spectrum of the modulated signal is the spectrum of the audio signal shifted to the frequency of the carrier.

modulated signal is equal to the audio spectrum *shifted* to the frequency of the carrier. This results in a modulated spectrum composed of three components: a **carrier wave**, an **upper sideband**, and a **lower sideband**.

These correspond to the three parts of the original audio signal: the DC component, the positive frequencies between 0.3 and 3 kHz, and the negative frequencies between -0.3 and -3 kHz, respectively. Even though the negative frequencies in the original audio signal are somewhat elusive and abstract, the resulting frequencies in the lower sideband are as real as you could want them to be. The ghosts have taken human form!

Communication engineers live and die by this type of frequency domain analysis. For example, consider the frequency spectrum for television transmission. A standard TV signal has a frequency spectrum from DC to 6 MHz. By using these frequency shifting techniques, 82 of these 6 MHz wide channels are stacked on top of each other. For instance, channel 3 is from 60 to 66 MHz, channel 4 is from 66 to 72 MHz, channel 83 is from 884 to 890 MHz, etc. The television receiver moves the desired channel back to the DC to 6 MHz band for display on the screen. This scheme is called **frequency domain multiplexing**.

The Discrete Time Fourier Transform

The Discrete Time Fourier Transform (DTFT) is the member of the Fourier transform family that operates on *aperiodic, discrete* signals. The best way to understand the DTFT is how it relates to the DFT. To start, imagine that you acquire an N sample signal, and want to find its frequency spectrum. By using the DFT, the signal can be decomposed into $N/2+1$ sine and cosine waves, with frequencies equally spaced between zero and one-half of the sampling rate. As discussed in the last chapter, padding the time domain signal with zeros makes the period of the time domain *longer*, as well as making the spacing between samples in the frequency domain *narrower*. As N approaches infinity, the time domain becomes *aperiodic*, and the frequency domain becomes a *continuous* signal. This is the DTFT, the Fourier transform that relates an *aperiodic, discrete* signal, with a *periodic, continuous* frequency spectrum.

The mathematics of the DTFT can be understood by starting with the synthesis and analysis equations for the DFT (Eqs. 8-2, 8-3 and 8-4), and taking N to infinity:

EQUATION 10-1
The DTFT analysis equation. In this relation, $x[n]$ is the time domain signal with n running from 0 to $N-1$. The frequency spectrum is held in: $Re\,X(\omega)$ and $Im\,X(\omega)$, with ω taking on values between 0 and π.

$$Re\,X(\omega) = \sum_{n=-\infty}^{+\infty} x[n]\cos(\omega n)$$

$$Im\,X(\omega) = -\sum_{n=-\infty}^{+\infty} x[n]\sin(\omega n)$$

EQUATION 10-2
The DTFT synthesis equation.

$$x[n] = \frac{1}{\pi}\int_0^\pi Re\,X(\omega)\cos(\omega n) - Im\,X(\omega)\sin(\omega n)\,d\omega$$

There are many subtle details in these relations. First, the time domain signal, $x[n]$, is still discrete, and therefore is represented by *brackets*. In comparison, the frequency domain signals, $Re\,X(\omega)$ & $Im\,X(\omega)$, are continuous, and are thus written with *parentheses*. Since the frequency domain is continuous, the synthesis equation must be written as an integral, rather than a summation.

As discussed in Chapter 8, frequency is represented in the DFT's frequency domain by one of three variables: k, an index that runs from 0 to $N/2$; f, the fraction of the sampling rate, running from 0 to 0.5; or ω, the fraction of the sampling rate expressed as a natural frequency, running from 0 to π. The spectrum of the DTFT is continuous, so either f or ω can be used. The common choice is ω, because it makes the equations shorter by eliminating the always present factor of 2π. Remember, when ω is used, the frequency spectrum extends from 0 to π, which corresponds to DC to one-half of the sampling rate. To make things even more complicated, many authors use Ω (an upper case omega) to represent this frequency in the DTFT, rather than ω (a lower case omega).

When calculating the inverse DFT, samples 0 and $N/2$ must be divided by two (Eq. 8-3) before the synthesis can be carried out (Eq. 8-2). This is not necessary with the DTFT. As you recall, this action in the DFT is related to the frequency spectrum being defined as a *spectral density*, i.e., amplitude per unit of bandwidth. When the spectrum becomes continuous, the special treatment of the end points disappear. However, there is still a normalization factor that must be included, the $2/N$ in the DFT (Eq. 8-3) becomes $1/\pi$ in the DTFT (Eq. 10-2). Some authors place these terms in front of the *synthesis* equation, while others place them in front of the *analysis* equation. Suppose you start with some time domain signal. After taking the Fourier transform, and then the Inverse Fourier transform, you want to end up with what you started. That is, the $1/\pi$ term (or the $2/N$ term) must be encountered somewhere along the way, either in the forward or in the inverse transform. Some authors even split the term between the two transforms by placing $1/\sqrt{\pi}$ in front of both.

Since the DTFT involves infinite summations and integrals, it cannot be calculated with a digital computer. Its main use is in theoretical problems as an alternative to the DFT. For instance, suppose you want to find the frequency response of a system from its impulse response. If the impulse response is known as an *array of numbers*, such as might be obtained from an experimental measurement or computer simulation, a DFT program is run on a computer. This provides the frequency spectrum as another *array of numbers*, equally spaced between 0 and 0.5 of the sampling rate.

In other cases, the impulse response might be know as an *equation*, such as a sinc function or an exponentially decaying sinusoid. The DTFT is used here to mathematically calculate the frequency domain as another *equation*, specifying the entire continuous curve between 0 and 0.5. While the DFT could also be used for this calculation, it would only provide an equation for *samples* of the frequency response, not the entire curve.

Parseval's Relation

Since the time and frequency domains are equivalent representations of the same signal, they must have the same *energy*. This is called Parseval's relation, and holds for all members of the Fourier transform family. For the DFT, Parseval's relation is expressed:

EQUATION 10-2
Parseval's relation. In this equation, $x[i]$ is a time domain signal with i running from 0 to N-1, and $X[k]$ is its *modified* frequency spectrum, with k running from 0 to $N/2$. The *modified* frequency spectrum is found by taking the DFT of the signal, and dividing the first and last frequencies (sample 0 and $N/2$) by the square-root of two.

$$\sum_{i=0}^{N-1} x[i]^2 \;=\; \frac{2}{N} \sum_{k=0}^{N/2} Mag\, X[k]^2$$

The left side of this equation is the total energy contained in the time domain signal, found by summing the energies of the N individual samples. Likewise, the right side is the energy contained in the frequency domain, found by summing the energies of the $N/2+1$ sinusoids. Remember from physics that *energy* is proportional to the *amplitude squared*. For example, the energy in a spring is proportional to the displacement squared, and the energy stored in a capacitor is proportional to the voltage squared. In Eq. 10-1, $X[f]$ is the frequency spectrum of $x[n]$, with one slight modification: the first and last frequency components, $X[0]$ & $X[N/2]$, have been divided by $\sqrt{2}$. This modification, along with the $2/N$ factor on the right side of the equation, accounts for several subtle details of calculating and summing energies.

To understand these corrections, start by finding the frequency domain of the signal by using the DFT. Next, convert the frequency domain into the amplitudes of the sinusoids needed to reconstruct the signal, as previously defined in Eq. 8-3. This is done by dividing the first and last points (sample 0 and $N/2$) by 2, and then dividing all of the points by $N/2$. While this provides the amplitudes of the sinusoids, they are expressed as a *peak* amplitude, not the *root-mean-square* (rms) amplitude needed for energy calculations. In a sinusoid, the peak amplitude is converted to rms by dividing by $\sqrt{2}$. This correction must be made to all of the frequency domain values, *except* sample 0 and $N/2$. This is because these two sinusoids are unique; one is a constant value, while the other alternates between two constant values. For these two special cases, the *peak* amplitude is already equal to the *rms* value. All of the values in the frequency domain are squared and then summed. The last step is to divide the summed value by N, to account for each sample in the frequency domain being converted into a sinusoid that covers N values in the time domain. Working through all of these details produces Eq. 10-1.

While Parseval's relation is interesting from the physics it describes (conservation of energy), it has few practical uses in DSP.

CHAPTER 11

Fourier Transform Pairs

For every *time domain* waveform there is a corresponding *frequency domain* waveform, and vice versa. For example, a rectangular pulse in the time domain coincides with a sinc function in the frequency domain. Duality provides that the reverse is also true; a rectangular pulse in the frequency domain matches a sinc function in the time domain. Waveforms that correspond to each other in this manner are called *Fourier transform pairs*. Several common pairs are presented in this chapter.

Delta Function Pairs

The delta function is a simple waveform, and has an equally simple Fourier transform pair. Figure 11-1a shows a delta function in the time domain, with its frequency spectrum in (b) and (c). The magnitude is a constant value, while the phase is entirely zero. As discussed in the last chapter, this can be understood by using the expansion/compression property. When the time domain is compressed until it becomes an impulse, the frequency domain is expanded until it becomes a constant value.

In (d) and (g), the time domain waveform is shifted four and eight samples to the right, respectively. As expected from the properties in the last chapter, shifting the time domain waveform does not affect the magnitude, but adds a linear component to the phase. The phase signals in this figure have not been *unwrapped*, and thus extend only from $-\pi$ to π. Also notice that the horizontal axes in the frequency domain run from -0.5 to 0.5. That is, they show the *negative* frequencies in the spectrum, as well as the *positive* ones. The negative frequencies are redundant information, but they are often included in DSP graphs and you should become accustomed to seeing them.

Figure 11-2 presents the same information as Fig. 11-1, but with the frequency domain in *rectangular form*. There are two lessons to be learned here. First, compare the polar and rectangular representations of the

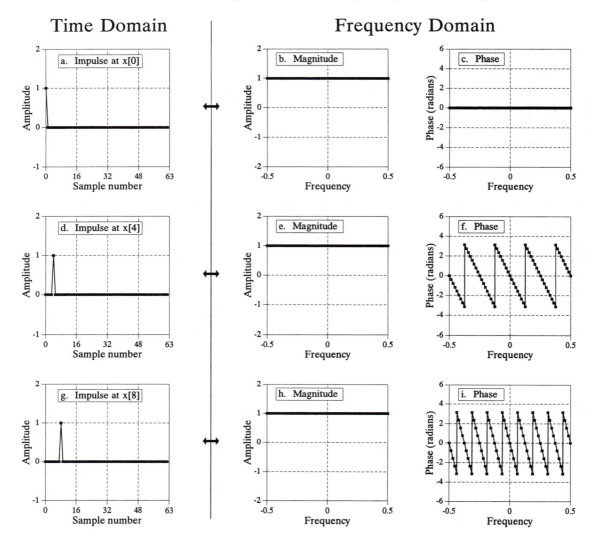

FIGURE 11-1
Delta function pairs in *polar form*. An impulse in the time domain corresponds to a constant magnitude and a linear phase in the frequency domain.

frequency domains. As is usually the case, the polar form is much easier to understand; the magnitude is nothing more than a constant, while the phase is a straight line. In comparison, the real and imaginary parts are sinusoidal oscillations that are difficult to attach a meaning to.

The second interesting feature in Fig. 11-2 is the *duality* of the DFT. In the conventional view, each sample in the DFT's frequency domain corresponds to a sinusoid in the time domain. However, the reverse of this is also true, each sample in the time domain corresponds to sinusoids in the frequency domain. Including the negative frequencies in these graphs allows the duality property to be more symmetrical. For instance, Figs. (d), (e), and

Time Domain

Frequency Domain

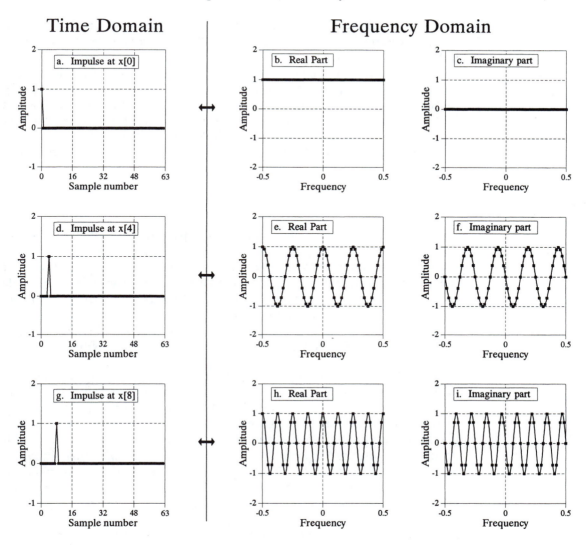

FIGURE 11-2

Delta function pairs in *rectangular form*. Each sample in the time domain results in a cosine wave in the real part, and a negative sine wave in the imaginary part of the frequency domain.

(f) show that an impulse at sample number four in the time domain results in four cycles of a cosine wave in the real part of the frequency spectrum, and four cycles of a negative sine wave in the imaginary part. As you recall, an impulse at sample number four in the real part of the frequency spectrum results in four cycles of a cosine wave in the time domain. Likewise, an impulse at sample number four in the imaginary part of the frequency spectrum results in four cycles of a negative sine wave being added to the time domain wave.

As mentioned in Chapter 8, this can be used as another way to calculate the DFT (besides correlating the time domain with sinusoids). Each sample in the time domain results in a cosine wave being added to the real part of the frequency domain, and a negative sine wave being added to the imaginary part. The *amplitude* of each sinusoid is given by the *amplitude* of the time domain sample. The *frequency* of each sinusoid is provided by the *sample number* of the time domain point. The algorithm involves: (1) stepping through each time domain sample, (2) calculating the sine and cosine waves that correspond to each sample, and (3) adding up all of the contributing sinusoids. The resulting program is nearly identical to the correlation method (Table 8-2), except that the outer and inner loops are exchanged.

The Sinc Function

Figure 11-4 illustrates a common transform pair: the *rectangular pulse* and the *sinc function*. The sinc function is defined as: $sinc(a) = sin(\pi a)/(\pi a)$, however, it is common to see the vague statement: "the sinc function is of the general form: $sin(x)/x$." In (a), the rectangular pulse is symmetrically centered on sample zero, making one-half of the pulse on the right of the graph and the other one-half on the left. This appears to the DFT as a single pulse because of the time domain periodicity. The DFT of this signal is shown in (b) and (c), with the *unwrapped* version in (d) and (e).

First look at the unwrapped spectrum, (d) and (e). The *unwrapped magnitude* is an oscillation that decreases in amplitude with increasing frequency. The phase is composed of all zeros, as you should expect for a time domain signal that is symmetrical around sample number zero. We are using the term *unwrapped magnitude* to indicate that it can have both positive and negative values. By definition, the *magnitude* must always be positive. This is shown in (b) and (c) where the magnitude is made all positive by introducing a phase shift of π at all frequencies where the unwrapped magnitude is negative in (d).

In (f), the signal is shifted so that it appears as one contiguous pulse, but is no longer centered on sample number zero. While this doesn't change the magnitude of the frequency domain, it does add a linear component to the phase, making it a jumbled mess. What does the frequency spectrum look like as real and imaginary parts ? Too confusing to even worry about.

An *N* point time domain signal that contains a unity amplitude rectangular pulse *M* points wide, has a DFT frequency spectrum given by:

EQUATION 11-1
DFT spectrum of a rectangular pulse. In this equation, N is the number of points in the time domain signal, all of which have a value of zero, except M adjacent points that have a value of one. The frequency spectrum is contained in $X[k]$, where k runs from 0 to $N/2$. To avoid the division by zero, use $X[0] = M$. The *sin* function uses radians, not degrees. This equation takes into account that the signal is *aliased*.

$$Mag \ X[k] = \left| \frac{\sin(\pi k M/N)}{\sin(\pi k/N)} \right|$$

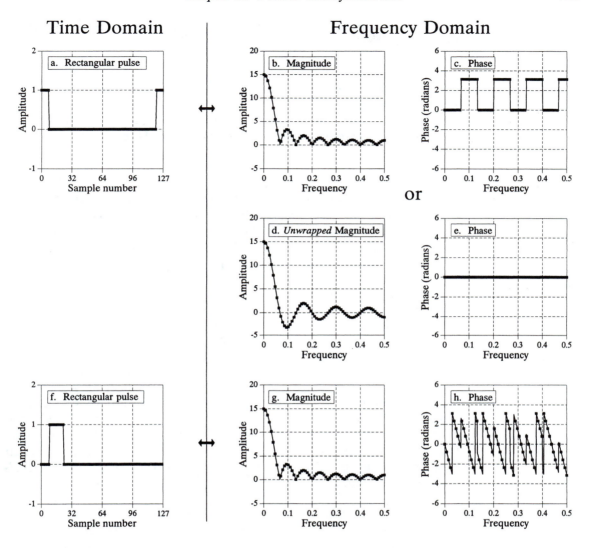

FIGURE 11-3
DFT of a rectangular pulse. A rectangular pulse in one domain corresponds to a sinc function in the other domain.

Alternatively, the DTFT can be used to express the frequency spectrum as a fraction of the sampling rate, f:

EQUATION 11-2
Equation 11-1 rewritten in terms of the sampling frequency. The parameter, f, is the fraction of the sampling rate, running continiously from 0 to 0.5. To avoid the division by zero, use $X(0)=M$.

$$Mag\ X(f) = \left| \frac{\sin(\pi f M)}{\sin(\pi f)} \right|$$

In other words, Eq. 11-1 provides $N/2+1$ *samples* in the frequency spectrum, while Eq. 11-2 provides the *continuous curve* that the samples lie on. These equations only provide the magnitude. The phase is determined solely by the left-right positioning of the time domain waveform, as discussed in the last chapter.

Notice in Fig. 11-3b that the amplitude of the oscillation does not decay to zero before a frequency of 0.5 is reached. As you should suspect, the waveform continues into the next period where it is *aliased*. This changes the shape of the frequency domain, an effect that is included in Eqs. 11-1 and 11-2.

It is often important to understand what the frequency spectrum looks like when aliasing *isn't* present. This is because *discrete signals* are often used to represent or model *continuous signals*, and continuous signals don't alias. To remove the aliasing in Eqs. 11-1 and 11-2, change the denominators from $sin(\pi kM/N)$ to $\pi kM/N$, and $sin(\pi f)$ to πf, respectively. Figure 11-4 shows the significance of this. The quantity πf can only run from 0 to 1.5708, since f can only run from 0 to 0.5. Over this range there isn't much difference between $sin(\pi f)$ and πf. At zero frequency they have the same value, and at a frequency of 0.5 there is only about a 36% difference. Without aliasing, the curve in Fig. 11-3b would show a slightly lower amplitude near the right side of the graph, and no change near the left side.

When the frequency spectrum of the rectangular pulse is *not* aliased (because the time domain signal is continuous, or because you are ignoring the aliasing), it is of the general form: $sin(x)/x$. This is given a special name in mathematics, the **sinc function** (pronounced "sink"). Mathematically this is written: $sinc(x) = sin(x)/x$. In other words, the sinc function is a sine wave that decays in amplitude as $1/x$. For continuous signals, the *rectangular pulse* and the *sinc function* are Fourier transform pairs. For discrete signals this is only an approximation, with the error being due to aliasing.

The sinc function has an annoying problem at $x = 0$, where $sin(x)/x$ becomes *zero* divided by *zero*. This is not a difficult mathematical problem; as x becomes very small, $sin(x)$ approaches the value of x (see Fig. 11-4).

FIGURE 11-4
Comparing x and $sin(x)$. The functions: $y(x) = x$, and $y(x) = sin(x)$ are similar for small values of x, and only differ by about 36% at 1.57 ($\pi/2$). This describes how aliasing distorts the frequency spectrum of the rectangular pulse from a pure sinc function.

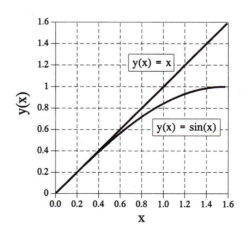

This turns the sinc function into x/x, which has a value of *one*. In other words, as x becomes smaller and smaller, the value of $sinc(x)$ approaches *one*, which includes $sinc(0) = 1$. Now try to tell your computer this! All it sees is a division by zero, causing it to complain and stop your program. The important point to remember is that your program must include special handling at $x = 0$ when calculating the sinc function.

A key trait of the sinc function is the location of the **zero crossings**. These occur at frequencies where an integer number of the sinusoid's cycles fit evenly into the rectangular pulse. For example, if the rectangular pulse is 20 points wide, the first zero in the frequency domain is at the frequency that makes one complete cycle in 20 points. The second zero is at the frequency that makes two complete cycles in 20 points, etc. This can be understood by remembering how the DFT is calculated by correlation. The amplitude of a frequency component is found by multiplying the time domain signal by a sinusoid and adding up the resulting samples. If the time domain waveform is a rectangular pulse of unity amplitude, this is the same as *adding* the sinusoid's samples that are within the rectangular pulse. If this summation occurs over an integral number of the sinusoid's cycles, the result will be zero.

The sinc function is widely used in DSP because it is the Fourier transform pair of a very simple waveform, the rectangular pulse. For example, the sinc function is used in *spectral analysis*, as discussed in Chapter 9. Consider the analysis of an infinitely long discrete signal. Since the DFT can only work with *finite* length signals, N samples are selected to represent the longer signal. The key here is that "selecting N samples from a longer signal" is the same as multiplying the longer signal by a rectangular pulse. The *ones* in the rectangular pulse retain the corresponding samples, while the *zeros* eliminate them. How does this affect the frequency spectrum of the signal? Multiplying the time domain by a rectangular pulse results in the frequency domain being *convolved* with a *sinc function*. This reduces the frequency spectrum's resolution, as previously shown in Fig. 9-5a.

Other Transform Pairs

Figure 11-5 (a) and (b) show the duality of the above: a rectangular pulse in the frequency domain corresponds to a sinc function (plus aliasing) in the time domain. Including the effects of aliasing, the time domain signal is given by:

EQUATION 11-3
Inverse DFT of the rectangular pulse. In the frequency domain, the pulse has an amplitude of one, and runs from sample number 0 through sample number M-1. The parameter N is the length of the DFT, and $x[i]$ is the time domain signal with i running from 0 to N-1. To avoid the division by zero, use $x[0] = (2M-1)/N$.

$$x[i] = \frac{1}{N} \frac{\sin(2\pi i(M-1/2)/N)}{\sin(\pi i/N)}$$

To eliminate the effects of aliasing from this equation, imagine that the frequency domain is so finely sampled that it turns into a continuous curve. This makes the time domain infinitely long with no periodicity. The DTFT is the Fourier transform to use here, resulting in the time domain signal being given by the relation:

EQUATION 11-4
Inverse DTFT of the rectangular pulse. In the frequency domain, the pulse has an amplitude of one, and runs from zero frequency to f_c, a value between 0 and 0.5. The time domain signal is held in $x[i]$ with i running from 0 to N-1. To avoid the division by zero, use $x[0] = 2f_c$.

$$x[i] = \frac{\sin(2\pi f_c i)}{i\pi}$$

This equation is very important in DSP, because the rectangular pulse in the frequency domain is the perfect *low-pass filter*. Therefore, the sinc function described by this equation is the filter kernel for the perfect low-pass filter. This is the basis for a very useful class of digital filters called the *windowed-sinc filters*, described in Chapter 15.

Figures (c) & (d) show that a triangular pulse in the time domain coincides with a sinc function *squared* (plus aliasing) in the frequency domain. This transform pair isn't as important as the *reason* it is true. A $2M-1$ point triangle in the time domain can be formed by convolving an M point rectangular pulse with itself. Since convolution in the time domain results in multiplication in the frequency domain, convolving a waveform with itself will *square* the frequency spectrum.

Is there a waveform that is its own Fourier Transform? The answer is yes, and there is *only* one: the Gaussian. Figure (e) shows a Gaussian curve, and (f) shows the corresponding frequency spectrum, also a Gaussian curve. This relationship is only true if you ignore aliasing. The relationship between the standard deviation of the time domain and frequency domain is given by: $2\pi\sigma_f = 1/\sigma_t$. While only one side of a Gaussian is shown in (f), the negative frequencies in the spectrum complete the full curve, with the center of symmetry at zero frequency.

Figure (g) shows what can be called a **Gaussian burst**. It is formed by multiplying a sine wave by a Gaussian. For example, (g) is a sine wave multiplied by the same Gaussian shown in (e). The corresponding frequency domain is a Gaussian centered somewhere other than zero frequency. As before, this transform pair is not as important as the *reason* it is true. Since the time domain signal is the multiplication of two signals, the frequency domain will be the convolution of the two frequency spectra. The frequency spectrum of the sine wave is a delta function centered at the frequency of the sine wave. The frequency spectrum of a Gaussian is a Gaussian centered at zero frequency. Convolving the two produces a Gaussian centered at the frequency of the sine wave. This should look familiar; it is identical to the procedure of *amplitude modulation* described in the last chapter.

Time Domain

Frequency Domain

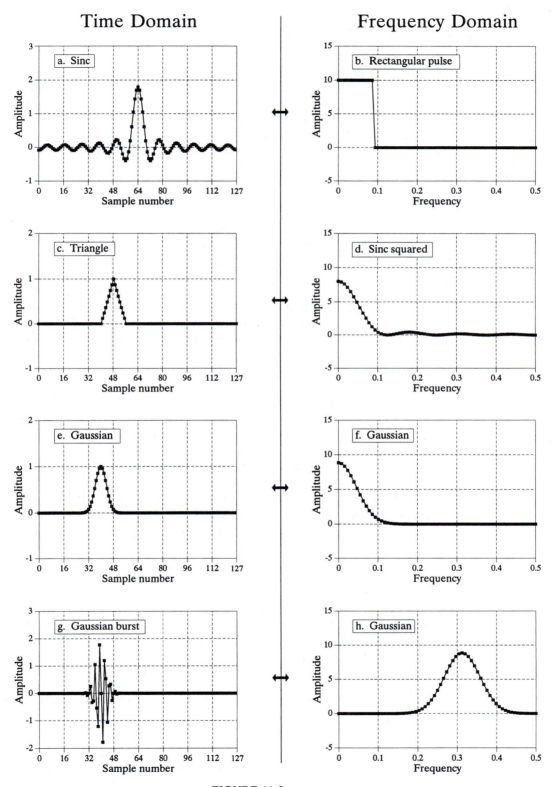

FIGURE 11-5
Common transform pairs.

Gibbs Effect

Figure 11-6 shows a time domain signal being synthesized from sinusoids. The signal being reconstructed is shown in the last graph, (h). Since this signal is 1024 points long, there will be 513 individual frequencies needed for a complete reconstruction. Figures (a) through (g) show what the reconstructed signal looks like if only *some* of these frequencies are used. For example, (f) shows a reconstructed signal using frequencies 0 through 100. This signal was created by taking the DFT of the signal in (h), setting frequencies 101 through 512 to a value of zero, and then using the Inverse DFT to find the resulting time domain signal.

As more frequencies are added to the reconstruction, the signal becomes closer to the final solution. The interesting thing is *how* the final solution is approached at the *edges* in the signal. There are three sharp edges in (h). Two are the edges of the rectangular pulse. The third is between sample numbers 1023 and 0, since the DFT views the time domain as periodic. When only some of the frequencies are used in the reconstruction, each edge shows *overshoot* and *ringing* (decaying oscillations). This overshoot and ringing is known as the **Gibbs effect,** after the mathematical physicist Josiah Gibbs, who explained the phenomenon in 1899.

Look closely at the overshoot in (e), (f), and (g). As more sinusoids are added, the *width* of the overshoot decreases; however, the *amplitude* of the overshoot remains about the same, roughly 9 percent. With discrete signals this is not a problem; the overshoot is eliminated when the last frequency is added. However, the reconstruction of continuous signals cannot be explained so easily. An infinite number of sinusoids must be added to synthesize a continuous signal. The problem is, the amplitude of the overshoot does not decrease as the number of sinusoids approaches infinity, it stays about the same 9%. Given this situation (and other arguments), it is reasonable to question if a summation of continuous sinusoids *can* reconstruct an edge. Remember the squabble between Lagrange and Fourier?

The critical factor in resolving this puzzle is that the *width* of the overshoot becomes smaller as more sinusoids are included. The overshoot is still present with an infinite number of sinusoids, but it has *zero* width. Exactly at the discontinuity the value of the reconstructed signal converges to the midpoint of the step. As shown by Gibbs, the summation converges to the signal in the sense that the *error* between the two has zero energy.

Problems related to the Gibbs effect are frequently encountered in DSP. For example, a low-pass filter is a *truncation* of the higher frequencies, resulting in overshoot and ringing at the edges in the *time domain*. Another common procedure is to truncate the ends of a time domain signal to prevent them from extending into neighboring periods. By duality, this distorts the edges in the *frequency domain*. These issues will resurface in future chapters on filter design.

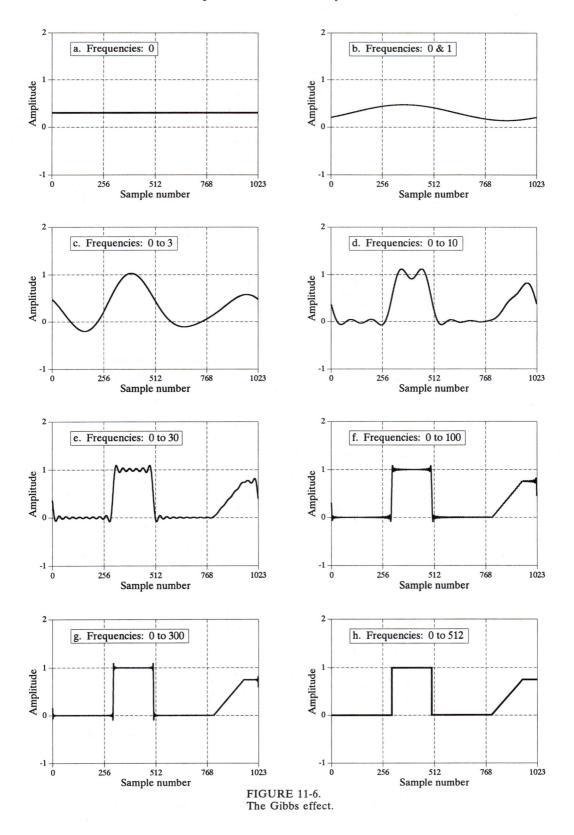

FIGURE 11-6.
The Gibbs effect.

Harmonics

If a signal is periodic with frequency f, the only frequencies composing the signal are integer multiples of f, i.e., f, $2f$, $3f$, $4f$, etc.   These frequencies are called **harmonics**. The **first harmonic** is f, the **second harmonic** is $2f$, the **third harmonic** is $3f$, and so forth.   The first harmonic (i.e., f) is also given a special name, the **fundamental frequency**. Figure 11-7 shows an

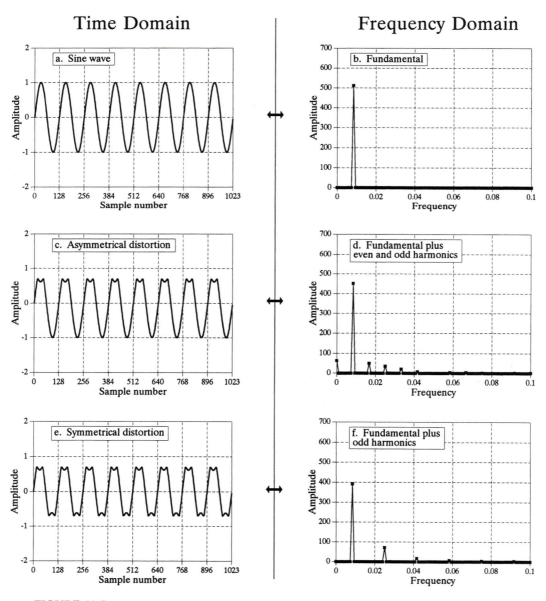

FIGURE 11-7
Example of harmonics. Asymmetrical distortion, shown in (c), results in even and odd harmonics, (d), while symmetrical distortion, shown in (e), produces only even harmonics, (f).

Time Domain

Frequency Domain

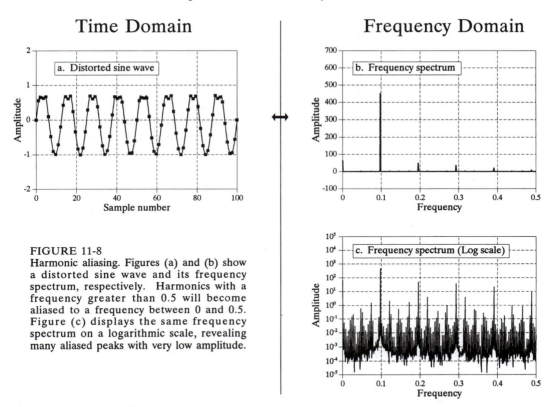

FIGURE 11-8
Harmonic aliasing. Figures (a) and (b) show a distorted sine wave and its frequency spectrum, respectively. Harmonics with a frequency greater than 0.5 will become aliased to a frequency between 0 and 0.5. Figure (c) displays the same frequency spectrum on a logarithmic scale, revealing many aliased peaks with very low amplitude.

example. Figure (a) is a pure sine wave, and (b) is its DFT, a single peak. In (c), the sine wave has been distorted by poking in the tops of the peaks. Figure (d) shows the result of this distortion in the frequency domain. Because the distorted signal is periodic with the same frequency as the original sine wave, the frequency domain is composed of the original peak plus harmonics. Harmonics can be of any amplitude; however, they usually become smaller as they increase in frequency. As with any signal, *sharp edges* result in *higher frequencies*. For example, consider a common TTL logic gate generating a 1 kHz square wave. The edges rise in a few nanoseconds, resulting in harmonics being generated to nearly 100 MHz, the *ten-thousandth* harmonic!

Figure (e) demonstrates a subtlety of harmonic analysis. If the signal is symmetrical around a horizontal axis, i.e., the top lobes are mirror images of the bottom lobes, all of the even harmonics will have a value of zero. As shown in (f), the only frequencies contained in the signal are the fundamental, the third harmonic, the fifth harmonic, etc.

All *continuous* periodic signals can be represented as a summation of harmonics, just as described. *Discrete* periodic signals have a problem that disrupts this simple relation. As you might have guessed, the problem is *aliasing*. Figure 11-8a shows a sine wave distorted in the same manner as before, by poking in the tops of the peaks. This waveform looks much less regular and smooth than in the previous example because the sine wave is at a much higher frequency, resulting in fewer samples per cycle. Figure (b) shows the frequency spectrum of this signal. As you would expect, you can identify the fundamental and harmonics. This example shows that harmonics can extend to frequencies greater than 0.5 of the sampling frequency, and will be *aliased* to frequencies somewhere between 0 and 0.5. You don't notice them in (b) because their amplitudes are too low. Figure (c) shows the frequency spectrum plotted on a logarithmic scale to reveal these low amplitude aliased peaks. At first glance, this spectrum looks like random noise. It isn't; this is a result of the many harmonics overlapping as they are aliased.

It is important to understand that this example involves distorting a signal *after* it has been digitally represented. If this distortion occurred in an analog signal, you would remove the offending harmonics with an antialias filter *before* digitization. Harmonic aliasing is only a problem when nonlinear operations are performed directly on a discrete signal. Even then, the amplitude of these aliased harmonics is often low enough that they can be ignored.

The concept of harmonics is also useful for another reason: it explains why the DFT views the time and frequency domains as *periodic*. In the frequency domain, an N point DFT consists of $N/2+1$ equally spaced frequencies. You can view the frequencies *between* these samples as (1) having a value of zero, or (2) not existing. Either way they don't contribute to the synthesis of the time domain signal. In other words, a *discrete* frequency spectrum consists of *harmonics*, rather than a continuous range of frequencies. This requires the time domain to be periodic with a frequency equal to the lowest sinusoid in the frequency domain, i.e., the fundamental frequency. Neglecting the DC value, the lowest frequency represented in the frequency domain makes one complete cycle every N samples, resulting in the time domain being periodic with a period of N. In other words, if one domain is *discrete*, the other domain must be *periodic*, and vice versa. This holds for all four members of the Fourier transform family. Since the DFT views both domains as discrete, it must also view both domains as periodic. The samples in each domain represent harmonics of the periodicity of the opposite domain.

Chirp Signals

Chirp signals are an ingenious way of handling a practical problem in echo location systems, such as radar and sonar. Figure 11-9 shows the frequency response of the chirp system. The magnitude has a constant value of one, while the phase is a parabola:

EQUATION 11-7
Phase of the chirp system.

$$Phase\ X[k] = \alpha k + \beta k^2$$

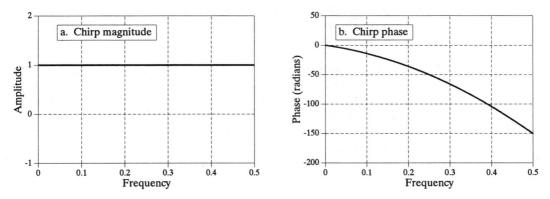

FIGURE 11-9
Frequency response of the chirp system. The magnitude is a constant, while the phase is a parabola.

The parameter α introduces a linear slope in the phase, that is, it simply shifts the impulse response left or right as desired. The parameter β controls the *curvature* of the phase. These two parameters must be chosen such that the phase at frequency 0.5 (i.e. k = N/2) is a multiple of 2π. Remember, whenever the phase is directly manipulated, frequency 0 and 0.5 must both have a phase of zero (or a multiple of 2π, which is the same thing).

Figure 11-10 shows an impulse entering a chirp system, and the impulse response exiting the system. The impulse response is an oscillatory burst that starts at a low frequency and changes to a high frequency as time progresses. This is called a *chirp* signal for a very simple reason: it sounds like the chirp of a bird when played through a speaker.

The key feature of the chirp system is that it is completely *reversible*. If you run the chirp signal through an *antichirp* system, the signal is again made into an impulse. This requires the antichirp system to have a magnitude of one, and the *opposite* phase of the chirp system. As discussed in the last

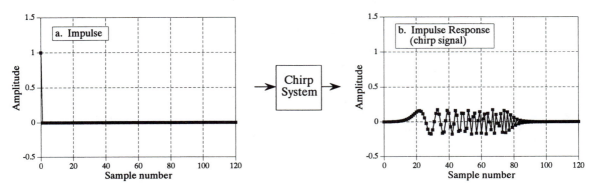

FIGURE 11-10
The chirp system. The impulse response of a chirp system is a chirp signal.

chapter, this means that the impulse response of the antichirp system is found by preforming a left-for-right flip of the chirp system's impulse response. Interesting, but what is it good for?

Consider how a radar system operates. A short burst of radio frequency energy is emitted from a directional antenna. Aircraft and other objects reflect some of this energy back to a radio receiver located next to the transmitter. Since radio waves travel at a constant rate, the elapsed time between the transmitted and received signals provides the distance to the target. This brings up the first requirement for the pulse: it needs to be as short as possible. For example, a 1 microsecond pulse provides a radio burst about 300 meters long. This means that the distance information we obtain with the system will have a resolution of about this same length. If we want better distance resolution, we need a shorter pulse.

The second requirement is obvious: if we want to detect objects farther away, you need more energy in your pulse. Unfortunately, *more energy* and *shorter pulse* are conflicting requirements. The electrical power needed to provide a pulse is equal to the energy of the pulse divided by the pulse length. Requiring both *more energy* and a *shorter pulse* makes electrical power handling a limiting factor in the system. The output stage of a radio transmitter can only handle so much power without destroying itself.

Chirp signals provide a way of breaking this limitation. Before the impulse reaches the final stage of the radio transmitter, it is passed through a chirp system. Instead of bouncing an impulse off the target aircraft, a chirp signal is used. After the chirp echo is received, the signal is passed through an antichirp system, restoring the signal to an impulse. This allows the portions of the system that measure distance to see short pulses, while the power handling circuits see long duration signals. This type of waveshaping is a fundamental part of modern radar systems.

The Fast Fourier Transform

There are several ways to calculate the Discrete Fourier Transform (DFT), such as solving simultaneous linear equations or the *correlation* method described in Chapter 8. The Fast Fourier Transform (FFT) is another method for calculating the DFT. While it produces the same result as the other approaches, it is incredibly more efficient, often reducing the computation time by *hundreds*. This is the same improvement as flying in a jet aircraft versus walking! If the FFT were not available, many of the techniques described in this book would not be practical. While the FFT only requires a few dozen lines of code, it is one of the most complicated algorithms in DSP. But don't despair! You can easily use published FFT routines without fully understanding the internal workings.

Real DFT Using the Complex DFT

J.W. Cooley and J.W. Tukey are given credit for bringing the FFT to the world in their paper: "An algorithm for the machine calculation of complex Fourier Series," *Mathematics Computation*, Vol. 19, 1965, pp 297-301. In retrospect, others had discovered the technique many years before. For instance, the great German mathematician Karl Friedrich Gauss (1777-1855) had used the method more than a century earlier. This early work was largely forgotten because it lacked the tool to make it practical: the *digital computer*. Cooley and Tukey are honored because they discovered the FFT at the right time, the beginning of the computer revolution.

The FFT is based on the *complex DFT*, a more sophisticated version of the *real DFT* discussed in the last four chapters. These transforms are named for the way each represents data, that is, using complex numbers or using real numbers. The term *complex* does not mean that this representation is difficult or complicated, but that a specific type of mathematics is used. Complex mathematics often *is* difficult and complicated, but that isn't where the name comes from. Chapter 29 discusses the complex DFT and provides the background needed to understand the details of the FFT algorithm. The

Real DFT

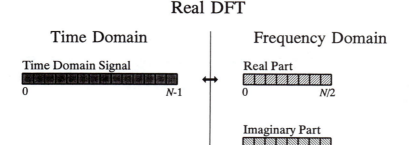

Complex DFT

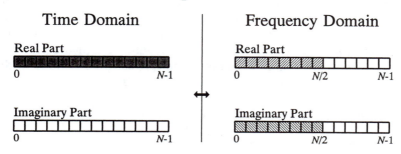

FIGURE 12-1
Comparing the real and complex DFTs. The real DFT takes an *N* point time domain signal and creates two *N*/2+1 point frequency domain signals. The complex DFT takes two *N* point time domain signals and creates two *N* point frequency domain signals. The shaded regions shows the values common to the two transforms.

topic of this chapter is simpler: how to use the FFT to calculate the real DFT, without drowning in a mire of advanced mathematics.

Since the FFT is an algorithm for calculating the complex DFT, it is important to understand how to transfer *real DFT* data into and out of the *complex DFT* format. Figure 12-1 compares how the real DFT and the complex DFT store data. The real DFT transforms an *N* point time domain signal into two *N*/2+1 point frequency domain signals. The time domain signal is called just that: the *time domain signal*. The two signals in the frequency domain are called the *real part* and the *imaginary part*, holding the amplitudes of the cosine waves and sine waves, respectively. This should be very familiar from past chapters.

In comparison, the complex DFT transforms two *N* point time domain signals into two *N* point frequency domain signals. The two time domain signals are called the *real part* and the *imaginary part*, just as are the frequency domain signals. In spite of their names, all of the values in these arrays are just ordinary numbers. (If you are familiar with complex numbers: the *j*'s are not included in the array values; they are a part of the *mathematics*. Recall that the operator, *Im*( ), returns a real number).

Suppose you have an *N* point signal, and need to calculate the *real DFT* by means of the *Complex DFT* (such as by using the FFT algorithm). First, move the *N* point signal into the real part of the complex DFT's time domain, and then set all of the samples in the imaginary part to *zero*. Calculation of the complex DFT results in a real and an imaginary signal in the frequency domain, each composed of *N* points. Samples 0 through *N*/2 of these signals correspond to the real DFT's spectrum.

As discussed in Chapter 10, the DFT's frequency domain is periodic when the negative frequencies are included (see Fig. 10-9). The choice of a single period is arbitrary; it can be chosen between -1.0 and 0, -0.5 and 0.5, 0 and 1.0, or any other one unit interval referenced to the sampling rate. The complex DFT's frequency spectrum includes the negative frequencies in the 0 to 1.0 arrangement. In other words, one full period stretches from sample 0 to sample *N*-1, corresponding with 0 to 1.0 times the sampling rate. The positive frequencies sit between sample 0 and *N*/2, corresponding with 0 to 0.5. The other samples, between *N*/2+1 and *N*-1, contain the negative frequency values (which are usually ignored).

Calculating a *real Inverse DFT* using a *complex Inverse DFT* is slightly harder. This is because you need to insure that the negative frequencies are loaded in the proper format. Remember, points 0 through *N*/2 in the complex DFT are the same as in the real DFT, for both the real and the imaginary parts. For the real part, point *N*/2+1 is the same as point *N*/2-1, point *N*/2+2 is the same as point *N*/2-2, etc. This continues to point *N*-1 being the same as point 1. The same basic pattern is used for the imaginary part, except the sign is changed. That is, point *N*/2+1 is the negative of point *N*/2-1, point *N*/2+2 is the negative of point *N*/2-2, etc. Notice that samples 0 and *N*/2 do not have a matching point in this duplication scheme. Use Fig. 10-9 as a guide to understanding this symmetry. In practice, you load the real DFT's frequency spectrum into samples 0 to *N*/2 of the complex DFT's arrays, and then use a subroutine to generate the negative frequencies between samples *N*/2+1 and *N*-1. Table 12-1 shows such a program. To check that the proper symmetry is present, after taking the inverse FFT, look at the imaginary part of the time domain. It will contain all zeros if everything is correct (except for a few parts-per-million of noise, using single precision calculations).

```
6000 'NEGATIVE FREQUENCY GENERATION
6010 'This subroutine creates the complex frequency domain from the real frequency domain.
6020 'Upon entry to this subroutine, N% contains the number of points in the signals, and
6030 'REX[ ] and IMX[ ] contain the real frequency domain in samples 0 to N%/2.
6040 'On return, REX[ ] and IMX[ ] contain the complex frequency domain in samples 0 to N%-1.
6050 '
6060 FOR K% = (N%/2+1) TO (N%-1)
6070   REX[K%] =  REX[N%-K%]
6080   IMX[K%] = -IMX[N%-K%]
6090 NEXT K%
6100 '
6110 RETURN
```

TABLE 12-1

How the FFT works

The FFT is a complicated algorithm, and its details are usually left to those that specialize in such things. This section describes the general operation of the FFT, but skirts a key issue: the use of *complex numbers*. If you have a background in complex mathematics, you can read between the lines to understand the true nature of the algorithm. Don't worry if the details elude you; few scientists and engineers that use the FFT could write the program from scratch.

In complex notation, the time and frequency domains each contain *one signal* made up of *N complex points*. Each of these complex points is composed of two numbers, the real part and the imaginary part. For example, when we talk about complex sample $X[42]$, it refers to the combination of $Re\,X[42]$ and $Im\,X[42]$. In other words, each complex variable holds two numbers. When two complex variables are multiplied, the four individual components must be combined to form the two components of the product (such as in Eq. 9-1). The following discussion on "*How the FFT works*" uses this jargon of complex notation. That is, the singular terms: *signal, point, sample,* and *value*, refer to the *combination* of the real part and the imaginary part.

The FFT operates by decomposing an N point time domain signal into N time domain signals each composed of a single point. The second step is to calculate the N frequency spectra corresponding to these N time domain signals. Lastly, the N spectra are synthesized into a single frequency spectrum.

Figure 12-2 shows an example of the time domain decomposition used in the FFT. In this example, a 16 point signal is decomposed through four

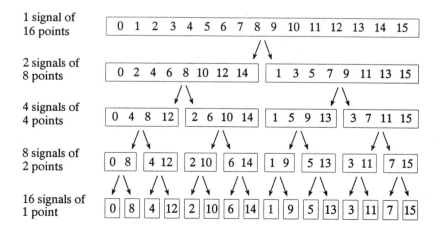

FIGURE 12-2
The FFT decomposition. An N point signal is decomposed into N signals each containing a single point. Each stage uses an *interlace decomposition*, separating the even and odd numbered samples.

Sample numbers in normal order			Sample numbers after bit reversal	
Decimal	*Binary*		*Decimal*	*Binary*
0	0000		0	0000
1	0001		8	1000
2	0010		4	0100
3	0011		12	1100
4	0100		2	0010
5	0101		10	1010
6	0110	$\Rightarrow$	6	0100
7	0111		14	1110
8	1000		1	0001
9	1001		9	1001
10	1010		5	0101
11	1011		13	1101
12	1100		3	0011
13	1101		11	1011
14	1110		7	0111
15	1111		15	1111

FIGURE 12-3
The FFT bit reversal sorting. The FFT time domain decomposition can be implemented by sorting the samples according to bit reversed order.

separate stages. The first stage breaks the 16 point signal into two signals each consisting of 8 points. The second stage decomposes the data into four signals of 4 points. This pattern continues until there are N signals composed of a single point. An **interlaced decomposition** is used each time a signal is broken in two, that is, the signal is separated into its even and odd numbered samples. The best way to understand this is by inspecting Fig. 12-2 until you grasp the pattern. There are $Log_2 N$ stages required in this decomposition, i.e., a 16 point signal requires 4 stages, a 512 point signal requires 7 stages, a 4096 point signal requires 12 stages, etc. Remember this value, $Log_2 N$; it will be referenced many times in this chapter.

Now that you understand the structure of the decomposition, it can be greatly simplified. The decomposition is nothing more than a *reordering* of the samples in the signal. Figure 12-3 shows the rearrangement pattern required. On the left, the sample numbers of the original signal are listed along with their binary equivalents. On the right, the rearranged sample numbers are listed, also along with their binary equivalents. The important idea is that the binary numbers are the *reversals* of each other. For example, sample 3 (0011) is exchanged with sample number 12 (1100). Likewise, sample number 14 (1110) is swapped with sample number 7 (0111), and so forth. The FFT time domain decomposition is usually carried out by a **bit reversal sorting** algorithm. This involves rearranging the order of the N time domain samples by counting in binary with the bits flipped left-for-right (such as in the far right column in Fig. 12-3).

The next step in the FFT algorithm is to find the frequency spectra of the 1 point time domain signals. Nothing could be easier; the frequency spectrum of a 1 point signal is equal to *itself*. This means that *nothing* is required to do this step. Although there is no work involved, don't forget that each of the 1 point signals is now a frequency spectrum, and not a time domain signal.

The last step in the FFT is to combine the *N* frequency spectra in the exact reverse order that the time domain decomposition took place. This is where the algorithm gets messy. Unfortunately, the bit reversal shortcut is not applicable, and we must go back one stage at a time. In the first stage, 16 frequency spectra (1 point each) are synthesized into 8 frequency spectra (2 points each). In the second stage, the 8 frequency spectra (2 points each) are synthesized into 4 frequency spectra (4 points each), and so on. The last stage results in the output of the FFT, a 16 point frequency spectrum.

Figure 12-4 shows how two frequency spectra, each composed of 4 points, are combined into a single frequency spectrum of 8 points. This synthesis must *undo* the interlaced decomposition done in the time domain. In other words, the frequency domain operation must correspond to the time domain procedure of *combining* two 4 point signals by interlacing. Consider two time domain signals, *abcd* and *efgh*. An 8 point time domain signal can be formed by two steps: dilute each 4 point signal with zeros to make it an

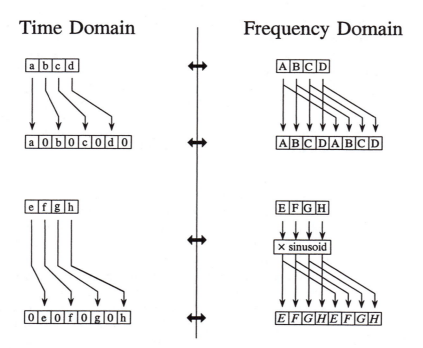

FIGURE 12-4
The FFT synthesis. When a time domain signal is diluted with zeros, the frequency domain is duplicated. If the time domain signal is also shifted by one sample during the dilution, the spectrum will additionally be multiplied by a sinusoid.

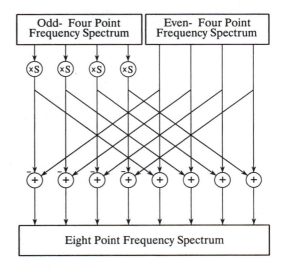

FIGURE 12-5
FFT synthesis flow diagram. This shows the method of combining two 4 point frequency spectra into a single 8 point frequency spectrum. The ×S operation means that the signal is multiplied by a sinusoid with an appropriately selected frequency.

8 point signal, and then add the signals together. That is, *abcd* becomes *a0b0c0d0*, and *efgh* becomes *0e0f0g0h*. Adding these two 8 point signals produces *aebfcgdh*. As shown in Fig. 12-4, diluting the time domain with zeros corresponds to a *duplication* of the frequency spectrum. Therefore, the frequency spectra are combined in the FFT by duplicating them, and then adding the duplicated spectra together.

In order to match up when added, the two time domain signals are diluted with zeros in a slightly different way. In one signal, the *odd points* are zero, while in the other signal, the *even points* are zero. In other words, one of the time domain signals (*0e0f0g0h* in Fig. 12-4) is shifted to the right by one sample. This time domain shift corresponds to multiplying the spectrum by a *sinusoid*. To see this, recall that a shift in the time domain is equivalent to convolving the signal with a shifted delta function. This multiplies the signal's spectrum with the spectrum of the shifted delta function. The spectrum of a shifted delta function is a sinusoid (see Fig 11-2).

Figure 12-5 shows a flow diagram for combining two 4 point spectra into a single 8 point spectrum. To reduce the situation even more, notice that Fig. 12-5 is formed from the basic pattern in Fig 12-6 repeated over and over.

FIGURE 12-6
The FFT butterfly. This is the basic calculation element in the FFT, taking two complex points and converting them into two other complex points.

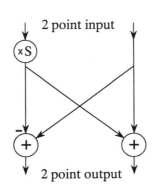

This simple flow diagram is called a **butterfly** due to its winged appearance. The butterfly is the basic computational element of the FFT, transforming two complex points into two other complex points.

Figure 12-7 shows the structure of the entire FFT. The time domain decomposition is accomplished with a bit reversal sorting algorithm. Transforming the decomposed data into the frequency domain involves *nothing* and therefore does not appear in the figure.

The frequency domain synthesis requires three loops. The outer loop runs through the Log_2N stages (i.e., each level in Fig. 12-2, starting from the bottom and moving to the top). The middle loop moves through each of the individual frequency spectra in the stage being worked on (i.e., each of the boxes on any one level in Fig. 12-2). The innermost loop uses the butterfly to calculate the points in each frequency spectra (i.e., looping through the samples inside any one box in Fig. 12-2). The overhead boxes in Fig. 12-7 determine the beginning and ending indexes for the loops, as well as calculating the sinusoids needed in the butterflies. Now we come to the heart of this chapter, the actual FFT programs.

Time Domain Data

FIGURE 12-7
Flow diagram of the FFT. This is based on three steps: (1) decompose an N point time domain signal into N signals each containing a single point, (2) find the spectrum of each of the N point signals (nothing required), and (3) synthesize the N frequency spectra into a single frequency spectrum.

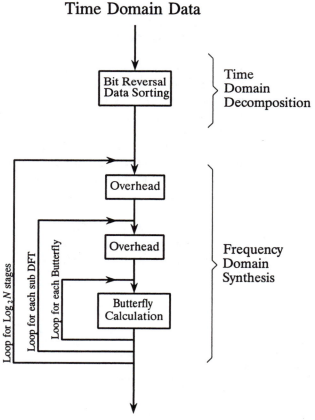

Frequency Domain Data

FFT Programs

As discussed in Chapter 8, the *real DFT* can be calculated by correlating the time domain signal with sine and cosine waves (see Table 8-2). Table 12-2 shows a program to calculate the *complex DFT* by the same method. In an apples-to-apples comparison, this is the program that the FFT improves upon.

Tables 12-3 and 12-4 show two different FFT programs, one in FORTRAN and one in BASIC. First we will look at the BASIC routine in Table 12-4. This subroutine produces exactly the same output as the correlation technique in Table 12-2, except it does it *much faster*. The block diagram in Fig. 12-7 can be used to identify the different sections of this program. Data are passed to this FFT subroutine in the arrays: REX[ ] and IMX[ ], each running from sample 0 to N-1. Upon return from the subroutine, REX[ ] and IMX[ ] are overwritten with the frequency domain data. This is another way that the FFT is highly optimized; the same arrays are used for the input, intermediate storage, and output. This efficient use of memory is important for designing fast hardware to calculate the FFT. The term **in-place computation** is used to describe this memory usage.

While all FFT programs produce the same numerical result, there are subtle variations in programming that you need to look out for. Several of these

```
5000 'COMPLEX DFT BY CORRELATION
5010 'Upon entry, N% contains the number of points in the DFT, and
5020 'XR[ ] and XI[ ] contain the real and imaginary parts of the time domain.
5030 'Upon return, REX[ ] and IMX[ ] contain the frequency domain data.
5040 'All signals run from 0 to N%-1.
5050 '
5060 PI = 3.14159265                    'Set constants
5070 '
5080 FOR K% = 0 TO N%-1                 'Zero REX[ ] and IMX[ ], so they can be used
5090   REX[K%] = 0                      'as accumulators during the correlation
5100   IMX[K%] = 0
5110 NEXT K%
5120 '
5130 FOR K% = 0 TO N%-1                 'Loop for each value in frequency domain
5140   FOR I% = 0 TO N%-1               'Correlate with the complex sinusoid, SR & SI
5150 '
5160    SR =  COS(2*PI*K%*I%/N%)        'Calculate complex sinusoid
5170    SI = -SIN(2*PI*K%*I%/N%)
5180    REX[K%] = REX[K%] + XR[I%]*SR - XI[I%]*SI
5190    IMX[K%] = IMX[K%] + XR[I%]*SI + XI[I%]*SR
5200 '
5210   NEXT I%
5220 NEXT K%
5230 '
5240 RETURN
```

TABLE 12-2

of these differences are illustrated by the FORTRAN program listed in Table 12-3. This program uses an algorithm called **decimation in frequency**, while the previously described algorithm is called **decimation in time**. In a decimation in frequency algorithm, the bit reversal sorting is done *after* the three nested loops. There are also FFT routines that completely eliminate the bit reversal sorting. None of these variations significantly improve the performance of the FFT, and you shouldn't worry about which one you are using.

The *important* differences between FFT algorithms concern how data are passed to and from the subroutines. In the BASIC program, data enter and leave the subroutine in the arrays REX[ ] and IMX[ ], with the samples running from index 0 to N-1. In the FORTRAN program, data are passed in the complex array $X(\ )$, with the samples running from 1 to N. Since this is an array of complex variables, each sample in X( ) consists of two numbers, a real part and an imaginary part. The length of the DFT must also be passed to these subroutines. In the BASIC program, the variable N% is used for this purpose. In comparison, the FORTRAN program uses the variable M, which is defined to equal $Log_2 N$. For instance, M will be

```
        SUBROUTINE  FFT(X,M)
        COMPLEX X(4096),U,S,T
        PI=3.14159265
        N=2**M
        DO 20 L=1,M
        LE=2**(M+1-L)
        LE2=LE/2
        U=(1.0,0.0)
        S=CMPLX(COS(PI/FLOAT(LE2)),-SIN(PI/FLOAT(LE2)))
        DO 20 J=1,LE2
        DO 10 I=J,N,LE
        IP=I+LE2
        T=X(I)+X(IP)
        X(IP)=(X(I)-X(IP))*U
10      X(I)=T
20      U=U*S
        ND2=N/2
        NM1=N-1
        J=1
        DO 50 I=1,NM1
        IF(I.GE.J) GO TO 30
        T=X(J)
        X(J)=X(I)
        X(I)=T
30      K=ND2
40      IF(K.GE.J) GO TO 50
        J=J-K
        K=K/2
        GO TO 40
50      J=J+K
        RETURN
        END
```

TABLE 12-3
The Fast Fourier Transform in FORTRAN. Data are passed to this subroutine in the variables $X(\ )$ and M. The integer, M, is the base two logarithm of the length of the DFT, i.e., $M = 8$ for a 256 point DFT, $M = 12$ for a 4096 point DFT, etc. The complex array, $X(\ )$, holds the time domain data upon entering the DFT. Upon return from this subroutine, $X(\ )$ is overwritten with the frequency domain data. Take note: this subroutine requires that the input and output signals run from $X(1)$ through $X(N)$, rather than the customary $X(0)$ through $X(N$-1$)$.

```
1000 'THE FAST FOURIER TRANSFORM
1010 'Upon entry, N% contains the number of points in the DFT, REX[ ] and
1020 'IMX[ ] contain the real and imaginary parts of the input. Upon return,
1030 'REX[ ] and IMX[ ] contain the DFT output. All signals run from 0 to N%-1.
1040 '
1050 PI = 3.14159265                        'Set constants
1060 NM1% = N%-1
1070 ND2% = N%/2
1080 M% = CINT(LOG(N%)/LOG(2))
1090 J% = ND2%
1100 '
1110 FOR I% = 1 TO N%-2                     'Bit reversal sorting
1120    IF I% >= J% THEN GOTO 1190
1130    TR = REX[J%]
1140    TI = IMX[J%]
1150    REX[J%] = REX[I%]
1160    IMX[J%] = IMX[I%]
1170    REX[I%] = TR
1180    IMX[I%] = TI
1190    K% = ND2%
1200    IF K% > J% THEN GOTO 1240
1210    J% = J%-K%
1220    K% = K%/2
1230    GOTO 1200
1240    J% = J%+K%
1250 NEXT I%
1260 '
1270 FOR L% = 1 TO M%                       'Loop for each stage
1280    LE% = CINT(2^L%)
1290    LE2% = LE%/2
1300    UR = 1
1310    UI = 0
1320    SR =  COS(PI/LE2%)                  'Calculate sine & cosine values
1330    SI  = -SIN(PI/LE2%)
1340    FOR J% = 1 TO LE2%                  'Loop for each sub DFT
1350       JM1% = J%-1
1360       FOR I% = JM1% TO NM1% STEP LE%   'Loop for each butterfly
1370          IP% = I%+LE2%
1380          TR = REX[IP%]*UR - IMX[IP%]*UI   'Butterfly calculation
1390          TI = REX[IP%]*UI + IMX[IP%]*UR
1400          REX[IP%] = REX[I%]-TR
1410          IMX[IP%] = IMX[I%]-TI
1420          REX[I%] = REX[I%]+TR
1430          IMX[I%] = IMX[I%]+TI
1440       NEXT I%
1450       TR = UR
1460       UR = TR*SR - UI*SI
1470       UI = TR*SI + UI*SR
1480    NEXT J%
1490 NEXT L%
1500 '
1510 RETURN
```

TABLE 12-4
The Fast Fourier Transform in BASIC.

8 for a 256 point DFT, 12 for a 4096 point DFT, etc. The point is, the programmer who writes an FFT subroutine has many options for interfacing with the host program. Arrays that run from 1 to N, such as in the FORTRAN program, are especially aggravating. Most of the DSP literature (including this book) explains algorithms assuming the arrays run from sample 0 to $N-1$. For instance, if the arrays run from 1 to N, the symmetry in the frequency domain is around points 1 and $N/2+1$, rather than points 0 and $N/2$,

Using the complex DFT to calculate the real DFT has another interesting advantage. The complex DFT is more symmetrical between the time and frequency domains than the real DFT. That is, the **duality** is stronger. Among other things, this means that the Inverse DFT is nearly identical to the Forward DFT. In fact, the easiest way to calculate an *Inverse FFT* is to calculate a *Forward FFT*, and then adjust the data. Table 12-5 shows a subroutine for calculating the Inverse FFT in this manner.

Suppose you copy one of these FFT algorithms into your computer program and start it running. How do you know if it is operating properly? Two tricks are commonly used for debugging. First, start with some arbitrary time domain signal, such as from a random number generator, and run it through the FFT. Next, run the resultant frequency spectrum through the Inverse FFT and compare the result with the original signal. They should be *identical*, except round-off noise (a few parts-per-million for single precision).

The second test of proper operation is that the signals have the correct *symmetry*. When the imaginary part of the time domain signal is composed of all zeros (the normal case), the frequency domain of the complex DFT will be symmetrical around samples 0 and $N/2$, as previously described.

```
2000 'INVERSE FAST FOURIER TRANSFORM SUBROUTINE
2010 'Upon entry, N% contains the number of points in the IDFT, REX[ ] and
2020 'IMX[ ] contain the real and imaginary parts of the complex frequency domain.
2030 'Upon return, REX[ ] and IMX[ ] contain the complex time domain.
2040 'All signals run from 0 to N%-1.
2050 '
2060 FOR K% = 0 TO N%-1              'Change the sign of IMX[ ]
2070   IMX[K%] = -IMX[K%]
2080 NEXT K%
2090 '
2100 GOSUB 1000                      'Calculate forward FFT  (Table 12-3)
2110 '
2120 FOR I% = 0 TO N%-1              'Divide the time domain by N% and
2130   REX[I%] = REX[I%]/N%          'change the sign of IMX[ ]
2140   IMX[I%] = -IMX[I%]/N%
2150 NEXT I%
2160 '
2170 RETURN
```

TABLE 12-5

Likewise, when this correct symmetry is present in the frequency domain, the Inverse DFT will produce a time domain that has an imaginary part composes of all zeros (plus round-off noise). These debugging techniques are essential for using the FFT; become familiar with them.

Speed and Precision Comparisons

When the DFT is calculated by correlation (as in Table 12-2), the program uses two nested loops, each running through N points. This means that the total number of operations is proportional to N *times* N. The time to complete the program is thus given by:

EQUATION 12-1
DFT execution time. The time required to calculate a DFT by correlation is proportional to the length of the DFT squared.

$$Execution\,Time \; = \; k_{DFT} N^2$$

where N is the number of points in the DFT and k_{DFT} is a constant of proportionality. If the sine and cosine values are calculated *within* the nested loops, k_{DFT} is equal to about 25 microseconds on a Pentium at 100 MHz. If you *precalculate* the sine and cosine values and store them in a look-up-table, k_{DFT} drops to about 7 microseconds. For example, a 1024 point DFT will require about 25 seconds, or nearly 25 milliseconds per point. That's slow!

Using this same strategy we can derive the execution time for the FFT. The time required for the bit reversal is negligible. In each of the $Log_2 N$ stages there are $N/2$ butterfly computations. This means the execution time for the program is approximated by:

EQUATION 12-2
FFT execution time. The time required to calculate a DFT using the FFT is proportional to N multiplied by the logarithm of N.

$$Execution\,Time \; = \; k_{FFT} N \log_2 N$$

The value of k_{FFT} is about 10 microseconds on a 100 MHz Pentium system. A 1024 point FFT requires about 70 milliseconds to execute, or 70 microseconds per point. This is more than 300 times faster than the DFT calculated by correlation!

Not only is $N Log_2 N$ less than N^2, it increases much more slowly as N becomes larger. For example, a 32 point FFT is about *ten* times faster than the correlation method. However, a 4096 point FFT is *one-thousand* times faster. For small values of N (say, 32 to 128), the FFT is important. For large values of N (1024 and above), the FFT is absolutely critical. Figure 12-8 compares the execution times of the two algorithms in a graphical form.

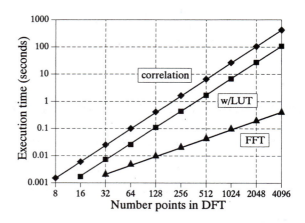

FIGURE 12-8
Execution times for calculating the DFT. The *correlation* method refers to the algorithm described in Table 12-2. This method can be made faster by precalculating the sine and cosine values and storing them in a look-up table (LUT). The FFT (Table 12-3) is the fastest algorithm when the DFT is greater than 16 points long. The times shown are for a Pentium processor at 100 MHz.

The FFT has another advantage besides raw speed. The FFT is calculated more *precisely* because the fewer number of calculations results in less round-off error. This can be demonstrated by taking the FFT of an arbitrary signal, and then running the frequency spectrum through an Inverse FFT. This reconstructs the original time domain signal, *except* for the addition of round-off noise from the calculations. A single number characterizing this noise can be obtained by calculating the standard deviation of the difference between the two signals. For comparison, this same procedure can be repeated using a DFT calculated by correlation, and a corresponding Inverse DFT. How does the round-off noise of the FFT compare to the DFT by correlation? See for yourself in Fig. 12-9.

Further Speed Increases

There are several techniques for making the FFT even faster; however, the improvements are only about 20-40%. In one of these methods, the time

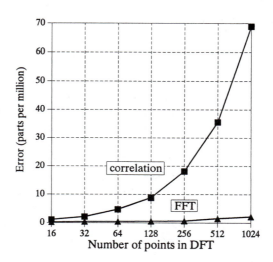

FIGURE 12-9
DFT precision. Since the FFT calculates the DFT *faster* than the correlation method, it also calculates it with less round-off error.

domain decomposition is stopped two stages early, when each signals is composed of only four points. Instead of calculating the last two stages, highly optimized code is used to jump directly into the frequency domain, using the simplicity of four point sine and cosine waves.

Another popular algorithm eliminates the wasted calculations associated with the imaginary part of the time domain being zero, and the frequency spectrum being symmetrical. In other words, the FFT is modified to calculate the *real DFT*, instead of the *complex DFT*. These algorithms are called the **real FFT** and the **real Inverse FFT** (or similar names). Expect them to be about 30% faster than the conventional FFT routines. Tables 12-6 and 12-7 show programs for these algorithms.

There are two small disadvantages in using the *real FFT*. First, the code is about twice as long. While your computer doesn't care, you must take the time to convert someone else's program to run on your computer. Second, debugging these programs is slightly harder because you cannot use symmetry as a check for proper operation. These algorithms *force* the imaginary part of the time domain to be zero, and the frequency domain to have left-right symmetry. For debugging, check that these programs produce the same output as the conventional FFT algorithms.

Figures 12-10 and 12-11 illustrate how the real FFT works. In Fig. 12-10, (a) and (b) show a time domain signal that consists of a pulse in the real part, and all zeros in the imaginary part. Figures (c) and (d) show the corresponding frequency spectrum. As previously described, the frequency domain's real part has an *even* symmetry around sample 0 and sample $N/2$, while the imaginary part has an *odd* symmetry around these same points.

```
4000 'INVERSE FFT FOR REAL SIGNALS
4010 'Upon entry, N% contains the number of points in the IDFT, REX[ ] and
4020 'IMX[ ] contain the real and imaginary parts of the frequency domain running from
4030 'index 0 to N%/2.  The remaining samples in REX[ ] and IMX[ ] are ignored.
4040 'Upon return, REX[ ] contains the real time domain, IMX[ ] contains zeros.
4050 '
4060 '
4070 FOR K% = (N%/2+1) TO (N%-1)            'Make frequency domain symmetrical
4080   REX[K%] =  REX[N%-K%]                '(as in Table 12-1)
4090   IMX[K%] = -IMX[N%-K%]
4100 NEXT K%
4110 '
4120 FOR K% = 0 TO N%-1                     'Add real and imaginary parts together
4130   REX[K%] =  REX[K%]+IMX[K%]
4140 NEXT K%
4150 '
4160 GOSUB 3000                             'Calculate forward real DFT (TABLE 12-6)
4170 '
4180 FOR I% = 0 TO N%-1                     'Add real and imaginary parts together
4190   REX[I%] = (REX[I%]+IMX[I%])/N%       'and divide the time domain by N%
4200   IMX[I%] = 0
4210 NEXT I%
4220 '
4230 RETURN
```

TABLE 12-6

Time Domain   Frequency Domain

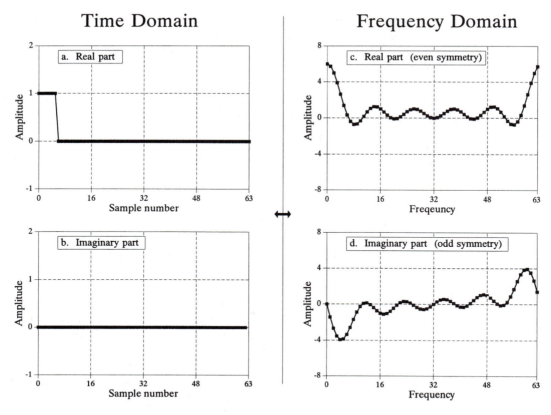

FIGURE 12-10
Real part symmetry of the DFT.

Now consider Fig. 12-11, where the pulse is in the imaginary part of the time domain, and the real part is all zeros. The symmetry in the frequency domain is *reversed*; the real part is odd, while the imaginary part is even. This situation will be discussed in Chapter 29. For now, take it for granted that this is how the complex DFT behaves.

What if there is a signal in *both parts* of the time domain? By additivity, the frequency domain will be the *sum* of the two frequency spectra. Now the key element: a frequency spectrum composed of these two types of symmetry can be perfectly separated into the two component signals. This is achieved by the *even/odd decomposition* discussed in Chapter 6. In other words, two DFT's can be calculated for the price of one. One of the signals is placed in the real part of the time domain, and the other signal is placed in the imaginary part. After calculating the complex DFT (via the FFT, of course), the spectra are separated using the even/odd decomposition. When two or more signals need to be passed through the FFT, this technique reduces the execution time by about 40%. The improvement isn't a full factor of two because of the calculation time required for the even/odd decomposition. This is a relatively simple technique with few pitfalls, nothing like writing an FFT routine from scratch.

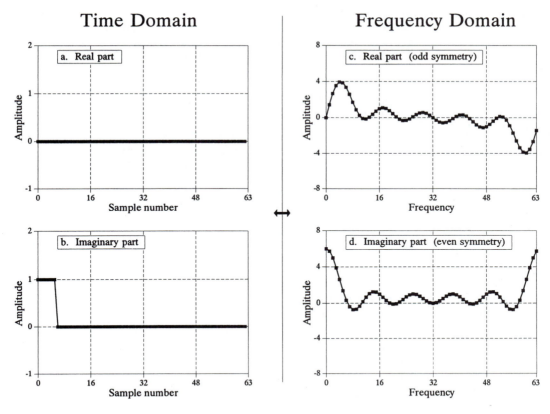

FIGURE 12-11
Imaginary part symmetry of the DFT.

The next step is to modify the algorithm to calculate a *single* DFT faster. It's ugly, but here is how it is done. The input signal is broken in half by using an interlaced decomposition. The $N/2$ even points are placed into the real part of the time domain signal, while the $N/2$ odd points go into the imaginary part. An $N/2$ point FFT is then calculated, requiring about one-half the time as an N point FFT. The resulting frequency domain is then separated by the even/odd decomposition, resulting in the frequency spectra of the two interlaced time domain signals. These two frequency spectra are then combined into a single spectrum, just as in the last synthesis stage of the FFT.

To close this chapter, consider that the *FFT* is to *Digital Signal Processing* what the *transistor* is to *electronics*. It is a foundation of the technology; everyone in the field knows its characteristics and how to use it. However, only a small number of specialists really understand the details of the internal workings.

```
3000 'FFT FOR REAL SIGNALS
3010 'Upon entry, N% contains the number of points in the DFT, REX[ ] contains
3020 'the real input signal, while values in IMX[ ] are ignored.  Upon return,
3030 'REX[ ] and IMX[ ] contain the DFT output. All signals run from 0 to N%-1.
3040 '
3050 NH% = N%/2-1                                    'Separate even and odd points
3060 FOR I% = 0 TO NH%
3070   REX(I%) = REX(2*I%)
3080   IMX(I%) = REX(2*I%+1)
3090 NEXT I%
3100 '
3110 N% = N%/2                                       'Calculate N%/2 point FFT
3120 GOSUB 1000                                      '(GOSUB 1000 is the FFT in Table 12-3)
3130 N% = N%*2
3140 '
3150 NM1% = N%-1                                     'Even/odd frequency domain decomposition
3160 ND2% = N%/2
3170 N4% = N%/4-1
3180 FOR I% = 1 TO N4%
3190   IM%  = ND2%-I%
3200   IP2% = I%+ND2%
3210   IPM% = IM%+ND2%
3220   REX(IP2%) =  (IMX(I%) + IMX(IM%))/2
3230   REX(IPM%) =  REX(IP2%)
3240   IMX(IP2%) = -(REX(I%) - REX(IM%))/2
3250   IMX(IPM%) = -IMX(IP2%)
3260   REX(I%)   =  (REX(I%) + REX(IM%))/2
3270   REX(IM%)  =  REX(I%)
3280   IMX(I%)   =  (IMX(I%) - IMX(IM%))/2
3290   IMX(IM%)  = -IMX(I%)
3300 NEXT I%
3310 REX(N%*3/4) = IMX(N%/4)
3320 REX(ND2%) = IMX(0)
3330 IMX(N%*3/4) = 0
3340 IMX(ND2%) = 0
3350 IMX(N%/4) = 0
3360 IMX(0) = 0
3370 '
3380 PI = 3.14159265                                 'Complete the last FFT stage
3390 L% = CINT(LOG(N%)/LOG(2))
3400 LE% = CINT(2^L%)
3410 LE2% = LE%/2
3420 UR = 1
3430 UI = 0
3440 SR =  COS(PI/LE2%)
3450 SI = -SIN(PI/LE2%)
3460 FOR J% = 1 TO LE2%
3470   JM1% = J%-1
3480   FOR I% = JM1% TO NM1% STEP LE%
3490     IP% = I%+LE2%
3500     TR = REX[IP%]*UR - IMX[IP%]*UI
3510     TI = REX[IP%]*UI + IMX[IP%]*UR
3520     REX[IP%] = REX[I%]-TR
3530     IMX[IP%] = IMX[I%]-TI
3540     REX[I%]  = REX[I%]+TR
3550     IMX[I%]  = IMX[I%]+TI
3560   NEXT I%
3570   TR = UR
3580   UR = TR*SR - UI*SI
3590   UI = TR*SI + UI*SR
3600 NEXT J%
3610 RETURN                          TABLE 12-7
```

CHAPTER 13

Continuous Signal Processing

Continuous signal processing is a parallel field to DSP, and most of the techniques are nearly identical. For example, both DSP and continuous signal processing are based on linearity, decomposition, convolution and Fourier analysis. Since continuous signals cannot be directly represented in digital computers, don't expect to find computer programs in this chapter. Continuous signal processing is based on *mathematics*; signals are represented as equations, and systems change one equation into another. Just as the *digital computer* is the primary tool used in DSP, *calculus* is the primary tool used in continuous signal processing. These techniques have been used for centuries, long before computers were developed.

The Delta Function

Continuous signals can be decomposed into scaled and shifted *delta functions*, just as done with discrete signals. The difference is that the continuous delta function is much more complicated and mathematically abstract than its discrete counterpart. Instead of defining the continuous delta function by what it *is*, we will define it by the *characteristics it has*.

A thought experiment will show how this works. Imagine an electronic circuit composed of linear components, such as resistors, capacitors and inductors. Connected to the input is a signal generator that produces various shapes of short *pulses*. The output of the circuit is connected to an oscilloscope, displaying the waveform produced by the circuit in response to each input pulse. The question we want to answer is: *how is the shape of the output pulse related to the characteristics of the input pulse?* To simplify the investigation, we will only use input pulses that are much shorter than the output. For instance, if the system responds in milliseconds, we might use input pulses only a few microseconds in length.

After taking many measurement, we come to three conclusions: First, the *shape* of the input pulse does not affect the shape of the output signal. This is illustrated in Fig. 13-1, where various shapes of short input pulses produce exactly the same shape of output pulse. Second, the shape of the output waveform is totally determined by the characteristics of the system, i.e., the value and configuration of the resistors, capacitors and inductors. Third, the *amplitude* of the output pulse is directly proportional to the *area* of the input pulse. For example, the output will have the same amplitude for inputs of: 1 volt for 1 microsecond, 10 volts for 0.1 microseconds, 1,000 volts for 1 nanosecond, etc. This relationship also allows for input pulses with *negative* areas. For instance, imagine the combination of a 2 volt pulse lasting 2 microseconds being quickly followed by a -1 volt pulse lasting 4 microseconds. The total area of the input signal is *zero*, resulting in the output doing *nothing*.

Input signals that are brief enough to have these three properties are called **impulses**. In other words, an impulse is any signal that is entirely zero except for a short *blip* of arbitrary shape. For example, an impulse to a microwave transmitter may have to be in the *picosecond* range because the electronics responds in *nanoseconds*. In comparison, a volcano that erupts for *years* may be a perfectly good impulse to geological changes that take *millennia*.

Mathematicians don't like to be limited by any particular system, and commonly use the term *impulse* to mean a signal that is short enough to be an impulse to *any possible* system. That is, a signal that is *infinitesimally* narrow. The **continuous delta function** is a normalized version of this type of impulse. Specifically, the continuous delta function is mathematically defined by three idealized characteristics: (1) the signal must be infinitesimally brief, (2) the pulse must occur at time zero, and (3) the pulse must have an area of one.

Since the delta function is defined to be infinitesimally narrow *and* have a fixed area, the amplitude is implied to be *infinite*. Don't let this bother you; it is completely unimportant. Since the amplitude is part of the *shape* of the impulse, you will never encounter a problem where the amplitude makes any difference, infinite or not. The delta function is a mathematical construct, not a real world signal. Signals in the real world that *act* as delta functions will always have a finite duration and amplitude.

Just as in the discrete case, the continuous delta function is given the mathematical symbol: $\delta(\ )$. Likewise, the output of a continuous system in response to a delta function is called the **impulse response**, and is often denoted by: $h(\ )$. Notice that parentheses, $(\ )$, are used to denote continuous signals, as compared to brackets, $[\ ]$, for discrete signals. This notation is used in this book and elsewhere in DSP, but isn't universal. Impulses are displayed in graphs as vertical arrows (see Fig. 13-1d), with the *length* of the arrow indicating the *area* of the impulse.

To better understand real world impulses, look into the night sky at a *planet* and a *star*, for instance, Mars and Sirius. Both appear about the same brightness and size to the unaided eye. The reason for this similarity is not

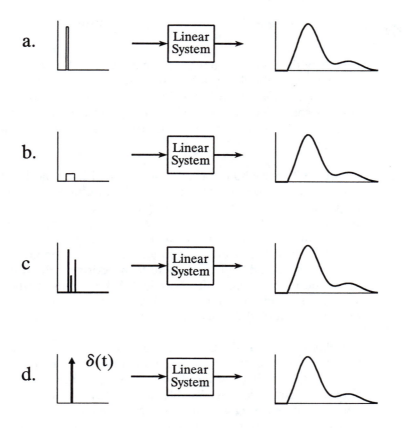

FIGURE 13-1
The continuous delta function. If the input to a linear system is brief compared to the resulting output, the shape of the output depends only on the characteristics of the system, and not the shape of the input. Such short input signals are called *impulses*. Figures a,b & c illustrate example input signals that are impulses for this particular system. The term *delta function* is used to describe a normalized impulse, i.e., one that occurs at $t = 0$ and has an area of one. The mathematical symbols for the delta function are shown in (d), a vertical arrow and $\delta(t)$.

obvious, since the viewing geometry is drastically different. Mars is about 6000 kilometers in diameter and 60 million kilometers from earth. In comparison, Sirius is about 300 times larger and over one-million times farther away. These dimensions should make Mars appear more than *three-thousand* times larger than Sirius. How is it possible that they look alike?

These objects look the same because they are small enough to be *impulses* to the human visual system. The perceived shape is the impulse response of the eye, not the actual image of the star or planet. This becomes obvious when the two objects are viewed through a small telescope; Mars appears as a dim disk, while Sirius still appears as a bright impulse. This is also the reason that stars twinkle while planets do not. The image of a star is small enough that it can be briefly blocked by particles or turbulence in the atmosphere, whereas the larger image of the planet is much less affected.

Convolution

Just as with discrete signals, the convolution of continuous signals can be viewed from the *input signal*, or the *output signal*. The input side viewpoint is the best *conceptual* description of how convolution operates. In comparison, the output side viewpoint describes the *mathematics* that must be used. These descriptions are virtually identical to those presented in Chapter 6 for discrete signals.

Figure 13-2 shows how convolution is viewed from the input side. An input signal, $x(t)$, is passed through a system characterized by an impulse response, $h(t)$, to produce an output signal, $y(t)$. This can be written in the familiar mathematical equation, $y(t) = x(t) * h(t)$. The input signal is divided into narrow columns, each short enough to act as an *impulse* to the system. In other words, the input signal is decomposed into an infinite number of scaled and shifted delta functions. Each of these impulses produces a scaled and shifted version of the impulse response in the output signal. The final output signal is then equal to the combined effect, i.e., the sum of all of the individual responses.

For this scheme to work, the width of the columns must be much shorter than the response of the system. Of course, mathematicians take this to the extreme by making the input segments *infinitesimally* narrow, turning the situation into a calculus problem. In this manner, the input viewpoint describes how a single point (or narrow region) in the input signal affects a larger portion of output signal.

In comparison, the output viewpoint examines how a single point in the output signal is determined by the various values from the input signal. Just as with discrete signals, each instantaneous value in the output signal is affected by a section of the input signal, weighted by the impulse response flipped left-for-right. In the discrete case, the signals are multiplied and *summed*. In the continuous case, the signals are multiplied and *integrated*. In equation form:

EQUATION 13-1
The convolution integral. This equation defines the meaning of: $y(t) = x(t) * h(t)$.

$$y(t) = \int_{-\infty}^{+\infty} x(\tau)\, h(t-\tau)\, d\tau$$

This equation is called the convolution integral, and is the twin of the convolution sum (Eq. 6-1) used with discrete signals. Figure 13-3 shows how this equation can be understood. The goal is to find an expression for calculating the value of the output signal at an arbitrary time, t. The first step is to change the independent variable used to move through the input signal and the impulse response. That is, we replace t with τ (a lower case

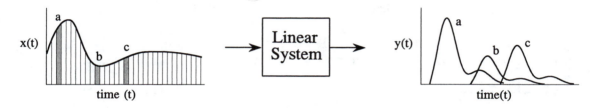

FIGURE 13-2
Convolution viewed from the input side. The input signal, $x(t)$, is divided into narrow segments, each acting as an impulse to the system. The output signal, $y(t)$, is the sum of the resulting scaled and shifted impulse responses. This illustration shows how three points in the input signal contribute to the output signal.

Greek tau). This makes $x(t)$ and $h(t)$ become $x(\tau)$ and $h(\tau)$, respectively. This change of variable names is needed because t is already being used to represent the point in the output signal being calculated. The next step is to flip the impulse response left-for-right, turning it into $h(-\tau)$. Shifting the flipped impulse response to the location t, results in the expression becoming $h(t-\tau)$. The input signal is then weighted by the flipped and shifted impulse response by multiplying the two, i.e., $x(\tau)\,h(t-\tau)$. The value of the output signal is then found by integrating this weighted input signal from negative to positive infinity, as described by Eq. 13-1.

If you have trouble understanding how this works, go back and review the same concepts for discrete signals in Chapter 6. Figure 13-3 is just another way of describing the convolution machine in Fig. 6-8. The only difference is that integrals are being used instead of summations. Treat this as an extension of what you already know, not something new.

An example will illustrate how continuous convolution is used in real world problems and the mathematics required. Figure 13-4 shows a simple continuous linear system: an electronic low-pass filter composed of a single resistor and a single capacitor. As shown in the figure, an impulse entering this system produces an output that quickly jumps to some value, and then exponentially decays toward zero. In other words, the impulse response of this simple electronic circuit is a *one-sided exponential*. Mathematically, the

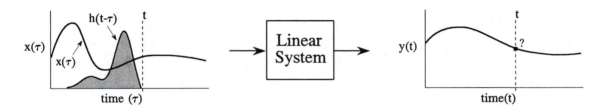

FIGURE 13-3
Convolution viewed from the output side. Each value in the output signal is influenced by many points from the input signal. In this figure, the output signal at time t is being calculated. The input signal, $x(\tau)$, is *weighted* (multiplied) by the flipped and shifted impulse response, given by $h(t-\tau)$. Integrating the weighted input signal produces the value of the output point, $y(t)$

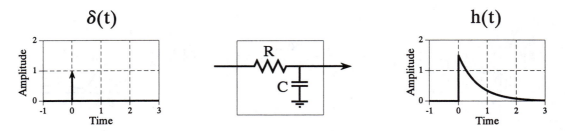

FIGURE 13-4
Example of a continuous linear system. This electronic circuit is a low-pass filter composed of a single resistor and capacitor. The impulse response of this system is a one-sided exponential.

impulse response of this system is broken into two sections, each represented by an equation:

$$h(t) = 0 \qquad \text{for t < 0}$$

$$h(t) = \alpha e^{-\alpha t} \qquad \text{for t} \geq 0$$

where $\alpha = 1/RC$ (R is in ohms, C is in farads, and t is in seconds). Just as in the discrete case, the continuous impulse response contains complete information about the system, that is, how it will react to all possible signals. To pursue this example further, Fig. 13-5 shows a square pulse entering the system, mathematically expressed by:

$$x(t) = 1 \qquad \text{for } 0 \leq t \leq 1$$

$$x(t) = 0 \qquad \text{otherwise}$$

Since both the input signal and the impulse response are completely known as mathematical expressions, the output signal, $y(t)$, can be calculated by evaluating the convolution integral of Eq. 13-1. This is complicated by the fact that both signals are defined by *regions* rather than a single

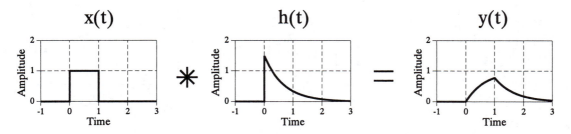

FIGURE 13-5
Example of continuous convolution. This figure illustrates a square pulse entering an RC low-pass filter (Fig. 13-4). The square pulse is convolved with the system's impulse response to produce the output.

a. No overlap
(t < 0)

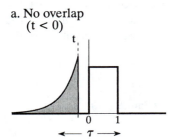

b. Partial overlap
(0 ≤ t ≤ 1)

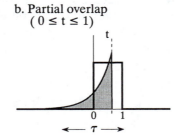

c. Full overlap
(t > 1)

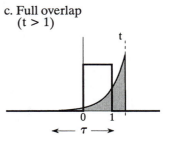

FIGURE 13-6

Calculating a convolution by segments. Since many continuous signals are defined by *regions*, the convolution calculation must be performed region-by-region. In this example, calculation of the output signal is broken into three sections: (a) no overlap, (b) partial overlap, and (c) total overlap, of the input signal and the shifted-flipped impulse response.

mathematical expression. This is very common in continuous signal processing. It is usually essential to draw a picture of how the two signals shift over each other for various values of t. In this example, Fig. 13-6a shows that the two signals do not overlap at all for $t < 0$. This means that the product of the two signals is zero at all locations along the τ axis, and the resulting output signal is:

$$y(t) = 0 \qquad \text{for } t < 0$$

A second case is illustrated in (b), where t is between 0 and 1. Here the two signals partially overlap, resulting in their product having nonzero values between $\tau = 0$ and $\tau = t$. Since this is the only nonzero region, it is the only section where the integral needs to be evaluated. This provides the output signal for $0 \le t \le 1$, given by:

$$y(t) = \int_{-\infty}^{\infty} x(\tau)\, h(t-\tau)\, d\tau \qquad \text{(start with Eq. 13-1)}$$

$$y(t) = \int_{0}^{t} 1 \cdot \alpha e^{-\alpha(t-\tau)}\, d\tau \qquad \text{(plug in the signals)}$$

$$y(t) = e^{-\alpha t} \left[e^{\alpha \tau} \right]\Big|_{0}^{t} \qquad \text{(evaluate the integral)}$$

$$y(t) = e^{-\alpha t} \left[e^{\alpha t} - 1 \right] \qquad \text{(reduce)}$$

$$y(t) = 1 - e^{-\alpha t} \qquad \text{for } 0 \le t \le 1$$

Figure (c) shows the calculation for the third section of the output signal, where t > 1. Here the overlap occurs between $\tau = 0$ and $\tau = 1$, making the calculation the same as for the second segment, except a change to the limits of integration:

$$y(t) = \int_0^1 1 \cdot \alpha e^{-\alpha(t-\tau)} \, d\tau \qquad \text{(plug into Eq. 13-1)}$$

$$y(t) = e^{-\alpha t} \left[e^{\alpha\tau} \right]\Big|_0^1 \qquad \text{(evaluate the integral)}$$

$$y(t) = \left[e^\alpha - 1 \right] e^{-\alpha t} \qquad \text{for } t > 1$$

The waveform in each of these three segments should agree with your knowledge of electronics: (1) The output signal must be zero until the input signal becomes nonzero. That is, the first segment is given by $y(t) = 0$ for $t < 0$. (2) When the step occurs, the RC circuit exponentially increases to match the input, according to the equation: $y(t) = 1 - e^{-\alpha t}$. (3) When the input is returned to zero, the output exponentially decays toward zero, given by the equation: $y(t) = k e^{-\alpha t}$ (where $k = e^\alpha - 1$, the voltage on the capacitor just before the discharge was started).

More intricate waveforms can be handled in the same way, although the mathematical complexity can rapidly become unmanageable. When faced with a nasty continuous convolution problem, you need to spend significant time evaluating *strategies* for solving the problem. If you start blindly evaluating integrals you are likely to end up with a mathematical mess. A common strategy is to break one of the signals into simpler additive components that can be *individually* convolved. Using the principles of linearity, the resulting waveforms can be added to find the answer to the original problem.

Figure 13-7 shows another strategy: modify one of the signals in some linear way, perform the convolution, and then undo the original modification. In this example the modification is the *derivative*, and it is undone by taking the *integral*. The derivative of a unit amplitude square pulse is two *impulses*, the first with an area of one, and the second with an area of negative one. To understand this, think about the opposite process of taking the integral of the two impulses. As you integrate past the first impulse, the integral rapidly increases from zero to one, i.e., a step function. After passing the negative impulse, the integral of the signal rapidly returns from one back to zero, completing the square pulse.

Taking the derivative simplifies this problem because convolution is easy when one of the signals is composed of impulses. Each of the two impulses in $x'(t)$ contributes a scaled and shifted version of the impulse response to

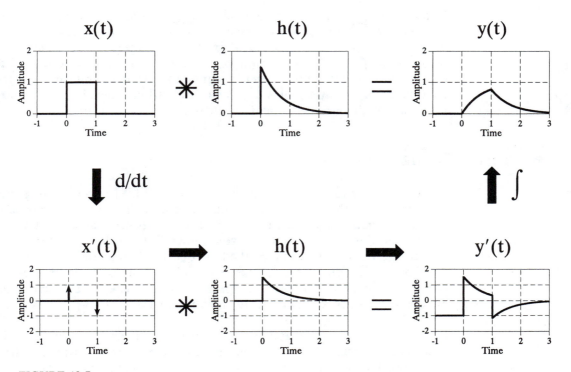

FIGURE 13-7
A strategy for convolving signals. Convolution problems can often be simplified by clever use of the rules governing linear systems. In this example, the convolution of two signals is simplified by taking the derivative of one of them. After performing the convolution, the derivative is undone by taking the integral.

the derivative of the output signal, $y'(t)$. That is, by inspection it is known that: $y'(t) = h(t) - h(t-1)$. The output signal, $y(t)$, can then be found by plugging in the exact equation for $h(t)$, and integrating the expression.

A slight nuisance in this procedure is that the DC value of the input signal is lost when the derivative is taken. This can result in an error in the DC value of the calculated output signal. The mathematics reflects this as the arbitrary constant that can be added during the integration. There is no systematic way of identifying this error, but it can usually be corrected by inspection of the problem. For instance, there is no DC error in the example of Fig. 13-7. This is known because the calculated output signal has the correct DC value when t becomes very large. If an error is present in a particular problem, an appropriate DC term is manually added to the output signal to complete the calculation.

This method also works for signals that can be reduced to impulses by taking the derivative *multiple* times. In the jargon of the field, these signals are called *piecewise polynomials*. After the convolution, the initial operation of multiple derivatives is undone by taking multiple integrals. The only catch is that the lost DC value must be found at each stage by finding the correct constant of integration.

Before starting a difficult continuous convolution problem, there is another approach that you should consider. Ask yourself the question: *Is a mathematical expression really needed for the output signal, or is a graph of the waveform sufficient?* If a graph is adequate, you may be better off to handle the problem with *discrete* techniques. That is, approximate the continuous signals by samples that can be directly convolved by a computer program. While not as mathematically pure, it can be much easier.

The Fourier Transform

The Fourier Transform for continuous signals is divided into two categories, one for signals that are *periodic*, and one for signals that are *aperiodic*. Periodic signals use a version of the Fourier Transform called the **Fourier Series**, and are discussed in the next section. The Fourier Transform used with aperiodic signals is simply called the **Fourier Transform**. This chapter describes these Fourier techniques using only *real* mathematics, just as the last several chapters have done for discrete signals. The more powerful use of *complex* mathematics will be reserved for Chapter 29.

Figure 13-8 shows an example of a continuous aperiodic signal and its frequency spectrum. The time domain signal extends from negative infinity to positive infinity, while each of the frequency domain signals extends from zero to positive infinity. This frequency spectrum is shown in rectangular form (real and imaginary parts); however, the polar form (magnitude and phase) is also used with continuous signals. Just as in the discrete case, the **synthesis equation** describes a recipe for constructing the time domain signal using the data in the frequency domain. In mathematical form:

$$x(t) = \frac{1}{\pi} \int_0^{+\infty} Re\, X(\omega)\cos(\omega t) - Im\, X(\omega)\sin(\omega t)\, d\omega$$

EQUATION 13-2
The Fourier transform synthesis equation. In this equation, $x(t)$ is the time domain signal being synthesized, and $Re\, X(\omega)$ & $Im\, X(\omega)$ are the real and imaginary parts of the frequency spectrum, respectively.

In words, the time domain signal is formed by adding (with the use of an integral) an infinite number of scaled sine and cosine waves. The real part of the frequency domain consists of the scaling factors for the cosine waves, while the imaginary part consists of the scaling factors for the sine waves. Just as with discrete signals, the synthesis equation is usually written with *negative* sine waves. Although the negative sign has no significance in this discussion, it is necessary to make the notation compatible with the complex mathematics described in Chapter 29. The key point to remember is that some authors put this negative sign in the equation, while others do not. Also notice that frequency is represented by the symbol, ω, a lower case

Time Domain

Frequency Domain

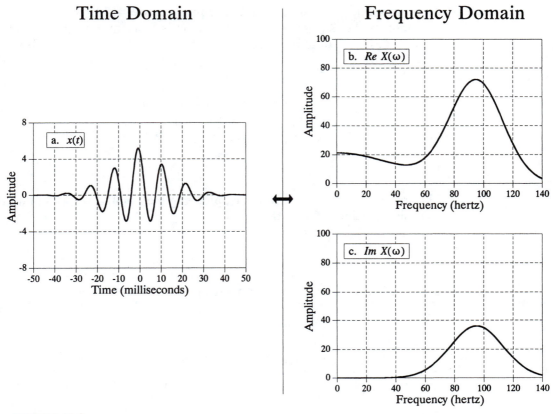

FIGURE 13-8
Example of the Fourier Transform. The time domain signal, $x(t)$, extends from negative to positive infinity. The frequency domain is composed of a real part, $Re\,X(\omega)$, and an imaginary part, $Im\,X(\omega)$, each extending from zero to positive infinity. The frequency axis in this illustration is labeled in *cycles per second* (hertz). To convert to natural frequency, multiply the numbers on the frequency axis by 2π.

Greek omega. As you recall, this notation is called the **natural frequency**, and has the units of radians per second. That is, $\omega = 2\pi f$, where f is the frequency in cycles per second (hertz). The natural frequency notation is favored by mathematicians and others doing signal processing by *solving equations*, because there are usually fewer symbols to write.

The **analysis equations** for continuous signals follow the same strategy as the discrete case: *correlation* with sine and cosine waves. The equations are:

EQUATION 13-3
The Fourier transform analysis equations. In this equation, $Re\,X(\omega)$ & $Im\,X(\omega)$ are the real and imaginary parts of the frequency spectrum, respectively, and $x(t)$ is the time domain signal being analyzed.

$$Re\,X(\omega) = \int_{-\infty}^{+\infty} x(t)\cos(\omega t)\,dt$$

$$Im\,X(\omega) = -\int_{-\infty}^{+\infty} x(t)\sin(\omega t)\,dt$$

As an example of using the analysis equations, we will find the frequency response of the RC low-pass filter. This is done by taking the Fourier transform of its impulse response, previously shown in Fig. 13-4, and described by:

$$h(t) = 0 \qquad \text{for } t < 0$$

$$h(t) = \alpha e^{-\alpha t} \qquad \text{for } t \geq 0$$

The frequency response is found by plugging the impulse response into the analysis equations. First, the real part:

$$Re\ H(\omega) = \int_{-\infty}^{+\infty} h(t) \cos(\omega t)\ dt \qquad \text{(start with Eq. 13-3)}$$

$$Re\ H(\omega) = \int_{0}^{+\infty} \alpha e^{-\alpha t} \cos(\omega t)\ dt \qquad \text{(plug in the signal)}$$

$$Re\ H(\omega) = \frac{\alpha e^{-\alpha t}}{\alpha^2 + \omega^2} \left[-\alpha \cos(\omega t) + \omega \sin(\omega t) \right] \Big|_{0}^{+\infty} \qquad \text{(evaluate)}$$

$$Re\ H(\omega) = \frac{\alpha^2}{\alpha^2 + \omega^2}$$

Using this same approach, the imaginary part of the frequency response is calculated to be:

$$Im\ H(\omega) = \frac{-\omega \alpha}{\alpha^2 + \omega^2}$$

Just as with discrete signals, the rectangular representation of the frequency domain is great for mathematical manipulation, but difficult for human understanding. The situation can be remedied by converting into polar notation with the standard relations: $Mag\ H(\omega) = [Re\ H(\omega)^2 + Im\ H(\omega)^2]^{1/2}$ and $Phase\ H(\omega) = \arctan[Re\ H(\omega)/Im\ H(\omega)]$. Working through the algebra provides the frequency response of the RC low-pass filter as magnitude and phase (i.e., polar form):

$$Mag\,H(\omega) \;=\; \frac{\alpha}{\left[\alpha^2 + \omega^2\right]^{1/2}}$$

$$Phase\,H(\omega) \;=\; \arctan\left[-\frac{\omega}{\alpha}\right]$$

Figure 13-9 shows graphs of these curves for a cutoff frequency of 1000 hertz (i.e., $\alpha = 2\pi\,1000$).

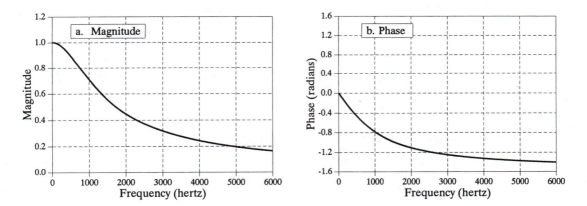

FIGURE 13-9
Frequency response of an RC low-pass filter. These curves were derived by calculating the Fourier transform of the impulse response, and then converting to polar form.

The Fourier Series

This brings us to the last member of the Fourier transform family: the *Fourier series*. The time domain signal used in the Fourier series is *periodic* and *continuous*. Figure 13-10 shows several examples of continuous waveforms that repeat themselves from negative to positive infinity. Chapter 11 showed that periodic signals have a frequency spectrum consisting of **harmonics**. For instance, if the time domain repeats at 1000 hertz, the frequency spectrum will contain a first harmonic at 1000 hertz, a second harmonic at 2000 hertz, a third harmonic at 3000 hertz, and so forth. The first harmonic, i.e., the frequency that the time domain repeats itself, is also called the **fundamental frequency.** This means that the frequency spectrum can be viewed in two ways: (1) the frequency spectrum is *continuous*, but zero at all frequencies except the harmonics, or (2) the frequency spectrum is *discrete*, and only *defined* at the harmonic frequencies. In other words, the frequencies between the harmonics can be thought of as having a value of zero, or simply not existing. The important point is that they do not contribute to forming the time domain signal.

The Fourier series **synthesis equation** creates a continuous periodic signal with a fundamental frequency, f, by adding scaled cosine and sine waves with frequencies: f, $2f$, $3f$, $4f$, etc. The amplitudes of the cosine waves are held in the variables: a_1, a_2, a_3, a_4, etc., while the amplitudes of the sine waves are held in: b_1, b_2, b_3, b_4, and so on. In other words, the "a" and "b" coefficients are the real and imaginary parts of the frequency spectrum, respectively. In addition, the coefficient a_0 is used to hold the DC value of the time domain waveform. This can be viewed as the amplitude of a cosine wave with zero frequency (a constant value). Sometimes a_0 is grouped with the other "a" coefficients, but it is often handled separately because it requires special calculations. There is no b_0 coefficient since a sine wave of zero frequency has a constant value of zero, and would be quite useless. The synthesis equation is written:

$$x(t) = a_0 + \sum_{n=1}^{\infty} a_n \cos(2\pi f t n) - \sum_{n=1}^{\infty} b_n \sin(2\pi f t n)$$

EQUATION 13-4
The Fourier series synthesis equation. Any periodic signal, $x(t)$, can be reconstructed from sine and cosine waves with frequencies that are multiples of the fundamental, f. The a_n and b_n coefficients hold the amplitudes of the cosine and sine waves, respectively.

The corresponding **analysis equations** for the Fourier series are usually written in terms of the *period* of the waveform, denoted by T, rather than the fundamental frequency, f (where $f = 1/T$). Since the time domain signal is periodic, the sine and cosine wave correlation only needs to be evaluated over a single period, i.e., $-T/2$ to $T/2$, 0 to T, $-T$ to 0, etc. Selecting different limits makes the mathematics different, but the final answer is always the same. The Fourier series analysis equations are:

$$a_0 = \frac{1}{T} \int_{-T/2}^{T/2} x(t)\, dt \qquad a_n = \frac{2}{T} \int_{-T/2}^{T/2} x(t) \cos\left(\frac{2\pi t n}{T}\right) dt$$

EQUATION 13-5
Fourier series analysis equations. In these equations, $x(t)$ is the time domain signal being decomposed, a_0 is the DC component, a_n & b_n hold the amplitudes of the cosine and sine waves, respectively, and T is the period of the signal, i.e., the reciprocal of the fundamental frequency.

$$b_n = \frac{-2}{T} \int_{-T/2}^{T/2} x(t) \sin\left(\frac{2\pi t n}{T}\right) dt$$

Time Domain

Frequency Domain

a. Pulse
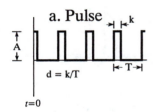
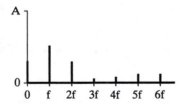

$$a_0 = A d$$
$$a_n = \frac{2A}{n\pi} \sin(n\pi d)$$
$$b_n = 0$$
($d = 0.27$ in this example)

b. Square
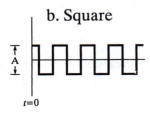
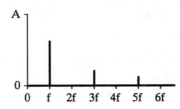

$$a_0 = 0$$
$$a_n = \frac{2A}{n\pi} \sin\left(\frac{n\pi}{2}\right)$$
$$b_n = 0$$
(all even harmonics are zero)

c. Triangle
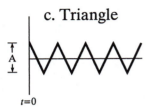
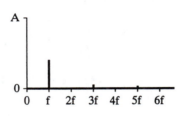

$$a_0 = 0$$
$$a_n = \frac{4A}{(n\pi)^2}$$
$$b_n = 0$$
(all even harmonics are zero)

d. Sawtooth
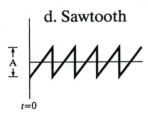
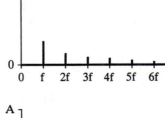

$$a_0 = 0$$
$$a_n = 0$$
$$b_n = \frac{A}{n\pi}$$

e. Rectified
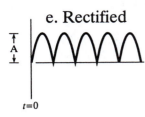
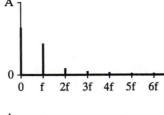

$$a_0 = 2A/\pi$$
$$a_n = \frac{-4A}{\pi(4n^2-1)}$$
$$b_n = 0$$

f. Cosine wave
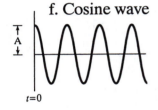
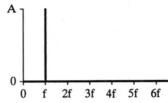

$$a_1 = A$$
(all other coefficients are zero)

FIGURE 13-10
Examples of the Fourier series. Six common time domain waveforms are shown, along with the equations to calculate their "a" and "b" coefficients.

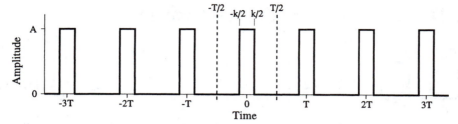

FIGURE 13-11
Example of calculating a Fourier series. This is a pulse train with a duty cycle of $d = k/T$. The Fourier series coefficients are calculated by correlating the waveform with cosine and sine waves over any full period. In this example, the period from $-T/2$ to $T/2$ is used.

Figure 13-11 shows an example of calculating a Fourier series using these equations. The time domain signal being analyzed is a *pulse train*, a square wave with unequal high and low durations. Over a single period from $-T/2$ to $T/2$, the waveform is given by:

$$x(t) = A \qquad \text{for } -k/2 \le t \le k/2$$

$$x(t) = 0 \qquad \text{otherwise}$$

The *duty cycle* of the waveform (the fraction of time that the pulse is "high") is thus given by $d = k/T$. The Fourier series coefficients can be found by evaluating Eq. 13-5. First, we will find the DC component, a_0:

$$a_0 = \frac{1}{T} \int_{-T/2}^{T/2} x(t)\, dt \qquad \text{(start with Eq. 13-5)}$$

$$a_0 = \frac{1}{T} \int_{-k/2}^{k/2} A\, dt \qquad \text{(plug in the signal)}$$

$$a_0 = \frac{A\,k}{T} \qquad \text{(evaluate the integral)}$$

$$a_0 = A\,d \qquad \text{(substitute: } d = k/T\text{)}$$

This result should make intuitive sense; the DC component is simply the average value of the signal. A similar analysis provides the "a" coefficients:

$$a_n = \frac{2}{T} \int_{-T/2}^{T/2} x(t) \cos\left(\frac{2\pi t n}{T}\right) dt \qquad \text{(start with Eq. 13-4)}$$

$$a_n = \frac{2}{T} \int_{-k/2}^{k/2} A \cos\left(\frac{2\pi t n}{T}\right) dt \qquad \text{(plug in the signal)}$$

$$a_n = \frac{2A}{T} \left[\frac{T}{2\pi n} \sin\left(\frac{2\pi t n}{T}\right) \right] \Bigg|_{-k/2}^{k/2} \qquad \text{(evaluate the integral)}$$

$$a_n = \frac{2A}{n\pi} \sin(\pi n d) \qquad \text{(reduce)}$$

The "*b*" coefficients are calculated in this same way; however, they all turn out to be *zero*. In other words, this waveform can be constructed using only cosine waves, with no sine waves being needed.

The "*a*" and "*b*" coefficients will change if the time domain waveform is shifted left or right. For instance, the "*b*" coefficients in this example will be zero *only* if one of the pulses is centered on $t = 0$. Think about it this way. If the waveform is *even* (i.e., symmetrical around $t = 0$), it will be composed solely of *even* sinusoids, that is, cosine waves. This makes all of the "*b*" coefficients equal to zero. If the waveform if *odd* (i.e., symmetrical but opposite in sign around $t = 0$), it will be composed of *odd* sinusoids, i.e., sine waves. This results in the "*a*" coefficients being zero. If the coefficients are converted to polar notation (say, M_n and θ_n coefficients), a shift in the time domain leaves the magnitude unchanged, but adds a linear component to the phase.

To complete this example, imagine a pulse train existing in an electronic circuit, with a frequency of 1 kHz, an amplitude of one volt, and a duty cycle of 0.2. The table in Fig. 13-12 provides the amplitude of each harmonic contained in this waveform. Figure 13-12 also shows the synthesis of the waveform using only the *first fourteen* of these harmonics. Even with this number of harmonics, the reconstruction is not very good. In mathematical jargon, the Fourier series *converges* very *slowly*. This is just another way of saying that sharp edges in the time domain waveform results in very high frequencies in the spectrum. Lastly, be sure and notice the overshoot at the sharp edges, i.e., the Gibbs effect discussed in Chapter 11.

An important application of the Fourier series is electronic **frequency multiplication**. Suppose you want to construct a very stable sine wave oscillator at 150 MHz. This might be needed, for example, in a radio transmitter operating at this frequency.   High stability calls for the circuit to be *crystal controlled*.   That is, the frequency of the oscillator is determined by a resonating quartz crystal that is a part of the circuit.   The problem is, quartz crystals only work to about 10 MHz.   The solution is to build a crystal controlled oscillator operating somewhere between 1 and 10 MHz, and then *multiply* the frequency to whatever you need.   This is accomplished by *distorting* the sine wave, such as by clipping the peaks with a diode, or running the waveform through a squaring circuit.   The harmonics in the distorted waveform are then isolated with band-pass filters.   This allows the frequency to be doubled, tripled, or multiplied by even higher integers numbers.   The most common technique is to use sequential stages of doublers and triplers to generate the required frequency multiplication, rather than just a single stage. The Fourier series is important to this type of design because it describes the *amplitude* of the multiplied signal, depending on the type of distortion and harmonic selected.

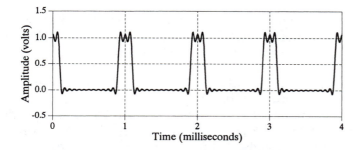

frequency	amplitude (volts)
DC	0.20000
1 kHz	0.37420
2 kHz	0.30273
3 kHz	0.20182
4 kHz	0.09355
5 kHz	0.00000
6 kHz	-0.06237
7 kHz	-0.08649
8 kHz	-0.07568
9 kHz	-0.04158
10 kHz	0.00000
11 kHz	0.03402
12 kHz	0.05046
⋮	
123 kHz	0.00492
124 kHz	0.00302
125 kHz	0.00000
126 kHz	-0.00297
⋮	
803 kHz	0.00075
804 kHz	0.00046
805 kHz	0.00000
806 kHz	-0.00046

FIGURE 13-12
Example of Fourier series synthesis. The waveform being constructed is a pulse train at 1 kHz, an amplitude of one volt, and a duty cycle of 0.2 (as illustrated in Fig. 13-11).   This table shows the amplitude of the harmonics, while the graph shows the reconstructed waveform using only the first fourteen harmonics.

Introduction to Digital Filters

Digital filters are used for two general purposes: (1) separation of signals that have been combined, and (2) restoration of signals that have been distorted in some way. Analog (electronic) filters can be used for these same tasks; however, digital filters can achieve far superior results. The most popular digital filters are described and compared in the next seven chapters. This introductory chapter describes the parameters you want to look for when learning about each of these filters.

Filter Basics

Digital filters are a very important part of DSP. In fact, their extraordinary performance is one of the key reasons that DSP has become so popular. As mentioned in the introduction, filters have two uses: signal *separation* and signal *restoration*. Signal separation is needed when a signal has been contaminated with interference, noise, or other signals. For example, imagine a device for measuring the electrical activity of a baby's heart (EKG) while still in the womb. The raw signal will likely be corrupted by the breathing and heartbeat of the mother. A filter might be used to separate these signals so that they can be individually analyzed.

Signal restoration is used when a signal has been distorted in some way. For example, an audio recording made with poor equipment may be filtered to better represent the sound as it actually occurred. Another example is the deblurring of an image acquired with an improperly focused lens, or a shaky camera.

These problems can be attacked with either analog or digital filters. Which is better? Analog filters are cheap, fast, and have a large dynamic range in both amplitude and frequency. Digital filters, in comparison, are vastly superior in the level of performance that can be achieved. For example, a low-pass digital filter presented in Chapter 16 has a gain of 1 +/- 0.0002 from DC to 1000 hertz, and a gain of less than 0.0002 for frequencies above

1001 hertz. The entire transition occurs within only 1 hertz. Don't expect this from an op amp circuit! Digital filters can achieve *thousands* of times better performance than analog filters. This makes a dramatic difference in how filtering problems are approached. With analog filters, the emphasis is on handling limitations of the electronics, such as the accuracy and stability of the resistors and capacitors. In comparison, digital filters are so good that the performance of the filter is frequently ignored. The emphasis shifts to the limitations of the *signals*, and the *theoretical* issues regarding their processing.

It is common in DSP to say that a filter's input and output signals are in the *time domain*. This is because signals are usually created by sampling at regular intervals of *time*. But this is not the only way sampling can take place. The second most common way of sampling is at equal intervals in *space*. For example, imagine taking simultaneous readings from an array of strain sensors mounted at one centimeter increments along the length of an aircraft wing. Many other domains are possible; however, time and space are by far the most common. When you see the term *time domain* in DSP, remember that it may actually refer to samples taken over time, or it may be a general reference to any domain that the samples are taken in.

As shown in Fig. 14-1, every linear filter has an **impulse response**, a **step response** and a **frequency response**. Each of these responses contains complete information about the filter, but in a different form. If one of the three is specified, the other two are fixed and can be directly calculated. All three of these representations are important, because they describe how the filter will react under different circumstances.

The most straightforward way to implement a digital filter is by *convolving* the input signal with the digital filter's *impulse response*. All possible linear filters can be made in this manner. (This should be obvious. If it isn't, you probably don't have the background to understand this section on filter design. Try reviewing the previous section on DSP fundamentals). When the *impulse response* is used in this way, filter designers give it a special name: the **filter kernel**.

There is also another way to make digital filters, called **recursion**. When a filter is implemented by convolution, each sample in the output is calculated by *weighting* the samples in the input, and adding them together. Recursive filters are an extension of this, using previously calculated values from the *output*, besides points from the *input*. Instead of using a filter kernel, recursive filters are defined by a set of **recursion coefficients**. This method will be discussed in detail in Chapter 19. For now, the important point is that all linear filters have an impulse response, even if you don't use it to implement the filter. To find the impulse response of a recursive filter, simply feed in an impulse, and see what comes out. The impulse responses of recursive filters are composed of sinusoids that exponentially decay in amplitude. In principle, this makes their impulse responses *infinitely long*. However, the amplitude eventually drops below the round-off noise of the system, and the remaining samples can be ignored. Because

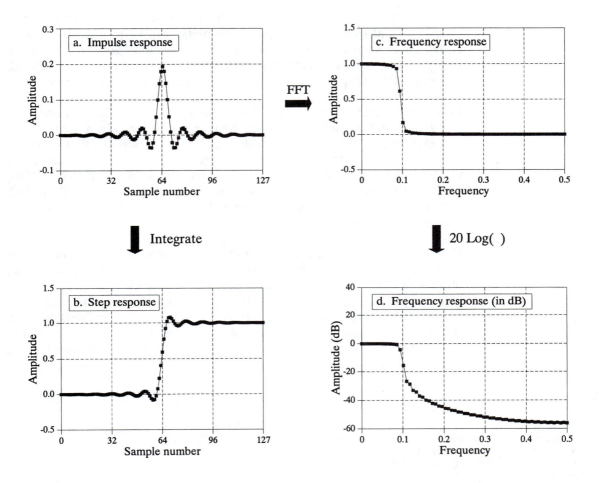

FIGURE 14-1

Filter parameters. Every linear filter has an impulse response, a step response, and a frequency response. The step response, (b), can be found by discrete integration of the impulse response, (a). The frequency response can be found from the impulse response by using the Fast Fourier Transform (FFT), and can be displayed either on a linear scale, (c), or in decibels, (d).

of this characteristic, recursive filters are also called **Infinite Impulse Response or IIR** filters. In comparison, filters carried out by convolution are called **Finite Impulse Response or FIR** filters.

As you know, the *impulse response* is the output of a system when the input is an *impulse*. In this same manner, the *step response* is the output when the input is a *step* (also called an *edge*, and an *edge response*). Since the step is the integral of the impulse, the step response is the integral of the impulse response. This provides two ways to find the step response: (1) feed a step waveform into the filter and see what comes out, or (2) integrate the impulse response. (To be mathematically correct: *integration* is used with continuous signals, while *discrete integration*, i.e., a running sum, is used with discrete signals). The frequency response can be found by taking the DFT (using the FFT algorithm) of the impulse response. This will be reviewed later in this chapter. The frequency response can be plotted on a linear vertical axis, such as in (c), or on a logarithmic scale (decibels), as shown in (d). The linear scale is best at showing the passband ripple and roll-off, while the decibel scale is needed to show the stopband attenuation.

Don't remember decibels? Here is a quick review. A **bel** (in honor of Alexander Graham Bell) means that the power is changed by a *factor of ten*. For example, an electronic circuit that has 3 bels of amplification produces an output signal with $10 \times 10 \times 10 = 1000$ times the power of the input. A **decibel (dB)** is one-tenth of a bel. Therefore, the decibel values of: -20dB, -10dB, 0dB, 10dB & 20dB, mean the power ratios: 0.01, 0.1, 1, 10, & 100, respectively. In other words, every *ten* decibels mean that the power has changed by a factor of ten.

Here's the catch: you usually want to work with a signal's *amplitude*, not its *power*. For example, imagine an amplifier with 20dB of gain. By definition, this means that the power in the signal has increased by a factor of 100. Since amplitude is proportional to the square-root of power, the amplitude of the output is 10 times the amplitude of the input. While 20dB means a factor of 100 in power, it only means a factor of 10 in amplitude. Every *twenty* decibels mean that the amplitude has changed by a factor of ten. In equation form:

EQUATION 14-1
Definition of decibels. Decibels are a way of expressing a *ratio* between two signals. Ratios of power (P_1 & P_2) use a different equation from ratios of amplitude (A_1 & A_2).

$$dB = 10 \log_{10} \frac{P_2}{P_1}$$

$$dB = 20 \log_{10} \frac{A_2}{A_1}$$

The above equations use the base 10 logarithm; however, many computer languages only provide a function for the base e logarithm (the natural log, written $\log_e x$ or $\ln x$). The natural log can be use by modifying the above equations: $dB = 4.342945 \log_e (P_2/P_1)$ and $dB = 8.685890 \log_e (A_2/A_1)$.

Since decibels are a way of expressing the ratio between two signals, they are ideal for describing the gain of a system, i.e., the ratio between the output and the input signal. However, engineers also use decibels to specify the amplitude (or power) of a *single* signal, by referencing it to some standard. For example, the term: **dBV** means that the signal is being referenced to a 1 volt rms signal. Likewise, **dBm** indicates a reference signal producing 1 mW into a 600 ohms load (about 0.78 volts rms).

If you understand nothing else about decibels, remember two things: First, -3dB means that the amplitude is reduced to 0.707 (and the power is therefore reduced to 0.5). Second, memorize the following conversions between decibels and *amplitude* ratios:

$$
\begin{aligned}
60\text{dB} &= 1000 \\
40\text{dB} &= 100 \\
20\text{dB} &= 10 \\
0\text{dB} &= 1 \\
-20\text{dB} &= 0.1 \\
-40\text{dB} &= 0.01 \\
-60\text{dB} &= 0.001
\end{aligned}
$$

How Information is Represented in Signals

The most important part of any DSP task is understanding how *information* is contained in the signals you are working with. There are many ways that information can be contained in a signal. This is especially true if the signal is manmade. For instance, consider all of the modulation schemes that have been devised: AM, FM, single-sideband, pulse-code modulation, pulse-width modulation, etc. The list goes on and on. Fortunately, there are only two ways that are common for information to be represented in naturally occurring signals. We will call these: **information represented in the time domain**, and **information represented in the frequency domain**.

Information represented in the time domain describes when something occurs and what the amplitude of the occurrence is. For example, imagine an experiment to study the light output from the sun. The light output is measured and recorded once each second. Each sample in the signal indicates what is happening at that instant, and the level of the event. If a solar flare occurs, the signal directly provides information on the time it occurred, the duration, the development over time, etc. Each sample contains information that is interpretable without reference to any other sample. Even if you have only one sample from this signal, you still know something about what you are measuring. This is the simplest way for information to be contained in a signal.

In contrast, information represented in the frequency domain is more indirect. Many things in our universe show periodic motion. For example, a wine glass struck with a fingernail will vibrate, producing a ringing sound; the pendulum of a grandfather clock swings back and forth; stars and planets rotate on their axis and revolve around each other, and so forth. By measuring the frequency, phase, and amplitude of this periodic motion, information can often be obtained about the system producing the motion. Suppose we sample the sound produced by the ringing wine glass. The fundamental frequency and harmonics of the periodic vibration relate to the mass and elasticity of the material. A single sample, in itself, contains no information about the periodic motion, and therefore no information about the wine glass. The information is contained in the *relationship* between many points in the signal.

This brings us to the importance of the step and frequency responses. The *step response* describes how information represented in the *time domain* is being modified by the system. In contrast, the *frequency response* shows how information represented in the *frequency domain* is being changed. This distinction is absolutely critical in filter design because it is not possible to optimize a filter for both applications. Good performance in the time domain results in poor performance in the frequency domain, and vice versa. If you are designing a filter to remove noise from an EKG signal (information represented in the time domain), the step response is the important parameter, and the frequency response is of little concern. If your task is to design a digital filter for a hearing aid (with the information in the frequency domain), the frequency response is all important, while the step response doesn't matter. Now let's look at what makes a filter optimal for time domain or frequency domain applications.

Time Domain Parameters

It may not be obvious why the step response is of such concern in time domain filters. You may be wondering why the impulse response isn't the important parameter. The answer lies in the way that the human mind understands and processes information. Remember that the step, impulse and frequency responses all contain identical information, just in different arrangements. The step response is useful in time domain analysis because it matches the way humans view the information contained in the signals.

For example, suppose you are given a signal of some unknown origin and asked to analyze it. The first thing you will do is divide the signal into regions of similar characteristics. You can't stop from doing this; your mind will do it automatically. Some of the regions may be smooth; others may have large amplitude peaks; others may be noisy. This segmentation is accomplished by identifying the points that separate the regions. This is where the step function comes in. The step function is the purest way of representing a division between two dissimilar regions. It can mark when an event starts, or when an event ends. It tells you that whatever is on the *left* is somehow different from whatever is on the *right*. This is how the human mind views time domain information: a group of step functions dividing the information into regions of similar characteristics. The step response, in turn, is important because it describes how the dividing lines are being modified by the filter.

The step response parameters that are important in filter design are shown in Fig. 14-2. To distinguish events in a signal, the duration of the step response must be shorter than the spacing of the events. This dictates that the step response should be as *fast* (the DSP jargon) as possible. This is shown in Figs. (a) & (b). The most common way to specify the **risetime** (more jargon) is to quote the number of samples between the 10% and 90% amplitude levels. Why isn't a very fast risetime always possible? There are many reasons, noise reduction, inherent limitations of the data acquisition system, avoiding aliasing, etc.

POOR

GOOD

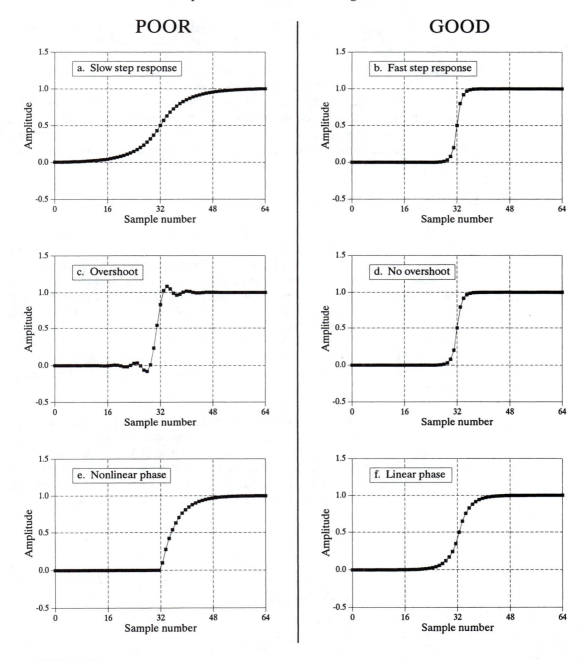

FIGURE 14-2
Parameters for evaluating *time domain* performance. The step response is used to measure how well a filter performs in the time domain. Three parameters are important: (1) transition speed (risetime), shown in (a) and (b), (2) overshoot, shown in (c) and (d), and (3) phase linearity (symmetry between the top and bottom halves of the step), shown in (e) and (f).

Figures (c) and (d) shows the next parameter that is important: **overshoot** in the step response. Overshoot must generally be eliminated because it changes the amplitude of samples in the signal; this is a basic distortion of the information contained in the time domain. This can be summed up in one question: Is the overshoot you observe in a signal coming from the thing you are trying to measure, or from the filter you have used?

Finally, it is often desired that the upper half of the step response be symmetrical with the lower half, as illustrated in (e) and (f). This symmetry is needed to make the *rising edges* look the same as the *falling edges*. This symmetry is called **linear phase**, because the frequency response has a phase that is a straight line (discussed in Chapter 19). Make sure you understand these three parameters; they are the key to evaluating time domain filters.

Frequency Domain Parameters

Figure 14-3 shows the four basic frequency responses. The purpose of these filters is to allow some frequencies to pass unaltered, while completely blocking other frequencies. The **passband** refers to those frequencies that are passed, while the **stopband** contains those frequencies that are blocked. The **transition band** is between. A **fast roll-off** means that the transition band is very narrow. The division between the passband and transition band is called the **cutoff frequency**. In analog filter design, the cutoff frequency is usually defined to be where the amplitude is reduced to 0.707 (i.e., -3dB). Digital filters are less standardized, and it is common to see 99%, 90%, 70.7%, and 50% amplitude levels defined to be the cutoff frequency.

Figure 14-4 shows three parameters that measure how well a filter performs in the frequency domain. To separate closely spaced frequencies, the filter must have a **fast roll-off**, as illustrated in (a) and (b). For the passband frequencies to move through the filter unaltered, there must be no **passband ripple**, as shown in (c) and (d). Lastly, to adequately block the stopband frequencies, it is necessary to have good **stopband attenuation**, displayed in (e) and (f).

FIGURE 14-3
The four common frequency responses. Frequency domain filters are generally used to pass certain frequencies (the *passband*), while blocking others (the *stopband*). Four responses are the most common: low-pass, high-pass, band-pass, and band-reject.

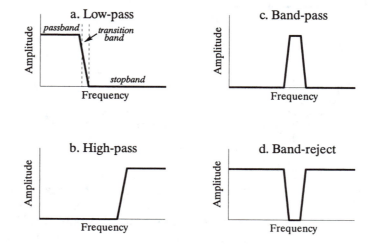

POOR

GOOD

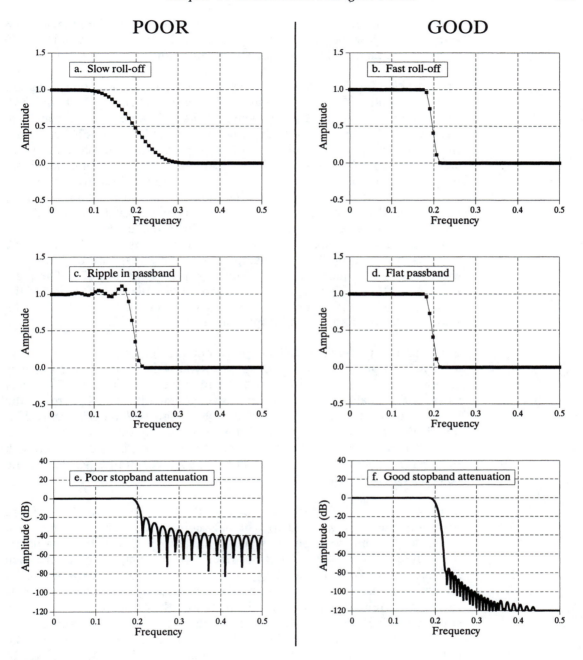

FIGURE 14-4

Parameters for evaluating *frequency domain* performance. The frequency responses shown are for low-pass filters. Three parameters are important: (1) roll-off sharpness, shown in (a) and (b), (2) passband ripple, shown in (c) and (d), and (3) stopband attenuation, shown in (e) and (f).

Why is there nothing about the *phase* in these parameters? First, the phase isn't important in most frequency domain applications. For example, the phase of an audio signal is almost completely random, and contains little useful information. Second, if the phase is important, it is very easy to make digital filters with a *perfect* phase response, i.e., all frequencies pass through the filter with a zero phase shift (also discussed in Chapter 19). In comparison, analog filters are ghastly in this respect.

Previous chapters have described how the DFT converts a system's impulse response into its frequency response. Here is a brief review. The quickest way to calculate the DFT is by means of the FFT algorithm presented in Chapter 12. Starting with a filter kernel N samples long, the FFT calculates the frequency spectrum consisting of an N point *real part* and an N point *imaginary part*. Only samples 0 to $N/2$ of the FFT's real and imaginary parts contain useful information; the remaining points are duplicates (negative frequencies) and can be ignored. Since the real and imaginary parts are difficult for humans to understand, they are usually converted into polar notation as described in Chapter 8. This provides the magnitude and phase signals, each running from sample 0 to sample $N/2$ (i.e., $N/2+1$ samples in each signal). For example, an impulse response of 256 points will result in a frequency response running from point 0 to 128. Sample 0 represents DC, i.e., zero frequency. Sample 128 represents one-half of the sampling rate. Remember, no frequencies higher than one-half of the sampling rate can appear in sampled data.

The number of samples used to represent the impulse response can be arbitrarily large. For instance, suppose you want to find the frequency response of a filter kernel that consists of 80 points. Since the FFT only works with signals that are a power of two, you need to add 48 zeros to the signal to bring it to a length of 128 samples. This *padding with zeros* does not change the impulse response. To understand why this is so, think about what happens to these added zeros when the input signal is convolved with the system's impulse response. The added zeros simply *vanish* in the convolution, and do not affect the outcome.

Taking this a step further, you could add *many* zeros to the impulse response to make it, say, 256, 512, or 1024 points long. The important idea is that longer impulse responses result in a closer spacing of the data points in the frequency response. That is, there are more samples spread between DC and one-half of the sampling rate. Taking this to the extreme, if the impulse response is padded with an *infinite* number of zeros, the data points in the frequency response are infinitesimally close together, i.e., a continuous line. In other words, the frequency response of a filter is really a *continuous* signal between DC and one-half of the sampling rate. The output of the DFT is a *sampling* of this continuous line. What length of impulse response should you use when calculating a filter's frequency response? As a first thought, try $N=1024$, but don't be afraid to change it if needed (such as insufficient resolution or excessive computation time).

Keep in mind that the "good" and "bad" parameters discussed in this chapter are only generalizations. Many signals don't fall neatly into categories. For example, consider an EKG signal contaminated with 60 hertz interference. The information is encoded in the *time domain*, but the interference is best dealt with in the *frequency domain*. The best design for this application is

Time Domain

Frequency Domain

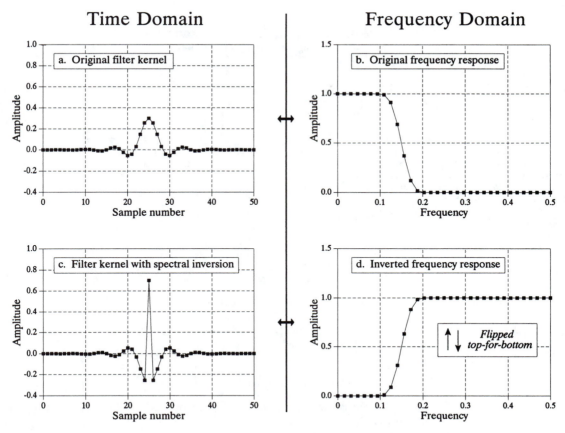

FIGURE 14-5
Example of spectral inversion. The low-pass filter kernel in (a) has the frequency response shown in (b). A high-pass filter kernel, (c), is formed by changing the sign of each sample in (a), and adding one to the sample at the center of symmetry. This action in the time domain *inverts* the frequency spectrum (i.e., flips it top-for-bottom), as shown by the high-pass frequency response in (d).

bound to have trade-offs, and might go against the conventional wisdom of this chapter. Remember the number one rule of education: *A paragraph in a book doesn't give you a license to stop thinking*.

High-Pass, Band-Pass and Band-Reject Filters

High-pass, band-pass and band-reject filters are designed by starting with a low-pass filter, and then converting it into the desired response. For this reason, most discussions on filter design only give examples of low-pass filters. There are two methods for the low-pass to high-pass conversion: **spectral inversion** and **spectral reversal**. Both are equally useful.

An example of *spectral inversion* is shown in 14-5. Figure (a) shows a low-pass filter kernel called a windowed-sinc (the topic of Chapter 16). This filter kernel is 51 points in length, although many of samples have a value so small that they appear to be zero in this graph. The corresponding

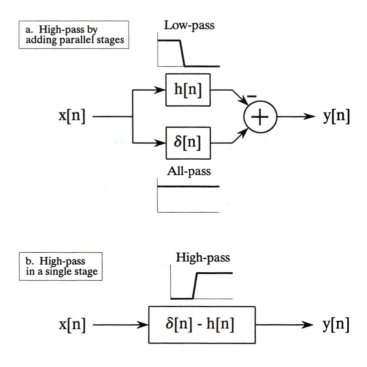

FIGURE 14-6
Block diagram of spectral inversion. In (a), the input signal, $x[n]$, is applied to two systems in parallel, having impulse responses of $h[n]$ and $\delta[n]$. As shown in (b), the combined system has an impulse response of $\delta[n]-h[n]$. This means that the frequency response of the combined system is the *inversion* of the frequency response of $h[n]$.

frequency response is shown in (b), found by adding 13 zeros to the filter kernel and taking a 64 point FFT. Two things must be done to change the low-pass filter kernel into a high-pass filter kernel. First, change the sign of each sample in the filter kernel. Second, add *one* to the sample at the center of symmetry. This results in the high-pass filter kernel shown in (c), with the frequency response shown in (d). Spectral inversion *flips* the frequency response *top-for-bottom*, changing the passbands into stopbands, and the stopbands into passbands. In other words, it changes a filter from low-pass to high-pass, high-pass to low-pass, band-pass to band-reject, or band-reject to band-pass.

Figure 14-6 shows why this two step modification to the time domain results in an inverted frequency spectrum. In (a), the input signal, $x[n]$, is applied to two systems in parallel. One of these systems is a low-pass filter, with an impulse response given by $h[n]$. The other system does *nothing* to the signal, and therefore has an impulse response that is a delta function, $\delta[n]$. The overall output, $y[n]$, is equal to the output of the all-pass system *minus* the output of the low-pass system. Since the low frequency components are subtracted from the original signal, only the high frequency components appear in the output. Thus, a high-pass filter is formed.

This could be performed as a two step operation in a computer program: run the signal through a low-pass filter, and then subtract the filtered signal from the original. However, the entire operation can be performed in a signal stage by combining the two filter kernels. As described in Chapter

Time Domain

Frequency Domain

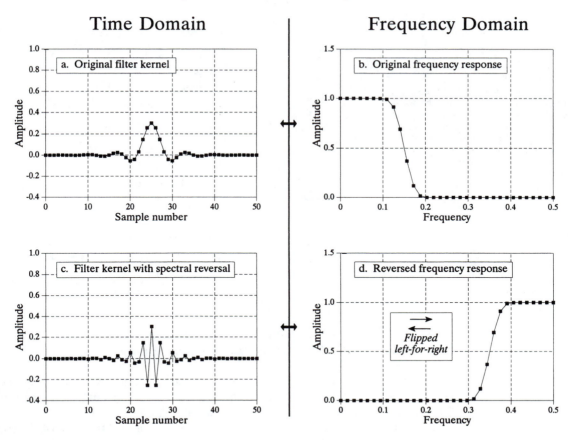

a. Original filter kernel

b. Original frequency response

c. Filter kernel with spectral reversal

d. Reversed frequency response

Flipped left-for-right

FIGURE 14-7
Example of spectral reversal. The low-pass filter kernel in (a) has the frequency response shown in (b). A high-pass filter kernel, (c), is formed by changing the sign of every other sample in (a). This action in the time domain results in the frequency domain being flipped *left-for-right*, resulting in the high-pass frequency response shown in (d).

7, parallel systems with added outputs can be combined into a single stage by adding their impulse responses. As shown in (b), the filter kernel for the high-pass filter is given by: $\delta[n] - h[n]$. That is, change the sign of all the samples, and then add one to the sample at the center of symmetry.

For this technique to work, the low-frequency components exiting the low-pass filter must have the same phase as the low-frequency components exiting the all-pass system. Otherwise a complete subtraction cannot take place. This places two restrictions on the method: (1) the original filter kernel must have left-right symmetry (i.e., a zero or linear phase), and (2) the impulse must be added at the center of symmetry.

The second method for low-pass to high-pass conversion, *spectral reversal*, is illustrated in Fig. 14-7. Just as before, the low-pass filter kernel in (a) corresponds to the frequency response in (b). The high-pass filter kernel, (c), is formed by *changing the sign of every other sample* in (a). As shown in (d), this flips the frequency domain *left-for-right*: 0 becomes 0.5 and 0.5

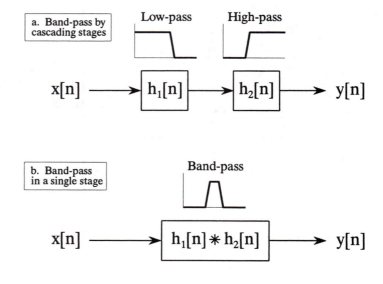

FIGURE 14-8
Deigning a band-pass filter. As shown in (a), a band-pass filter can be formed by cascading a low-pass filter and a high-pass filter. This can be reduced to a single stage, shown in (b). The filter kernel of the single stage is equal to the *convolution* of the low-pass and high-pass filter kernels.

becomes 0. The cutoff frequency of the example low-pass filter is 0.15, resulting in the cutoff frequency of the high-pass filter being 0.35.

Changing the sign of every other sample is equivalent to multiplying the filter kernel by a sinusoid with a frequency of 0.5. As discussed in Chapter 10, this has the effect of *shifting* the frequency domain by 0.5. Look at (b) and imagine the negative frequencies between -0.5 and 0 that are of mirror image of the frequencies between 0 and 0.5. The frequencies that appear in (d) are the negative frequencies from (b) shifted by 0.5.

Lastly, Figs. 14-8 and 14-9 show how low-pass and high-pass filter kernels can be combined to form band-pass and band-reject filters. In short, *adding* the filter kernels produces a *band-reject* filter, while *convolving* the filter kernels produces a *band-pass* filter. These are based on the way cascaded and parallel systems are be combined, as discussed in Chapter 7. Multiple combination of these techniques can also be used. For instance, a band-pass filter can be designed by adding the two filter kernels to form a band-pass filter, and then use *spectral inversion* or *spectral reversal* as previously described. All these techniques work very well with few surprises.

Filter Classification

Table 14-1 summarizes how digital filters are classified by their *use* and by their *implementation*. The use of a digital filter can be broken into three categories: *time domain*, *frequency domain* and *custom*.   As previously described, time domain filters are used when the information is encoded in the shape of the signal's waveform. Time domain filtering is used for such actions as: smoothing, DC removal, waveform shaping, etc. In contrast, frequency domain filters are used when the information is contained in the

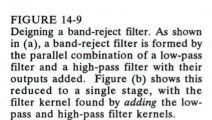

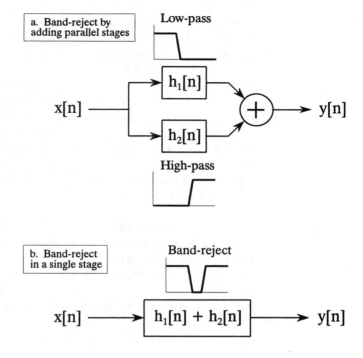

FIGURE 14-9
Deigning a band-reject filter. As shown in (a), a band-reject filter is formed by the parallel combination of a low-pass filter and a high-pass filter with their outputs added. Figure (b) shows this reduced to a single stage, with the filter kernel found by *adding* the low-pass and high-pass filter kernels.

amplitude, frequency, and phase of the component sinusoids. The goal of these filters is to separate one band of frequencies from another. Custom filters are used when a special action is required by the filter, something more elaborate than the four basic responses (high-pass, low-pass, band-pass and band-reject). For instance, Chapter 17 describes how custom filters can be used for *deconvolution*, a way of counteracting an unwanted convolution.

FILTER IMPLEMENTED BY:

FILTER USED FOR:		Convolution *Finite Impulse Response (FIR)*	Recursion *Infinite Impulse Response (IIR)*
	Time Domain *(smoothing, DC removal)*	Moving average (Ch. 15)	Single pole (Ch. 19)
	Frequency Domain *(separating frequencies)*	Windowed-sinc (Ch. 16)	Chebyshev (Ch. 20)
	Custom *(Deconvolution)*	FIR custom (Ch. 17)	Iterative design (Ch. 26)

Table 14-1
Filter classification. Filters can be divided by their *use*, and how they are *implemented*.

Digital filters can be implemented in two ways, by *convolution* (also called *finite impulse response* or *FIR*) and by *recursion* (also called *infinite impulse response* or *IIR*). Filters carried out by convolution can have far better performance than filters using recursion, but execute much more slowly.

The next six chapters describe digital filters according to the classifications in Table 14-1. First, we will look at filters carried out by convolution. The *moving average* (Chapter 15) is used in the time domain, the *windowed-sinc* (Chapter 16) is used in the frequency domain, and *FIR custom* (Chapter 17) is used when something special is needed. To finish the discussion of FIR filters, Chapter 18 presents a technique called FFT convolution. This is an algorithm for increasing the speed of convolution, allowing FIR filters to execute faster.

Next, we look at recursive filters. The *single pole* recursive filter (Chapter 19) is used in the time domain, while the *Chebyshev* (Chapter 20) is used in the frequency domain. Recursive filters having a custom response are designed by *iterative techniques*. For this reason, we will delay their discussion until Chapter 26, where they will be presented with another type of iterative procedure: the neural network.

As shown in Table 14-1, *convolution* and *recursion* are rival techniques; you must use one or the other for a particular application. How do you choose? Chapter 21 presents a head-to-head comparison of the two, in both the time and frequency domains.

CHAPTER 15

Moving Average Filters

The moving average is the most common filter in DSP, mainly because it is the easiest digital filter to understand and use. In spite of its simplicity, the moving average filter is *optimal* for a common task: reducing random noise while retaining a sharp step response. This makes it the premier filter for time domain encoded signals. However, the moving average is the *worst* filter for frequency domain encoded signals, with little ability to separate one band of frequencies from another. Relatives of the moving average filter include the Gaussian, Blackman, and multiple-pass moving average. These have slightly better performance in the frequency domain, at the expense of increased computation time.

Implementation by Convolution

As the name implies, the moving average filter operates by averaging a number of points from the input signal to produce each point in the output signal. In equation form, this is written:

EQUATION 15-1
Equation of the moving average filter. In this equation, $x[\ ]$ is the input signal, $y[\ ]$ is the output signal, and M is the number of points used in the moving average. This equation only uses points on *one side* of the output sample being calculated.

$$y[i] \ = \ \frac{1}{M} \sum_{j=0}^{M-1} x[i+j]$$

Where $x[\ ]$ is the input signal, $y[\ ]$ is the output signal, and M is the number of points in the average. For example, in a 5 point moving average filter, point 80 in the output signal is given by:

$$y[80] \ = \ \frac{x[80] + x[81] + x[82] + x[83] + x[84]}{5}$$

As an alternative, the group of points from the input signal can be chosen *symmetrically* around the output point:

$$y[80] = \frac{x[78] + x[79] + x[80] + x[81] + x[82]}{5}$$

This corresponds to changing the summation in Eq. 15-1 from: $j = 0$ to $M-1$, to: $j = -(M-1)/2$ to $(M-1)/2$. For instance, in an 11 point moving average filter, the index, j, can run from 0 to 11 (one side averaging) or -5 to 5 (symmetrical averaging). Symmetrical averaging requires that M be an *odd* number. Programming is slightly easier with the points on only one side; however, this produces a relative shift between the input and output signals.

You should recognize that the moving average filter is a *convolution* using a very simple filter kernel. For example, a 5 point filter has the filter kernel: $\cdots 0, 0, 1/5, 1/5, 1/5, 1/5, 1/5, 0, 0 \cdots$. That is, the moving average filter is a convolution of the input signal with a *rectangular pulse* having an area of *one*. Table 15-1 shows a program to implement the moving average filter.

```
100 'MOVING AVERAGE FILTER
110 'This program filters 5000 samples with a 101 point moving
120 'average filter, resulting in 4900 samples of filtered data.
130 '
140 DIM X[4999]                    'X[ ] holds the input signal
150 DIM Y[4999]                    'Y[ ] holds the output signal
160 '
170 GOSUB XXXX                     'Mythical subroutine to load X[ ]
180 '
190 FOR I% = 50 TO 4949           'Loop for each point in the output signal
200   Y[I%] = 0                    'Zero, so it can be used as an accumulator
210   FOR J% = -50 TO 50           'Calculate the summation
220     Y[I%] = Y[I%] + X(I%+J%]
230   NEXT J%
240   Y[I%] = Y[I%]/101            'Complete the average by dividing
250 NEXT I%
260 '
270 END
```

TABLE 15-1

Noise Reduction vs. Step Response

Many scientists and engineers feel guilty about using the moving average filter. Because it is so very simple, the moving average filter is often the first thing tried when faced with a problem. Even if the problem is completely solved, there is still the feeling that something more should be done. This situation is truly ironic. Not only is the moving average filter very good for many applications, it is *optimal* for a common problem, reducing random white noise while keeping the sharpest step response.

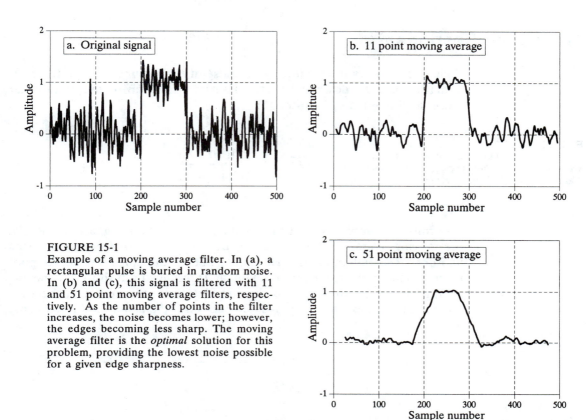

FIGURE 15-1

Example of a moving average filter. In (a), a rectangular pulse is buried in random noise. In (b) and (c), this signal is filtered with 11 and 51 point moving average filters, respectively. As the number of points in the filter increases, the noise becomes lower; however, the edges becoming less sharp. The moving average filter is the *optimal* solution for this problem, providing the lowest noise possible for a given edge sharpness.

Figure 15-1 shows an example of how this works. The signal in (a) is a pulse buried in random noise. In (b) and (c), the smoothing action of the moving average filter decreases the amplitude of the random noise (good), but also reduces the sharpness of the edges (bad). Of all the possible linear filters that could be used, the moving average produces the lowest noise for a given edge sharpness. The amount of noise reduction is equal to the square-root of the number of points in the average. For example, a 100 point moving average filter reduces the noise by a factor of 10.

To understand why the moving average if the best solution, imagine we want to design a filter with a fixed edge sharpness. For example, let's assume we fix the edge sharpness by specifying that there are eleven points in the rise of the step response. This requires that the filter kernel have eleven points. The optimization question is: how do we choose the eleven values in the filter kernel to minimize the noise on the output signal? Since the noise we are trying to reduce is random, none of the input points is special; each is just as noisy as its neighbor. Therefore, it is useless to give preferential treatment to any one of the input points by assigning it a larger coefficient in the filter kernel. The lowest noise is obtained when all the input samples are treated equally, i.e., the moving average filter. (Later in this chapter we show that other filters are essentially *as* good. The point is, no filter is *better* than the simple moving average).

Frequency Response

Figure 15-2 shows the frequency response of the moving average filter. It is mathematically described by the Fourier transform of the rectangular pulse, as discussed in Chapter 11:

EQUATION 15-2
Frequency response of an M point moving average filter. The frequency, f, runs between 0 and 0.5. For $f = 0$, use: $H[f] = 1$

$$H[f] = \frac{\sin(\pi f M)}{M \sin(\pi f)}$$

The roll-off is very slow and the stopband attenuation is ghastly. Clearly, the moving average filter cannot separate one band of frequencies from another. Remember, good performance in the time domain results in poor performance in the frequency domain, and vice versa. In short, the moving average is an exceptionally good *smoothing filter* (the action in the time domain), but an exceptionally bad *low-pass filter* (the action in the frequency domain).

FIGURE 15-2
Frequency response of the moving average filter. The moving average is a very poor low-pass filter, due to its slow roll-off and poor stopband attenuation. These curves are generated by Eq. 15-2.

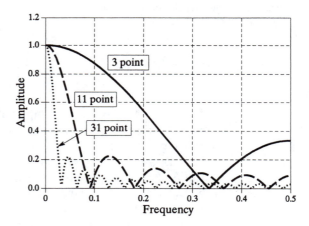

Relatives of the Moving Average Filter

In a perfect world, filter designers would only have to deal with time domain *or* frequency domain encoded information, but never a mixture of the two in the same signal. Unfortunately, there are some applications where both domains are simultaneously important. For instance, television signals fall into this nasty category. Video information is encoded in the time domain, that is, the shape of the waveform corresponds to the patterns of brightness in the image. However, during transmission the video signal is treated according to its frequency composition, such as its total bandwidth, how the carrier waves for sound & color are added, elimination & restoration of the DC component, etc. As another example, electro-magnetic interference is best understood in the frequency domain, even if

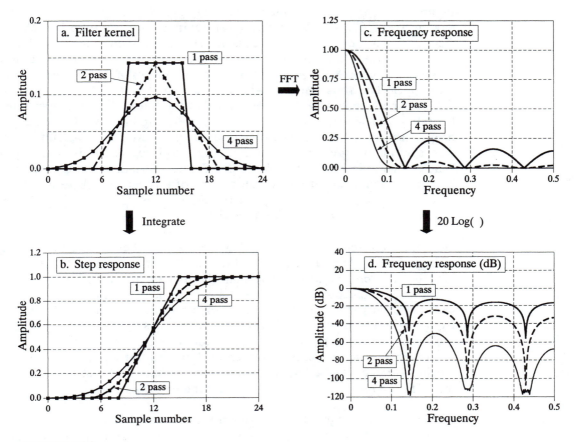

FIGURE 15-3
Characteristics of multiple-pass moving average filters. Figure (a) shows the filter kernels resulting from passing a seven point moving average filter over the data once, twice and four times. Figure (b) shows the corresponding step responses, while (c) and (d) show the corresponding frequency responses.

the signal's information is encoded in the time domain. For instance, the temperature monitor in a scientific experiment might be contaminated with 60 hertz from the power lines, 30 kHz from a switching power supply, or 1320 kHz from a local AM radio station. Relatives of the moving average filter have better frequency domain performance, and can be useful in these mixed domain applications.

Multiple-pass moving average filters involve passing the input signal through a moving average filter two or more times. Figure 15-3a shows the overall filter kernel resulting from one, two and four passes. Two passes are equivalent to using a *triangular* filter kernel (a rectangular filter kernel convolved with itself). After four or more passes, the equivalent filter kernel looks like a *Gaussian* (recall the Central Limit Theorem). As shown in (b), multiple passes produce an "s" shaped step response, as compared to the straight line of the single pass. The frequency responses in (c) and (d) are given by Eq. 15-2 *multiplied* by itself for each pass. That is, each time domain convolution results in a multiplication of the frequency spectra.

Figure 15-4 shows the frequency response of two other relatives of the moving average filter. When a pure **Gaussian** is used as a filter kernel, the frequency response is also a Gaussian, as discussed in Chapter 11. The Gaussian is important because it is the impulse response of many natural and manmade systems. For example, a brief pulse of light entering a long fiber optic transmission line will exit as a Gaussian pulse, due to the different paths taken by the photons within the fiber. The Gaussian filter kernel is also used extensively in *image processing* because it has unique properties that allow fast two-dimensional convolutions (see Chapter 24). The second frequency response in Fig. 15-4 corresponds to using a **Blackman window** as a filter kernel. (The term *window* has no meaning here; it is simply part of the accepted name of this curve). The exact shape of the Blackman window is given in Chapter 16 (Eq. 16-2, Fig. 16-2); however, it looks much like a Gaussian.

How are these relatives of the moving average filter better than the moving average filter itself? Three ways: First, and most important, these filters have better *stopband attenuation* than the moving average filter. Second, the filter kernels *taper* to a smaller amplitude near the ends. Recall that each point in the output signal is a weighted sum of a group of samples from the input. If the filter kernel tapers, samples in the input signal that are farther away are given less weight than those close by. Third, the step responses are *smooth* curves, rather than the abrupt straight line of the moving average. These last two are usually of limited benefit, although you might find applications where they are genuine advantages.

The moving average filter and its relatives are all about the same at reducing random noise while maintaining a sharp step response. The ambiguity lies in how the *risetime* of the step response is measured. If the risetime is measured from 0% to 100% of the step, the moving average filter is the best you can do, as previously shown. In comparison, measuring the risetime from 10% to 90% makes the Blackman window *better* than the moving average filter. The point is, this is just theoretical squabbling; consider these filters equal in this parameter.

The biggest difference in these filters is *execution speed*. Using a recursive algorithm (described next), the moving average filter will run like lightning in your computer. In fact, it is the *fastest* digital filter available. Multiple passes of the moving average will be correspondingly slower, but still very quick. In comparison, the Gaussian and Blackman filters are excruciatingly slow, because they must use convolution. Think a factor of ten times the number of points in the filter kernel (based on multiplication being about 10 times slower than addition). For example, expect a 100 point Gaussian to be 1000 times slower than a moving average using recursion.

Recursive Implementation

A tremendous advantage of the moving average filter is that it can be implemented with an algorithm that is very fast. To understand this

FIGURE 15-4
Frequency response of the Blackman window and Gaussian filter kernels. Both these filters provide better stopband attenuation than the moving average filter. This has no advantage in removing random noise from time domain encoded signals, but it can be useful in mixed domain problems. The disadvantage of these filters is that they must use *convolution*, a terribly slow algorithm.

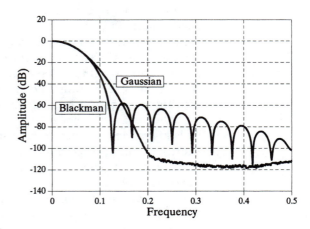

algorithm, imagine passing an input signal, $x[\ ]$, through a seven point moving average filter to form an output signal, $y[\ ]$. Now look at how two adjacent output points, $y[50]$ and $y[51]$, are calculated:

$$y[50] = x[47] + x[48] + x[49] + x[50] + x[51] + x[52] + x[53]$$

$$y[51] = x[48] + x[49] + x[50] + x[51] + x[52] + x[53] + x[54]$$

These are nearly the same calculation; points $x[48]$ through $x[53]$ must be added for $y[50]$, and again for $y[51]$. If $y[50]$ has already been calculated, the most *efficient* way to calculate $y[51]$ is:

$$y[51] = y[50] + x[54] - x[47]$$

Once $y[51]$ has been found using $y[50]$, then $y[52]$ can be calculated from sample $y[51]$, and so on. After the first point is calculated in $y[\ ]$, all of the other points can be found with only a single addition and subtraction per point. This can be expressed in the equation:

EQUATION 15-3
Recursive implementation of the moving average filter. In this equation, $x[\ ]$ is the input signal, $y[\ ]$ is the output signal, M is the number of points in the moving average (an odd number). Before this equation can be used, the first point in the signal must be calculated using a standard summation.

$$y[i] = y[i-1] + x[i+p] - x[i-q]$$

where: $p = (M-1)/2$
$$q = p + 1$$

Notice that this equation use two sources of data to calculate each point in the output: points from the input *and* previously calculated points from the output. This is called a **recursive** equation, meaning that the result of one calculation is used in *future* calculations. (The term "recursive" also has other meanings, especially in computer science). Chapter 19 discusses a variety of recursive filters in more detail. Be aware that the moving average recursive filter is very different from typical recursive filters. In particular, most recursive filters have an infinitely long impulse response (IIR), composed of sinusoids and exponentials. The impulse response of the moving average is a rectangular pulse (finite impulse response, or FIR).

This algorithm is faster than other digital filters for several reasons. First, there are only two computations per point, regardless of the length of the filter kernel. Second, addition and subtraction are the only math operations needed, while most digital filters require time-consuming multiplication. Third, the indexing scheme is very simple. Each index in Eq. 15-3 is found by adding or subtracting integer constants that can be calculated before the filtering starts (i.e., p and q). Forth, the entire algorithm can be carried out with integer representation. Depending on the hardware used, integers can be more than an order of magnitude faster than floating point.

Surprisingly, integer representation works *better* than floating point with this algorithm, in addition to being *faster*. The round-off error from floating point arithmetic can produce unexpected results if you are not careful. For example, imagine a 10,000 sample signal being filtered with this method. The last sample in the filtered signal contains the accumulated error of 10,000 additions and 10,000 subtractions. This appears in the output signal as a drifting offset. Integers don't have this problem because there is no round-off error in the arithmetic. If you must use floating point with this algorithm, the program in Table 15-2 shows how to use a double precision accumulator to eliminate this drift.

```
100 'MOVING AVERAGE FILTER IMPLEMENTED BY RECURSION
110 'This program filters 5000 samples with a 101 point moving
120 'average filter, resulting in 4900 samples of filtered data.
130 'A double precision accumulator is used to prevent round-off drift.
140 '
150 DIM X[4999]                 'X[ ] holds the input signal
160 DIM Y[4999]                 'Y[ ] holds the output signal
170 DEFDBL  ACC                 'Define the variable ACC to be double precision
180 '
190 GOSUB XXXX                  'Mythical subroutine to load X[ ]
200 '
210 ACC = 0                     'Find Y[50] by averaging points X[0] to X[100]
220 FOR I% = 0 TO 100
230   ACC = ACC + X[I%]
240 NEXT I%
250 Y[[50] = ACC/101
260 '                           'Recursive moving average filter (Eq. 15-3)
270 FOR I% = 51 TO 4949
280   ACC = ACC + X[I%+50] - X[I%-51]
290   Y[I%] = ACC
300 NEXT I%
310 '
320 END
```

TABLE 15-2

Windowed-Sinc Filters

Windowed-sinc filters are used to separate one band of frequencies from another. They are very stable, produce few surprises, and can be pushed to incredible performance levels. These exceptional frequency domain characteristics are obtained at the expense of poor performance in the time domain, including excessive ripple and overshoot in the step response. When carried out by standard convolution, windowed-sinc filters are easy to program, but slow to execute. Chapter 18 shows how the FFT can be used to dramatically improve the computational speed of these filters.

Strategy of the Windowed-Sinc

Figure 16-1 illustrates the idea behind the windowed-sinc filter. In (a), the frequency response of the *ideal* low-pass filter is shown. All frequencies below the cutoff frequency, f_C, are passed with unity amplitude, while all higher frequencies are blocked. The passband is perfectly flat, the attenuation in the stopband is infinite, and the transition between the two is infinitesimally small.

Taking the Inverse Fourier Transform of this ideal frequency response produces the ideal filter kernel (impulse response) shown in (b). As previously discussed (see Chapter 11, Eq. 11-4), this curve is of the general form: $\sin(x)/x$, called the **sinc function**, given by:

$$h[i] = \frac{\sin(2\pi f_c i)}{i\pi}$$

Convolving an input signal with this filter kernel provides a *perfect* low-pass filter. The problem is, the sinc function continues to both negative and positive infinity without dropping to zero amplitude. While this infinite length is not a problem for *mathematics*, it is a show stopper for *computers*.

To get around this problem, we will make two modifications to the sinc function in (b), resulting in the waveform shown in (c). First, it is truncated to $M+1$ points, symmetrically chosen around the main lobe, where M is an even number. All samples outside these $M+1$ points are set to zero, or simply ignored. Second, the entire sequence is shifted to the right so that it runs from 0 to M. This allows the filter kernel to be represented using only *positive* indexes. While many programming languages allow *negative* indexes, they are a nuisance to use. The sole effect of this $M/2$ shift in the filter kernel is to shift the output signal by the same amount.

Since the modified filter kernel is only an approximation to the ideal filter kernel, it will not have an ideal frequency response. To find the frequency response that is obtained, the Fourier transform can be taken of the signal in (c), resulting in the curve in (d). It's a mess! There is excessive ripple in the passband and poor attenuation in the stopband (recall the Gibbs effect discussed in Chapter 11). These problems result from the abrupt discontinuity at the ends of the truncated sinc function. Increasing the length of the filter kernel does not reduce these problems; the discontinuity is significant no matter how long M is made.

Fortunately, there is a simple method of improving this situation. Figure (e) shows a smoothly tapered curve called a **Blackman window**. Multiplying the truncated-sinc, (c), by the Blackman window, (e), results in the **windowed-sinc** filter kernel shown in (f). The idea is to reduce the abruptness of the truncated ends and thereby improve the frequency response. Figure (g) shows this improvement. The passband is now flat, and the stopband attenuation is so good it cannot be seen in this graph.

Several different windows are available, most of them named after their original developers in the 1950s. Only two are worth using, the **Hamming window** and the **Blackman window** These are given by:

EQUATION 16-1
The Hamming window. These windows run from $i = 0$ to M, for a total of $M+1$ points.

$$w[i] = 0.54 - 0.46 \cos(2\pi i/M)$$

EQUATION 16-2
The Blackman window.

$$w[i] = 0.42 - 0.5 \cos(2\pi i/M) + 0.08 \cos(4\pi i/M)$$

Figure 16-2a shows the shape of these two windows for $M = 50$ (i.e., 51 total points in the curves). Which of these two windows should you use? It's a trade-off between parameters. As shown in Fig. 16-2b, the Hamming window has about a 20% faster *roll-off* than the Blackman. However,

FIGURE 16-1 (facing page)
Derivation of the windowed-sinc filter kernel. The frequency response of the ideal low-pass filter is shown in (a), with the corresponding filter kernel in (b), a sinc function. Since the sinc is infinitely long, it must be truncated to be used in a computer, as shown in (c). However, this truncation results in undesirable changes in the frequency response, (d). The solution is to multiply the truncated-sinc with a smooth window, (e), resulting in the windowed-sinc filter kernel, (f). The frequency response of the windowed-sinc, (g), is smooth and well behaved. These figures are not to scale.

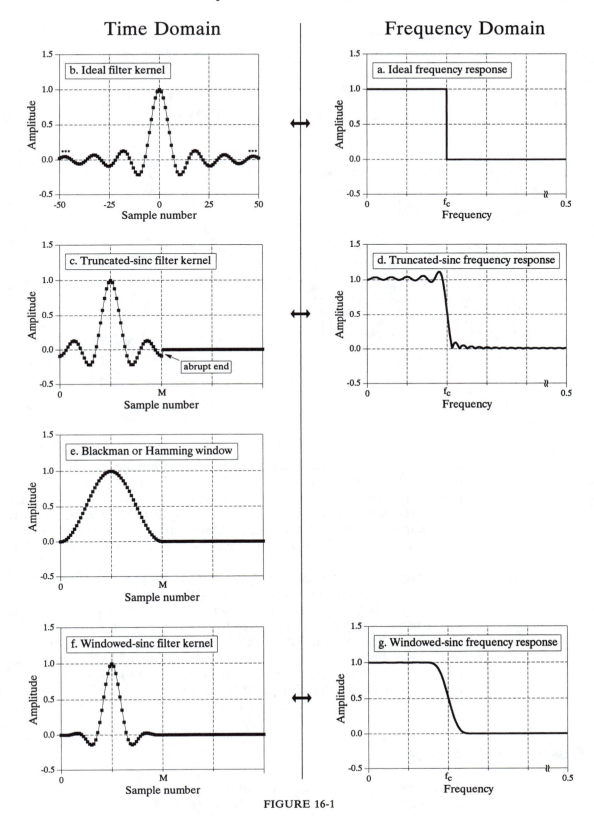

FIGURE 16-1

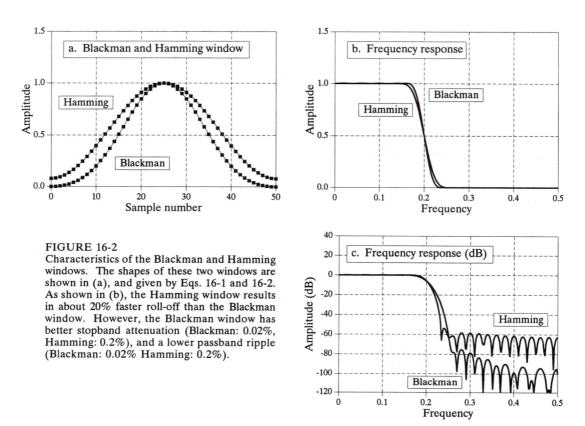

FIGURE 16-2
Characteristics of the Blackman and Hamming
windows. The shapes of these two windows are
shown in (a), and given by Eqs. 16-1 and 16-2.
As shown in (b), the Hamming window results
in about 20% faster roll-off than the Blackman
window. However, the Blackman window has
better stopband attenuation (Blackman: 0.02%,
Hamming: 0.2%), and a lower passband ripple
(Blackman: 0.02% Hamming: 0.2%).

(c) shows that the Blackman has a better *stopband attenuation*. To be exact,
the stopband attenuation for the Blackman is -74dB (~0.02%), while the
Hamming is only -53dB (~0.2%). Although it cannot be seen in these
graphs, the Blackman has a *passband ripple* of only about 0.02%, while the
Hamming is typically 0.2%. In general, the Blackman should be your first
choice; a slow roll-off is easier to handle than poor stopband attenuation.

There are other windows you might hear about, although they fall short of
the Blackman and Hamming. The **Bartlett window** is a triangle, using
straight lines for the taper. The **Hanning window,** also called the **raised
cosine window**, is given by: $w[i] = 0.5 - 0.5\cos(2\pi i / M)$. These two windows
have about the same roll-off speed as the Hamming, but worse stopband
attenuation (Bartlett: -25dB or 5.6%, Hanning -44dB or 0.63%). You might
also hear of a **rectangular window**. This is the same as *no* window, just a
truncation of the tails (such as in 16-1c). While the roll-off is ~2.5 times
faster than the Blackman, the stopband attenuation is only -21dB (8.9%).

Designing the Filter

To design a windowed-sinc, two parameters must be selected: the cutoff
frequency, f_C, and the length of the filter kernel, M. The cutoff frequency

is expressed as a fraction of the sampling rate, and therefore must be between 0 and 0.5. The value for M sets the *roll-off* according to the approximation:

EQUATION 16-3
Filter length vs. roll-off. The length of the filter kernel, M, determines the transition bandwidth of the filter, BW. This is only an approximation since roll-off depends on the particular window being used.

$$M \approx \frac{4}{BW}$$

where BW is the width of the transition band, measured from where the curve just barely leaves one, to where it almost reaches zero (say, 99% to 1% of the curve). The transition bandwidth is also expressed as a fraction of the sampling frequency, and must between 0 and 0.5. Figure 16-3a shows an example of how this approximation is used. The three curves shown are generated from filter kernels with: M = 20, 40, and 200. From Eq. 16-3, the transition bandwidths are: BW = 0.2, 0.1, and 0.02, respectively. Figure (b) shows that the shape of the frequency response does not depend on the cutoff frequency selected.

Since the time required for a convolution is proportional to the length of the signals, Eq. 16-3 expresses a trade-off between *computation time* (the value of M) and *filter sharpness* (the value of BW). For instance, the 20% slower roll-off of the Blackman window (as compared with the Hamming) can be compensated for by using a filter kernel 20% longer. In other words, it could be said that the Blackman window is 20% slower to execute that an equivalent roll-off Hamming window. This is important because the execution speed of windowed-sinc filters is already terribly slow.

As also shown in Fig. 16-3b, the cutoff frequency of the windowed-sinc filter is measured at the *one-half amplitude* point. Why use 0.5 instead of the

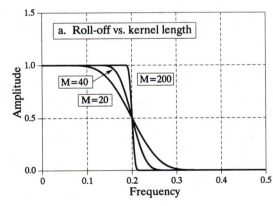

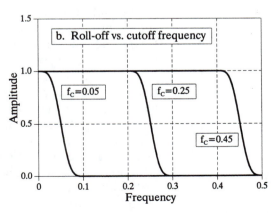

FIGURE 16-3
Filter length vs. roll-off of the windowed-sinc filter. As shown in (a), for M = 20, 40, and 200, the transition bandwidths are BW = 0.2, 0.1, and 0.02 of the sampling rate, respectively. As shown in (b), the shape of the frequency response does not change with different cutoff frequencies. In (b), M = 60.

standard 0.707 (-3dB) used in analog electronics and other digital filters? This is because the windowed-sinc's frequency response is *symmetrical* between the passband and the stopband. For instance, the Hamming window results in a passband ripple of 0.2%, and an *identical* stopband attenuation (i.e., ripple in the stopband) of 0.2%. Other filters do not show this symmetry, and therefore have no advantage in using the one-half amplitude point to mark the cutoff frequency. As shown later in this chapter, this symmetry makes the windowed-sinc ideal for *spectral inversion*.

After f_C and M have been selected, the filter kernel is calculated from the relation:

$$h[i] = K \frac{\sin(2\pi f_c(i-M/2))}{i-M/2} \left[0.42 - 0.5\cos\left(\frac{2\pi i}{M}\right) + 0.08\cos\left(\frac{4\pi i}{M}\right) \right]$$

EQUATION 16-4
The windowed-sinc filter kernel. The cutoff frequency, f_C, is expressed as a fraction of the sampling rate, a value between 0 and 0.5. The length of the filter kernel is determined by M, which must be an even integer. The sample number i, is an integer that runs from 0 to M, resulting in $M+1$ total points in the filter kernel. The constant, K, is chosen to provide unity gain at zero frequency. To avoid a divide-by-zero error, for $i = M/2$, use $h[i] = 2\pi f_c K$.

Don't be intimidated by this equation! Based on the previous discussion, you should be able to identify three components: the *sinc function*, the *M/2 shift*, and the *Blackman window*. For the filter to have unity gain at DC, the constant K must be chosen such that the sum of all the samples is equal to one. In practice, ignore K during the calculation of the filter kernel, and then *normalize* all of the samples as needed. The program listed in Table 16-1 shows how this is done. Also notice how the calculation is handled at the center of the sinc, $i = M/2$, which involves a division by zero.

This equation may be long, but it is easy to use; simply type it into your computer program and forget it. Let the computer handle the calculations. If you find yourself trying to evaluate this equation by hand, you are doing something very very wrong.

Let's be specific about where the filter kernel described by Eq. 16-4 is located in your computer array. As an example, M will be chosen to be 100. Remember, M must be an even number. The first point in the filter kernel is in array location 0, while the last point is in array location 100. This means that the entire signal is 101 points long. The center of symmetry is at point 50, i.e., $M/2$. The 50 points to the left of point 50 are symmetrical with the 50 points to the right. Point 0 is the same value as point 100, and point 49 is the same as point 51. If you must have a specific number of samples in the filter kernel, such as to use the FFT, simply add zeros to one end or the other. For example, with $M = 100$, you could make samples 101 through 127 equal to zero, resulting in a filter kernel 128 points long.

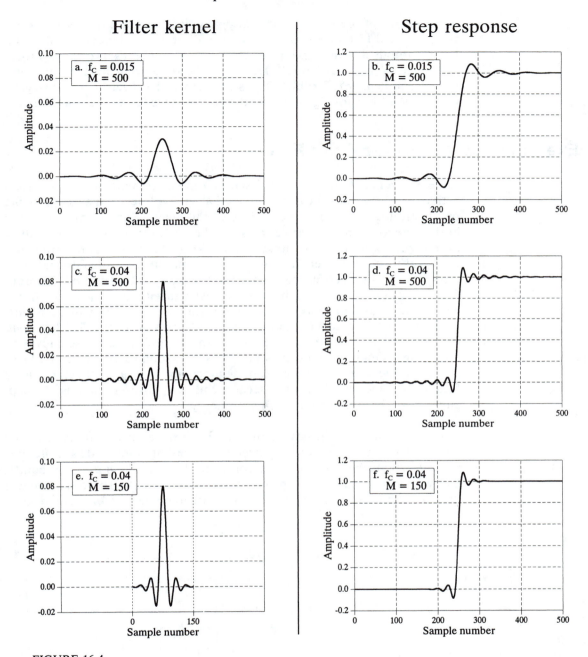

FIGURE 16-4
Example filter kernels and the corresponding step responses. The frequency of the sinusoidal oscillation is approximately equal to the cutoff frequency, f_C, while M determines the kernel length.

Figure 16-4 shows examples of windowed-sinc filter kernels, and their corresponding step responses. The samples at the beginning and end of the filter kernels are so small that they can't even be seen in the graphs. Don't make the mistake of thinking they are unimportant! These samples may be small in value; however, they collectively have a large effect on the performance of the filter. This is also why floating point representation is typically used to implement windowed-sinc filters. Integers usually don't have enough dynamic range to capture the large variation of values contained in the filter kernel. How does the windowed-sinc filter perform in the time domain? Terrible! The step response has overshoot and ringing; this is *not* a filter for signals with information encoded in the time domain.

Examples of Windowed-Sinc Filters

An electroencephalogram, or EEG, is a measurement of the electrical activity of the brain. It can be detected as millivolt level signals appearing on electrodes attached to the surface of the head. Each nerve cell in the brain generates small electrical pulses. The EEG is the combined result of an enormous number of these electrical pulses being generated in a (hopefully) coordinated manner. Although the relationship between thought and this electrical coordination is very poorly understood, different frequencies in the EEG can be identified with specific mental states. If you close your eyes and relax, the predominant EEG pattern will be a slow oscillation between about 7 and 12 hertz. This waveform is called the *alpha rhythm*, and is associated with contentment and a decreased level of attention. Opening your eyes and looking around causes the EEG to change to the *beta rhythm*, occurring between about 17 and 20 hertz. Other frequencies and waveforms are seen in children, different depths of sleep, and various brain disorders such as epilepsy.

In this example, we will assume that the EEG signal has been amplified by analog electronics, and then digitized at a sampling rate of 100 samples per second. Acquiring data for 50 seconds produces a signal of 5,000 points. Our goal is to separate the alpha from the beta rhythms. To do this, we will design a digital low-pass filter with a cutoff frequency of 14 hertz, or 0.14

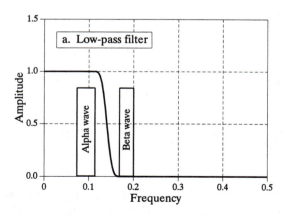
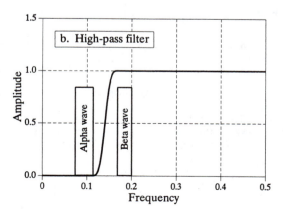

FIGURE 16-5
Example of windowed-sinc filters. The alpha and beta rhythms in an EEG are separated by low-pass and high-pass filters with $M = 100$. The program to implement the low-pass filter is shown in Table 16-1. The program for the high-pass filter is identical, except for a *spectral inversion* of the low-pass filter kernel.

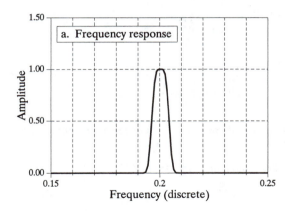 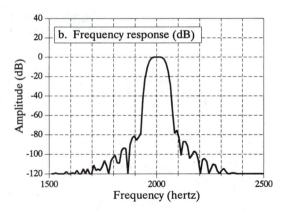

FIGURE 16-6

Example of a windowed-sinc band-pass filter. This filter was designed for a sampling rate of 10 kHz. When referenced to the analog signal, the center frequency of the passband is at 2 kHz, the passband is 80 hertz, and the transition bands are 50 hertz. The windowed-sinc uses 801 points in the filter kernel to achieve this roll-off, and a Blackman window for good stopband attenuation. Figure (a) shows the resulting frequency response on a linear scale, while (b) shows it in decibels. The frequency axis in (a) is expressed as a fraction of the sampling frequency, while (b) is expressed in terms of the analog signal before digitization.

of the sampling rate. The transition bandwidth will be set at 4 hertz, or 0.04 of the sampling rate. From Eq. 16-3, the filter kernel needs to be about 101 points long, and we will arbitrarily choose to use a Hamming window. The program in Table 16-1 shows how the filter is carried out. The frequency response of the filter, obtained by taking the Fourier Transform of the filter kernel, is shown in Fig. 16-5.

In a second example, we will design a *band-pass filter* to isolate a *signaling tone* in an audio signal, such as when a button on a telephone is pressed. We will assume that the signal has been digitized at 10 kHz, and the goal is to isolate an 80 hertz band of frequencies centered on 2 kHz. In terms of the sampling rate, we want to block all frequencies below 0.196 and above 0.204 (corresponding to 1960 hertz and 2040 hertz, respectively). To achieve a transition bandwidth of 50 hertz (0.005 of the sampling rate), we will make the filter kernel 801 points long, and use a Blackman window. Table 16-2 contains a program for calculating the filter kernel, while Fig. 16-6 shows the frequency response. The design involves several steps. First, *two* low-pass filters are designed, one with a cutoff at 0.196, and the other with a cutoff at 0.204. This second filter is then *spectrally inverted*, making it a high-pass filter (see Chapter 14, Fig. 14-6). Next, the two filter kernels are added, resulting in a band-reject filter (see Fig. 14-8). Finally, another *spectral inversion* makes this into the desired band-pass filter.

Pushing it to the Limit

The windowed-sinc filter can be pushed to incredible performance levels without nasty surprises. For instance, suppose you need to isolate a 1 *millivolt* signal riding on a 120 *volt* power line. The low-pass filter will need

```
100 'LOW-PASS WINDOWED-SINC FILTER
110 'This program filters 5000 samples with a 101 point windowed-sinc filter,
120 'resulting in 4900 samples of filtered data.
130 '
140 DIM X[4999]                    'X[ ] holds the input signal
150 DIM Y[4999]                    'Y[ ] holds the output signal
160 DIM H[100]                     'H[ ] holds the filter kernel
170 '
180 PI = 3.14159265
190 FC = .14                       'Set the cutoff frequency (between 0 and 0.5)
200 M% = 100                       'Set filter length (101 points)
210 '
220 GOSUB XXXX                     'Mythical subroutine to load X[ ]
230 '
240 '                             'Calculate the low-pass filter kernel via Eq. 16-4
250 FOR I% = 0 TO 100
260    IF (I%-M%/2) = 0 THEN H[I%] = 2*PI*FC
270    IF (I%-M%/2) <> 0  THEN H[I%] = SIN(2*PI*FC * (I%-M%/2)) / (I%-M%/2)
280    H[I%] = H[I%] * (0.54 - 0.46*COS(2*PI*I%/M%) )
290 NEXT I%
300 '
310 SUM = 0                        'Normalize the low-pass filter kernel for
320 FOR I% = 0 TO 100              'unity gain at DC
330    SUM = SUM + H[I%]
340 NEXT I%
350 '
360 FOR I% = 0 TO 100
370    H[I%] = H[I%] / SUM
380 NEXT I%
390 '
400 FOR J% = 100 TO 4999          'Convolve the input signal & filter kernel
410    Y[J%] = 0
420    FOR I% = 0 TO 100
430       Y[J%] = Y[J%] + X[J%-I%] * H[I%]
440    NEXT I%
450 NEXT J%
460 '
470 END
```

TABLE 16-1

a stopband attenuation of at least -120dB (one part in one-million for those that refuse to learn decibels). As previously shown, the Blackman window only provides -74dB (one part in five-thousand). Fortunately, greater stopband attenuation is easy to obtain. The input signal can be filtered using a conventional windowed-sinc filter kernel, providing an intermediate signal. The intermediate signal can then be passed through the filter a second time, further increasing the stopband attenuation to -148dB (1 part in 30 million, wow!). It is also possible to combine the two stages into a single filter. The kernel of the combined filter is equal to the *convolution* of the filter kernels of the two stages. This also means that convolving any filter kernel *with itself* results in a filter kernel with a much improved stopband attenuation. The price you pay is a longer filter kernel and a slower roll-off. Figure 16-7a shows the frequency response of a 201 point low-pass filter, formed by convolving a 101 point Blackman windowed-sinc with itself. Amazing performance! (If you really need more than -100dB of stopband attenuation, you should use double precision. Single precision

```
100 'BAND-PASS WINDOWED-SINC FILTER
110 'This program calculates an 801 point band-pass filter kernel
120 '
130 DIM A[800]                      'A[ ] workspace for the lower cutoff
140 DIM B[800]                      'B[ ] workspace for the upper cutoff
150 DIM H[800]                      'H[ ] holds the final filter kernel
160 '
170 PI = 3.1415926
180 M% = 800                        'Set filter kernel length (801 points)
190 '
200 '                               'Calculate the first low-pass filter kernel via Eq. 16-4,
210 FC = 0.196                      'with a cutoff frequency of 0.196, store in A[ ]
220 FOR I% = 0 TO 800
230    IF (I%-M%/2) = 0 THEN A[I%] = 2*PI*FC
240    IF (I%-M%/2) <> 0 THEN A[I%] = SIN(2*PI*FC * (I%-M%/2)) / (I%-M%/2)
250    A[I%] = A[I%] * (0.42 - 0.5*COS(2*PI*I%/M%) + 0.08*COS(4*PI*I%/M%))
260 NEXT I%
270 '
280 SUM = 0                         'Normalize the first low-pass filter kernel for
290 FOR I% = 0 TO 800               'unity gain at DC
300    SUM = SUM + A[I%]
310 NEXT I%
320 '
330 FOR I% = 0 TO 800
340    A[I%] = A[I%] / SUM
350 NEXT I%
360 '                               'Calculate the second low-pass filter kernel via Eq. 16-4,
370 FC = 0.204                      'with a cutoff frequency of 0.204, store in B[ ]
380 FOR I% = 0 TO 800
390    IF (I%-M%/2) = 0 THEN B[I%] = 2*PI*FC
400    IF (I%-M%/2) <> 0 THEN B[I%] = SIN(2*PI*FC * (I%-M%/2)) / (I%-M%/2)
410    B[I%] = B[I%] * (0.42 - 0.5*COS(2*PI*I%/M%) + 0.08*COS(4*PI*I%/M%))
420 NEXT I%
430 '
440 SUM = 0                         'Normalize the second low-pass filter kernel for
450 FOR I% = 0 TO 800               'unity gain at DC
460    SUM = SUM + B[I%]
470 NEXT I%
480 '
490 FOR I% = 0 TO 800
500    B[I%] = B[I%] / SUM
510 NEXT I%
520 '
530 FOR I% = 0 TO 800               'Change the low-pass filter kernel in B[ ] into a high-pass
540    B[I%] = - B[I%]              'filter kernel using spectral inversion (as in Fig. 14-5)
550 NEXT I%
560 B[400] =  B[400] + 1
570 '
580 '
590 FOR I% = 0 TO 800               'Add the low-pass filter kernel in A[ ], to the high-pass
600    H[I%] = A[I%] + B[I%]        'filter kernel in B[ ], to form a band-reject filter kernel
610 NEXT I%                         'stored in H[ ] (as in Fig. 14-8)
620 '
630 FOR I% = 0 TO 800               'Change the band-reject filter kernel into a band-pass
640    H[I%] = -H[I%]               'filter kernel by using spectral inversion
650 NEXT I%
660 H[400] =  H[400] + 1
670 '                               'The band-pass filter kernel now resides in H[ ]
680 END
```

TABLE 16-2

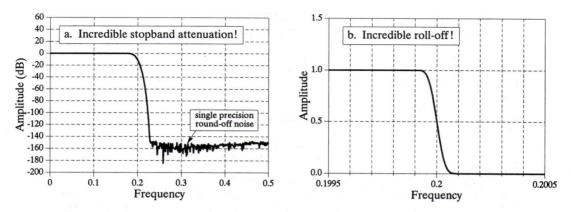

FIGURE 16-7
The incredible performance of the windowed-sinc filter. Figure (a) shows the frequency response of a windowed-sinc filter with increased stopband attenuation. This is achieved by convolving a windowed-sinc filter kernel with itself. Figure (b) shows the very rapid roll-off a 32,001 point windowed-sinc filter.

round-off noise on signals in the *passband* can erratically appear in the *stopband* with amplitudes in the -100dB to -120dB range).

Figure 16-7b shows another example of the windowed-sinc's incredible performance: a low-pass filter with 32,001 points in the kernel. The frequency response appears as expected, with a roll-off of 0.000125 of the sampling rate. How good is this filter? Try building an analog electronic filter that passes signals from DC to 1000 hertz with less than a 0.02% variation, and blocks all frequencies above 1001 hertz with less than 0.02% residue. Now that's a filter! If you really want to be impressed, remember that both the filters in Fig. 16-7 use *single precision*. Using *double precision* allows these performance levels to be extended by a *million* times.

The strongest limitation of the windowed-sinc filter is the *execution time*; it can be unacceptably long if there are many points in the filter kernel and standard convolution is used. A high-speed algorithm for this filter (FFT convolution) is presented in Chapter 18. Recursive filters (Chapter 19) also provide good frequency separation and are a reasonable alternative to the windowed-sinc filter.

Is the windowed-sinc the optimal filter kernel for separating frequencies? No, filter kernels resulting from more sophisticated techniques can be better. But beware! Before you jump into this very mathematical field, you should consider exactly what you hope to gain. The windowed-sinc will provide *any* level of performance that you could possibly need. What the advanced filter design methods may provide is a slightly shorter filter kernel for a given level of performance. This, in turn, may mean a slightly faster execution speed. Be warned that you may get little return for the effort expended.

CHAPTER 17

Custom Filters

Most filters have one of the four standard frequency responses: low-pass, high-pass, band-pass or band-reject. This chapter presents a general method of designing digital filters with an *arbitrary* frequency response, tailored to the needs of your particular application. DSP excels in this area, solving problems that are far above the capabilities of analog electronics. Two important uses of custom filters are discussed in this chapter: *deconvolution*, a way of restoring signals that have undergone an unwanted convolution, and *optimal filtering*, the problem of separating signals with overlapping frequency spectra. This is DSP at its best.

Arbitrary Frequency Response

The approach used to derive the windowed-sinc filter in the last chapter can also be used to design filters with virtually *any* frequency response. The only difference is how the desired response is moved from the frequency domain into the time domain. In the windowed-sinc filter, the frequency response and the filter kernel are both represented by *equations*, and the conversion between them is made by evaluating the *mathematics* of the Fourier transform. In the method presented here, both signals are represented by *arrays of numbers*, with a *computer program* (the FFT) being used to find one from the other.

Figure 17-1 shows an example of how this works. The frequency response we want the filter to produce is shown in (a). To say the least, it is very irregular and would be virtually impossible to obtain with analog electronics. This ideal frequency response is *defined* by an array of numbers that have been selected, not some mathematical equation. In this example, there are 513 samples spread between 0 and 0.5 of the sampling rate. More points could be used to better represent the desired frequency response, while a smaller number may be needed to reduce the computation time during the filter design. However, these concerns are usually small, and 513 is a good length for most applications.

Besides the desired *magnitude* array shown in (a), there must be a corresponding *phase* array of the same length. In this example, the phase of the desired frequency response is entirely *zero* (this array is not shown in Fig. 17-1). Just as with the magnitude array, the phase array can be loaded with any arbitrary curve you would like the filter to produce. However, remember that the first and last samples (i.e., 0 and 512) of the phase array must have a value of *zero* (or a multiple of 2π, which is the same thing). The frequency response can also be specified in rectangular form by defining the array entries for the *real* and *imaginary parts*, instead of using the magnitude and phase.

The next step is to take the Inverse DFT to move the filter into the time domain. The quickest way to do this is to convert the frequency domain to rectangular form, and then use the Inverse FFT. This results in a 1024 sample signal running from 0 to 1023, as shown in (b). This is the impulse response that corresponds to the frequency response we want; however, it is not suitable for use as a filter kernel (more about this shortly). Just as in the last chapter, it needs to be *shifted*, *truncated*, and *windowed*. In this example, we will design the filter kernel with $M = 40$, i.e., 41 points running from sample 0 to sample 40. Table 17-1 shows a computer program that converts the signal in (b) into the filter kernel shown in (c). As with the windowed-sinc filter, the points near the ends of the filter kernel are so small that they appear to be zero when plotted. Don't make the mistake of thinking they can be deleted!

```
100 'CUSTOM FILTER DESIGN
110 'This program converts an aliased 1024 point impulse response into an M+1 point
120 'filter kernel (such as Fig. 17-1b being converted into Fig. 17-1c)
130 '
140 DIM REX[1023]               'REX[ ] holds the signal being converted
150 DIM T[1023]                 'T[ ] is a temporary storage buffer
160 '
170 PI = 3.14159265
180 M% = 40                     'Set filter kernel length (41 total points)
190 '
200 GOSUB XXXX                  'Mythical subroutine to load REX[ ] with impulse response
210 '
220 FOR I% = 0 TO 1023          'Shift (rotate) the signal M/2 points to the right
230    INDEX% = I% + M%/2
240    IF INDEX% > 1023 THEN INDEX% = INDEX%-1024
250    T[INDEX%] = REX[I%]
260 NEXT I%
270 '
280 FOR I% = 0 TO 1023
290    REX[I%] = T[I%]
300 NEXT I%
310 '                           'Truncate and window the signal
320 FOR I% = 0 TO 1023
330    IF I% <= M% THEN REX[I%] = REX[I%] * (0.54 - 0.46 * COS(2*PI*I%/M%))
340    IF I% > M%  THEN REX[I%] = 0
350 NEXT I%
360 '                           'The filter kernel now resides in REX[0] to REX[40]
370 END
```

TABLE 17-1

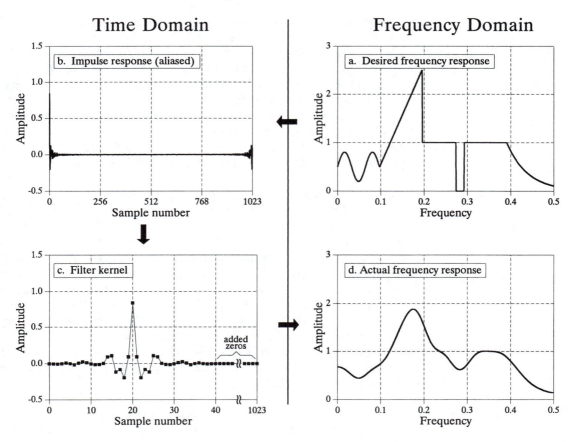

FIGURE 17-1

Example of FIR filter design. Figure (a) shows the desired frequency response, with 513 samples running between 0 to 0.5 of the sampling rate. Taking the Inverse DFT results in (b), an *aliased* impulse response composed of 1024 samples. To form the filter kernel, (c), the aliased impulse response is truncated to *M*+1 samples, shifted to the right by *M*/2 samples, and multiplied by a Hamming or Blackman window. In this example, *M* is 40. The program in Table 17-1 shows how this is done. The filter kernel is tested by padding it with zeros and taking the DFT, providing the actual frequency response of the filter, (d).

The last step is to *test* the filter kernel. This is done by taking the DFT (using the FFT) to find the actual frequency response, as shown in (d). To obtain better resolution in the frequency domain, pad the filter kernel with zeros before the FFT. For instance, using 1024 total samples (41 in the filter kernel, plus 983 zeros), results in 513 samples between 0 and 0.5.

As shown in Fig. 17-2, the length of the filter kernel determines how well the *actual* frequency response matches the *desired* frequency response. The exceptional performance of FIR digital filters is apparent; virtually any frequency response can be obtained if a long enough filter kernel is used.

This is the entire design method; however, there is a subtle *theoretical* issue that needs to be clarified. Why isn't it possible to directly use the impulse response shown in 17-1b as the filter kernel? After all, if (a) is the Fourier transform of (b), wouldn't convolving an input signal with (b) produce the *exact* frequency response we want? The answer is *no*, and here's why.

When designing a custom filter, the desired frequency response is defined by the values in an array. Now consider this: what does the frequency response do *between* the specified points? For simplicity, two cases can be imagined, one "good" and one "bad." In the "good" case, the frequency response is a smooth curve between the defined samples. In the "bad" case, there are wild fluctuations between. As luck would have it, the impulse response in (b) corresponds to the "bad" frequency response. This can be shown by padding it with a large number of zeros, and then taking the DFT. The frequency response obtained by this method will show the erratic behavior between the originally defined samples, and look just awful.

To understand this, imagine that we force the frequency response to be what we want by defining it at an infinite number of points between 0 and 0.5. That is, we create a continuous curve. The inverse DTFT is then used to find the impulse response, which will be *infinite* in length. In other words, the "good" frequency response corresponds to something that cannot be represented in a computer, an infinitely long impulse response. When we represent the frequency spectrum with $N/2 + 1$ samples, only N points are provided in the time domain, making it unable to correctly contain the signal. The result is that the infinitely long impulse response wraps up (aliases) into the N points. When this aliasing occurs, the frequency response changes from "good" to "bad." Fortunately, windowing the N point impulse response greatly reduces this aliasing, providing a smooth curve between the frequency domain samples.

Designing a digital filter to produce a given frequency response is quite simple. The hard part is finding what frequency response to use. Let's look at some strategies used in DSP to design custom filters.

Deconvolution

Unwanted convolution is an inherent problem in transferring analog information. For instance, all of the following can be modeled as a convolution: image blurring in a shaky camera, echoes in long distance telephone calls, the finite bandwidth of analog sensors and electronics, etc. Deconvolution is the process of filtering a signal to compensate for an undesired convolution. The goal of deconvolution is to recreate the signal as it existed *before* the convolution took place. This usually requires the characteristics of the convolution (i.e., the impulse or frequency response) to be known. This can be distinguished from **blind deconvolution**, where the characteristics of the parasitic convolution are *not* known. Blind deconvolution is a much more difficult problem that has no general solution, and the approach must be tailored to the particular application.

Deconvolution is nearly impossible to understand in the *time domain*, but quite straightforward in the *frequency domain*. Each sinusoid that composes the original signal can be changed in amplitude and/or phase as it passes through the undesired convolution. To extract the original signal, the deconvolution filter must *undo* these amplitude and phase changes. For

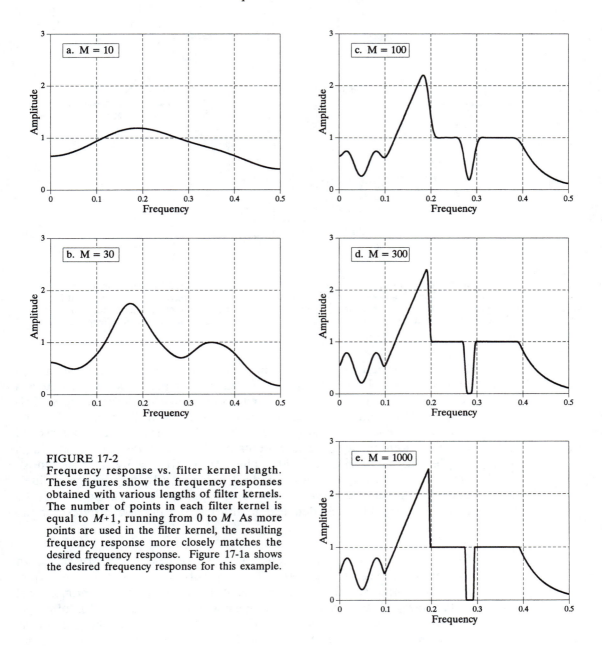

FIGURE 17-2
Frequency response vs. filter kernel length. These figures show the frequency responses obtained with various lengths of filter kernels. The number of points in each filter kernel is equal to $M+1$, running from 0 to M. As more points are used in the filter kernel, the resulting frequency response more closely matches the desired frequency response. Figure 17-1a shows the desired frequency response for this example.

example, if the convolution changes a sinusoid's amplitude by 0.5 with a 30 degree phase shift, the deconvolution filter must amplify the sinusoid by 2.0 with a -30 degree phase change.

The example we will use to illustrate deconvolution is a *gamma ray detector*. As illustrated in Fig. 17-3, this device is composed of two parts, a *scintillator* and a *light detector*. A scintillator is a special type of transparent material, such as sodium iodide or bismuth germanate. These compounds change the energy in each gamma ray into a brief burst of visible light. This light

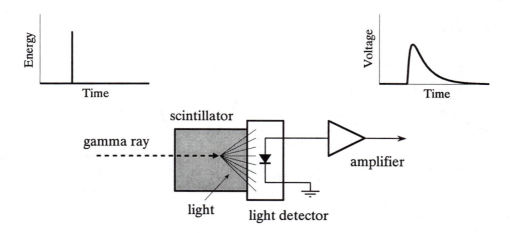

FIGURE 17-3
Example of an unavoidable convolution. A gamma ray detector can be formed by mounting a *scintillator* on a *light detector*. When a gamma ray strikes the scintillator, its energy is converted into a pulse of light. This pulse of light is then converted into an electronic signal by the light detector. The gamma ray is an *impulse*, while the output of the detector (i.e., the *impulse response*) resembles a one-sided exponential.

is then converted into an electronic signal by a light detector, such as a photodiode or photomultiplier tube. Each pulse produced by the detector resembles a *one-sided exponential*, with some rounding of the corners. This shape is determined by the characteristics of the scintillator used. When a gamma ray deposits its energy into the scintillator, nearby atoms are excited to a higher energy level. These atoms randomly *deexcite*, each producing a single photon of visible light. The net result is a light pulse whose amplitude decays over a few hundred nanoseconds (for sodium iodide). Since the arrival of each gamma ray is an *impulse*, the output pulse from the detector (i.e., the one-sided exponential) is the *impulse response* of the system.

Figure 17-4a shows pulses generated by the detector in response to randomly arriving gamma rays. The information we would like to extract from this output signal is the *amplitude* of each pulse, which is proportional to the *energy* of the gamma ray that generated it. This is useful information because the energy can tell interesting things about where the gamma ray has been. For example, it may provide medical information on a patient, tell the age of a distant galaxy, detect a bomb in airline luggage, etc.

Everything would be fine if only an occasional gamma ray were detected, but this is usually not the case. As shown in (a), two or more pulses may overlap, shifting the measured amplitude. One answer to this problem is to *deconvolve* the detector's output signal, making the pulses narrower so that less pile-up occurs. Ideally, we would like each pulse to resemble the original impulse. As you may suspect, this isn't possible and we must settle for a pulse that is finite in length, but significantly shorter than the detected pulse. This goal is illustrated in Fig. 17-4b.

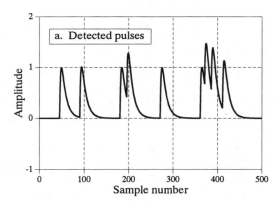
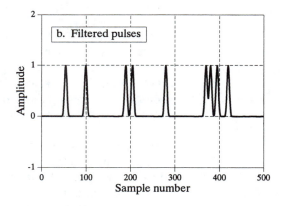

FIGURE 17-4

Example of deconvolution. Figure (a) shows the output signal from a gamma ray detector in response to a series of randomly arriving gamma rays. The deconvolution filter is designed to convert (a) into (b), by reducing the width of the pulses. This minimizes the amplitude shift when pulses land on top of each other.

Even though the detector signal has its information encoded in the *time domain*, much of our analysis must be done in the *frequency domain*, where the problem is easier to understand. Figure 17-5a is the signal produced by the detector (something we know). Figure (c) is the signal we wish to have (also something we know). This desired pulse was arbitrarily selected to be the same shape as a Blackman window, with a length about one-third that of the original pulse. Our goal is to find a filter kernel, (e), that when convolved with the signal in (a), produces the signal in (c). In equation form: if $a * e = c$, and given a and c, find e.

If these signals were combined by addition or multiplication instead of convolution, the solution would be easy: *subtraction* is used to "de-add" and *division* is used to "de-multiply". Convolution is different; there is not a simple inverse operation that can be called "deconvolution." Convolution is too messy to be undone by directly manipulating the time domain signals.

Fortunately, this problem is simpler in the frequency domain. Remember, *convolution* in one domain corresponds with *multiplication* in the other domain. Again referring to the signals in Fig. 17-5: if $b \times f = d$, and given b and d, find f. This is an easy problem to solve: the frequency response of the filter, (f), is the frequency spectrum of the desired pulse, (d), *divided* by the frequency spectrum of the detected pulse, (b). Since the detected pulse is asymmetrical, it will have a *nonzero* phase. This means that a *complex* division must be used (that is, a magnitude & phase divided by another magnitude & phase). In case you have forgotten, Chapter 9 defines how to perform a complex division of one spectrum by another. The required filter kernel, (e), is then found from the frequency response by the custom filter method (IDFT, shift, truncate, & multiply by a window).

There are limits to the improvement that deconvolution can provide. In other words, if you get greedy, things will fall apart. Getting greedy in this example means trying to make the desired pulse excessively narrow. Let's look at what happens. If the desired pulse is made narrower, its frequency spectrum must contain more high frequency components. Since these high frequency components are at a very low amplitude in the detected pulse, the filter must have a very high gain at these frequencies. For instance, (f) shows that some frequencies must be multiplied by a factor of *three* to achieve the desired pulse in (c). If the desired pulse is made narrower, the gain of the deconvolution filter will be even greater at high frequencies.

The problem is, small errors are very unforgiving in this situation. For instance, if some frequency is amplified by 30, when only 28 is required, the deconvolved signal will probably be a mess. When the deconvolution is pushed to greater levels of performance, the characteristics of the unwanted convolution must be understood with greater *accuracy* and *precision*. There are always unknowns in real world applications, caused by such villains as: electronic noise, temperature drift, variation between devices, etc. These unknowns set a limit on how well deconvolution will work.

Even if the unwanted convolution is *perfectly* understood, there is still a factor that limits the performance of deconvolution: *noise*. For instance, most unwanted convolutions take the form of a low-pass filter, reducing the amplitude of the high frequency components in the signal. Deconvolution corrects this by amplifying these frequencies. However, if the amplitude of these components falls below the inherent noise of the system, the information contained in these frequencies is lost. No amount of signal processing can retrieve it. It's gone forever. Adios! Goodbye! Sayonara! Trying to reclaim this data will only amplify the noise. As an extreme case, the amplitude of some frequencies may be completely reduced to *zero*. This not only obliterates the information, it will try to make the deconvolution filter have *infinite* gain at these frequencies. The solution: design a less aggressive deconvolution filter and/or place limits on how much gain is allow at any of the frequencies.

How far can you go? How greedy is too greedy? This depends totally on the problem you are attacking. If the signal is well behaved and has low noise, a significant improvement can probably be made (think a factor of 5-10). If the signal changes over time, isn't especially well understood, or is noisy, you won't do nearly as well (think a factor of 1-2). Successful deconvolution involves a great deal of testing. If it works at some level, try going farther; you will know when it falls apart. No amount of theoretical work will allow you to bypass this iterative process.

Deconvolution can also be applied to *frequency domain* encoded signals. A classic example is the restoration of old recordings of the famous opera singer, Enrico Caruso (1873-1921). These recordings were made with very primitive equipment by modern standards. The most significant problem is the *resonances* of the long tubular recording horn used to gather the sound. Whenever the singer happens to hit one of these resonance frequencies, the loudness of the recording abruptly increases. Digital deconvolution has improved the subjective quality of these recordings by

Time Domain

Frequency Domain

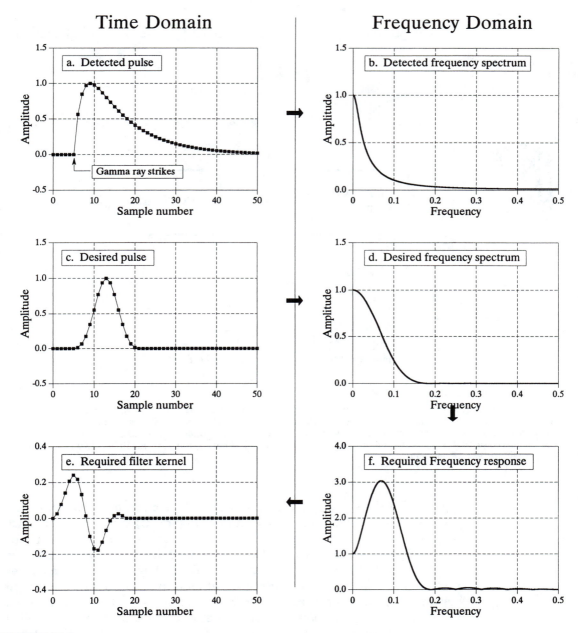

FIGURE 17-5
Example of deconvolution in the time and frequency domains. The impulse response of the example gamma ray detector is shown in (a), while the desired impulse response is shown in (c). The frequency spectra of these two signals are shown in (b) and (d), respectively. The filter that changes (a) into (c) has a frequency response, (f), equal to (b) divided by (d). The filter kernel of this filter, (e), is then found from the frequency response using the custom filter design method (inverse DFT, truncation, windowing). Only the magnitudes of the frequency domain signals are shown in this illustration; however, the phases are nonzero and must also be used.

reducing the loud spots in the music. We will only describe the general method; for a detailed description, see the original paper: T. Stockham, T. Cannon, and R. Ingebretsen, "Blind Deconvolution Through Digital Signal Processing", *Proc. IEEE*, vol. 63, Apr. 1975, pp. 678-692.

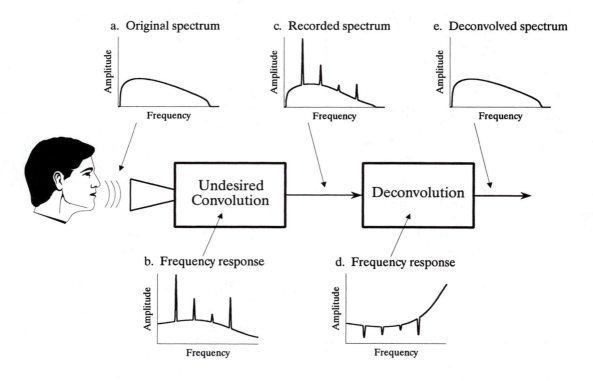

a. Original spectrum

c. Recorded spectrum

e. Deconvolved spectrum

Undesired Convolution

Deconvolution

b. Frequency response

d. Frequency response

FIGURE 17-6

Deconvolution of old phonograph recordings. The frequency spectrum produced by the original singer is illustrated in (a). Resonance peaks in the primitive equipment, (b), produce distortion in the recorded frequency spectrum, (c). The frequency response of the deconvolution filter, (d), is designed to counteracts the undesired convolution, restoring the original spectrum, (e). These graphs are for illustrative purposes only; they are not actual signals.

Figure 17-6 shows the general approach. The frequency spectrum of the original audio signal is illustrated in (a). Figure (b) shows the frequency response of the recording equipment, a relatively smooth curve except for several sharp resonance peaks. The spectrum of the recorded signal, shown in (c), is equal to the true spectrum, (a), multiplied by the uneven frequency response, (b). The goal of the deconvolution is to *counteract* the undesired convolution. In other words, the frequency response of the deconvolution filter, (d), must be the *inverse* of (b). That is, each peak in (b) is cancelled by a corresponding dip in (d). If this filter were perfectly designed, the resulting signal would have a spectrum, (e), identical to that of the original. Here's the catch: the original recording equipment has long been discarded, and its frequency response, (b), is a mystery. In other words, this is a *blind deconvolution* problem; given only (c), how can we determine (d)?

Blind deconvolution problems are usually attacked by making an estimate or assumption about the unknown parameters. To deal with this example, the *average spectrum* of the original music is assumed to match the *average spectrum* of the same music performed by a present day singer using modern equipment. The *average spectrum* is found by the techniques of Chapter 9:

break the signal into a large number of segments, take the DFT of each segment, convert into polar form, and then average the magnitudes together. In the simplest case, the unknown frequency response is taken as the average spectrum of the old recording, divided by the average spectrum of the modern recording. (The method used by Stockham et al. is based on a more sophisticated technique called *homomorphic* processing, providing a better estimate of the characteristics of the recording system).

Optimal Filters

Figure 17-7a illustrates a common filtering problem: trying to extract a waveform (in this example, an exponential pulse) buried in random noise. As shown in (b), this problem is no easier in the frequency domain. The signal has a spectrum composed mainly of low frequency components. In comparison, the spectrum of the noise is *white* (the same amplitude at all frequencies). Since the spectra of the signal and noise *overlap*, it is not clear how the two can best be separated. In fact, the real question is how to define what "best" means. We will look at three filters, each of which is "best" (optimal) in a different way. Figure 17-8 shows the filter kernel and frequency response for each of these filters. Figure 17-9 shows the result of using these filters on the example waveform of Fig. 17-7a.

The **moving average filter** is the topic of Chapter 15. As you recall, each output point produced by the moving average filter is the average of a certain number of points from the input signal. This makes the filter kernel a rectangular pulse with an amplitude equal to the reciprocal of the number of points in the average. The moving average filter is optimal in the sense that it provides the fastest step response for a given noise reduction.

The **matched filter** was previously discussed in Chapter 7. As shown in Fig. 17-8a, the filter kernel of the matched filter is the same as the target signal

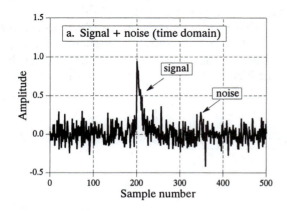
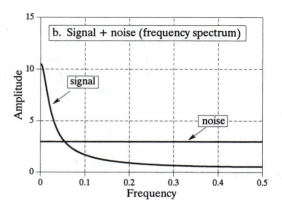

FIGURE 17-7

Example of optimal filtering. In (a), an exponential pulse buried in random noise. The frequency spectra of the pulse and noise are shown in (b). Since the signal and noise overlap in both the time and frequency domains, the best way to separate them isn't obvious.

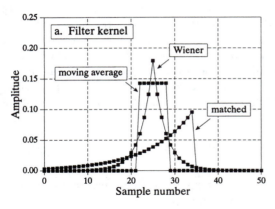 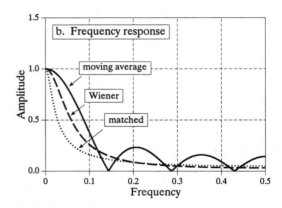

FIGURE 17-8
Example of optimal filters. In (a), three filter kernels are shown, each of which is optimal in some sense. The corresponding frequency responses are shown in (b). The moving average filter is designed to have a rectangular pulse for a filter kernel. In comparison, the filter kernel of the matched filter looks like the signal being detected. The Wiener filter is designed in the frequency domain, based on the relative amounts of signal and noise present at each frequency.

being detected, except it has been flipped left-for-right. The idea behind the matched filter is *correlation*, and this flip is required to perform *correlation* using *convolution*. The amplitude of each point in the output signal is a measure of how well the filter kernel *matches* the corresponding section of the input signal. Recall that the output of a matched filter does not necessarily look like the signal being detected. This doesn't really matter; if a matched filter is used, the shape of the target signal must already be known. The matched filter is optimal in the sense that the top of the peak is farther above the noise than can be achieved with any other linear filter (see Fig. 17-9b).

The **Wiener filter** (named after the optimal estimation theory of Norbert Wiener) separates signals based on their frequency spectra. As shown in Fig. 17-7b, at some frequencies there is mostly signal, while at others there is mostly noise. It seems logical that the "mostly signal" frequencies should be passed through the filter, while the "mostly noise" frequencies should be blocked. The Wiener filter takes this idea a step further; the gain of the filter *at each frequency* is determined by the relative amount of signal and noise *at that frequency*:

EQUATION 17-1
The Wiener filter. The frequency response, represented by $H[f]$, is determined by the frequency spectra of the noise, $N[f]$, and the signal, $S[f]$. Only the magnitudes are important; all of the phases are zero.

$$H[f] = \frac{S[f]^2}{S[f]^2 + N[f]^2}$$

This relation is used to convert the spectra in Fig. 17-7b into the Wiener filter's frequency response in Fig. 17-8b. The Wiener filter is optimal in the sense that it maximizes the ratio of the signal power to the noise power

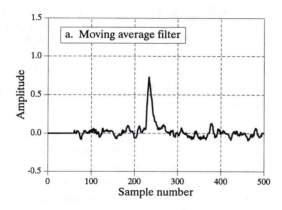

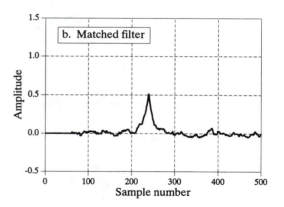

FIGURE 17-9
Example of using three optimal filters. These signals result from filtering the waveform in Fig. 17-7 with the filters in Fig. 17-8. Each of these three filters is optimal *in some sense*. In (a), the moving average filter results in the sharpest edge response for a given level of random noise reduction. In (b), the matched filter produces a peak that is farther above the residue noise than provided by any other filter. In (c), the Wiener filter optimizes the signal-to-noise ratio.

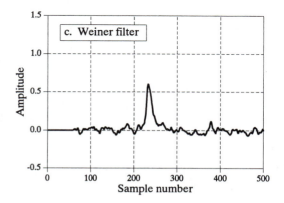

(over the length of the signal, not at each individual point). An appropriate filter kernel is designed from the Wiener frequency response using the custom method.

While the ideas behind these optimal filters are mathematically elegant, they often fail in practicality. This isn't to say they should never be used. The point is, don't hear the word "optimal" and stop thinking. Let's look at several reasons why you might *not* want to use them.

First, the difference between the signals in Fig. 17-9 is very unimpressive. In fact, if you weren't told what parameters were being optimized, you probably couldn't tell by looking at the signals. This is usually the case for problems involving overlapping frequency spectra. The small amount of extra performance obtained from an optimal filter may not be worth the the increased program complexity, the extra design effort, or the longer execution time.

Second: The Wiener and matched filters are completely determined by the characteristics of the problem. Other filters, such as the windowed-sinc and moving average, can be tailored to your liking. Optimal filter advocates would claim that this diddling can only reduce the effectiveness of the filter.

This is very arguable. Remember, each of these filters is optimal in one specific way (i.e., "in some sense"). This is seldom sufficient to claim that the entire problem has been optimized, especially if the resulting signals are interpreted by a human observer. For instance, a biomedical engineer might use a Wiener filter to maximize the signal-to-noise ratio in an electro-cardiogram. However, it is not obvious that this also optimizes a physician's ability to detect irregular heart activity by looking at the signal.

Third: The Wiener and matched filter must be carried out by *convolution*, making them extremely slow to execute. Even with the speed improvements discussed in the next chapter (FFT convolution), the computation time can be excessively long. In comparison, *recursive* filters (such as the moving average or others presented in Chapter 19) are much faster, and may provide an acceptable level of performance.

CHAPTER 18

FFT Convolution

This chapter presents two important DSP techniques, the *overlap-add method*, and *FFT convolution*. The overlap-add method is used to break long signals into smaller segments for easier processing. FFT convolution uses the overlap-add method together with the Fast Fourier Transform, allowing signals to be convolved by multiplying their frequency spectra. For filter kernels longer than about 64 points, FFT convolution is faster than standard convolution, while producing exactly the same result.

The Overlap-Add Method

There are many DSP applications where a long signal must be filtered in *segments*. For instance, high fidelity digital *audio* requires a data rate of about 5 Mbytes/min, while digital *video* requires about 500 Mbytes/min. With data rates this high, it is common for computers to have insufficient memory to simultaneously hold the entire signal to be processed. There are also systems that process segment-by-segment because they operate in *real time*. For example, telephone signals cannot be delayed by more than a few hundred milliseconds, limiting the amount of data that are available for processing at any one instant. In still other applications, the *processing* may require that the signal be segmented. An example is FFT convolution, the main topic of this chapter.

The overlap-add method is based on the fundamental technique in DSP: (1) decompose the signal into simple components, (2) process each of the components in some useful way, and (3) recombine the processed components into the final signal. Figure 18-1 shows an example of how this is done for the overlap-add method. Figure (a) is the signal to be filtered, while (b) shows the filter kernel to be used, a windowed-sinc low-pass filter. Jumping to the bottom of the figure, (i) shows the filtered signal, a smoothed version of (a). The key to this method is how the *lengths* of these signals are affected by the convolution. When an *N* sample signal is convolved with an M sample filter kernel, the output signal is $N+M-1$ samples long. For instance, the input signal, (a), is 300 samples (running from 0 to 299), the filter kernel, (b), is 101 samples (running from 0 to 100), and the output signal, (i), is 400 samples (running from 0 to 399).

In other words, when an N sample signal is filtered, it will be *expanded* by $M-1$ points *to the right*. (This is assuming that the filter kernel runs from index 0 to M. If negative indexes are used in the filter kernel, the expansion will also be to the *left*). In (a), zeros have been added to the signal between sample 300 and 399 to illustrate where this expansion will occur. Don't be confused by the small values at the ends of the output signal, (i). This is simply a result of the windowed-sinc filter kernel having small values near its ends. All 400 samples in (i) are nonzero, even though some of them are too small to be seen in the graph.

Figures (c), (d) and (e) show the decomposition used in the overlap-add method. The signal is broken into segments, with each segment having 100 samples from the original signal. In addition, 100 zeros are added to the right of each segment. In the next step, each segment is individually filtered by convolving it with the filter kernel. This produces the output segments shown in (f), (g), and (h). Since each input segment is 100 samples long, and the filter kernel is 101 samples long, each output segment will be 200 samples long. The important point to understand is that the 100 zeros were added to each input segment to allow for the expansion during the convolution.

Notice that the expansion results in the output segments *overlapping* each other. These overlapping output segments are added to give the output signal, (i). For instance, samples 200 to 299 in (i) are found by adding the corresponding samples in (g) and (h). The overlap-add method produces exactly the same output signal as direct convolution. The disadvantage is a much greater program complexity to keep track of the overlapping samples.

FFT Convolution

FFT convolution uses the principle that *multiplication* in the frequency domain corresponds to *convolution* in the time domain. The input signal is transformed into the frequency domain using the DFT, multiplied by the frequency response of the filter, and then transformed back into the time domain using the Inverse DFT. This basic technique was known since the days of Fourier; however, no one really cared. This is because the time required to calculate the DFT was *longer* than the time to directly calculate the convolution. This changed in 1965 with the development of the Fast Fourier Transform (FFT). By using the FFT algorithm to calculate the DFT, convolution via the frequency domain can be *faster* than directly convolving the time domain signals. The final result is the same; only the number of calculations has been changed by a more efficient algorithm. For this reason, FFT convolution is also called **high-speed convolution**.

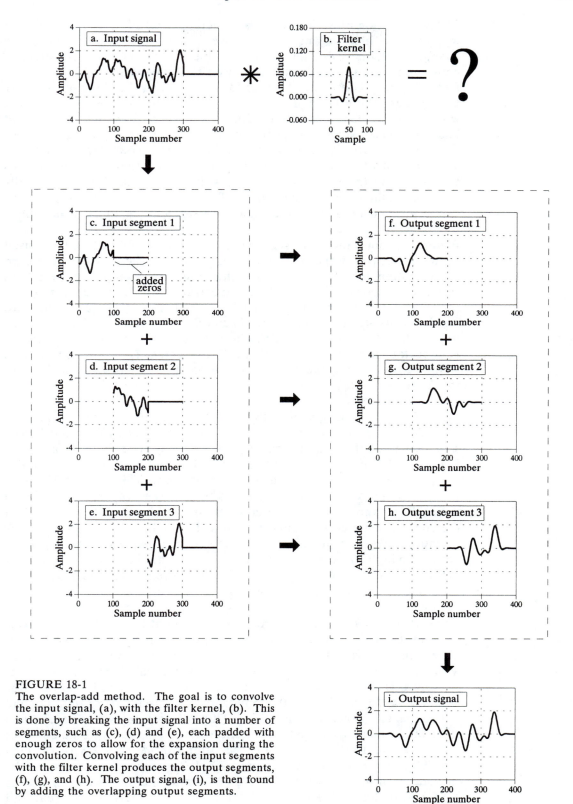

FIGURE 18-1
The overlap-add method. The goal is to convolve
the input signal, (a), with the filter kernel, (b). This
is done by breaking the input signal into a number of
segments, such as (c), (d) and (e), each padded with
enough zeros to allow for the expansion during the
convolution. Convolving each of the input segments
with the filter kernel produces the output segments,
(f), (g), and (h). The output signal, (i), is then found
by adding the overlapping output segments.

FFT convolution uses the overlap-add method shown in Fig. 18-1; only the way that the input segments are converted into the output segments is changed. Figure 18-2 shows an example of how an input segment is converted into an output segment by FFT convolution. To start, the frequency response of the filter is found by taking the DFT of the filter kernel, using the FFT. For instance, (a) shows an example filter kernel, a windowed-sinc band-pass filter. The FFT converts this into the real and imaginary parts of the frequency response, shown in (b) & (c). These frequency domain signals may not *look* like a band-pass filter because they are in rectangular form. Remember, polar form is usually best for humans to understand the frequency domain, while rectangular form is normally best for mathematical calculations. These real and imaginary parts are stored in the computer for use when each segment is being calculated.

Figure (d) shows the input segment to being processed. The FFT is used to find its frequency spectrum, shown in (e) & (f). The frequency spectrum of the output segment, (h) & (i) is then found by multiplying the filter's frequency response, (b) & (c), by the spectrum of the input segment, (e) & (f). Since these spectra consist of real and imaginary parts, they are multiplied according to Eq. 9-1 in Chapter 9. The Inverse FFT is then used to find the output segment, (g), from its frequency spectrum, (h) & (i). It is important to recognize that this output segment is exactly the same as would be obtained by the direct convolution of the input segment, (d), and the filter kernel, (a).

The FFTs must be long enough that *circular convolution* does not take place (also described in Chapter 9). This means that the FFT should be the same length as the output segment, (g). For instance, in the example of Fig. 18-2, the filter kernel contains 129 points and each segment contains 128 points, making output segment 256 points long. This calls for 256 point FFTs to be used. This means that the filter kernel, (a), must be padded with 127 zeros to bring it to a total length of 256 points. Likewise, each of the input segments, (d), must be padded with 128 zeros. As another example, imagine you need to convolve a very long signal with a filter kernel having 600 samples. One alternative would be to use segments of 425 points, and 1024 point FFTs. Another alternative would be to use segments of 1449 points, and 2048 point FFTs.

Table 18-1 shows an example program to carry out FFT convolution. This program filters a 10 million point signal by convolving it with a 400 point filter kernel. This is done by breaking the input signal into 16000 segments, with each segment having 625 points. When each of these segments is convolved with the filter kernel, an output segment of $625+400-1 = 1024$ points is produced. Thus, 1024 point FFTs are used. After defining and initializing all the arrays (lines 130 to 230), the first step is to calculate and store the frequency response of the filter (lines 250 to 310). Line 260 calls a mythical subroutine that loads the filter kernel into XX[0] through XX[399], and sets XX[400] through XX[1023] to a value of zero. The subroutine in line 270 is the FFT, transforming the 1024 samples held in XX[ ] into the 513 samples held in REX[ ] & IMX[ ], the real and

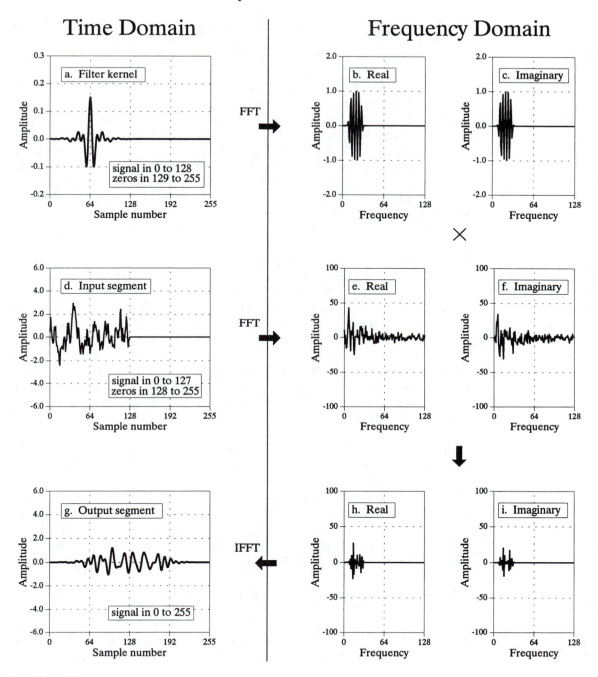

FIGURE 18-2
FFT convolution. The filter kernel, (a), and the signal segment, (d), are converted into their respective spectra, (b) & (c) and (d) & (e), via the FFT. These spectra are multiplied, resulting in the spectrum of the output segment, (h) & (i). The Inverse FFT then finds the output segment, (g).

imaginary parts of the frequency response. These values are transferred into the arrays REFR[ ] & IMFR[ ] (for: REal and IMaginary Frequency Response), to be used later in the program.

The FOR-NEXT loop between lines 340 and 580 controls how the 16000 segments are processed. In line 360, a subroutine loads the next segment to be processed into XX[0] through XX[624], and sets XX[625] through XX[1023] to a value of zero. In line 370, the FFT subroutine is used to find this segment's frequency spectrum, with the real part being placed in the 513 points of REX[ ], and the imaginary part being placed in the 513 points of IMX[ ]. Lines 390 to 430 show the multiplication of the segment's frequency spectrum, held in REX[ ] & IMX[ ], by the filter's frequency response, held in REFR[ ] and IMFR[ ]. The result of the multiplication is stored in REX[ ] & IMX[ ], overwriting the data previously there. Since this is now the frequency spectrum of the output segment, the IFFT can be used to find the output segment. This is done by the mythical IFFT subroutine in line 450, which transforms the 513 points held in REX[ ] & IMX[ ] into the 1024 points held in XX[ ], the output segment.

Lines 470 to 550 handle the overlapping of the segments. Each output segment is divided into two sections. The first 625 points (0 to 624) need to be combined with the overlap from the *previous* output segment, and then written to the output signal. The last 399 points (625 to 1023) need to be saved so that they can overlap with the *next* output segment.

To understand this, look back at Fig 18-1. Samples 100 to 199 in (g) need to be combined with the overlap from the *previous* output segment, (f), and can then be moved to the output signal (i). In comparison, samples 200 to 299 in (g) need to be saved so that they can be combined with the *next* output segment, (h).

Now back to the program. The array OLAP[ ] is used to hold the 399 samples that overlap from one segment to the next. In lines 470 to 490 the 399 values in this array (from the previous output segment) are added to the output segment currently being worked on, held in XX[ ]. The mythical subroutine in line 550 then outputs the 625 samples in XX[0] to XX[624] to the file holding the output signal. The 399 samples of the current output segment that need to be held over to the next output segment are then stored in OLAP[ ] in lines 510 to 530.

After all 0 to 15999 segments have been processed, the array, OLAP[ ], will contain the 399 samples from segment 15999 that should overlap segment 16000. Since segment 16000 doesn't exist (or can be viewed as containing all zeros), the 399 samples are written to the output signal in line 600. This makes the length of the output signal $16000 \times 625 + 399 = 10,000,399$ points. This matches the length of input signal, plus the length of the filter kernel, minus 1.

Speed Improvements

When is FFT convolution faster than standard convolution? The answer depends on the length of the filter kernel, as shown in Fig. 18-3. The time

```
100 'FFT CONVOLUTION
110 'This program convolves a 10 million point signal with a 400 point filter kernel.  The input
120 'signal is broken into 16000 segments, each with 625 points.  1024 point FFTs are used.
130 '
130 '                               'INITIALIZE THE ARRAYS
140 DIM XX[1023]                    'the time domain signal (for the FFT)
150 DIM REX[512]                    'real part of the frequency domain (for the FFT)
160 DIM IMX[512]                    'imaginary part of the frequency domain (for the FFT)
170 DIM REFR[512]                   'real part of the filter's frequency response
180 DIM IMFR[512]                   'imaginary part of the filter's frequency response
190 DIM OLAP[398]                   'holds the overlapping samples from segment to segment
200 '
210 FOR I% = 0 TO 398              'zero the array holding the overlapping samples
220   OLAP[I%] = 0
230 NEXT I%
240 '
250 '                               'FIND & STORE THE FILTER'S FREQUENCY RESPONSE
260 GOSUB XXXX                      'Mythical subroutine to load the filter kernel into XX[ ]
270 GOSUB XXXX                      'Mythical FFT subroutine: XX[ ] --> REX[ ] & IMX[ ]
280 FOR F% = 0 TO 512              'Save the frequency response in REFR[ ] & IMFR[ ]
290   REFR[F%] = REX[F%]
300   IMFR[F%] = IMX[F%]
310 NEXT F%
320 '
330 '                               'PROCESS EACH OF THE 16000 SEGMENTS
340 FOR SEGMENT% = 0 TO 15999
350 '
360   GOSUB XXXX                    'Mythical subroutine to load the next input segment into XX[ ]
370   GOSUB XXXX                    'Mythical FFT subroutine:  XX[ ] -->  REX[ ] & IMX[ ]
380   '
390   FOR F% = 0 TO 512            'Multiply the frequency spectrum by the frequency response
400     TEMP     = REX[F%]*REFR[F%] - IMX[F%]*IMFR[F%]
410     IMX[F%] = REX[F%]*IMFR[F%] + IMX[F%]*REFR[F%]
420     REX[F%] = TEMP
430   NEXT F%
440   '
450   GOSUB XXXX                    'Mythical IFFT subroutine:  REX[ ] & IMX[ ] -->  XX[ ]
460   '
470   FOR I% = 0 TO 398            'Add the last segment's overlap to this segment
480     XX[I%] = XX[I%] + OLAP[I%]
490   NEXT I%
500   '
510   FOR I% = 625 TO 1023         'Save the samples that will overlap the next segment
520     OLAP[I%-625] = XX[I%]
530   NEXT I%
540   '
550   GOSUB XXXX                    'Mythical subroutine to output the 625 samples stored
560   '                             'in XX[0] to XX[624]
570   '
580 NEXT SEGMENT%
590 '
600 GOSUB XXXX                      'Mythical subroutine to output all 399 samples in OLAP[ ]
610 END
```

TABLE 18-1

for standard convolution is directly proportional to the number of points in the filter kernel. In comparison, the time required for FFT convolution increases very slowly, only as the *logarithm* of the number of points in the

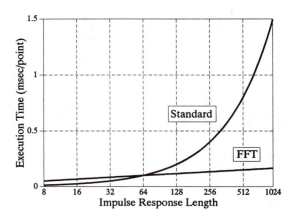

FIGURE 18-3
Execution times for FFT convolution. FFT convolution is faster than the standard method when the filter kernel is longer than about 60 points. These execution times are for a 100 MHz Pentium, using single precision floating point.

filter kernel.   The crossover occurs when the filter kernel has about 40 to 80 samples (depending on the particular hardware used).

The important idea to remember: filter kernels shorter than about 60 points can be implemented faster with standard convolution, and the execution time is proportional to the kernel length. Longer filter kernels can be implemented faster with FFT convolution. With FFT convolution, the filter kernel can be made as long as you like, with very little penalty in execution time. For instance, a 16,000 point filter kernel only requires about *twice* as long to execute as one with only 64 points.

The *speed* of the convolution also dictates the *precision* of the calculation (just as described for the FFT in Chapter 12). This is because the round-off error in the output signal depends on the total number of calculations, which is directly proportional to the computation time. If the output signal is calculated *faster*, it will also be calculated more *precisely*. For instance, imagine convolving a signal with a 1000 point filter kernel, with single precision floating point. Using standard convolution, the typical round-off noise can be expected to be about 1 part in 20,000 (from the guidelines in Chapter 4). In comparison, FFT convolution can be expected to be an order of magnitude *faster*, and an order of magnitude more *precise* (i.e., 1 part in 200,000).

Keep FFT convolution tucked away for when you have a large amount of data to process and need an extremely long filter kernel. Think in terms of a *million* sample signal and a *thousand* point filter kernel. Anything less won't justify the extra programming effort. Don't want to write your own FFT convolution routine? Look in software libraries and packages for prewritten code. Start with this book's web site (see the copyright page).

CHAPTER 19

Recursive Filters

Recursive filters are an efficient way of achieving a long impulse response, without having to perform a long convolution. They execute very rapidly, but have less performance and flexibility than other digital filters. Recursive filters are also called *Infinite Impulse Response* (IIR) filters, since their impulse responses are composed of decaying exponentials. This distinguishes them from digital filters carried out by convolution, called *Finite Impulse Response* (FIR) filters. This chapter is an introduction to how recursive filters operate, and how simple members of the family can be designed. Chapters 20, 26 and 31 present more sophisticated design methods.

The Recursive Method

To start the discussion of recursive filters, imagine that you need to extract information from some signal, $x[\ ]$. Your need is so great that you hire an old mathematics professor to process the data for you. The professor's task is to filter $x[\ ]$ to produce $y[\ ]$, which hopefully contains the information you are interested in. The professor begins his work of calculating each point in $y[\ ]$ according to some algorithm that is locked tightly in his over-developed brain. Part way through the task, a most unfortunate event occurs. The professor begins to babble about analytic singularities and fractional transforms, and other demons from a mathematician's nightmare. It is clear that the professor has lost his mind. You watch with anxiety as the professor, and your algorithm, are taken away by several men in white coats.

You frantically review the professor's notes to find the algorithm he was using. You find that he had completed the calculation of points $y[0]$ through $y[27]$, and was about to start on point $y[28]$. As shown in Fig. 19-1, we will let the variable, n, represent the point that is currently being calculated. This means that $y[n]$ is sample 28 in the output signal, $y[n-1]$ is sample 27, $y[n-2]$ is sample 26, etc. Likewise, $x[n]$ is point 28 in the input signal, $x[n-1]$ is point 27, etc. To understand the algorithm being used, we ask ourselves: "What information was available to the professor to calculate $y[n]$, the sample currently being worked on?"

The most obvious source of information is the *input signal*, that is, the values: $x[n]$, $x[n-1]$, $x[n-2]$, $\cdots$. The professor could have been multiplying each point in the input signal by a coefficient, and adding the products together:

$$y[n] = a_0 x[n] + a_1 x[n-1] + a_2 x[n-2] + a_3 x[n-3] + \cdots$$

You should recognize that this is nothing more than simple convolution, with the coefficients: $a_0, a_1, a_2, \cdots$, forming the convolution kernel. If this was all the professor was doing, there wouldn't be much need for this story, or this chapter. However, there is another source of information that the professor had access to: the *previously* calculated values of the output signal, held in: $y[n-1]$, $y[n-2]$, $y[n-3]$, $\cdots$. Using this additional information, the algorithm would be in the form:

$$y[n] = a_0 x[n] + a_1 x[n-1] + a_2 x[n-2] + a_3 x[n-3] + \cdots$$
$$+ b_1 y[n-1] + b_2 y[n-2] + b_3 y[n-3] + \cdots$$

EQUATION 19-1
The recursion equation. In this equation, $x[\ ]$ is the input signal, $y[\ ]$ is the output signal, and the a's and b's are coefficients.

In words, each point in the output signal is found by multiplying the values from the input signal by the "a" coefficients, multiplying the previously calculated values from the output signal by the "b" coefficients, and adding the products together. Notice that there isn't a value for b_0, because this corresponds to the sample being calculated. Equation 19-1 is called the **recursion equation**, and filters that use it are called **recursive filters**. The "a" and "b" values that define the filter are called the **recursion coefficients**. In actual practice, no more than about a dozen recursion coefficients can be used or the filter becomes unstable (i.e., the output continually increases or oscillates). Table 19-1 shows an example recursive filter program.

Recursive filters are useful because they *bypass* a longer convolution. For instance, consider what happens when a delta function is passed through a recursive filter. The output is the filter's *impulse response*, and will typically be a sinusoidal oscillation that exponentially decays. Since this impulse response in infinitely long, recursive filters are often called *infinite impulse response* (IIR) filters. In effect, recursive filters *convolve* the input signal with a very long filter kernel, although only a few coefficients are involved.

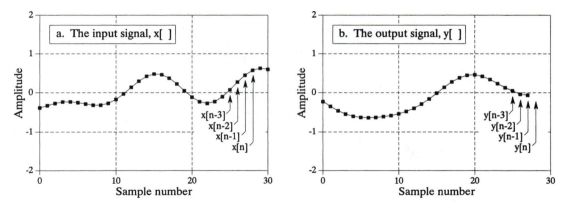

FIGURE 19-1
Recursive filter notation. The output sample being calculated, $y[n]$, is determined by the values from the input signal, $x[n]$, $x[n-1]$, $x[n-2]$, ⋯, as well as the *previously* calculated values in the output signal, $y[n-1]$, $y[n-2]$, $y[n-3]$, ⋯. These figures are shown for $n = 28$.

The relationship between the recursion coefficients and the filter's response is given by a mathematical technique called the **z-transform**, the topic of Chapter 31. For example, the z-transform can be used for such tasks as: converting between the recursion coefficients and the frequency response, combining cascaded and parallel stages into a single filter, designing recursive systems that mimic analog filters, etc. Unfortunately, the z-transform is very mathematical, and more complicated than most DSP *users* are willing to deal with. This is the realm of those that specialize in DSP.

There are three ways to find the recursion coefficients without having to understand the z-transform. First, this chapter provides design equations for several types of simple recursive filters. Second, Chapter 20 provides a "cookbook" computer program for designing the more sophisticated *Chebyshev* low-pass and high-pass filters. Third, Chapter 26 describes an iterative method for designing recursive filters with an *arbitrary* frequency response.

```
100 'RECURSIVE FILTER
110 '
120 DIM X[499]                'holds the input signal
130 DIM Y[499]                'holds the filtered output signal
140 '
150 GOSUB XXXX                'Mythical subroutine to calculate the recursion
160 '                         'coefficients: A0, A1, A2, B1, B2
170 '
180 GOSUB XXXX                'Mythical subroutine to load X[ ] with the input data
190 '
200 FOR I% = 2 TO 499
210   Y[I%] = A0*X[I%] + A1*X[I%-1] + A2*X[I%-2] + B1*Y[I%-1] + B2*Y[I%-2]
220 NEXT I%
230 '
240 END
```

TABLE 19-1

Digital Filter

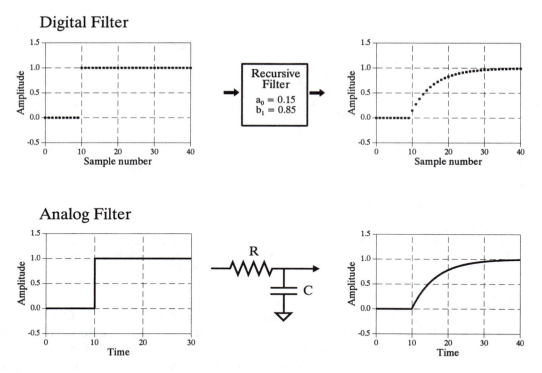

Analog Filter

FIGURE 19-2
Single pole low-pass filter. Digital recursive filters can mimic analog filters composed of resistors and capacitors. As shown in this example, a single pole low-pass recursive filter smoothes the edge of a step input, just as an electronic RC filter.

Single Pole Recursive Filters

Figure 19-2 shows an example of what is called a **single pole** low-pass filter. This recursive filter uses just two coefficients, $a_0 = 0.15$ and $b_1 = 0.85$. For this example, the input signal is a step function. As you should expect for a low-pass filter, the output is a smooth rise to the steady state level. This figure also shows something that ties into your knowledge of electronics. This low-pass recursive filter is completely analogous to an electronic low-pass filter composed of a single resistor and capacitor.

The beauty of the recursive method is in its ability to create a wide variety of responses by changing only a few parameters. For example, Fig. 19-3 shows a filter with three coefficients: $a_0 = 0.93$, $a_1 = -0.93$ and $b_1 = 0.86$. As shown by the similar step responses, this digital filter mimics an electronic RC high-pass filter.

These single pole recursive filters are definitely something you want to keep in your DSP toolbox. You can use them to process digital signals just as you would use RC networks to process analog electronic signals. This includes everything you would expect: DC removal, high-frequency noise suppression, wave shaping, smoothing, etc. They are easy to program, fast

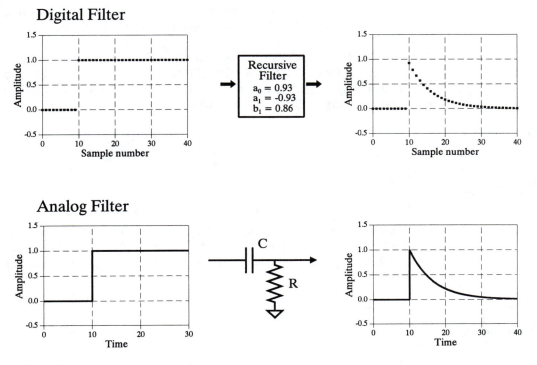

FIGURE 19-3

Single pole high-pass filter. Proper coefficient selection can also make the recursive filter mimic an electronic RC high-pass filter. These single pole recursive filters can be used in DSP just as you would use RC circuits in analog electronics.

to execute, and produce few surprises. The coefficients are found from these simple equations:

EQUATION 19-2
Single pole low-pass filter. The filter's response is controlled by the parameter, x, a value between zero and one.

$$a_0 = 1-x$$
$$b_1 = x$$

EQUATION 19-3
Single pole high-pass filter.

$$a_0 = (1+x)/2$$
$$a_1 = -(1+x)/2$$
$$b_1 = x$$

The characteristics of these filters are controlled by the parameter, x, a value between zero and one. Physically, x is the amount of *decay* between adjacent samples. For instance, x is 0.86 in Fig. 19-3, meaning that the value of each sample in the output signal is 0.86 the value of the sample before it. The higher the value of x, the slower the decay. Notice that the filter becomes *unstable* if x is made greater than one. That is, any nonzero value on the input will make the output increase until an overflow occurs.

The value for x can be directly specified, or found from the desired *time constant* of the filter. Just as $R{\times}C$ is the number of seconds it takes an RC circuit to decay to 36.8% of its final value, d is the number of samples it takes for a recursive filter to decay to this same level:

EQUATION 19-4
Time constant of single pole filters. This equation relates the amount of decay between samples, x, with the filter's time constant, d, the number of samples for the filter to decay to 36.8%.

$$x = e^{-1/d}$$

For instance, a sample-to-sample decay of $x = 0.86$ corresponds to a time constant of $d = 6.63$ samples (as shown in Fig 19-3). There is also a fixed relationship between x and the -3dB *cutoff frequency*, f_C, of the digital filter:

EQUATION 19-5
Cutoff frequency of single pole filters. The amount of decay between samples, x, is related to the cutoff frequency of the filter, f_C, a value between 0 and 0.5.

$$x = e^{-2\pi f_C}$$

This provides three ways to find the "a" and "b" coefficients, starting with the time constant, the cutoff frequency, or just directly picking x.

Figure 19-4 shows an example of using single pole recursive filters. In (a), the original signal is a smooth curve, except a burst of a high frequency sine wave. Figure (b) shows the signal after passing through low-pass and high-pass filters. The signals have been separated fairly well, but not perfectly, just as if simple RC circuits were used on an analog signal.

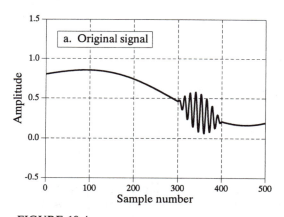
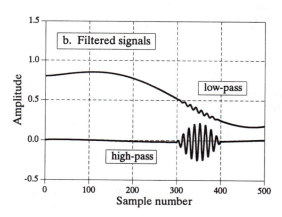

FIGURE 19-4
Example of single pole recursive filters. In (a), a high frequency burst rides on a slowly varying signal. In (b), single pole low-pass and high-pass filters are used to separate the two components. The low-pass filter uses $x = 0.95$, while the high-pass filter is for $x = 0.86$.

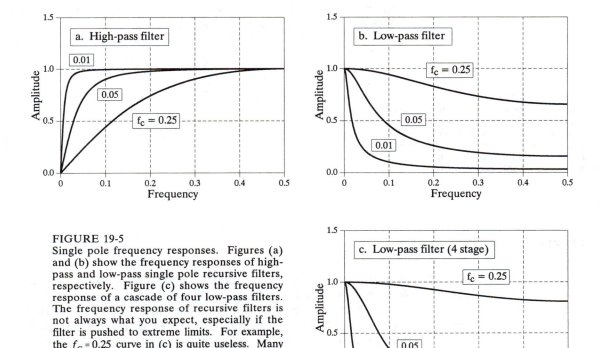

FIGURE 19-5
Single pole frequency responses. Figures (a) and (b) show the frequency responses of high-pass and low-pass single pole recursive filters, respectively. Figure (c) shows the frequency response of a cascade of four low-pass filters. The frequency response of recursive filters is not always what you expect, especially if the filter is pushed to extreme limits. For example, the $f_C = 0.25$ curve in (c) is quite useless. Many factors are to blame, including: aliasing, round-off noise, and the nonlinear phase response.

Figure 19-5 shows the frequency responses of various single pole recursive filters. These curves are obtained by passing a delta function through the filter to find the filter's impulse response. The FFT is then used to convert the impulse response into the frequency response. In principle, the impulse response is infinitely long; however, it decays below the single precision round-off noise after about 15 to 20 time constants. For example, when the time constant of the filter is $d = 6.63$ samples, the impulse response can be contained in about 128 samples.

The key feature in Fig. 19-5 is that single pole recursive filters have little ability to separate one band of frequencies from another. In other words, they perform well in the time domain, and poorly in the frequency domain. The frequency response can be improved slightly by cascading several stages. This can be accomplished in two ways. First, the signal can be passed through the filter several times. Second, the z-transform can be used to find the recursion coefficients that combine the cascade into a single stage. Both ways work and are commonly used. Figure (c) shows the frequency response of a cascade of four low-pass filters. Although the stopband attenuation is significantly improved, the roll-off is still terrible. If you need better performance in the frequency domain, look at the Chebyshev filters of the next chapter.

The four stage low-pass filter is comparable to the Blackman and Gaussian filters (relatives of the moving average, Chapter 15), but with a much faster execution speed. The design equations for a four stage low-pass filter are:

EQUATION 19-6
Four stage low-pass filter. These equations provide the "a" and "b" coefficients for a cascade of four single pole low-pass filters. The relationship between x and the cutoff frequency of this filter is given by Eq. 19-5, with the 2π replaced by 14.445.

$$a_0 = (1-x)^4$$
$$b_1 = 4x$$
$$b_2 = -6x^2$$
$$b_3 = 4x^3$$
$$b_4 = -x^4$$

Narrow-band Filters

A common need in electronics and DSP is to isolate a narrow band of frequencies from a wider bandwidth signal. For example, you may want to eliminate 60 hertz interference in an instrumentation system, or isolate the signaling tones in a telephone network. Two types of frequency responses are available: the *band-pass* and the *band-reject* (also called a **notch filter**). Figure 19-6 shows the frequency response of these filters, with the recursion coefficients provided by the following equations:

EQUATION 19-7
Band-pass filter. An example frequency response is shown in Fig. 19-6a. To use these equations, first select the center frequency, f, and the bandwidth, BW. Both of these are expressed as a fraction of the sampling rate, and therefore in the range of 0 to 0.5. Next, calculate R, and then K, and then the recursion coefficients.

$$a_0 = 1-K$$
$$a_1 = 2(K-R)\cos(2\pi f)$$
$$a_2 = R^2-K$$
$$b_1 = 2R\cos(2\pi f)$$
$$b_2 = -R^2$$

EQUATION 19-8
Band-reject filter. This filter is commonly called a notch filter. Example frequency responses are shown in Fig. 19-6b.

$$a_0 = K$$
$$a_1 = -2K\cos(2\pi f)$$
$$a_2 = K$$
$$b_1 = 2R\cos(2\pi f)$$
$$b_2 = -R^2$$

where:

$$K = \frac{1-2R\cos(2\pi f)+R^2}{2-2\cos(2\pi f)}$$

$$R = 1-3BW$$

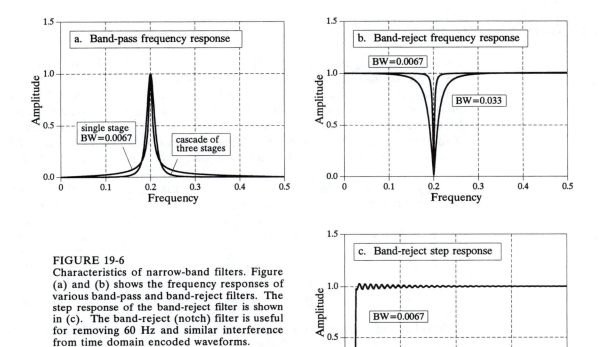

FIGURE 19-6

Characteristics of narrow-band filters. Figure (a) and (b) shows the frequency responses of various band-pass and band-reject filters. The step response of the band-reject filter is shown in (c). The band-reject (notch) filter is useful for removing 60 Hz and similar interference from time domain encoded waveforms.

Two parameters must be selected before using these equations: f, the center frequency, and BW, the bandwidth (measured at an amplitude of 0.707). Both of these are expressed as a fraction of the sampling frequency, and therefore must be between 0 and 0.5. From these two specified values, calculate the intermediate variables: R and K, and then the recursion coefficients.

As shown in (a), the band-pass filter has relatively large *tails* extending from the main peak. This can be improved by cascading several stages. Since the design equations are quite long, it is simpler to implement this cascade by filtering the signal several times, rather than trying to find the coefficients needed for a single filter.

Figure (b) shows examples of the band-reject filter. The narrowest bandwidth that can be obtain with single precision is about 0.0003 of the sampling frequency. When pushed beyond this limit, the attenuation of the notch will degrade. Figure (c) shows the step response of the band-reject filter. There is noticeable overshoot and ringing, but its amplitude is quite small. This allows the filter to remove narrowband interference (60 Hz and the like) with only a minor distortion to the time domain waveform.

Phase Response

There are three types of *phase response* that a filter can have: **zero phase**, **linear phase**, and **nonlinear phase**. An example of each of these is shown in Figure 19-7. As shown in (a), the *zero phase* filter is characterized by an impulse response that is symmetrical around sample zero. The actual shape doesn't matter, only that the negative numbered samples are a mirror image of the positive numbered samples. When the Fourier transform is taken of this symmetrical waveform, the phase will be entirely zero, as shown in (b).

The disadvantage of the zero phase filter is that it requires the use of negative indexes, which can be inconvenient to work with. The linear phase filter is a way around this. The impulse response in (d) is identical to that shown in (a), except it has been shifted to use only positive numbered samples. The impulse response is still symmetrical between the left and right; however, the location of symmetry has been shifted from zero. This shift results in the phase, (e), being a *straight line*, accounting for the name: *linear phase*. The slope of this straight line is directly proportional to the amount of the shift. Since the shift in the impulse response does nothing but produce an identical shift in the output signal, the linear phase filter is equivalent to the zero phase filter for most purposes.

Figure (g) shows an impulse response that is *not* symmetrical between the left and right. Correspondingly, the phase, (h), is *not* a straight line. In other words, it has a *nonlinear phase*. Don't confuse the terms: *nonlinear and linear phase* with the concept of *system linearity* discussed in Chapter 5. Although both use the word *linear*, they are not related.

Why does anyone care if the phase is linear or not? Figures (c), (f), and (i) show the answer. These are the **pulse responses** of each of the three filters. The pulse response is nothing more than a positive going step response followed by a negative going step response. The pulse response is used here because it displays what happens to both the rising and falling edges in a signal. Here is the important part: zero and linear phase filters have left and right edges that look the *same*, while nonlinear phase filters have left and right edges that look *different*. Many applications cannot tolerate the left and right edges looking different. One example is the display of an oscilloscope, where this difference could be misinterpreted as a feature of the signal being measured. Another example is in video processing. Can you imagine turning on your TV to find the left ear of your favorite actor looking different from his right ear?

It is easy to make an FIR (finite impulse response) filter have a linear phase. This is because the impulse response (filter kernel) is directly *specified* in the design process. Making the filter kernel have left-right symmetry is all that is required. This is not the case with IIR (recursive) filters, since the recursion coefficients are what is specified, not the impulse response. The impulse response of a recursive filter is *not* symmetrical between the left and right, and therefore has a *nonlinear* phase.

Zero Phase Filter

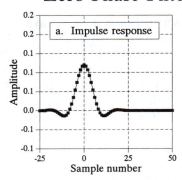

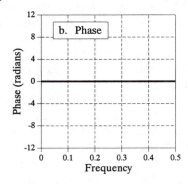

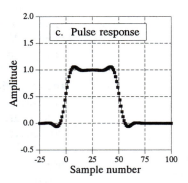

Linear Phase Filter

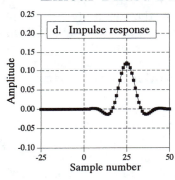

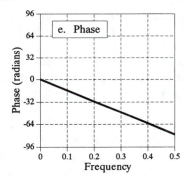

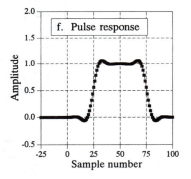

Nonlinear Phase Filter

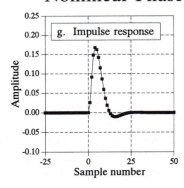

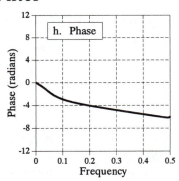

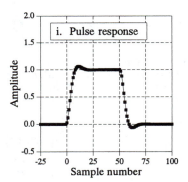

FIGURE 19-7
Zero, linear, and nonlinear phase filters. A *zero phase* filter has an impulse response that has left-right symmetry around sample number zero, as in (a). This results in a frequency response that has a phase composed entirely of zeros, as in (b). Zero phase impulse responses are desirable because their step responses are symmetrical between the top and bottom, making the left and right edges of pulses look the same, as is shown in (c). *Linear phase* filters have left-right symmetry, but not around sample zero, as illustrated in (d). This results in a phase that is linear, that is, a straight line, as shown in (e). The linear phase pulse response, shown in (f), has all the advantages of the zero phase pulse response. In comparison, the impulse responses of *nonlinear phase* filters are not symmetrical between the left and right, as in (g), and the phases are not a straight line, as in (h). The worst part is that the left and right edges of the pulse response are not the same, as shown in (i).

Analog electronic circuits have this same problem with the phase response. Imagine a circuit composed of resistors and capacitors sitting on your desk. If the input has always been zero, the output will also have always been zero. When an impulse is applied to the input, the capacitors quickly charge to some value and then begin to exponentially decay through the resistors. The impulse response (i.e., the output signal) is a combination of these various decaying exponentials. The impulse response *cannot* be symmetrical, because the output was zero before the impulse, and the exponential decay never quite reaches a value of zero again. Analog filter designers attack this problem with the **Bessel filter**, presented in Chapter 3. The Bessel filter is designed to have as linear phase as possible; however, it is far below the performance of digital filters. The ability to provide an *exact* linear phase is a clear advantage of digital filters.

Fortunately, there is a simple way to modify recursive filters to obtain a *zero phase*. Figure 19-8 shows an example of how this works. The input signal to be filtered is shown in (a). Figure (b) shows the signal after it has been filtered by a single pole low-pass filter. Since this is a nonlinear phase filter, the left and right edges do not look the same; they are inverted versions of each other. As previously described, this recursive filter is implemented by starting at sample 0 and working toward sample 150, calculating each sample along the way.

Now, suppose that instead of moving from sample 0 toward sample 150, we start at sample 150 and move toward sample 0. In other words, each sample in the output signal is calculated from input and output samples to the *right* of the sample being worked on. This means that the recursion equation, Eq. 19-1, is changed to:

$$y[n] = a_0 x[n] + a_1 x[n+1] + a_2 x[n+2] + a_3 x[n+3] + \cdots$$
$$+ b_1 y[n+1] + b_2 y[n+2] + b_3 y[n+3] + \cdots$$

EQUATION 19-9
The *reverse* recursion equation. This is the same as Eq. 19-1, except the signal is filtered from left-to-right, instead of right-to-left.

Figure (c) shows the result of this **reverse filtering**. This is analogous to passing an analog signal through an electronic RC circuit while running time *backwards*. !esrevinu eht pu-wercs nac lasrever emit -noituaC

Filtering in the reverse direction does not produce any benefit in itself; the filtered signal still has left and right edges that do not look alike. The magic happens when forward and reverse filtering are *combined*. Figure (d) results from filtering the signal in the forward direction and then filtering again in the reverse direction. Voila! This produces a *zero phase* recursive filter. In fact, *any* recursive filter can be converted to zero phase with this **bidirectional filtering** technique. The only penalty for this improved performance is a factor of two in execution time and program complexity.

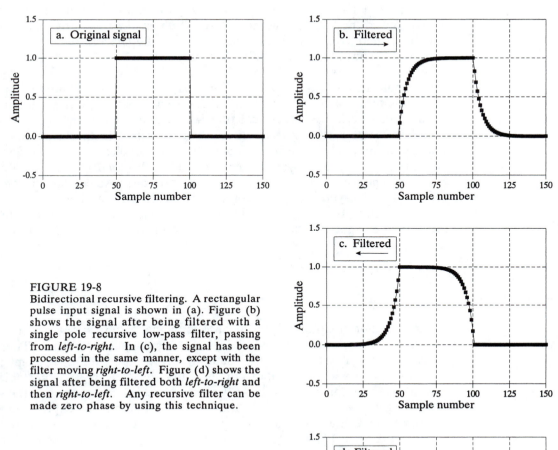

FIGURE 19-8
Bidirectional recursive filtering. A rectangular pulse input signal is shown in (a). Figure (b) shows the signal after being filtered with a single pole recursive low-pass filter, passing from *left-to-right*. In (c), the signal has been processed in the same manner, except with the filter moving *right-to-left*. Figure (d) shows the signal after being filtered both *left-to-right* and then *right-to-left*. Any recursive filter can be made zero phase by using this technique.

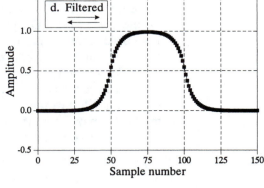

How do you find the impulse and frequency responses of the overall filter? The magnitude of the frequency response is the same for each direction, while the phases are opposite in sign. When the two directions are combined, the magnitude becomes *squared*, while the phase cancels to *zero*. In the time domain, this corresponds to convolving the original impulse response with a left-for-right flipped version of itself. For instance, the impulse response of a single pole low-pass filter is a one-sided exponential. The impulse response of the corresponding bidirectional filter is a one-sided exponential that decays to the right, convolved with a one-sided exponential that decays to the left. Going through the mathematics, this turns out to be a double-sided exponential that decays both to the left and right, with the same decay constant as the original filter.

Some applications only have a portion of the signal in the computer at a particular time, such as systems that alternately input and output data on a continuing basis. Bidirectional filtering can be used in these cases by combining it with the overlap-add method described in the last chapter. When you come to the question of how long the impulse response is, don't say "infinite." If you do, you will need to pad each signal segment with an *infinite* number of zeros. Remember, the impulse response can be truncated when it has decayed below the round-off noise level, i.e., about 15 to 20 time constants. Each segment will need to be padded with zeros on both the left and right to allow for the expansion during the bidirectional filtering.

Using Integers

Single precision floating point is ideal to implement these simple recursive filters. The use of integers is possible, but it is much more difficult. There are two main problems. First, the round-off error from the limited number of bits can degrade the response of the filter, or even make it unstable. Second, the fractional values of the recursion coefficients must be handled with integer math. One way to attack this problem is to express each coefficient as a fraction. For example, 0.15 becomes 19/128. Instead of multiplying by 0.15, you first multiply by 19 and then divide by 128. Another way is to replace the multiplications with look-up tables. For example, a 12 bit ADC produces samples with a value between 0 and 4095. Instead of multiplying each sample by 0.15, you pass the samples through a look-up table that is 4096 entries long. The value obtained from the look-up table is equal to 0.15 times the value entering the look-up table. This method is very fast, but it does require extra memory; a separate look-up table is needed for each coefficient. Before you try either of these integer methods, make sure the recursive algorithm for the moving average filter will not suit your needs. It *loves* integers.

Chebyshev Filters

Chebyshev filters are used to separate one band of frequencies from another. Although they cannot match the performance of the windowed-sinc filter, they are more than adequate for many applications. The primary attribute of Chebyshev filters is their speed, typically more than an order of magnitude faster than the windowed-sinc. This is because they are carried out by *recursion* rather than *convolution*. The design of these filters is based on a mathematical technique called the *z-transform*, discussed in Chapter 31. This chapter presents the information needed to *use* Chebyshev filters without wading through a mire of advanced mathematics.

The Chebyshev and Butterworth Responses

The Chebyshev response is a mathematical strategy for achieving a faster *roll-off* by allowing *ripple* in the frequency response. Analog and digital filters that use this approach are called *Chebyshev filters*. For instance, analog Chebyshev filters were used in Chapter 3 for analog-to-digital and digital-to-analog conversion. These filters are named from their use of the *Chebyshev polynomials*, developed by the Russian mathematician Pafnuti Chebyshev (1821-1894). This name has been translated from Russian and appears in the literature with different spellings, such as: Chebychev, Tschebyscheff, Tchebysheff and Tchebichef.

Figure 20-1 shows the frequency response of low-pass Chebyshev filters with passband ripples of: 0%, 0.5% and 20%. As the ripple increases (bad), the roll-off becomes sharper (good). The Chebyshev response is an optimal trade-off between these two parameters. When the ripple is set to 0%, the filter is called a **maximally flat** or **Butterworth filter** (after S. Butterworth, a British engineer who described this response in 1930). A ripple of 0.5% is a often good choice for digital filters. This matches the typical precision and accuracy of the analog electronics that the signal has passed through.

The Chebyshev filters discussed in this chapter are called **type 1** filters, meaning that the ripple is only allowed in the *passband*. In comparison,

FIGURE 20-1
The Chebyshev response. Chebyshev filters achieve a faster roll-off by allowing ripple in the passband. When the ripple is set to 0%, it is called a *maximally flat* or *Butterworth* filter. Consider using a ripple of 0.5% in your designs; this passband unflatness is so small that it cannot be seen in this graph, but the roll-off is greatly improved over the Butterworth.

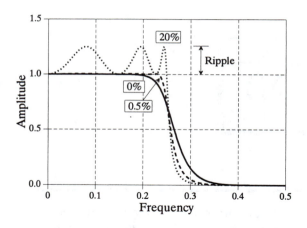

type 2 Chebyshev filters have ripple only in the *stopband*. Type 2 filters are seldom used, and we won't discuss them. There is, however, an important design called the **elliptic filter**, which has ripple in *both* the passband and the stopband. Elliptic filters provide the fastest roll-off for a given number of poles, but are much harder to design. We won't discuss the elliptic filter here, but be aware that it is frequently the first choice of professional filter designers, both in analog electronics and DSP. If you need this level of performance, buy a software package for designing digital filters.

Designing the Filter

You must select four parameters to design a Chebyshev filter: (1) a high-pass or low-pass response, (2) the cutoff frequency, (3) the percent ripple in the passband, and (4) the number of poles. Just what is a *pole*? Here are two answers. If you don't like one, maybe the other will help:

Answer 1- The Laplace transform and z-transform are mathematical ways of breaking an impulse response into sinusoids and decaying exponentials. This is done by expressing the system's characteristics as one complex polynomial divided by another complex polynomial. The roots of the numerator are called *zeros*, while the roots of the denominator are called *poles*. Since poles and zeros can be complex numbers, it is common to say they have a "location" in the complex plane. Elaborate systems have more poles and zeros than simple ones. Recursive filters are designed by first selecting the location of the poles and zeros, and then finding the appropriate recursion coefficients (or analog components). For example, Butterworth filters have poles that lie on a *circle* in the complex plane, while in a Chebyshev filter they lie on an *ellipse*. This is the topic of Chapters 30 and 31.

Answer 2- Poles are containers filled with magic powder. The more poles in a filter, the better the filter works.

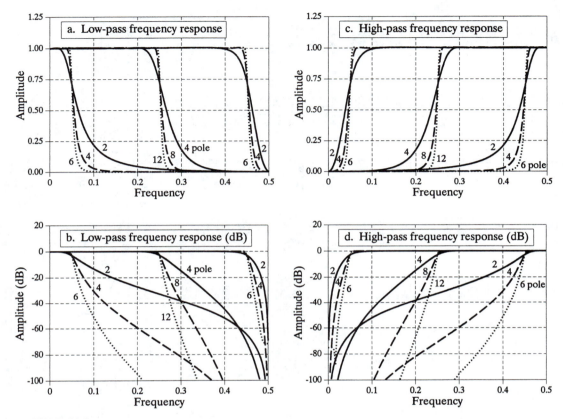

FIGURE 20-2
Chebyshev frequency responses. Figures (a) and (b) show the frequency responses of low-pass Chebyshev filters with 0.5% ripple, while (c) and (d) show the corresponding high-pass filter responses.

Kidding aside, the point is that you can use these filters very effectively without knowing the nasty mathematics behind them. Filter design is a specialty. In actual practice, more engineers, scientists and programmers think in terms of answer 2, than answer 1.

Figure 20-2 shows the frequency response of several Chebyshev filters with 0.5% ripple. For the method used here, the number of poles must be *even*. The cutoff frequency of each filter is measured where the amplitude crosses 0.707 (-3dB). Filters with a cutoff frequency near 0 or 0.5 have a sharper roll-off than filters in the center of the frequency range. For example, a two pole filter at $f_C = 0.05$ has about the same roll-off as a four pole filter at $f_C = 0.25$. This is fortunate; fewer poles can be used near 0 and 0.5 because of round-off noise. More about this later.

There are two ways of finding the recursion coefficients without using the z-transform. First, the cowards way: use a table. Tables 20-1 and 20-2 provide the recursion coefficients for low-pass and high-pass filters with 0.5% passband ripple. If you only need a quick and dirty design, copy the appropriate coefficients into your program, and you're done.

f_C	2 Pole	4 Pole	6 Pole

0.01

2 Pole:
```
a0=  8.663387E-04
a1=  1.732678E-03   b1=  1.919129E+00
a2=  8.663387E-04   b2= -9.225943E-01
```
4 Pole: (!! Unstable !!)
```
a0=  4.149425E-07
a1=  1.659770E-06   b1=  3.893453E+00
a2=  2.489655E-06   b2= -5.688233E+00
a3=  1.659770E-06   b3=  3.695783E+00
a4=  4.149425E-07   b4= -9.010106E-01
```
6 Pole: (!! Unstable !!)
```
a0=  1.391351E-10
a1=  8.348109E-10   b1=  5.883343E+00
a2=  2.087027E-09   b2= -1.442798E+01
a3=  2.782703E-09   b3=  1.887786E+01
a4=  2.087027E-09   b4= -1.389914E+01
a5=  8.348109E-10   b5=  5.459909E+00
a6=  1.391351E-10   b6= -8.939932E-01
```

0.025

2 Pole:
```
a0=  5.112374E-03
a1=  1.022475E-02   b1=  1.797154E+00
a2=  5.112374E-03   b2= -8.176033E-01
```
4 Pole:
```
a0=  1.504626E-05
a1=  6.018503E-05   b1=  3.725385E+00
a2=  9.027754E-05   b2= -5.226004E+00
a3=  6.018503E-05   b3=  3.270902E+00
a4=  1.504626E-05   b4= -7.705239E-01
```
6 Pole: (!! Unstable !!)
```
a0=  3.136210E-08
a1=  1.881726E-07   b1=  5.691653E+00
a2=  4.704314E-07   b2= -1.353172E+01
a3=  6.272419E-07   b3=  1.719986E+01
a4=  4.704314E-07   b4= -1.232689E+01
a5=  1.881726E-07   b5=  4.722721E+00
a6=  3.136210E-08   b6= -7.556340E-01
```

0.05

2 Pole:
```
a0=  1.868823E-02
a1=  3.737647E-02   b1=  1.593937E+00
a2=  1.868823E-02   b2= -6.686903E-01
```
4 Pole:
```
a0=  2.141509E-04
a1=  8.566037E-04   b1=  3.425455E+00
a2=  1.284906E-03   b2= -4.479272E+00
a3=  8.566037E-04   b3=  2.643718E+00
a4=  2.141509E-04   b4= -5.933269E-01
```
6 Pole:
```
a0=  1.771089E-06
a1=  1.062654E-05   b1=  5.330512E+00
a2=  2.656634E-05   b2= -1.196611E+01
a3=  3.542179E-05   b3=  1.447067E+01
a4=  2.656634E-05   b4= -9.937710E+00
a5=  1.062654E-05   b5=  3.673283E+00
a6=  1.771089E-06   b6= -5.707561E-01
```

0.075

2 Pole:
```
a0=  3.869430E-02
a1=  7.738860E-02   b1=  1.392667E+00
a2=  3.869430E-02   b2= -5.474446E-01
```
4 Pole:
```
a0=  9.726342E-04
a1=  3.890537E-03   b1=  3.103944E+00
a2=  5.835806E-03   b2= -3.774453E+00
a3=  3.890537E-03   b3=  2.111238E+00
a4=  9.726342E-04   b4= -4.562908E-01
```
6 Pole:
```
a0=  1.797538E-05
a1=  1.078523E-04   b1=  4.921746E+00
a2=  2.696307E-04   b2= -1.035734E+01
a3=  3.595076E-04   b3=  1.189764E+01
a4=  2.696307E-04   b4= -7.854533E+00
a5=  1.078523E-04   b5=  2.822109E+00
a6=  1.797538E-05   b6= -4.307710E-01
```

0.1

2 Pole:
```
a0=  6.372802E-02
a1=  1.274560E-01   b1=  1.194365E+00
a2=  6.372802E-02   b2= -4.492774E-01
```
4 Pole:
```
a0=  2.780755E-03
a1=  1.112302E-02   b1=  2.764031E+00
a2=  1.668453E-02   b2= -3.122854E+00
a3=  1.112302E-02   b3=  1.664554E+00
a4=  2.780755E-03   b4= -3.502232E-01
```
6 Pole:
```
a0=  9.086148E-05
a1=  5.451688E-04   b1=  4.470118E+00
a2=  1.362922E-03   b2= -8.755594E+00
a3=  1.817229E-03   b3=  9.543712E+00
a4=  1.362922E-03   b4= -6.079376E+00
a5=  5.451688E-04   b5=  2.140062E+00
a6=  9.086148E-05   b6= -3.247363E-01
```

0.15

2 Pole:
```
a0=  1.254285E-01
a1=  2.508570E-01   b1=  8.070778E-01
a2=  1.254285E-01   b2= -3.087918E-01
```
4 Pole:
```
a0=  1.180009E-02
a1=  4.720034E-02   b1=  2.039039E+00
a2=  7.080051E-02   b2= -2.012961E+00
a3=  4.720034E-02   b3=  9.897915E-01
a4=  1.180009E-02   b4= -2.046700E-01
```
6 Pole:
```
a0=  8.618665E-04
a1=  5.171199E-03   b1=  3.455239E+00
a2=  1.292800E-02   b2= -5.754735E+00
a3=  1.723733E-02   b3=  5.645387E+00
a4=  1.292800E-02   b4= -3.394902E+00
a5=  5.171199E-03   b5=  1.177469E+00
a6=  8.618665E-04   b6= -1.836195E-01
```

0.2

2 Pole:
```
a0=  1.997396E-01
a1=  3.994792E-01   b1=  4.291048E-01
a2=  1.997396E-01   b2= -2.280633E-01
```
4 Pole:
```
a0=  3.224554E-02
a1=  1.289821E-01   b1=  1.265912E+00
a2=  1.934732E-01   b2= -1.203878E+00
a3=  1.289821E-01   b3=  5.405908E-01
a4=  3.224554E-02   b4= -1.185538E-01
```
6 Pole:
```
a0=  4.187408E-03
a1=  2.512445E-02   b1=  2.315806E+00
a2=  6.281112E-02   b2= -3.293726E+00
a3=  8.374816E-02   b3=  2.904862E+00
a4=  6.281112E-02   b4= -1.694128E+00
a5=  2.512445E-02   b5=  6.021426E-01
a6=  4.187408E-03   b6= -1.029147E-01
```

0.25

2 Pole:
```
a0=  2.858110E-01
a1=  5.716221E-01   b1=  5.423258E-02
a2=  2.858110E-01   b2= -1.974768E-01
```
4 Pole:
```
a0=  7.015301E-02
a1=  2.806120E-01   b1=  4.541481E-01
a2=  4.209180E-01   b2= -7.417536E-01
a3=  2.806120E-01   b3=  2.361222E-01
a4=  7.015301E-02   b4= -7.096476E-02
```
6 Pole:
```
a0=  1.434449E-02
a1=  8.606697E-02   b1=  1.076052E+00
a2=  2.151674E-01   b2= -1.662847E+00
a3=  2.868899E-01   b3=  1.191063E+00
a4=  2.151674E-01   b4= -7.403087E-01
a5=  8.606697E-02   b5=  2.752158E-01
a6=  1.434449E-02   b6= -5.722251E-02
```

0.3

2 Pole:
```
a0=  3.849163E-01
a1=  7.698326E-01   b1= -3.249116E-01
a2=  3.849163E-01   b2= -2.147536E-01
```
4 Pole:
```
a0=  1.335566E-01
a1=  5.342263E-01   b1= -3.904486E-01
a2=  8.013394E-01   b2= -6.784138E-01
a3=  5.342263E-01   b3= -1.412021E-02
a4=  1.335566E-01   b4= -5.392238E-02
```
6 Pole:
```
a0=  3.997487E-02
a1=  2.398492E-01   b1= -2.441152E-01
a2=  5.996231E-01   b2= -1.130306E+00
a3=  7.994975E-01   b3=  1.063167E-01
a4=  5.996231E-01   b4= -3.463299E-01
a5=  2.398492E-01   b5=  8.882992E-02
a6=  3.997487E-02   b6= -3.278741E-02
```

0.35

2 Pole:
```
a0=  5.001024E-01
a1=  1.000205E+00   b1= -7.158993E-01
a2=  5.001024E-01   b2= -2.845103E-01
```
4 Pole:
```
a0=  2.340973E-01
a1=  9.363892E-01   b1= -1.263672E+00
a2=  1.404584E+00   b2= -1.080487E+00
a3=  9.363892E-01   b3= -3.276296E-01
a4=  2.340973E-01   b4= -7.376791E-02
```
6 Pole:
```
a0=  9.792321E-02
a1=  5.875393E-01   b1= -1.627573E+00
a2=  1.468848E+00   b2= -1.955020E+00
a3=  1.958464E+00   b3= -1.075051E+00
a4=  1.468848E+00   b4= -5.106501E-01
a5=  5.875393E-01   b5= -7.239843E-02
a6=  9.792321E-02   b6= -2.639193E-02
```

0.40

2 Pole:
```
a0=  6.362308E-01
a1=  1.272462E+00   b1= -1.125379E+00
a2=  6.362308E-01   b2= -4.195441E-01
```
4 Pole:
```
a0=  3.896966E-01
a1=  1.558787E+00   b1= -2.161179E+00
a2=  2.338180E+00   b2= -2.033992E+00
a3=  1.558787E+00   b3= -8.789098E-01
a4=  3.896966E-01   b4= -1.610655E-01
```
6 Pole:
```
a0=  2.211834E-01
a1=  1.327100E+00   b1= -3.058672E+00
a2=  3.317751E+00   b2= -4.390465E+00
a3=  4.423668E+00   b3= -3.523254E+00
a4=  3.317751E+00   b4= -1.684185E+00
a5=  1.327100E+00   b5= -4.414881E-01
a6=  2.211834E-01   b6= -5.767513E-02
```

0.45

2 Pole:
```
a0=  8.001101E-01
a1=  1.600220E+00   b1= -1.556269E+00
a2=  8.001101E-01   b2= -6.441713E-01
```
4 Pole:
```
a0=  6.291693E-01
a1=  2.516677E+00   b1= -3.077062E+00
a2=  3.775016E+00   b2= -3.641323E+00
a3=  2.516677E+00   b3= -1.949229E+00
a4=  6.291693E-01   b4= -3.990945E-01
```
6 Pole:
```
a0=  4.760635E-01
a1=  2.856381E+00   b1= -4.522403E+00
a2=  7.140952E+00   b2= -8.676844E+00
a3=  9.521270E+00   b3= -9.007512E+00
a4=  7.140952E+00   b4= -5.328429E+00
a5=  2.856381E+00   b5= -1.702543E+00
a6=  4.760635E-01   b6= -2.303303E-01
```

TABLE 20-1
Low-pass Chebyshev filters (0.5% ripple)

f_C	2 Pole		4 Pole		6 Pole	
0.01	a0= 9.567529E-01 a1= -1.913506E+00 a2= 9.567529E-01	b1= 1.911437E+00 b2= -9.155749E-01	a0= 9.121579E-01 (!! Unstable !!) a1= -3.648632E+00 b1= 3.815952E+00 a2= 5.472947E+00 b2= -5.465026E+00 a3= -3.648632E+00 b3= 3.481295E+00 a4= 9.121579E-01 b4= -8.322529E-01		a0= 8.630195E-01 (!! Unstable !!) a1= -5.178118E+00 b1= 5.705102E+00 a2= 1.294529E+01 b2= -1.356935E+01 a3= -1.726039E+01 b3= 1.722231E+01 a4= 1.294529E+01 b4= -1.230214E+01 a5= -5.178118E+00 b5= 4.689218E+00 a6= 8.630195E-01 b6= -7.451429E-01	
0.025	a0= 8.950355E-01 a1= -1.790071E+00 a2= 8.950355E-01	b1= 1.777932E+00 b2= -8.022106E-01	a0= 7.941874E-01 a1= -3.176750E+00 b1= 3.538919E+00 a2= 4.765125E+00 b2= -4.722213E+00 a3= -3.176750E+00 b3= 2.814036E+00 a4= 7.941874E-01 b4= -6.318300E-01		a0= 6.912863E-01 (!! Unstable !!) a1= -4.147718E+00 b1= 5.261399E+00 a2= 1.036929E+01 b2= -1.157800E+01 a3= -1.382573E+01 b3= 1.363599E+01 a4= 1.036929E+01 b4= -9.063840E+00 a5= -4.147718E+00 b5= 3.223738E+00 a6= 6.912863E-01 b6= -4.793541E-01	
0.05	a0= 8.001102E-01 a1= -1.600220E+00 a2= 8.001102E-01	b1= 1.556269E+00 b2= -6.441715E-01	a0= 6.291694E-01 a1= -2.516678E+00 b1= 3.077062E+00 a2= 3.775016E+00 b2= -3.641324E+00 a3= -2.516678E+00 b3= 1.949230E+00 a4= 6.291694E-01 b4= -3.990947E-01		a0= 4.760636E-01 a1= -2.856382E+00 b1= 4.522403E+00 a2= 7.140954E+00 b2= -8.676846E+00 a3= -9.521272E+00 b3= 9.007515E+00 a4= 7.140954E+00 b4= -5.328431E+00 a5= -2.856382E+00 b5= 1.702544E+00 a6= 4.760636E-01 b6= -2.303304E-01	
0.075	a0= 7.142028E-01 a1= -1.428406E+00 a2= 7.142028E-01	b1= 1.338264E+00 b2= -5.185469E-01	a0= 4.965350E-01 a1= -1.986140E+00 b1= 2.617304E+00 a2= 2.979210E+00 b2= -2.749252E+00 a3= -1.986140E+00 b3= 1.325548E+00 a4= 4.965350E-01 b4= -2.524546E-01		a0= 3.259100E-01 a1= -1.955460E+00 b1= 3.787397E+00 a2= 4.888651E+00 b2= -6.288362E+00 a3= -6.518201E+00 b3= 5.747801E+00 a4= 4.888651E+00 b4= -3.041570E+00 a5= -1.955460E+00 b5= 8.808669E-01 a6= 3.259100E-01 b6= -1.122464E-01	
0.1	a0= 6.362307E-01 a1= -1.272461E+00 a2= 6.362307E-01	b1= 1.125379E+00 b2= -4.195440E-01	a0= 3.896966E-01 a1= -1.558786E+00 b1= 2.161179E+00 a2= 2.338179E+00 b2= -2.033991E+00 a3= -1.558786E+00 b3= 8.789094E-01 a4= 3.896966E-01 b4= -1.610655E-01		a0= 2.211833E-01 a1= -1.327100E+00 b1= 3.058671E+00 a2= 3.317750E+00 b2= -4.390464E+00 a3= -4.423667E+00 b3= 3.523252E+00 a4= 3.317750E+00 b4= -1.684184E+00 a5= -1.327100E+00 b5= 4.414878E-01 a6= 2.211833E-01 b6= -5.767508E-02	
0.15	a0= 5.001024E-01 a1= -1.000205E+00 a2= 5.001024E-01	b1= 7.158993E-01 b2= -2.845103E-01	a0= 2.340973E-01 a1= -9.363892E-01 b1= 1.263672E+00 a2= 1.404584E+00 b2= -1.080487E+00 a3= -9.363892E-01 b3= 3.276296E-01 a4= 2.340973E-01 b4= -7.376791E-02		a0= 9.792321E-02 a1= -5.875393E-01 b1= 1.627573E+00 a2= 1.468848E+00 b2= -1.955020E+00 a3= -1.958464E+00 b3= 1.075051E+00 a4= 1.468848E+00 b4= -5.106501E-01 a5= -5.875393E-01 b5= 7.239843E-02 a6= 9.792321E-02 b6= -2.639193E-02	
0.2	a0= 3.849163E-01 a1= -7.698326E-01 a2= 3.849163E-01	b1= 3.249116E-01 b2= -2.147536E-01	a0= 1.335566E-01 a1= -5.342262E-01 b1= 3.904484E-01 a2= 8.013393E-01 b2= -6.784138E-01 a3= -5.342262E-01 b3= 1.412016E-02 a4= 1.335566E-01 b4= -5.392238E-02		a0= 3.997486E-02 a1= -2.398492E-01 b1= 2.441149E-01 a2= 5.996230E-01 b2= -1.130306E+00 a3= -7.994973E-01 b3= -1.063169E-01 a4= 5.996230E-01 b4= -3.463299E-01 a5= -2.398492E-01 b5= -8.882996E-02 a6= 3.997486E-02 b6= -3.278741E-02	
0.25	a0= 2.858111E-01 a1= -5.716222E-01 a2= 2.858111E-01	b1= -5.423243E-02 b2= -1.974768E-01	a0= 7.015302E-02 a1= -2.806121E-01 b1= -4.541478E-01 a2= 4.209180E-01 b2= -7.417535E-01 a3= -2.806121E-01 b3= -2.361221E-01 a4= 7.015302E-02 b4= -7.096475E-02		a0= 1.434450E-02 a1= -8.606701E-02 b1= -1.076051E+00 a2= 2.151675E-01 b2= -1.662847E+00 a3= -2.868900E-01 b3= -1.191062E+00 a4= 2.151675E-01 b4= -7.403085E-01 a5= -8.606701E-02 b5= -2.752156E-01 a6= 1.434450E-02 b6= -5.722250E-02	
0.3	a0= 1.997396E-01 a1= -3.994792E-01 a2= 1.997396E-01	b1= -4.291049E-01 b2= -2.280633E-01	a0= 3.224553E-02 a1= -1.289821E-01 b1= -1.265912E+00 a2= 1.934732E-01 b2= -1.203878E+00 a3= -1.289821E-01 b3= -5.405908E-01 a4= 3.224553E-02 b4= -1.185538E-01		a0= 4.187407E-03 a1= -2.512444E-02 b1= -2.315806E+00 a2= 6.281111E-02 b2= -3.293726E+00 a3= -8.374815E-02 b3= -2.904827E+00 a4= 6.281111E-02 b4= -1.694129E+00 a5= -2.512444E-02 b5= -6.021426E-01 a6= 4.187407E-03 b6= -1.029147E-01	
0.35	a0= 1.254285E-01 a1= -2.508570E-01 a2= 1.254285E-01	b1= -8.070777E-01 b2= -3.087918E-01	a0= 1.180009E-02 a1= -4.720035E-02 b1= -2.039039E+00 a2= 7.080051E-02 b2= -2.012961E+00 a3= -4.720035E-02 b3= -9.897915E-01 a4= 1.180009E-02 b4= -2.046700E-01		a0= 8.618665E-04 a1= -5.171200E-03 b1= -3.455239E+00 a2= 1.292800E-02 b2= -5.754734E+00 a3= -1.723733E-02 b3= -5.645387E+00 a4= 1.292800E-02 b4= -3.394902E+00 a5= -5.171200E-03 b5= -1.177469E+00 a6= 8.618665E-04 b6= -1.836195E-01	
0.40	a0= 6.372801E-02 a1= -1.274560E-01 a2= 6.372801E-02	b1= -1.194365E+00 b2= -4.492774E-01	a0= 2.780754E-03 a1= -1.112302E-02 b1= -2.764031E+00 a2= 1.668453E-02 b2= -3.122854E+00 a3= -1.112302E-02 b3= -1.664554E+00 a4= 2.780754E-03 b4= -3.502233E-01		a0= 9.086141E-05 a1= -5.451685E-04 b1= -4.470118E+00 a2= 1.362921E-03 b2= -8.755595E+00 a3= -1.817228E-03 b3= -9.543712E+00 a4= 1.362921E-03 b4= -6.079377E+00 a5= -5.451685E-04 b5= -2.140062E+00 a6= 9.086141E-05 b6= -3.247363E-01	
0.45	a0= 1.868823E-02 a1= -3.737647E-02 a2= 1.868823E-02	b1= -1.593937E+00 b2= -6.686903E-01	a0= 2.141509E-04 a1= -8.566037E-04 b1= -3.425455E+00 a2= 1.284906E-03 b2= -4.479272E+00 a3= -8.566037E-04 b3= -2.643718E+00 a4= 2.141509E-04 b4= -5.933269E-01		a0= 1.771089E-06 a1= -1.062654E-05 b1= -5.330512E+00 a2= 2.656634E-05 b2= -1.196611E+01 a3= -3.542179E-05 b3= -1.447067E+01 a4= 2.656634E-05 b4= -9.937710E+00 a5= -1.062654E-05 b5= -3.673283E+00 a6= 1.771089E-06 b6= -5.707561E-01	

TABLE 20-2
High-pass Chebyshev filters (0.5% ripple)

There are two problems with using tables to design digital filters. First, tables have a limited choice of parameters. For instance, Table 20-1 only provides 12 different cutoff frequencies, a maximum of 6 poles per filter, and *no* choice of passband ripple. Without the ability to select parameters from a continuous range of values, the filter design cannot be *optimized*. Second, the coefficients must be manually transferred from the table into the program. This is very time consuming and will discourage you from trying alternative values.

Instead of using tabulated values, consider including a subroutine in your program that *calculates* the coefficients. Such a program is shown in Table 20-4. The good news is that the program is relatively simple in structure. After the four filter parameters are entered, the program spits out the "a" and "b" coefficients in the arrays A[ ] and B[ ]. The bad news is that the program calls the subroutine in Table 20-5. At first glance this subroutine is really ugly. Don't despair; it isn't as bad as it seems! There is one simple branch in line 1120. Everything else in the subroutine is straightforward number crunching. Six variables enter the routine, five variables leave the routine, and fifteen temporary variables (plus indexes) are used within. Table 20-5 provides two sets of test data for debugging this subroutine. Chapter 31 discusses the operation of this program in detail.

Step Response Overshoot

Butterworth and Chebyshev filters have an overshoot of 5 to 30% in their step responses, becoming larger as the number of poles is increased. Figure 20-3a shows the step response for two example Chebyshev filters. Figure (b) shows something that is unique to digital filters and has no counterpart in analog electronics: the amount of overshoot in the step response depends to a small degree on the cutoff frequency of the filter. The excessive overshoot and ringing in the step response results from the Chebyshev filter being optimized for the *frequency domain* at the expense of the *time domain*.

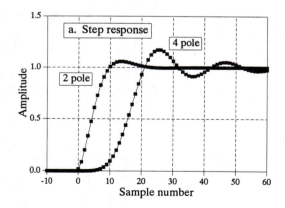
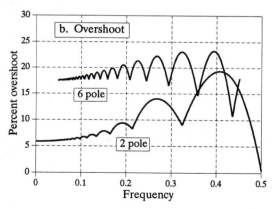

FIGURE 20-3

Chebyshev step response. The overshoot in the Chebyshev filter's step response is 5% to 30%, depending on the number of poles, as shown in (a), and the cutoff frequency, as shown in (b). Figure (a) is for a cutoff frequency of 0.05, and may be scaled to other cutoff frequencies.

Stability

The main limitation of digital filters carried out by convolution is *execution time*. It is possible to achieve nearly any filter response, provided you are willing to wait for the result. Recursive filters are just the opposite. They run like lightning; however, they are limited in performance. For example, consider a 6 pole, 0.5% ripple, low-pass filter with a 0.01 cutoff frequency. The recursion coefficients for this filter can be obtained from Table 20-1:

```
a0=  1.391351E-10
a1=  8.348109E-10    b1=   5.883343E+00
a2=  2.087027E-09    b2=  -1.442798E+01
a3=  2.782703E-09    b3=   1.887786E+01
a4=  2.087027E-09    b4=  -1.389914E+01
a5=  8.348109E-10    b5=   5.459909E+00
a6=  1.391351E-10    b6=  -8.939932E-01
```

Look carefully at these coefficients. The "b" coefficients have an absolute value of about *ten*. Using single precision, the round-off noise on each of these numbers is about one ten-millionth of the value, i.e., 10^{-6}. Now look at the "a" coefficients, with a value of about 10^{-9}. Something is obviously wrong here. The contribution from the input signal (via the "a" coefficients) will be 1000 times smaller than the *noise* from the previously calculated output signal (via the "b" coefficients). This filter won't work! In short, round-off noise limits the number of poles that can be used in a filter. The actual number will depend slightly on the ripple and if it is a high or low-pass filter. The approximate numbers for single precision are:

TABLE 20-3 The maximum number of poles for single precision.	Cutoff frequency	0.02	0.05	0.10	0.25	0.40	0.45	0.48
	Maximum poles	4	6	10	20	10	6	4

The filter's performance will start to degrade as this limit is approached; the step response will show more overshoot, the stopband attenuation will be poor, and the frequency response will have excessive ripple. If the filter is pushed too far, or there is an error in the coefficients, the output will probably oscillate until an overflow occurs.

There are two ways of extending the maximum number of poles that can be used. First, use double precision. This requires using double precision in the coefficient calculation as well (including the value for *pi*).

The second method is to implement the filter in *stages*. For example, a six pole filter starts out as a cascade of three stages of two poles each. The program in Table 20-4 combines these three stages into a single set of recursion coefficients for easier programming. However, the filter is more stable if carried out as the original three separate stages. This requires knowing the "a" and "b" coefficients for each of the stages. These can

```
100 'CHEBYSHEV FILTER-  RECURSION COEFFICIENT CALCULATION
110 '
120                                    'INITIALIZE VARIABLES
130 DIM A[22]                         'holds the "a" coefficients upon program completion
140 DIM B[22]                         'holds the "b" coefficients upon program completion
150 DIM TA[22]                        'internal use for combining stages
160 DIM TB[22]                        'internal use for combining stages
170 '
180 FOR I% = 0 TO 22
190   A[I%] = 0
200   B[I%] = 0
210 NEXT I%
220 '
230 A[2] = 1
240 B[2] = 1
250 PI = 3.14159265
260                                    'ENTER THE FOUR FILTER PARAMETERS
270 INPUT "Enter cutoff frequency  (0 to .5):      ", FC
280 INPUT "Enter  0 for LP,  1 for HP filter:      ", LH
290 INPUT "Enter percent ripple    (0 to 29):      ", PR
300 INPUT "Enter number of poles (2,4,...20):      ", NP
310 '
320 FOR P% = 1 TO NP/2                'LOOP FOR EACH POLE-PAIR
330   '
340   GOSUB 1000                      'The subroutine in TABLE 20-5
350   '
360   FOR I% = 0 TO 22                'Add coefficients to the cascade
370     TA[I%] = A[I%]
380     TB[I%] = B[I%]
390   NEXT I%
400   '
410   FOR I% = 2 TO 22
420     A[I%] = A0*TA[I%] + A1*TA[I%-1] + A2*TA[I%-2]
430     B[I%] =      TB[I%] - B1*TB[I%-1] - B2*TB[I%-2]
440   NEXT I%
450   '
460 NEXT P%
470 '
480 B[2] = 0                          'Finish combining coefficients
490 FOR I% = 0 TO 20
500   A[I%] = A[I%+2]
510   B[I%] = -B[I%+2]
520 NEXT I%
530 '
540 SA = 0                            'NORMALIZE THE GAIN
550 SB = 0
560 FOR I% = 0 TO 20
570   IF LH = 0 THEN SA     = SA   +  A[I%]
580   IF LH = 0 THEN SB     = SB   +  B[I%]
590   IF LH = 1 THEN SA     = SA   +  A[I%] * (-1) ^ I%
600   IF LH = 1 THEN SB     = SB   +  B[I%] * (-1) ^ I%
610 NEXT I%
620 '
630 GAIN = SA / (1 - SB)
640 '
650 FOR I% = 0 TO 20
660   A[I%] = A[I%] / GAIN
670 NEXT I%
680 '                                 'The final recursion coefficients are in A[ ] and B[ ]
690 END
```

TABLE 20-4

```
1000 'THIS SUBROUTINE IS CALLED FROM TABLE 20-4, LINE 340
1010 '
1020 '  Variables entering subroutine:       PI, FC, LH, PR, HP, P%
1030 '  Variables exiting subroutine:        A0, A1, A2, B1, B2
1040 '  Variables used internally:           RP, IP, ES, VX, KX, T, W, M, D, K,
1050 '                                       X0, X1, X2, Y1, Y2
1060 '
1070 '                                       'Calculate the pole location on the unit circle
1080 RP = -COS(PI/(NP*2) + (P%-1) * PI/NP)
1090 IP =   SIN(PI/(NP*2) + (P%-1) * PI/NP)
1100 '
1110 '                                       'Warp from a circle to an ellipse
1120 IF PR = 0 THEN GOTO 1210
1130 ES = SQR( (100 / (100-PR))^2 -1 )
1140 VX = (1/NP) * LOG( (1/ES) + SQR( (1/ES^2) + 1) )
1150 KX = (1/NP) * LOG( (1/ES) + SQR( (1/ES^2) - 1) )
1160 KX = (EXP(KX) + EXP(-KX))/2
1170 RP = RP * ( (EXP(VX) - EXP(-VX) ) /2 ) / KX
1180 IP  = IP * ( (EXP(VX) + EXP(-VX) ) /2 ) / KX
1190 '
1200 '                                       's-domain to z-domain conversion
1210 T  = 2 * TAN(1/2)
1220 W  = 2*PI*FC
1230 M  = RP^2 + IP^2
1240 D = 4 - 4*RP*T + M*T^2
1250 X0 = T^2/D
1260 X1 = 2*T^2/D
1270 X2 = T^2/D
1280 Y1 = (8 - 2*M*T^2)/D
1290 Y2 = (-4 - 4*RP*T - M*T^2)/D
1300 '
1310 '                                       'LP TO LP, or LP TO HP transform
1320 IF LH = 1 THEN K = -COS(W/2 + 1/2) / COS(W/2 - 1/2)
1330 IF LH = 0 THEN K =  SIN(1/2 - W/2) / SIN(1/2 + W/2)
1340 D = 1 + Y1*K - Y2*K^2
1350 A0 = (X0 - X1*K + X2*K^2)/D
1360 A1 = (-2*X0*K + X1 + X1*K^2 - 2*X2*K)/D
1370 A2 = (X0*K^2 - X1*K + X2)/D
1380 B1 = (2*K + Y1 + Y1*K^2 - 2*Y2*K)/D
1390 B2 = (-K^2 - Y1*K + Y2)/D
1400 IF LH = 1 THEN A1 = -A1
1410 IF LH = 1 THEN B1 = -B1
1420 '
1430 RETURN
```

TABLE 20-5

TABLE 20-4 and 20-5
Program to calculate the "a" and "b" coefficients for Chebyshev recursive filters. In lines 270-300, four parameters are entered into the program. The cutoff frequency, FC, is expressed as a fraction of the sampling frequency, and therefore must be in the range: 0 to 0.5. The variable, LH, is set to a value of *one* for a high-pass filter, and *zero* for a low-pass filter. The value entered for PR must be in the range of 0 to 29, corresponding to 0 to 29% ripple in the filter's frequency response. The number of poles in the filter, entered in the variable NP, must be an even integer between 2 and 20. At the completion of the program, the "a" and "b" coefficients are stored in the arrays A[ ] and B[ ] (a_0 = A[0], a_1 = A[1], etc.). TABLE 20-5 is a subroutine called from line 340 of the main program. Six variables are passed to this subroutine, and five variables are returned. Table 20-6 (next page) contains two sets of data to help debug this subroutine. The functions: COS and SIN, use radians, not degrees. The function: LOG is the natural (base *e*) logarithm. Declaring all floating point variables (including the value of π) to be double precision will allow more poles to be used. Tables 20-1 and 20-2 were generated with this program and can be used to test for proper operation. Chapter 31 describes the mathematical operation of this program.

DATA SET 1   DATA SET 2

Enter the subroutine with these values:

FC	=	0.1		FC	=	0.1
LH	=	0		LH	=	1
PR	=	0		PR	=	10
NP	=	4		NP	=	4
P%	=	1		P%	=	2
PI	=	3.141592		PI	=	3.141592

These values should be present at line 1200:

RP	=	-0.923879		RP	=	-0.136178
IP	=	0.382683		IP	=	0.933223
ES	=	not used		ES	=	0.484322
VX	=	not used		VX	=	0.368054
KX	=	not used		KX	=	1.057802

These values should be present at line 1310:

T	=	1.092605		T	=	1.092605
W	=	0.628318		W	=	0.628318
M	=	1.000000		M	=	0.889450
D	=	9.231528		D	=	5.656972
X0	=	0.129316		X0	=	0.211029
X1	=	0.258632		X1	=	0.422058
X2	=	0.129316		X2	=	0.211029
Y1	=	0.607963		Y1	=	1.038784
Y2	=	-0.125227		Y2	=	-0.789584

These values should be return to the main program:

A0	=	0.061885		A0	=	0.922919
A1	=	0.123770		A1	=	-1.845840
A2	=	0.061885		A2	=	0.922919
B1	=	1.048600		B1	=	1.446913
B2	=	-0.296140		B2	=	-0.836653

TABLE 20-6
Debugging data. This table contains two sets of data for debugging the
subroutine listed in Table 20-5.

be obtained from the program in Table 20-4. The subroutine in Table 20-5
is called once for each stage in the cascade. For example, it is called three
times for a six pole filter. At the completion of the subroutine, five
variables are return to the main program: *A0, A1, A2, B1, & B2*. These are
the recursion coefficients for the two pole stage being worked on, and can
be used to implement the filter in stages.

CHAPTER
21

Filter Comparison

Decisions, decisions, decisions! With all these filters to choose from, how do you know which to use? This chapter is a head-to-head competition between filters; we'll select champions from each side and let them fight it out. In the first match, *digital* filters are pitted against *analog* filters to see which technology is best. In the second round, the windowed-sinc is matched against the Chebyshev to find the king of the *frequency domain* filters. In the final battle, the moving average fights the single pole filter for the *time domain* championship. Enough talk; let the competition begin!

Match #1: Analog vs. Digital Filters

Most digital signals originate in analog electronics. If the signal needs to be filtered, is it better to use an analog filter before digitization, or a digital filter after? We will answer this question by letting two of the best contenders deliver their blows.

The goal will be to provide a low-pass filter at 1 kHz. Fighting for the analog side is a six pole Chebyshev filter with 0.5 dB (6%) ripple. As described in Chapter 3, this can be constructed with 3 op amps, 12 resistors, and 6 capacitors. In the digital corner, the windowed-sinc is warming up and ready to fight. The analog signal is digitized at a 10 kHz sampling rate, making the cutoff frequency 0.1 on the digital frequency scale. The length of the windowed-sinc will be chosen to be 129 points, providing the same 90% to 10% roll-off as the analog filter. Fair is fair. Figure 21-1 shows the frequency and step responses for these two filters.

Let's compare the two filters blow-by-blow. As shown in (a) and (b), the analog filter has a 6% ripple in the passband, while the digital filter is perfectly flat (within 0.02%). The analog designer might argue that the ripple can be *selected* in the design; however, this misses the point. The flatness achievable with analog filters is limited by the accuracy of their resistors and capacitors. Even if a Butterworth response is designed (i.e., 0% ripple), filters of this complexity will have a residue ripple of, perhaps, 1%. On the other hand, the flatness of digital filters is primarily limited by round-off error, making them *hundreds* of times flatter than their analog counterparts. Score one point for the digital filter.

Next, look at the frequency response on a log scale, as shown in (c) and (d). Again, the digital filter is clearly the victor in both *roll-off* and *stopband attenuation*. Even if the analog performance is improved by adding additional stages, it still can't compare to the digital filter. For instance, imagine that you need to improve these two parameters by a factor of 100. This can be done with simple modifications to the windowed-sinc, but is virtually impossible for the analog circuit. Score two more for the digital filter.

The step response of the two filters is shown in (e) and (f). The digital filter's step response is symmetrical between the lower and upper portions of the step, i.e., it has a linear phase. The analog filter's step response is *not* symmetrical, i.e., it has a nonlinear phase. One more point for the digital filter. Lastly, the analog filter overshoots about 20% on one side of the step. The digital filter overshoots about 10%, but on both sides of the step. Since both are bad, no points are awarded.

In spite of this beating, there are still many applications where analog filters should, or must, be used. This is not related to the actual performance of the filter (i.e., what goes in and what comes out), but to the general advantages that analog circuits have over digital techniques. The first advantage is *speed*: digital is slow; analog is fast. For example, a personal computer can only filter data at about 10,000 samples per second, using FFT convolution. Even simple op amps can operate at 100 kHz to 1 MHz, 10 to 100 times as fast as the digital system!

The second inherent advantage of analog over digital is *dynamic range*. This comes in two flavors. **Amplitude dynamic range** is the ratio between the largest signal that can be passed through a system, and the inherent noise of the system. For instance, a 12 bit ADC has a saturation level of 4095, and an rms quantization noise of 0.29 digital numbers, for a dynamic range of about 14000. In comparison, a standard op amp has a saturation voltage of about 20 volts and an internal noise of about 2 microvolts, for a dynamic range of about *ten million*. Just as before, a simple op amp devastates the digital system.

The other flavor is **frequency dynamic range**. For example, it is easy to design an op amp circuit to simultaneously handle frequencies between 0.01 Hz and 100 kHz (seven decades). When this is tried with a digital system, the computer becomes swamped with data. For instance, sampling at 200 kHz, it takes 20 million points to capture one complete cycle at 0.01 Hz. You may have noticed that the frequency response of digital filters is almost always plotted on a *linear* frequency scale, while analog filters are usually displayed with a *logarithmic* frequency. This is because digital filters need

Analog Filter
(6 pole 0.5dB Chebyshev)

Digital Filter
(129 point windowed-sinc)

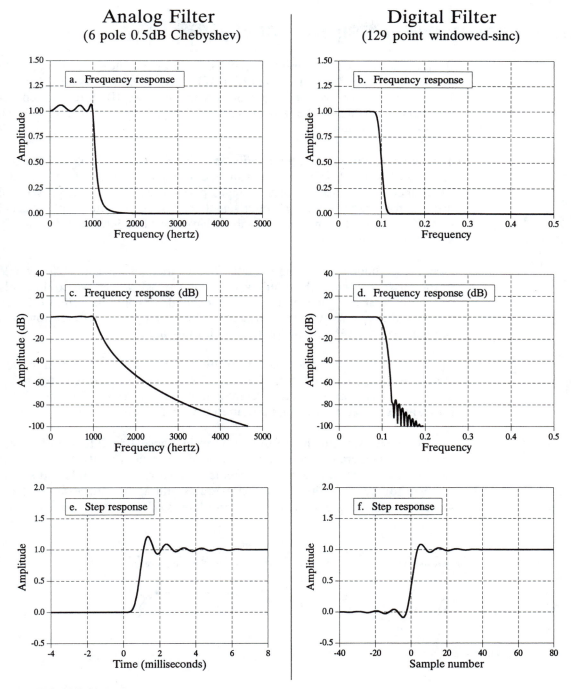

FIGURE 21-1
Comparison of analog and digital filters. Digital filters have better performance in many areas, such as: passband ripple, (a) vs. (b), roll-off and stopband attenuation, (c) vs. (d), and step response symmetry, (e) vs. (f). The digital filter in this example has a cutoff frequency of 0.1 of the 10 kHz sampling rate. This provides a fair comparison to the 1 kHz cutoff frequency of the analog filter.

a linear scale to show their exceptional filter performance, while analog filters need the logarithmic scale to show their huge dynamic range.

Match #2: Windowed-Sinc vs. Chebyshev

Both the windowed-sinc and the Chebyshev filters are designed to separate one band of frequencies from another. The windowed-sinc is an FIR filter implemented by *convolution*, while the Chebyshev is an IIR filter carried out by *recursion*. Which is the best digital filter in the frequency domain? We'll let them fight it out.

The recursive filter contender will be a 0.5% ripple, 6 pole Chebyshev low-pass filter. A fair comparison is complicated by the fact that the Chebyshev's frequency response changes with the cutoff frequency. We will use a cutoff frequency of 0.2, and select the windowed-sinc's filter kernel to be 51 points. This makes both filters have the same 90% to 10% roll-off, as shown in Fig. 21-2(a).

Now the pushing and shoving begins. The recursive filter has a 0.5% ripple in the passband, while the windowed-sinc is flat. However, we could easily set the recursive filter ripple to 0% if needed. No points. Figure 21-2b shows that the windowed-sinc has a much better stopband attenuation than the Chebyshev. One point for the windowed-sinc.

Figure 21-3 shows the step response of the two filters. Both are bad, as you should expect for frequency domain filters. The recursive filter has a nonlinear phase, but this can be corrected with bidirectional filtering. Since both filters are so ugly in this parameter, we will call this a draw.

So far, there isn't much difference between these two filters; either will work when moderate performance is needed. The heavy hitting comes over two critical issues: *maximum performance* and *speed*. The windowed-sinc is a powerhouse, while the Chebyshev is quick and agile. Suppose you have a really tough frequency separation problem, say, needing to isolate a 100

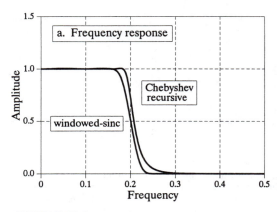
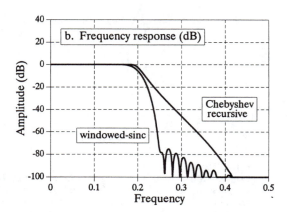

FIGURE 21-2
Windowed-sinc and Chebyshev frequency responses. Frequency responses are shown for a 51 point windowed-sinc filter and a 6 pole, 0.5% ripple Chebyshev recursive filter. The windowed-sinc has better stopband attenuation, but either will work in moderate performance applications. The cutoff frequency of both filters is 0.2, measured at an amplitude of 0.5 for the windowed-sinc, and 0.707 for the recursive.

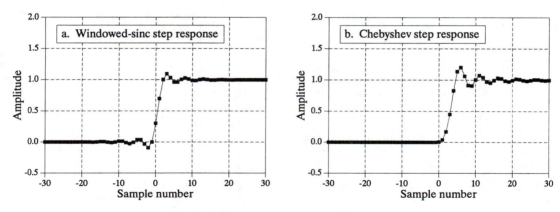

FIGURE 21-3
Windowed--sinc and Chebyshev step responses. The step responses are shown for a 51 point windowed-sinc filter and a 6 pole, 0.5% ripple Chebyshev recursive filter. Each of these filters has a cutoff frequency of 0.2. The windowed-sinc has a slightly better step response because it has less overshoot and a zero phase.

millivolt signal at 61 hertz that is riding on a 120 volt power line at 60 hertz. Figure 21-4 shows how these two filters compare when you need maximum performance. The recursive filter is a 6 pole Chebyshev with 0.5% ripple. This is the maximum number of poles that can be used at a 0.05 cutoff frequency with single precision. The windowed-sinc uses a 1001 point filter kernel, formed by convolving a 501 point windowed-sinc filter kernel with itself. As shown in Chapter 16, this provides greater stopband attenuation.

How do these two filters compare when maximum performance is needed? The windowed-sinc crushes the Chebyshev! Even if the recursive filter were improved (more poles, multistage implementation, double precision, etc.), it is still no match for the FIR performance. This is especially impressive when you consider that the windowed-sinc has only begun to fight. There are strong limits on the maximum performance that recursive filters can provide. The windowed-sinc, in contrast, can be pushed to incredible levels. This is, of course, provided you are willing to wait for the result. Which brings up the second critical issue: *speed*.

FIGURE 21-4
Maximum performance of FIR and IIR filters. The frequency response of the windowed-sinc can be virtually any shape needed, while the Chebyshev recursive filter is very limited. This graph compares the frequency response of a six pole Chebyshev recursive filter with a 1001 point windowed-sinc filter.

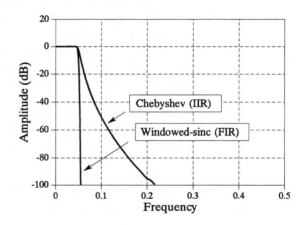

FIGURE 21-5
Comparing FIR and IIR execution speeds. These curves shows the relative execution times for a windowed-sinc filter compared with an equivalent six pole Chebyshev recursive filter. Curves are shown for implementing the FIR filter by both the standard and the FFT convolution algorithms. The windowed-sinc execution time rises at low and high frequencies because the filter kernel must be made longer to keep up with the greater performance of the recursive filter at these frequencies. In general, IIR filters are an order of magnitude faster than FIR filters of comparable performance.

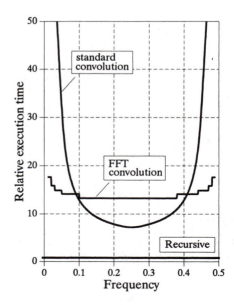

Comparing these filters for speed is like racing a Ferrari against a go-cart. Figure 21-5 shows how much longer the windowed-sinc takes to execute, compared to a six pole recursive filter. Since the recursive filter has a faster roll-off at low and high frequencies, the length of the windowed-sinc kernel must be made *longer* to match the performance (i.e., to keep the comparison fair). This accounts for the increased execution time for the windowed-sinc near frequencies 0 and 0.5. The important point is that FIR filters can be expected to be about an order of magnitude slower than comparable IIR filters (go-cart: 15 mph, Ferrari: 150 mph).

Match #3: Moving Average vs. Single Pole

Our third competition will be a battle of the time domain filters. The first fighter will be a nine point moving average filter. Its opponent for today's match will be a single pole recursive filter using the bidirectional technique. To achieve a comparable frequency response, the single pole filter will use a sample-to-sample decay of $x = 0.70$. The battle begins in Fig. 21-6 where the frequency response of each filter is shown. Neither one is very impressive, but of course, frequency separation isn't what these filters are used for. No points for either side.

Figure 21-7 shows the step responses of the filters. In (a), the moving average step response is a straight line, the most rapid way of moving from one level to another. In (b), the recursive filter's step response is smoother, which may be better for some applications. One point for each side.

These filters are quite equally matched in terms of performance and often the choice between the two is made on personal preference. However,

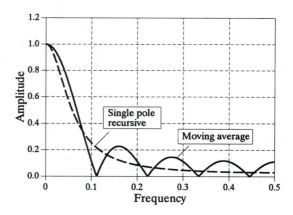

FIGURE 21-6
Moving average and single pole frequency responses. Both of these filters have a poor frequency response, as you should expect for time domain filters.

there are two cases where one filter has a slight edge over the other. These are based on the trade-off between *development* time and *execution* time. In the first instance, you want to reduce development time and are willing to accept a slower filter. For example, you might have a one time need to filter a few thousand points. Since the entire program runs in only a few seconds, it is pointless to spend time optimizing the algorithm. Floating point will almost certainly be used. The choice is to use the moving average filter carried out by convolution, or a single pole recursive filter. The winner here is the recursive filter. It will be slightly easier to program and modify, and will execute much faster.

The second case is just the opposite; your filter must operate as fast as possible and you are willing to spend the extra development time to get it. For instance, this filter might be a part of a commercial product, with the potential to be run *millions* of times. You will probably use integers for the highest possible speed. Your choice of filters will be the moving average

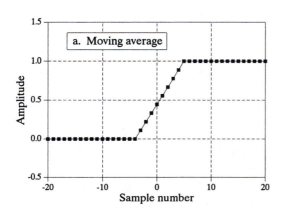

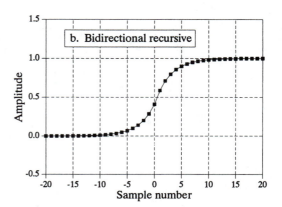

FIGURE 21-7
Step responses of the moving average and the bidirectional single pole filter. The moving average step response occurs over a smaller number of samples, while the single pole filter's step response is smoother.

carried out by *recursion*, or the single pole recursive filter implemented with look-up tables or integer math. The winner is the moving average filter. It will execute faster and not be susceptible to the development and execution problems of integer arithmetic.

CHAPTER 22

Audio Processing

Audio processing covers many diverse fields, all involved in presenting sound to human listeners. Three areas are prominent: (1) *high fidelity music reproduction*, such as in audio compact discs, (2) *voice telecommunications*, another name for telephone networks, and (3) *synthetic speech*, where computers generate and recognize human voice patterns. While these applications have different goals and problems, they are linked by a common umpire: the human ear. Digital Signal Processing has produced revolutionary changes in these and other areas of audio processing.

Human Hearing

The human ear is an exceedingly complex organ. To make matters even more difficult, the information from *two* ears is combined in a perplexing neural network, the human brain. Keep in mind that the following is only a brief overview; there are many subtle effects and poorly understood phenomena related to human hearing.

Figure 22-1 illustrates the major structures and processes that comprise the human ear. The *outer ear* is composed of two parts, the visible flap of skin and cartilage attached to the side of the head, and the *ear canal*, a tube about 0.5 cm in diameter extending about 3 cm into the head. These structures direct environmental sounds to the sensitive *middle and inner ear* organs located safely inside of the skull bones. Stretched across the end of the ear canal is a thin sheet of tissue called the *tympanic membrane* or *ear drum*. Sound waves striking the tympanic membrane cause it to vibrate. The middle ear is a set of small bones that transfer this vibration to the *cochlea* (inner ear) where it is converted to neural impulses. The cochlea is a liquid filled tube roughly 2 mm in diameter and 3 cm in length. Although shown straight in Fig. 22-1, the cochlea is curled up and looks like a small snail shell. In fact, *cochlea* is derived from the Greek word for *snail*.

When a sound wave tries to pass from air into liquid, only a small fraction of the sound is transmitted through the interface, while the remainder of the energy is reflected. This is because air has a *low* mechanical impedance (low acoustic pressure and high particle velocity resulting from low density and high compressibility), while liquid has a *high* mechanical impedance. In less technical terms, it requires more effort to wave your hand in water than it does to wave it in air. This difference in mechanical impedance results in most of the sound being reflected at an air/liquid interface.

The middle ear is an *impedance matching* network that increases the fraction of sound energy entering the liquid of the inner ear. For example, fish do not have an ear drum or middle ear, because they have no need to hear in air. Most of the impedance conversion results from the difference in *area* between the ear drum (receiving sound from the air) and the *oval window* (transmitting sound into the liquid, see Fig. 22-1). The ear drum has an area of about 60 (mm)^2, while the oval window has an area of roughly 4 (mm)^2. Since pressure is equal to force divided by area, this difference in area increases the sound wave pressure by about 15 times.

Contained within the cochlea is the *basilar membrane*, the supporting structure for about 12,000 sensory cells forming the *cochlear nerve*. The basilar membrane is stiffest near the oval window, and becomes more flexible toward the opposite end, allowing it to act as a *frequency spectrum analyzer*. When exposed to a high frequency signal, the basilar membrane resonates where it is stiff, resulting in the excitation of nerve cells close to the oval window. Likewise, low frequency sounds excite nerve cells at the far end of the basilar membrane. This makes specific fibers in the cochlear nerve respond to specific frequencies. This organization is called the **place principle**, and is preserved throughout the auditory pathway into the brain.

Another information encoding scheme is also used in human hearing, called the **volley principle**. Nerve cells transmit information by generating brief electrical pulses called *action potentials*. A nerve cell on the basilar membrane can encode audio information by producing an action potential in response to each cycle of the vibration. For example, a 200 hertz sound wave can be represented by a neuron producing 200 action potentials per second. However, this only works at frequencies below about 500 hertz, the maximum rate that neurons can produce action potentials. The human ear overcomes this problem by allowing several nerve cells to take turns performing this single task. For example, a 3000 hertz tone might be represented by *ten* nerve cells alternately firing at 300 times per second. This extends the range of the volley principle to about 4 kHz, above which the place principle is exclusively used.

Table 22-1 shows the relationship between sound intensity and perceived loudness. It is common to express sound intensity on a logarithmic scale, called **decibel SPL** (Sound Power Level). On this scale, 0 dB SPL is a sound wave power of 10^{-16} watts/cm^2, about the weakest sound detectable by the human ear. Normal speech is at about 60 dB SPL, while painful damage to the ear occurs at about 140 dB SPL.

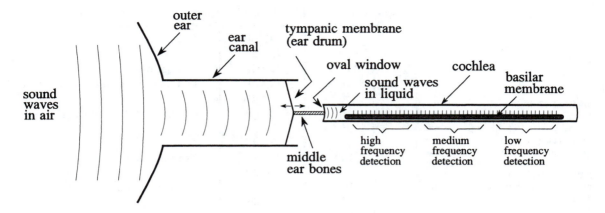

FIGURE 22-1
Functional diagram of the human ear. The outer ear collects sound waves from the environment and channels them to the tympanic membrane (ear drum), a thin sheet of tissue that vibrates in synchronization with the air waveform. The middle ear bones (hammer, anvil and stirrup) transmit these vibrations to the oval window, a flexible membrane in the fluid filled cochlea. Contained within the cochlea is the basilar membrane, the supporting structure for about 12,000 nerve cells that form the cochlear nerve. Due to the varying stiffness of the basilar membrane, each nerve cell only responses to a narrow range of audio frequencies, making the ear a frequency spectrum analyzer.

The difference between the loudest and faintest sounds that humans can hear is about 120 dB, a range of one-million in amplitude. Listeners can detect a *change* in loudness when the signal is altered by about 1 dB (a 12% change in amplitude). In other words, there are only about 120 levels of loudness that can be perceived from the faintest whisper to the loudest thunder. The sensitivity of the ear is amazing; when listening to very weak sounds, the ear drum vibrates less than the diameter of a single molecule!

The perception of loudness relates roughly to the sound power to an exponent of 1/3. For example, if you increase the sound power by a factor of *ten*, listeners will report that the loudness has increased by a factor of about *two* ($10^{1/3} \approx 2$). This is a major problem for eliminating undesirable environmental sounds, for instance, the beefed-up stereo in the next door apartment. Suppose you diligently cover 99% of your wall with a perfect soundproof material, missing only 1% of the surface area due to doors, corners, vents, etc. Even though the sound power has been reduced to only 1% of its former value, the perceived loudness has only dropped to about $0.01^{1/3} \approx 0.2$, or 20%.

The range of human hearing is generally considered to be 20 Hz to 20 kHz, but it is far more sensitive to sounds between 1 kHz and 4 kHz. For example, listeners can detect sounds as low as 0 dB SPL at 3 kHz, but require 40 dB SPL at 100 hertz (an amplitude increase of 100). Listeners can tell that two tones are different if their frequencies differ by more than about 0.3% at 3 kHz. This increases to 3% at 100 hertz. For comparison, adjacent keys on a piano differ by about 6% in frequency.

Watts/cm^2		Decibels SPL	Example sound
10^{-2}		140 dB	Pain
10^{-3}		130 dB	
10^{-4}		120 dB	Discomfort
10^{-5}		110 dB	Jack hammers and rock concerts
10^{-6}		100 dB	
10^{-7}	Louder	90 dB	OSHA limit for industrial noise
10^{-8}		80 dB	
10^{-9}		70 dB	
10^{-10}		60 dB	Normal conversation
10^{-11}	Softer	50 dB	
10^{-12}		40 dB	Weakest audible at 100 hertz
10^{-13}		30 dB	
10^{-14}		20 dB	Weakest audible at 10kHz
10^{-15}		10 dB	
10^{-16}		0 dB	Weakest audible at 3 kHz
10^{-17}		-10 dB	
10^{-18}		-20 dB	

TABLE 22-1
Units of sound intensity. Sound intensity is expressed as power per unit area (such as watts/cm^2), or more commonly on a logarithmic scale called *decibels SPL*. As this table shows, human hearing is the most sensitive between 1 kHz and 4 kHz.

The primary advantage of having *two* ears is the ability to identify the *direction* of the sound. Human listeners can detect the difference between two sound sources that are placed as little as three degrees apart, about the width of a person at 10 meters. This directional information is obtained in two separate ways. First, frequencies above about 1 kHz are strongly *shadowed* by the head. In other words, the ear nearest the sound receives a stronger signal than the ear on the opposite side of the head. The second clue to directionality is that the ear on the far side of the head hears the sound slightly *later* than the near ear, due to its greater distance from the source. Based on a typical head size (about 22 cm) and the speed of sound (about 340 meters per second), an angular discrimination of three degrees requires a timing precision of about 30 microseconds. Since this timing requires the volley principle, this clue to directionality is predominately used for sounds less than about 1 kHz.

Both these sources of directional information are greatly aided by the ability to turn the head and observe the change in the signals. An interesting sensation occurs when a listener is presented with exactly the same sounds to both ears, such as listening to monaural sound through headphones. The brain concludes that the sound is coming from the center of the listener's head!

While human hearing can determine the *direction* a sound is from, it does poorly in identifying the *distance* to the sound source. This is because there are few clues available in a sound wave that can provide this information. Human hearing weakly perceives that high frequency sounds are nearby, while low frequency sounds are distant. This is because sound waves dissipate their higher frequencies as they propagate long distances. Echo content is another weak clue to distance, providing a perception of the room size. For example, sounds in a large auditorium will contain echoes at about 100 millisecond intervals, while 10 milliseconds is typical for a small office. Some species have solved this ranging problem by using *active sonar*. For example, bats and dolphins produce clicks and squeaks that reflect from nearby objects. By measuring the interval between transmission and echo, these animals can locate objects with about 1 cm resolution. Experiments have shown that some humans, particularly the blind, can also use active echo localization to a small extent.

Timbre

The perception of a continuous sound, such as a note from a musical instrument, is often divided into three parts: **loudness**, **pitch**, and **timbre** (pronounced "timber"). *Loudness* is a measure of sound wave intensity, as previously described. *Pitch* is the frequency of the fundamental component in the sound, that is, the frequency with which the waveform repeats itself. While there are subtle effects in both these perceptions, they are a straightforward match with easily characterized physical quantities.

Timbre is more complicated, being determined by the *harmonic content* of the signal. Figure 22-2 illustrates two waveforms, each formed by adding a 1 kHz sine wave with an amplitude of *one*, to a 3 kHz sine wave with an amplitude of *one-half*. The difference between the two waveforms is that the one shown in (b) has the higher frequency *inverted* before the addition. Put another way, the third harmonic (3 kHz) is phase shifted by 180 degrees compared to the first harmonic (1 kHz). In spite of the very different time domain waveforms, these two signals sound *identical*. This is because hearing is based on the *amplitude* of the frequencies, and is very insensitive to their *phase*. The *shape* of the time domain waveform is only indirectly related to hearing, and usually not considered in audio systems.

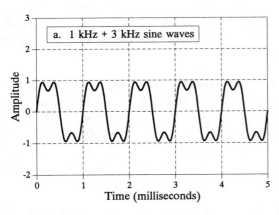
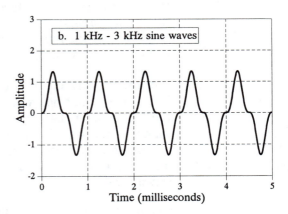

FIGURE 22-2
Phase detection of the human ear. The human ear is very insensitive to the relative phase of the component sinusoids. For example, these two waveforms would sound identical, because the *amplitudes* of their components are the same, even though their relative *phases* are different.

The ear's insensitivity to phase can be understood by examining how sound propagates through the environment. Suppose you are listening to a person speaking across a small room. Much of the sound reaching your ears is reflected from the walls, ceiling and floor. Since sound propagation depends on frequency (such as: attenuation, reflection, and resonance), different frequencies will reach your ear through different paths. This means that the relative phase of each frequency will change as you move about the room. Since the ear disregards these phase variations, you perceive the voice as *unchanging* as you move position. From a physics standpoint, the phase of an audio signal becomes randomized as it propagates through a complex environment. Put another way, the ear is insensitive to phase because it contains little useful information.

However, it cannot be said that the ear is completely deaf to the phase. This is because a phase change can rearrange the *time sequence* of an audio signal. An example is the chirp system (Chapter 11) that changes an impulse into a much longer duration signal. Although they differ only in their phase, the ear can distinguish between the two sounds because of their difference in duration. For the most part, this is just a curiosity, not something that happens in the normal listening environment.

Suppose that we ask a violinist to play a note, say, the *A* below middle *C*. When the waveform is displayed on an oscilloscope, it appear much as the sawtooth shown in Fig. 22-3a. This is a result of the sticky rosin applied to the fibers of the violinist's bow. As the bow is drawn across the string, the waveform is formed as the string sticks to the bow, is pulled back, and eventually breaks free. This cycle repeats itself over and over resulting in the sawtooth waveform.

Figure 22-3b shows how this sound is perceived by the ear, a frequency of 220 hertz, plus harmonics at 440, 660, 880 hertz, etc. If this note were played on another instrument, the waveform would *look* different; however, the ear would still hear a frequency of 220 hertz plus the harmonics. Since the two instruments produce the same fundamental frequency for this note, they sound similar, and are said to have identical *pitch*. Since the relative amplitude of the *harmonics* is different, they will not sound identical, and will be said to have different *timbre*.

It is often said that timbre is determined by the shape of the waveform. This is true, but slightly misleading. The perception of timbre results from the ear detecting harmonics. While harmonic content is determined by the shape of the waveform, the insensitivity of the ear to phase makes the relationship very one-sided. That is, a particular waveform will have only one timbre, while a particular timbre has an infinite number of possible waveforms.

The ear is very accustomed to hearing a fundamental plus harmonics. If a listener is presented with the combination of a 1 kHz and 3 kHz sine wave, they will report that it sounds natural and pleasant. If sine waves of 1 kHz and 3.1 kHz are used, it will sound objectionable.

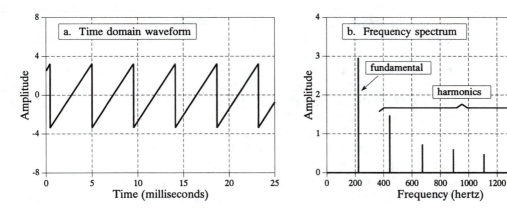

FIGURE 22-3

Violin waveform. A bowed violin produces a sawtooth waveform, as illustrated in (a). The sound *heard by the ear* is shown in (b), the fundamental frequency plus harmonics.

This is the basis of the standard musical scale, as illustrated by the piano keyboard in Fig. 22-4. Striking the farthest left key on the piano produces a fundamental frequency of 27.5 hertz, plus harmonics at 55, 110, 220, 440, 880 hertz, etc. (there are also harmonics between these frequencies, but they aren't important for this discussion). These harmonics correspond to the fundamental frequency produced by other keys on the keyboard. Specifically, every *seventh* white key is a harmonic of the far left key. That is, the eighth key from the left has a fundamental frequency of 55 hertz, the 15th key has a fundamental frequency of 110 hertz, etc. Being harmonics of each other, these keys sound similar when played, and are harmonious when played in unison. For this reason, they are *all* called the note, *A*. In this same manner, the white key immediate right of each *A* is called a *B*, and *they* are all harmonics of each other. This pattern repeats for the seven notes: *A*, *B*, *C*, *D*, *E*, *F*, and *G*.

The term **octave** means a *factor of two in frequency*. On the piano, one octave occurs every seven white keys, and the entire keyboard spans a little over seven octaves. The range of human hearing is generally quoted as 20 hertz to 20 kHz, corresponding to about ½ octave to the left, and two

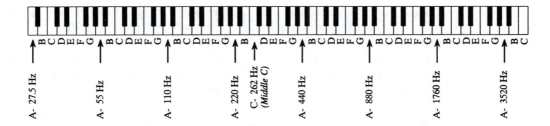

FIGURE 22-4

The Piano keyboard. The keyboard of the piano is a *logarithmic* frequency scale, with the fundamental frequency doubling every seven white keys. These white keys are the notes: *A, B, C, D, E, F and G*.

octaves to the right of the piano keyboard. Since octaves are based on doubling the frequency every fixed number of keys, they are a *logarithmic* representation of frequency. This is important because audio information is generally distributed in this same way. For example, as much audio information is carried in the octave between 50 hertz and 100 hertz, as in the octave between 10 kHz and 20 kHz. Even though the piano only covers about 20% of the frequencies that humans can hear (4 kHz out of 20 kHz), it can produce more than 70% of the audio information that humans can perceive (7 out of 10 octaves). Likewise, the highest frequency a human can detect drops from about 20 kHz to 10 kHz over the course of an adult's lifetime. However, this is only a loss of about 10% of the hearing ability (one octave out of ten). As shown next, this logarithmic distribution of information directly affects the required *sampling rate* of audio signals.

Sound Quality vs. Data Rate

When designing a digital audio system there are two questions that need to be asked: (1) how good does it need to sound? and (2) what data rate can be tolerated? The answer to these questions usually results in one of three categories. First, **high fidelity music,** where sound quality is of the greatest importance, and almost any data rate will be acceptable. Second, **telephone communication**, requiring natural sounding speech *and* a low data rate to reduce the system cost. Third, **compressed speech**, where reducing the data rate is very important and some unnaturalness in the sound quality can be tolerated. This includes military communication, cellular telephones, and digitally stored speech for voice mail and multimedia.

Table 22-2 shows the tradeoff between sound quality and data rate for these three categories. High fidelity music systems sample fast enough (44.1 kHz), and with enough precision (16 bits), that they can capture virtually all of the sounds that humans are capable of hearing. This magnificent sound quality comes at the price of a high data rate, 44.1 kHz × 16 bits = 706k bits/sec. This is pure brute force.

Whereas music requires a bandwidth of 20 kHz, natural sounding speech only requires about 3.2 kHz. Even though the frequency range has been reduced to only 16% (3.2 kHz out of 20 kHz), the signal still contains 80% of the original sound information (8 out of 10 octaves). Telecommunication systems typically operate with a sampling rate of about 8 kHz, allowing natural sounding speech, but greatly reduced music quality. You are probably already familiar with this difference in sound quality: FM radio stations broadcast with a bandwidth of almost 20 kHz, while AM radio stations are limited to about 3.2 kHz. Voices sound normal on the AM stations, but the music is weak and unsatisfying.

Voice-only systems also reduce the precision from 16 bits to 12 bits per sample, with little noticeable change in the sound quality. This can be reduced to only 8 bits per sample if the quantization step size is made unequal. This is a widespread procedure called **companding**, and will be

Sound Quality Required	Bandwidth	Sampling rate	Number of bits	Data rate (bits/sec)	Comments
High fidelity music (compact disc)	5 Hz to 20 kHz	44.1 kHz	16 bit	706k	Satisfies even the most picky audiophile. Better than human hearing.
Telephone quality speech	200 Hz to 3.2 kHz	8 kHz	12 bit	96k	Good speech quality, but very poor for music.
(with companding)	200 Hz to 3.2 kHz	8 kHz	8 bit	64k	Nonlinear ADC reduces the data rate by 50%. A very common technique.
Speech encoded by Linear Predictive Coding	200 Hz to 3.2 kHz	8 kHz	12 bit	4k	DSP speech compression technique. Very low data rates, poor voice quality.

TABLE 22-2
Audio data rate vs. sound quality. The sound quality of a digitized audio signal depends on its *data rate*, the product of its sampling rate and number of bits per sample. This can be broken into three categories, high fidelity music (706 kbits/sec), telephone quality speech (64 kbits/sec), and compressed speech (4 kbits/sec).

discussed later in this chapter. An 8 kHz sampling rate, with an ADC precision of 8 bits per sample, results in a data rate of 64k bits/sec. This is the *brute force* data rate for natural sounding speech. Notice that speech requires less than 10% of the data rate of high fidelity music.

The data rate of 64k bits/sec represents the straightforward application of sampling and quantization theory to audio signals. Techniques for lowering the data rate further are based on *compressing* the data stream by removing the inherent redundancies in speech signals. Data compression is the topic of Chapter 27. One of the most efficient ways of compressing an audio signal is **Linear Predictive Coding (LPC)**, of which there are several variations and subgroups. Depending on the speech quality required, LPC can reduce the data rate to as little as 2-6k bits/sec. We will revisit LPC later in this chapter with *speech synthesis*.

High Fidelity Audio

Audiophiles demand the utmost sound quality, and all other factors are treated as secondary. If you had to describe the mindset in one word, it would be: *overkill*. Rather than just matching the abilities of the human ear, these systems are designed to *exceed* the limits of hearing. It's the only way to be sure that the reproduced music is pristine. Digital audio was brought to the world by the **compact laser disc**, or **CD**. This was a revolution in music; the sound quality of the CD system far exceeds older systems, such as records and tapes. DSP has been at the forefront of this technology.

Figure 22-5 illustrates the surface of a compact laser disc, such as viewed through a high power microscope. The main surface is shiny (reflective of light), with the digital information stored as a series of dark pits burned on the surface with a laser. The information is arranged in a single track that spirals from the outside to the inside, the same as a phonograph record. The rotation of the CD is changed from about 210 to 480 rpm as the information is read from the outside to the inside of the spiral, making the scanning velocity a constant 1.2 meters per second. (In comparison, phonograph records spin at a fixed rate, such as 33, 45 or 78 rpm). During playback, an optical sensor detects if the surface is reflective or nonreflective, generating the corresponding binary information.

As shown by the geometry in Fig. 22-5, the CD stores about 1 bit per $(\mu m)^2$, corresponding to 1 million bits per $(mm)^2$, and 15 billion bits per disk. This is about the same feature size used in integrated circuit manufacturing, and for a good reason. One of the properties of light is that it cannot be focused to smaller than about one-half wavelength, or 0.3 μm. Since both integrated circuits and laser disks are created by optical means, the fuzziness of light below 0.3 μm limits how small of features can be used.

Figure 22-6 shows a block diagram of a typical compact disc playback system. The raw data rate is 4.3 million bits per second, corresponding to 1 bit each 0.28 μm of track length. However, this is in conflict with the specified geometry of the CD; each pit must be no shorter than 0.8 μm, and no longer than 3.5 μm. In other words, each binary *one* must be part of a group of 3 to 13 *ones*. This has the advantage of reducing the error rate due to the optical pickup, but how do you force the binary data to comply with this strange bunching?

The answer is an encoding scheme called **eight-to-fourteen modulation** **(EFM)**. Instead of directly storing a byte of data on the disc, the 8 bits are passed through a look-up table that pops out 14 bits. These 14 bits have the desired bunching characteristics, and are stored on the laser disc. Upon playback, the binary values read from the disc are passed through the inverse of the EFM look-up table, resulting in each 14 bit group being turned back into the correct 8 bits.

FIGURE 22-5
Compact disc surface. Micron size pits are burned into the surface of the CD to represent ones and zeros. This results in a data density of 1 bit per μm^2, or one million bits per mm^2. The pit depth is 0.16 μm.

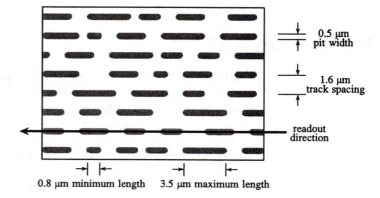

0.5 μm
pit width

1.6 μm
track spacing

readout
direction

0.8 μm minimum length   3.5 μm maximum length

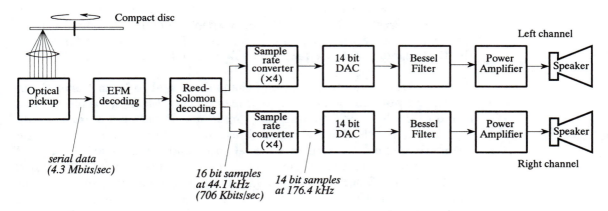

FIGURE 22-6
Compact disc playback block diagram. The digital information is retrieved from the disc with an optical sensor, corrected for EFM and Reed-Solomon encoding, and converted to stereo analog signals.

In addition to EFM, the data are encoded in a format called **two-level Reed-Solomon coding**. This involves combining the left and right stereo channels along with data for error detection and correction. Digital errors detected during playback are either: *corrected* by using the redundant data in the encoding scheme, *concealed* by interpolating between adjacent samples, or *muted* by setting the sample value to zero. These encoding schemes result in the data rate being *tripled*, i.e., 1.4 Mbits/sec for the stereo audio signals versus 4.3 Mbits/sec stored on the disc.

After decoding and error correction, the audio signals are represented as 16 bit samples at a 44.1 kHz sampling rate. In the simplest system, these signals could be run through a 16 bit DAC, followed by a low-pass analog filter. However, this would require high performance analog electronics to pass frequencies below 20 kHz, while rejecting all frequencies above 22.05 kHz, ½ of the sampling rate. A more common method is to use a **multirate** technique, that is, convert the digital data to a higher sampling rate before the DAC. A factor of four is commonly used, converting from 44.1 kHz to 176.4 kHz. This is called **interpolation**, and can be explained as a two step process (although it may not actually be carried out this way). First, three samples with a value of zero are placed between the original samples, producing the higher sampling rate. In the frequency domain, this has the effect of duplicating the 0 to 22.05 kHz spectrum three times, at 22.05 to 44.1 kHz, 41 to 66.15 kHz, and 66.15 to 88.2 kHz. In the second step, an efficient *digital* filter is used to remove the newly added frequencies.

The sample rate increase makes the sampling interval smaller, resulting in a smoother signal being generated by the DAC. The signal still contains frequencies between 20 Hz and 20 kHz; however, the Nyquist frequency has been increased by a factor of four. This means that the analog filter only needs to pass frequencies below 20 kHz, while blocking frequencies above 88.2 kHz. This is usually done with a three pole Bessel filter. Why use a *Bessel* filter if the ear is insensitive to phase? Overkill, remember?

Since there are four times as many samples, the number of bits per sample can be reduced from 16 bits to 14 bits, without degrading the sound quality. The $\sin(x)/x$ correction needed to compensate for the zeroth order hold of the DAC can be part of either the analog or digital filter.

Audio systems with more than one channel are said to be in **stereo** (from the Greek word for *solid*, or *three-dimensional*). Multiple channels send sound to the listener from different directions, providing a more accurate reproduction of the original music. Music played through a monaural (one channel) system often sounds artificial and bland. In comparison, a good stereo reproduction makes the listener feel as if the musicians are only a few feet away. Since the 1960s, high fidelity music has used two channels (left and right), while motion pictures have used four channels (left, right, center, and surround). In early stereo recordings (say, the Beatles or the Mamas And The Papas), individual singers can often be heard in only one channel or the other. This rapidly progressed into a more sophisticated **mix-down**, where the sound from many microphones in the recording studio is combined into the two channels. Mix-down is an art, aimed at providing the listener with the perception of *being there*.

The four channel sound used in motion pictures is called **Dolby Stereo**, with the home version called **Dolby Surround Pro Logic**. ("Dolby" and "Pro Logic" are trademarks of Dolby Laboratories Licensing Corp.). The four channels are encoded into the standard left and right channels, allowing regular two-channel stereo systems to reproduce the music. A Dolby decoder is used during playback to recreate the four channels of sound. The left and right channels, from speakers placed on each side of the movie or television screen, is similar to that of a regular two-channel stereo system. The speaker for the center channel is usually placed directly above or below the screen. Its purpose is to reproduce speech and other visually connected sounds, keeping them firmly centered on the screen, regardless of the seating position of the viewer/listener. The surround speakers are placed to the left and right of the listener, and may involve as many as twenty speakers in a large auditorium. The surround channel only contains midrange frequencies (say, 100 Hz to 7 kHz), and is *delayed* by 15 to 30 milliseconds. This delay makes the listener perceive that speech is coming from the screen, and not the sides. That is, the listener hears the speech coming from the front, followed by a delayed version of the speech coming from the sides. The listener's mind interprets the delayed signal as a reflection from the walls, and ignores it.

Companding

The data rate is important in telecommunication because it is directly proportional to the *cost* of transmitting the signal. Saving bits is the same as saving money. **Companding** is a common technique for reducing the data rate of audio signals by making the quantization levels *unequal*. As previously mentioned, the loudest sound that can be tolerated (120 dB SPL) is about one-million times the amplitude of the weakest sound that can be detected (0 dB SPL). However, the ear cannot distinguish between sounds that are closer than about 1 dB (12% in amplitude) apart. In other words, there are only about 120 different loudness levels that can be detected, spaced logarithmically over an amplitude range of one-million.

This is important for digitizing audio signals. If the quantization levels are equally spaced, 12 bits must be used to obtain telephone quality speech. However, only 8 bits are required if the quantization levels are made *unequal*, matching the characteristics of human hearing. This is quite intuitive: if the signal is small, the levels need to be very close together; if the signal is large, a larger spacing can be used.

Companding can be carried out in three ways: (1) run the analog signal through a nonlinear circuit before reaching a linear 8 bit ADC, (2) use an 8 bit ADC that internally has unequally spaced steps, or (3) use a linear 12 bit ADC followed by a digital look-up table (12 bits in, 8 bits out). Each of these three options requires the same nonlinearity, just in a different place: an analog circuit, an ADC, or a digital circuit.

Two nearly identical standards are used for companding curves: **μ255 law** (also called **mu law**), used in North America, and **"A" law**, used in Europe. Both use a logarithmic nonlinearity, since this is what converts the spacing detectable by the human ear into a linear spacing. In equation form, the curves used in μ255 law and "A" law are given by:

EQUATION 22-1
Mu law companding. This equation provides the nonlinearity for μ255 law companding. The constant, μ, has a value of 255, accounting for the name of this standard.

$$y = \frac{\ln(1+\mu x)}{\ln(1+\mu)} \qquad \text{for } 0 \le x \le 1$$

$$y = \frac{1+\ln(Ax)}{1+\ln(A)} \qquad \text{for } 1/A \le x \le 1$$

EQUATION 22-2
"A" law companding. The constant, A, has a value of 87.6.

$$y = \frac{Ax}{1+\ln(A)} \qquad \text{for } 0 \le x \le 1/A$$

Figure 22-7 graphs these equations for the input variable, x, being between -1 and +1, resulting in the output variable also assuming values between -1 and +1. Equations 22-1 and 22-2 only handle positive input values; portions of the curves for negative input values are found from symmetry. As shown in (a), the curves for μ255 law and "A" law are nearly identical. The only significant difference is near the origin, shown in (b), where μ255 law is a smooth curve, and "A" law switches to a straight line.

Producing a stable nonlinearity is a difficult task for analog electronics. One method is to use the logarithmic relationship between current and

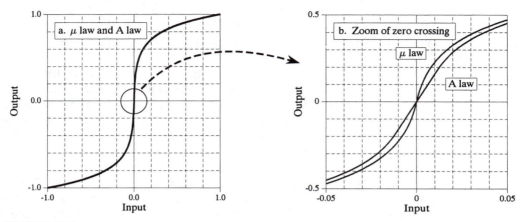

FIGURE 22-7
Companding curves. The $\mu255$ law and "A" law companding curves are nearly identical, differing only near the origin. Companding increases the amplitude when the signal is small, and decreases it when it is large.

voltage across a *pn* diode junction, and then add circuitry to correct for the ghastly temperature drift. Most companding circuits take another strategy: approximate the nonlinearity with a group of straight lines. A typical scheme is to approximate the logarithmic curve with a group of 16 straight segments, called **cords**. The first bit of the 8 bit output indicates if the input is positive or negative. The next three bits identify which of the 8 positive or 8 negative cords is used. The last four bits break each cord into 16 equally spaced increments. As with most integrated circuits, companding chips have sophisticated and proprietary internal designs. Rather than worrying about what goes on inside of the chip, pay the most attention to the pinout and the specification sheet.

Speech Synthesis and Recognition

Computer generation and recognition of speech are formidable problems; many approaches have been tried, with only mild success. This is an active area of DSP research, and will undoubtedly remain so for many years to come. You will be very disappointed if you are expecting this section to describe how to build speech synthesis and recognition circuits. Only a brief introduction to the typical approaches can be presented here. Before starting, it should be pointed out that most commercial products that produce human sounding speech do not *synthesize* it, but merely play back a digitally recorded segment from a human speaker. This approach has great sound quality, but it is limited to the prerecorded words and phrases.

Nearly all techniques for speech synthesis and recognition are based on the model of human speech production shown in Fig. 22-8. Most human speech sounds can be classified as either **voiced** or **fricative**. Voiced sounds occur when air is forced from the lungs, through the vocal cords, and out of the mouth and/or nose. The vocal cords are two thin flaps of tissue

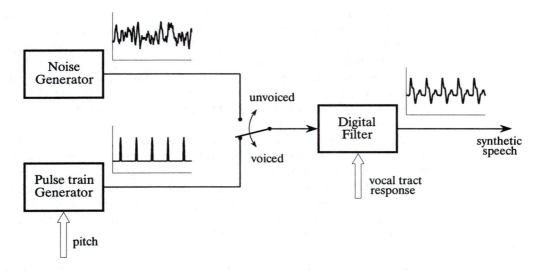

FIGURE 22-8
Human speech model. Over a short segment of time, about 2 to 40 milliseconds, speech can be modeled by three parameters: (1) the selection of either a periodic or a noise excitation, (2) the pitch of the periodic excitation, and (3) the coefficients of a recursive linear filter mimicking the vocal tract response.

stretched across the air flow, just behind the Adam's apple. In response to varying muscle tension, the vocal cords vibrate at frequencies between 50 and 1000 Hz, resulting in periodic puffs of air being injected into the throat. Vowels are an example of voiced sounds. In Fig. 22-8, voiced sounds are represented by the pulse train generator, with the pitch (i.e., the fundamental frequency of the waveform) being an adjustable parameter.

In comparison, *fricative* sounds originate as random noise, not from vibration of the vocal cords. This occurs when the air flow is nearly blocked by the tongue, lips, and/or teeth, resulting in air turbulence near the constriction. Fricative sounds include: *s*, *f*, *sh*, *z*, *v*, and *th*. In the model of Fig. 22-8, fricatives are represented by a *noise generator*.

Both these sound sources are modified by the acoustic cavities formed from the tongue, lips, mouth, throat, and nasal passages. Since sound propagation through these structures is a linear process, it can be represented as a linear filter with an appropriately chosen impulse response. In most cases, a *recursive* filter is used in the model, with the recursion coefficients specifying the filter's characteristics. Because the acoustic cavities have dimensions of several centimeters, the frequency response is primarily a series of resonances in the kilohertz range. In the jargon of audio processing, these resonance peaks are called the **format frequencies**. By changing the relative position of the tongue and lips, the format frequencies can be changed in both frequency and amplitude.

Figure 22-9 shows a common way to display speech signals, the **voice spectrogram**, or **voiceprint**. The audio signal is broken into short segments, say 2 to 40 milliseconds, and the FFT used to find the frequency spectrum of each segment. These spectra are placed side-by-side, and converted into a grayscale image (low amplitude becomes light, and high amplitude becomes dark). This provides a graphical way of observing how the frequency content of speech changes with time. The segment length is chosen as a tradeoff between *frequency resolution* (favored by longer segments) and *time resolution* (favored by shorter segments).

As demonstrated by the *a* in *rain*, voiced sounds have a periodic time domain waveform, shown in (a), and a frequency spectrum that is a series of regularly spaced harmonics, shown in (b). In comparison, the *s* in *storm*, shows that fricatives have a noisy time domain signal, as in (c), and a noisy spectrum, displayed in (d). These spectra also show the shaping by the format frequencies for both sounds. Also notice that the time-frequency display of the word *rain* looks similar both times it is spoken.

Over a short period, say 25 milliseconds, a speech signal can be approximated by specifying three parameters: (1) the selection of either a periodic or random noise excitation, (2) the frequency of the periodic wave (if used), and (3) the coefficients of the digital filter used to mimic the vocal tract response. Continuous speech can then be synthesized by continually updating these three parameters about 40 times a second. This approach was responsible for one the early commercial successes of DSP: the *Speak & Spell*, a widely marketed electronic learning aid for children. The sound quality of this type of speech synthesis is poor, sounding very mechanical and not quite human. However, it requires a very low data rate, typically only a few kbits/sec.

This is also the basis for the **linear predictive coding (LPC)** method of speech compression. Digitally recorded human speech is broken into short segments, and each is characterized according to the three parameters of the model. This typically requires about a dozen bytes per segment, or 2 to 6 kbytes/sec. The segment information is transmitted or stored as needed, and then reconstructed with the speech synthesizer.

Speech recognition algorithms take this a step further by trying to recognize patterns in the extracted parameters. This typically involves comparing the segment information with templates of previously stored sounds, in an attempt to identify the spoken words. The problem is, this method does not work very well. It is useful for some applications, but is far below the capabilities of human listeners. To understand why speech recognition is so difficult for computers, imagine someone unexpectedly speaking the following sentence:

Larger run medical buy dogs fortunate almost when.

Of course, you will not understand the meaning of this sentence, because it has none. More important, you will probably not even understand all of the individual words that were spoken. This is basic to the way that humans

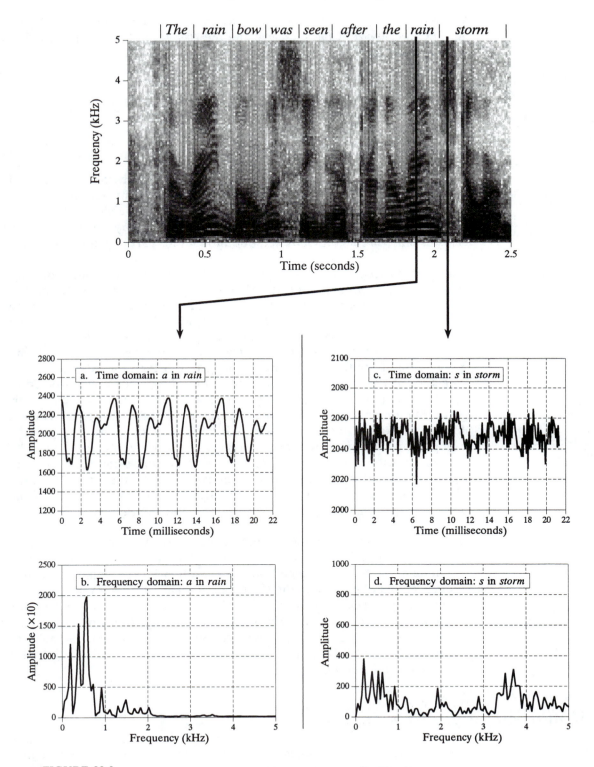

FIGURE 22-9
Voice spectrogram. The spectrogram of the phrase: *"The rainbow was seen after the rain storm."* Figures (a) and (b) shows the time and frequency signals for the voiced *a* in *rain*. Figures (c) and (d) show the time and frequency signals for the fricative *s* in *storm*.

perceive and understand speech. Words are recognized by their sounds, but also by the *context* of the sentence, and the *expectations* of the listener. For example, imagine hearing the two sentences:

The child wore a <u>spider ring</u> on Halloween.

He was an American <u>spy during</u> the war.

Even if exactly the same sounds were produced to convey the underlined words, listeners *hear* the correct words for the context. From your accumulated knowledge about the world, you know that children don't wear secret agents, and people don't become spooky jewelry during wartime. This usually isn't a conscious act, but an inherent part of human hearing.

Most speech recognition algorithms rely only on the sound of the individual words, and not on their context. They attempt to *recognize words*, but not to *understand speech*. This places them at a tremendous disadvantage compared to human listeners. Three annoyances are common in speech recognition systems: (1) The recognized speech must have distinct pauses between the words. This eliminates the need for the algorithm to deal with phrases that sound alike, but are composed of different words (i.e., *spider ring* and *spy during*). This is slow and awkward for people accustomed to speaking in an overlapping flow. (2) The vocabulary is often limited to only a few hundred words. This means that the algorithm only has to search a limited set to find the best match. As the vocabulary is made larger, the recognition time and error rate both increase. (3) The algorithm must be *trained* on each speaker. This requires each person using the system to speak each word to be recognized, often needing to be repeated five to ten times. This personalized database greatly increases the accuracy of the word recognition, but it is inconvenient and time consuming.

The prize for developing a successful speech recognition technology is enormous. Speech is the quickest and most efficient way for humans to communicate. Speech recognition has the potential of replacing writing, typing, keyboard entry, and the electronic control provided by switches and knobs. It just needs to work a little better to become accepted by the commercial marketplace. Progress in speech recognition will likely come from the areas of artificial intelligence and neural networks as much as through DSP itself. Don't think of this as a technical *difficulty*; think of it as a technical *opportunity*.

Nonlinear Audio Processing

Digital filtering can improve audio signals in many ways. For instance, *Wiener filtering* can be used to separate frequencies that are mainly signal, from frequencies that are mainly noise (see Chapter 17). Likewise, *deconvolution* can compensate for an undesired convolution, such as in the restoration of old recordings (also discussed in Chapter 17). These types of linear techniques are the backbone of DSP. Several *nonlinear* techniques are also useful for audio processing. Two will be briefly described here.

The first nonlinear technique is used for reducing wideband noise in speech signals. This type of noise includes: magnetic tape hiss, electronic noise in analog circuits, wind blowing by microphones, cheering crowds, etc. Linear filtering is of little use, because the frequencies in the noise completely overlap the frequencies in the voice signal, both covering the range from 200 hertz to 3.2 kHz. How can two signals be separated when they overlap in both the time domain *and* the frequency domain?

Here's how it is done. In a short segment of speech, the amplitude of the frequency components are greatly *unequal*. As an example, Fig. 22-10a illustrates the frequency spectrum of a 16 millisecond segment of speech (i.e., 128 samples at an 8 kHz sampling rate). Most of the signal is contained in a few large amplitude frequencies. In contrast, (b) illustrates the spectrum when only random noise is present; it is very irregular, but more uniformly distributed at a low amplitude.

Now the key concept: if both signal and noise are present, the two can be partially separated by looking at the *amplitude* of each frequency. If the amplitude is large, it is probably mostly signal, and should therefore be retained. If the amplitude is small, it can be attributed to mostly noise, and should therefore be discarded, i.e., set to zero. Mid-size frequency components are adjusted in some smooth manner between the two extremes.

Another way to view this technique is as a *time varying Wiener filter*. As you recall, the frequency response of the Wiener filter passes frequencies that are mostly signal, and rejects frequencies that are mostly noise. This

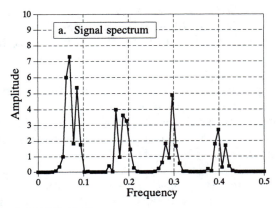

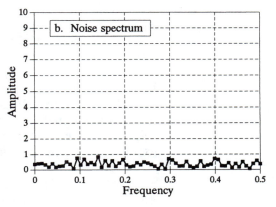

FIGURE 22-10
Spectra of speech and noise. While the frequency spectra of speech and noise generally overlap, there is some separation if the signal segment is made short enough. Figure (a) illustrates the spectrum of a 16 millisecond speech segment, showing many frequencies carry little speech information, *in this particular segment*. Figure (b) illustrates the spectrum of a random noise source; all the components have a small amplitude. (These graphs are not of real signals, but illustrations to show the noise reduction technique).

requires a knowledge of the signal and noise spectra *beforehand*, so that the filter's frequency response can be determined. This nonlinear technique uses the same idea, except that the Wiener filter's frequency response is recalculated for each segment, based on the spectrum *of that segment*. In other words, the filter's frequency response changes from segment-to-segment, as determined by the characteristics of the signal itself.

One of the difficulties in implementing this (and other) nonlinear techniques is that the overlap-add method for filtering long signals is not valid. Since the frequency response changes, the time domain waveform of each segment will no longer align with the neighboring segments. This can be overcome by remembering that audio information is encoded in frequency patterns that change over time, and not in the shape of the time domain waveform. A typical approach is to divide the original time domain signal into *overlapping* segments. After processing, a smooth window is applied to each of the over-lapping segments before they are recombined. This provides a smooth transition of the frequency spectrum from one segment to the next.

The second nonlinear technique is called **homomorphic** signal processing. This term literally means: *the same structure*. Addition is not the only way that noise and interference can be combined with a signal of interest; multiplication and convolution are also common means of mixing signals together. If signals are combined in a nonlinear way (i.e., anything other than addition), they cannot be separated by linear filtering. Homomorphic techniques attempt to separate signals combined in a nonlinear way by making the problem *become* linear. That is, the problem is converted to the *same structure* as a linear system.

For example, consider an audio signal transmitted via an AM radio wave. As atmospheric conditions change, the received amplitude of the signal increases and decreases, resulting in the loudness of the received audio signal slowly changing over time. This can be modeled as the audio signal, represented by $a[\ ]$, being *multiplied* by a slowly varying signal, $g[\ ]$, that represents the changing gain. This problem is usually handled in an electronic circuit called an *automatic gain control* (AGC), but it can also be corrected with nonlinear DSP.

As shown in Fig. 22-11, the input signal, $a[\ ] \times g[\ ]$, is passed through the logarithm function. From the identity, $\log(x+y) = \log x + \log y$, this results in two signals that are combined by addition, i.e., $\log a[\ ] + \log g[\ ]$. In other words, the *logarithm* is the homomorphic transform that turns the nonlinear problem of *multiplication* into the linear problem of *addition*.

Next, the added signals are separated by a conventional linear filter, that is, some frequencies are passed, while others are rejected. For the AGC, the gain signal, $g[\ ]$, will be composed of very low frequencies, far below the 200 hertz to 3.2 kHz band of the voice signal. The logarithm of these signals will have more complicated spectra, but the idea is the same: a high-pass filter is used to eliminate the varying gain component from the signal.

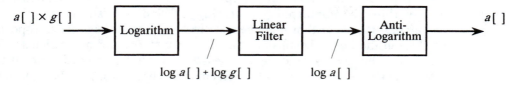

FIGURE 22-11
Homomorphic separation of multiplied signals. Taking the logarithm of the input signal transforms components that are *multiplied* into components that are *added*. These components can then be separated by linear filtering, and the effect of the logarithm undone.

In effect, $\log a[\ ] + \log g[\ ]$ is converted into $\log a[\ ]$. In the last step, the logarithm is undone by using the exponential function (the anti-logarithm, or e^x), producing the desired output signal, $a[\ ]$

Figure 22-12 shows a homomorphic system for separating signals that have been *convolved*. An application where this has proven useful is in removing echoes from audio signals. That is, the audio signal is convolved with an impulse response consisting of a delta function plus a shifted and scaled delta function. The homomorphic transform for convolution is composed of two stages, the *Fourier transform*, changing the convolution into a multiplication, followed by the *logarithm*, turning the multiplication into an addition. As before, the signals are then separated by linear filtering, and the homomorphic transform undone.

An interesting twist in Fig. 22-12 is that the linear filtering is dealing with frequency domain signals in the same way that time domain signals are usually processed. In other words, the time and frequency domains have been swapped from their normal use. For example, if FFT convolution were used to carry out the linear filtering stage, the "spectra" being multiplied would be in the *time domain*. This role reversal has given birth to a strange jargon. For instance, *cepstrum* (a rearrangement of *spectrum*) is the Fourier transform of the logarithm of the Fourier transform. Likewise, there are *long-pass* and *short-pass* filters, rather than low-pass and high-pass filters. Some authors even use *Quefrency Alanysis* and *liftering*.

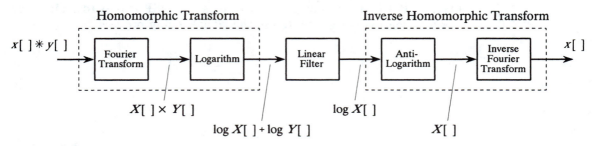

FIGURE 22-12
Homomorphic separation of convolved signals. Components that have been *convolved* are converted into components that are *added* by taking the Fourier transform followed by the logarithm. After linear filtering to separate the added components, the original steps are undone.

Keep in mind that these are simplified descriptions of sophisticated DSP algorithms; homomorphic processing is filled with subtle details. For example, the logarithm must be able to handle both negative and positive values in the input signal, since this is a characteristic of audio signals. This requires the use of the *complex logarithm*, a more advanced concept than the logarithm used in everyday science and engineering. When the linear filtering is restricted to be a *zero phase* filter, the complex log is found by taking the simple logarithm of the absolute value of the signal. After passing through the zero phase filter, the sign of the original signal is reapplied to the filtered signal.

Another problem is *aliasing* that occurs when the logarithm is taken. For example, imagine digitizing a continuous *sine wave*. In accordance with the sampling theorem, two or more samples per cycle is sufficient. Now consider digitizing the logarithm of this continuous sine wave. The sharp corners require many more samples per cycle to capture the waveform, i.e., to prevent aliasing. The required sampling rate can easily be 100 times as great after the log, as before. Further, it doesn't matter if the logarithm is applied to the continuous signal, or to its digital representation; the result is the same. Aliasing will result unless the sampling rate is high enough to capture the sharp corners produced by the nonlinearity. The result is that audio signals may need to be sampled at 100 kHz or more, instead of only the standard 8 kHz.

Even if these details are handled, there is no guarantee that the linearized signals *can* be separated by the linear filter. This is because the spectra of the linearized signals can overlap, even if the spectra of the original signals do not. For instance, imagine adding two sine waves, one at 1 kHz, and one at 2 kHz. Since these signals do not overlap in the frequency domain, they can be completely separated by linear filtering. Now imagine that these two sine waves are multiplied. Using homomorphic processing, the log is taken of the combined signal, resulting in the log of one sine wave plus the log of the other sine wave. The problem is, the logarithm of a sine wave contains many harmonics. Since the harmonics from the two signals overlap, their complete separation is not possible.

In spite of these obstacles, homomorphic processing teaches an important lesson: signals should be processed in a manner *consistent* with how they are formed. Put another way, the first step in any DSP task is to understand how information is represented in the signals being process.

chapter 23

Image Formation & Display

Images are a description of how a parameter varies over a surface. For example, standard visual images result from light intensity variations across a two-dimensional plane. However, light is not the only parameter used in scientific imaging. For example, an image can be formed of the *temperature* of an integrated circuit, *blood velocity* in a patient's artery, *x-ray emission* from a distant galaxy, *ground motion* during an earthquake, etc. These exotic images are usually converted into conventional pictures (i.e., light images), so that they can be evaluated by the human eye. This first chapter on image processing describes how digital images are formed and presented to human observers.

Digital Image Structure

Figure 23-1 illustrates the structure of a digital image. This example image is of the planet Venus, acquired by microwave radar from an orbiting space probe. Microwave imaging is necessary because the dense atmosphere blocks visible light, making standard photography impossible. The image shown is represented by 40,000 samples arranged in a two-dimensional array of 200 columns by 200 rows. Just as with one-dimensional signals, these rows and columns can be numbered 0 through 199, or 1 through 200. In imaging jargon, each sample is called a **pixel**, a contraction of the phrase: *picture element*. Each *pixel* in this example is a single number between 0 and 255. When the image was acquired, this number related to the amount of microwave energy being reflected from the corresponding location on the planet's surface. To display this as a *visual image*, the value of each pixel is converted into a **grayscale**, where 0 is black, 255 is white, and the intermediate values are shades of gray.

Images have their information encoded in the **spatial domain**, the image equivalent of the time domain. In other words, features in images are represented by *edges*, not *sinusoids*. This means that the spacing and number of pixels are determined by how small of features need to be seen, rather than by the formal constraints of the sampling theorem. Aliasing *can* occur in images, but it is generally thought of as a nuisance rather than a major problem. For instance, pinstriped suits look terrible on television because the repetitive pattern is greater than the Nyquist frequency. The aliased frequencies appear as light and dark bands that move across the clothing as the person changes position.

A "typical" digital image is composed of about 500 rows by 500 columns. This is the image quality encountered in television, personnel computer applications, and general scientific research. Images with fewer pixels, say 250 by 250, are regarded as having unusually poor resolution. This is frequently the case with new imaging modalities; as the technology matures, more pixels are added. These low resolution images look noticeably unnatural, and the individual pixels can often be seen. On the other end, images with more than 1000 by 1000 pixels are considered exceptionally good. This is the quality of the best computer graphics, high-definition television, and 35 mm motion pictures. There are also applications needing even higher resolution, requiring several thousand pixels per side: digitized x-ray images, space photographs, and glossy advertisements in magazines.

The strongest motivation for using lower resolution images is that there are *fewer* pixels to handle. This is not trivial; one of the most difficult problems in image processing is managing massive amounts of data. For example, one second of digital audio requires about eight *kilobytes*. In comparison, one second of television requires about eight *Megabytes*. Transmitting a 500 by 500 pixel image over a 33.6 kbps modem requires nearly a minute! Jumping to an image size of 1000 by 1000 *quadruples* these problems.

It is common for 256 **gray levels** (quantization levels) to be used in image processing, corresponding to a single byte per pixel. There are several reasons for this. First, a single byte is convenient for data management, since this is how computers usually store data. Second, the large number of pixels in an image compensate to a certain degree for a limited number of quantization steps. For example, imagine a group of adjacent pixels alternating in value between digital numbers (DN) 145 and 146. The human eye perceives the region as a brightness of 145.5. In other words, images are very *dithered*. Third, and most important, a brightness step size of 1/256 (0.39%) is smaller than the eye can perceive. An image presented to a human observer will not be improved by using more than 256 levels.

However, some images need to be stored with more than 8 bits per pixel. Remember, most of the images encountered in DSP represent nonvisual parameters. The acquired image may be able to take advantage of more quantization levels to properly capture the subtle details of the signal. The point of this is, don't expect to human eye to see all the information contained in these finely spaced levels. We will consider ways around this problem during a later discussion of brightness and contrast.

The value of each pixel in the digital image represents a small *region* in the continuous image being digitized. For example, imagine that the Venus

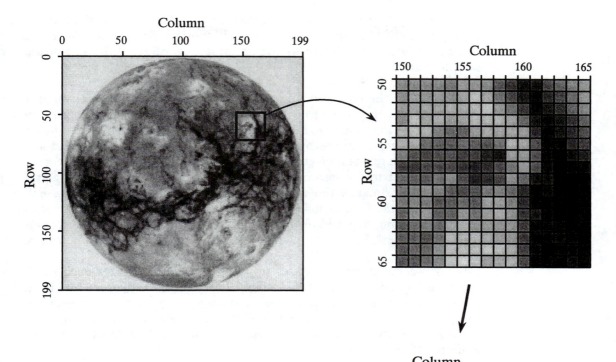

FIGURE 23-1
Digital image structure. This example image is the planet Venus, as viewed in reflected microwaves. Digital images are represented by a two-dimensional array of numbers, each called a *pixel*. In this image, the array is 200 rows by 200 columns, with each pixel a number between 0 to 255. When this image was acquired, the value of each pixel corresponded to the level of reflected microwave energy. A *grayscale* image is formed by assigning each of the 0 to 255 values to varying shades of gray.

Column

150					155					160					165
183	183	181	184	177	200	200	189	159	135	94	105	160	174	191	196
186	195	190	195	191	205	216	206	174	153	112	80	134	157	174	196
194	196	198	201	206	209	215	216	199	175	140	77	106	142	170	186
184	212	200	204	201	202	214	214	214	205	173	102	84	120	134	159
202	215	203	179	165	165	199	207	202	208	197	129	73	112	131	146
203	208	166	159	160	168	166	157	174	211	204	158	69	79	127	143
174	149	143	151	156	148	146	123	118	203	208	162	81	58	101	125
143	137	147	153	150	140	121	133	157	184	203	164	94	56	66	80
164	165	159	179	188	159	126	134	150	199	174	119	100	41	41	58
173	187	193	181	167	151	162	182	192	175	129	60	88	47	37	50
172	184	179	153	158	172	163	207	205	188	127	63	56	43	42	55
156	191	196	159	167	195	178	203	214	201	143	101	69	38	44	52
154	163	175	165	207	211	197	201	201	199	138	79	76	67	51	53
144	150	143	162	215	212	211	209	197	198	133	71	69	77	63	53
140	151	150	185	215	214	210	210	211	209	135	80	45	69	66	60
135	143	151	179	213	216	214	191	201	205	138	61	59	61	77	63

probe takes samples every 10 meters along the planet's surface as it orbits overhead. This defines a square **sample spacing** and **sampling grid,** with each pixel representing a 10 meter by 10 meter area. Now, imagine what happens in a single microwave reflection measurement. The space probe emits a highly focused burst of microwave energy, striking the surface in, for example, a circular area 15 meters in diameter. Each pixel therefore contains information about this circular area, regardless of the size of the sampling grid.

This region of the continuous image that contributes to the pixel value is called the **sampling aperture**. The size of the sampling aperture is often related to the inherent capabilities of the particular imaging system being used. For example, microscopes are limited by the quality of the optics and the wavelength of light, electronic cameras are limited by random electron diffusion in the image sensor, and so on. In most cases, the sampling grid is made approximately the same as the sampling aperture of the system. Resolution in the final digital image will be limited primary by the larger of the two, the sampling grid or the sampling aperture. We will return to this topic in Chapter 25 when discussing the spatial resolution of digital images.

Color is added to digital images by using three numbers for each pixel, representing the intensity of the three primary colors: red, green and blue. Mixing these three colors generates all possible colors that the human eye can perceive. A single byte is frequently used to store each of the color intensities, allowing the image to capture a total of $256 \times 256 \times 256 = 16.8$ million different colors.

Color is very important when the goal is to present the viewer with a true picture of the world, such as in television and still photography. However, this is usually not how images are used in science and engineering. The purpose here is to analyze a two-dimensional signal by using the human visual system as a *tool*. Black and white images are sufficient for this.

Cameras and Eyes

The structure and operation of the eye is very similar to an electronic camera, and it is natural to discuss them together. Both are based on two major components: a lens assembly, and an imaging sensor. The lens assembly captures a portion of the light emanating from an object, and focus it onto the imaging sensor. The imaging sensor then transforms the pattern of light into a video signal, either electronic or neural.

Figure 23-2 shows the operation of the lens. In this example, the image of an ice skater is focused onto a screen. The term *focus* means there is a one-to-one match of every point on the ice skater with a corresponding point on the screen. For example, consider a 1 mm × 1 mm region on the tip of the toe. In bright light, there are roughly 100 trillion photons of light striking this one square millimeter area each second. Depending on the characteristics of the surface, between 1 and 99 percent of these incident light photons will be reflected in random directions. Only a small portion of these reflected photons will pass through the lens. For example, only about one-millionth of the reflected light will pass through a one centimeter diameter lens located 3 meters from the object.

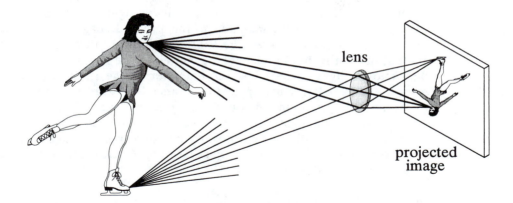

FIGURE 23-2
Focusing by a lens. A lens gathers light expanding from a point source, and force it to return to a point at another location. This allows a lens to project an image onto a surface.

Refraction in the lens changes the direction of the individual photons, depending on the location and angle they strike the glass/air interface. These direction changes cause light expanding from a single point to return to a single point on the projection screen. All of the photons that reflect from the toe *and* pass through the lens are brought back together at the "toe" in the projected image. In a similar way, a portion of the light coming from *any* point on the object will pass through the lens, and be focused to a corresponding point in the projected image.

Figures 23-3 and 23-4 illustrate the major structures in an electronic camera and the human eye, respectively. Both are light tight enclosures with a lens mounted at one end and an image sensor at the other. The camera is filled with air, while the eye is filled with a transparent liquid. Each lens system has two adjustable parameters: **focus** and **iris diameter**.

If the lens is not properly focused, each point on the object will project to a circular region on the imaging sensor, causing the image to be blurry. In the camera, focusing is achieved by physically moving the lens toward or away from the imaging sensor. In comparison, the eye contains two lenses, a bulge on the front of the eyeball called the cornea, and an adjustable lens inside the eye. The cornea does most of the light refraction, but is fixed in shape and location. Adjustment to the focusing is accomplished by the inner lens, a flexible structure that can be deformed by the action of the *ciliary muscles*. As these muscles contract, the lens flattens to bring the object into a sharp focus.

In both systems, the *iris* is used to control how much of the lens is exposed to light, and therefore the brightness of the image projected onto the imaging sensor. The iris of the eye is formed from opaque muscle tissue that can be contracted to make the *pupil* (the light opening) larger. The iris in a camera is a mechanical assembly that performs the same function.

The parameters in optical systems interact in many unexpected ways. For example, consider how the amount of available light and the sensitivity of the light sensor affects the *sharpness* of the acquired image. This is because the *iris diameter* and the *exposure time* are adjusted to transfer the proper amount of light from the scene being viewed to the image sensor. If more than enough light is available, the diameter of the iris can be reduced, resulting in a greater *depth-of-field* (the range of distance from the camera where an object remains in focus). A greater depth-of-field provides a sharper image when objects are at various distances. In addition, an abundance of light allows the exposure time to be reduced, resulting in less blur from camera shaking and object motion. Optical systems are full of these kinds of trade-offs.

An adjustable iris is necessary in both the camera and eye because the range of light intensities in the environment is much larger than can be directly handled by the light sensors. For example, the difference in light intensities between sunlight and moonlight is about one-million. Adding to this that reflectance can vary between 1% and 99%, results in a light intensity range of almost *one-hundred million*.

The **dynamic range** of an electronic camera is typically 300 to 1000, defined as the largest signal that can be measured, divided by the inherent noise of the device. Put another way, the maximum signal produced is 1 volt, and the rms noise in the dark is about 1 millivolt. Typical camera lenses have an iris that change the area of the light opening by a factor of about 300. This results in a typical electronic camera having a dynamic range of a few hundred thousand. Clearly, the same camera and lens assembly used in bright sunlight will be useless on a dark night.

In comparison, the eye operates over a dynamic range that nearly covers the large environmental variations. Surprisingly, the iris is not the main way that this tremendous dynamic range is achieved. From dark to light, the area of the pupil only changes by a factor of about 20. The light detecting nerve cells gradually adjust their sensitivity to handle the remaining dynamic range. For instance, it takes several minutes for your eyes to adjust to the low light after walking into a dark movie theater.

One way that DSP can improve images is by reducing the dynamic range an observer is required to view. That is, we do not want very light and very dark areas in the same image. A reflection image is formed from *two* image signals: the two-dimensional pattern of how the scene is *illuminated*, multiplied by the two-dimensional pattern of *reflectance* in the scene. The pattern of reflectance has a dynamic range of less than 100, because all ordinary materials reflect between 1% and 99% of the incident light. This is where most of the *image information* is contained, such as where objects are located in the scene and what their surface characteristics are. In comparison, the illumination signal depends on the light sources around the objects, but not on the objects themselves. The illumination signal can have a dynamic range of millions, although 10 to 100 is more typical within a single image. The illumination signal carries little interesting information,

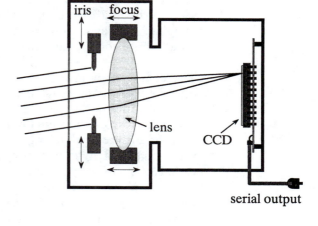

FIGURE 23-3
Diagram of an electronic camera. Focusing is achieved by moving the lens toward or away from the imaging sensor. The amount of light reaching the sensor is controlled by the iris, a mechanical device that changes the effective diameter of the lens. The most common imaging sensor in present day cameras is the CCD, a two-dimensional array of light sensitive elements.

FIGURE 23-4
Diagram of the human eye. The eye is a liquid filled sphere about 3 cm in diameter, enclosed by a tough outer case called the sclera. Focusing is mainly provided by the cornea, a fixed lens on the front of the eye. The focus is adjusted by contracting muscles attached to a flexible lens within the eye. The amount of light entering the eye is controlled by the iris, formed from opaque muscle tissue covering a portion of the lens. The rear hemisphere of the eye contains the retina, a layer of light sensitive nerve cells that converts the image to a neural signal in the optic nerve.

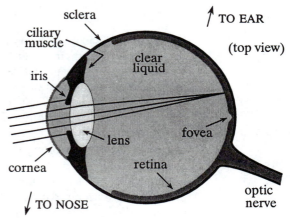

but can degrade the final image by increasing its dynamic range. DSP can improve this situation by suppressing the illumination signal, allowing the reflectance signal to dominate the image. The next chapter presents an approach for implementing this algorithm.

The light sensitive surface that covers the rear of the eye is called the **retina**. As shown in Fig. 23-5, the retina can be divided into three main layers of specialized nerve cells: one for converting light into neural signals, one for image processing, and one for transferring information to the optic nerve leading to the brain. In nearly all animals, these layers are seemingly *backward*. That is, the light sensitive cells are in last layer, requiring light to pass through the other layers before being detected.

There are two types of cells that detect light: **rods** and **cones**, named for their physical appearance under the microscope. The rods are specialized in operating with very little light, such as under the nighttime sky. Vision appears very *noisy* in near darkness, that is, the image appears to be filled with a continually changing grainy pattern. This results from the image signal being very weak, and is not a limitation of the eye. There is so little light entering the eye, the random detection of individual photons can be seen. This is called *statistical noise*, and is encountered in all low-light imaging, such as military night vision systems. Chapter 25 will revisit this topic. Since rods cannot detect color, low-light vision is in black and white.

The cone receptors are specialized in distinguishing color, but can only operate when a reasonable amount of light is present. There are three types of cones in the eye: red sensitive, green sensitive, and blue sensitive. This results from their containing different *photopigments*, chemicals that absorbs different wavelengths (colors) of light. Figure 23-6 shows the wavelengths of light that trigger each of these three receptors. This is called **RGB encoding**, and is how color information leaves the eye through the optic nerve. The human perception of color is made more complicated by neural processing in the lower levels of the brain. The RGB encoding is converted into another encoding scheme, where colors are classified as: red *or* green, blue *or* yellow, and light *or* dark.

RGB encoding is an important limitation of human vision; the wavelengths that exist in the environment are lumped into only three broad categories. In comparison, specialized cameras can separate the optical spectrum into hundreds or thousands of individual colors. For example, these might be used to classify cells as cancerous or healthy, understand the physics of a distant star, or see camouflaged soldiers hiding in a forest. Why is the eye so limited in detecting color? Apparently, all humans need for survival is to find a *red* apple, among the *green* leaves, silhouetted against the *blue* sky.

Rods and cones are roughly 3 μm wide, and are closely packed over the entire 3 cm by 3 cm surface of the retina. This results in the retina being composed of an array of roughly 10,000 × 10,000 = 100 million receptors. In comparison, the optic nerve only has about one-million nerve fibers that connect to these cells. On the average, each optic nerve fiber is connected to roughly 100 light receptors through the connecting layer. In addition to consolidating information, the connecting layer enhances the image by sharpening edges and suppressing the illumination component of the scene. This biological image processing will be discussed in the next chapter.

Directly in the center of the retina is a small region called the **fovea** (Latin for *pit*), which is used for high resolution vision (see Fig. 23-4). The fovea is different from the remainder of the retina in several respects. First, the optic nerve and interconnecting layers are pushed to the side of the fovea, allowing the receptors to be more directly exposed to the incoming light. This results in the fovea appearing as a small depression in the retina. Second, only cones are located in the fovea, and they are more tightly packed that in the remainder of the retina. This absence of rods in the fovea explains why night vision is often better when looking to the *side* of an object, rather than directly at it. Third, each optic nerve fiber is influenced by only a few cones, proving good localization ability. The fovea is surprisingly small. At normal reading distance, the fovea only sees about a 1 mm diameter area, less than the size of a single letter! The resolution is equivalent to about a 20×20 grid of pixels within this region.

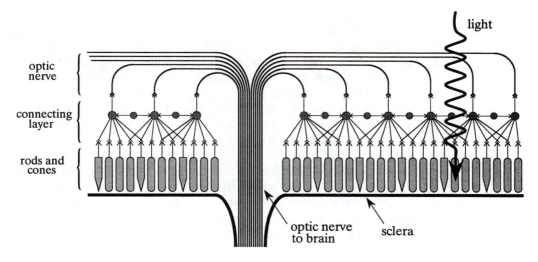

FIGURE 23-5
The human retina. The retina contains three principle layers: (1) the rod and cone light receptors, (2) an intermediate layer for data reduction and image processing, and (3) the optic nerve fibers that lead to the brain. The structure of these layers is seemingly *backward*, requiring light to pass through the other layers before reaching the light receptors.

Human vision overcomes the small size of the fovea by jerky eye movements called **saccades**. These abrupt motions allow the high resolution fovea to rapidly scan the field of vision for pertinent information. In addition, saccades present the rods and cones with a continually changing pattern of light. This is important because of the natural ability of the retina to adapt to changing levels of light intensity. In fact, if the eye is forced to remain fixed on the same scene, detail and color begin to fade in a few seconds.

The most common image sensor used in electronic cameras is the **charge coupled device** (**CCD**). The CCD is an integrated circuit that replaced most vacuum tube cameras in the 1980s, just as transistors replaced vacuum tube amplifiers twenty years before. The heart of the CCD is a thin wafer of

FIGURE 23-6
Spectral response of the eye. The three types of cones in the human eye respond to different sections of the optical spectrum, roughly corresponding to red, green, and blue. Combinations of these three form all colors that humans can perceive. The cones do not have enough sensitivity to be used in low-light environments, where the rods are used to detect the image. This is why colors are difficult to perceive at night.

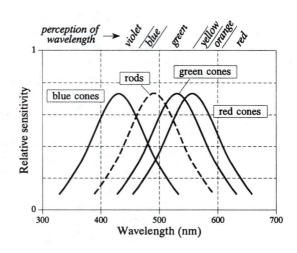

silicon, typically about 1 cm square. As shown by the cross-sectional view in Fig. 23-7, the backside is coated with a thin layer of metal connected to ground potential. The topside is covered with a thin electrical insulator, and a repetitive pattern of electrodes. The most common type of CCD is the **three phase readout**, where every third electrode is connected together. The silicon used is called *p-type*, meaning it has an excess of positive charge carriers called *holes*. For this discussion, a hole can be thought of as a positively charged particle that is free to move around in the silicon. Holes are represented in this figure by the "+" symbol.

In (a), +10 volts is applied to one of the three phases, while the other two are held at 0 volts. This causes the holes to move away from every third electrode, since positive charges are repelled by a positive voltage. This forms a region under these electrodes called a **well**, a shortened version of the physics term: *potential well*.

Each well in the CCD is a very efficient light sensor. As shown in (b), a single photon of light striking the silicon converts its energy into the formation of two charged particles, one electron, and one hole. The hole moves away, leaving the electron stuck in the well, held by the positive voltage on the electrode. Electrons in this illustration are represented by the "-" symbol. During the **integration period**, the pattern of light striking the CCD is transferred into a pattern of charge within the CCD wells. Dimmer light sources require longer integration periods. For example, the integration period for standard television is 1/60th of a second, while astrophotography can accumulate light for many hours.

Readout of the electronic image is quite clever; the accumulated electrons in each well are *pushed* to the output amplifier. As shown in (c), a positive voltage is placed on *two* of the phase lines. This results in each well expanding to the right. As shown in (d), the next step is to remove the voltage from the first phase, causing the original wells to collapse. This leaves the accumulated electrons in one well to the right of where they started. By repeating this pulsing sequence among the three phase lines, the accumulated electrons are pushed to the right until they reach a **charge sensitive amplifier**. This is a fancy name for a capacitor followed by a unity gain buffer. As the electrons are pushed from the last well, they flow onto the capacitor where they produce a voltage. To achieve high sensitivity, the capacitors are made extremely small, usually less than 1 pF. This capacitor and amplifier are an integral part of the CCD, and are made on the same piece of silicon. The signal leaving the CCD is a sequence of voltage levels proportional to the amount of light that has fallen on sequential wells.

Figure 23-8 shows how the two-dimensional image is read from the CCD. After the integration period, the charge accumulated in each well is moved up the column, one row at a time. For example, all the wells in row 15 are first moved into row 14, then row 13, then row 12, etc. Each time the rows are moved up, all the wells in row number 1 are transferred into the **horizontal register**. This is a group of specialized CCD wells that rapidly move the charge in a horizontal direction to the charge sensitive amplifier.

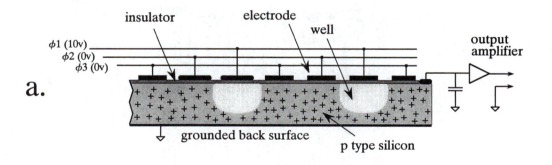

a.

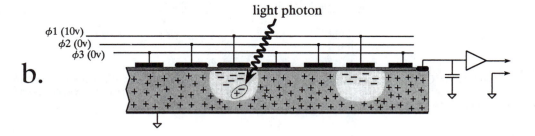

b.

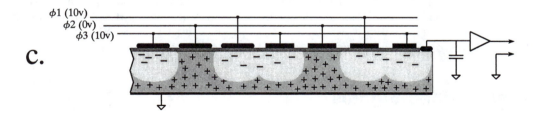

c.

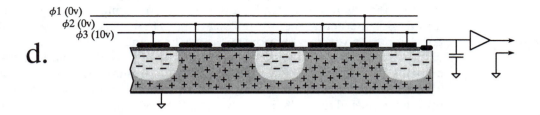

d.

FIGURE 23-7

Operation of the charge coupled device (CCD). As shown in this cross-sectional view, a thin sheet of p-type silicon is covered with an insulating layer and an array of electrodes. The electrodes are connected in groups of three, allowing three separate voltages to be applied: $\phi 1$, $\phi 2$, and $\phi 3$. When a positive voltage is applied to an electrode, the holes (i.e., the positive charge carriers indicated by the "+") are pushed away. This results in an area depleted of holes, called a *well*. Incoming light generates holes and electrons, resulting in an accumulation of electrons confined to each well (indicated by the "-"). By manipulating the three electrode voltages, the electrons in each well can be moved to the edge of the silicon where a charge sensitive amplifier converts the charge into a voltage.

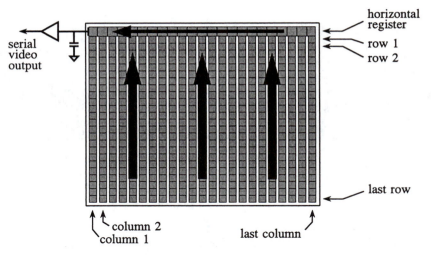

FIGURE 23-8
Architecture of the CCD. The imaging wells of the CCD are arranged in columns. During readout, the charge from each well is moved up the column into a horizontal register. The horizontal register is then readout into the charge sensitive preamplifier.

Notice that this architecture converts a two-dimensional array into a serial data stream in a particular sequence. The first pixel to be read is at the top-left corner of the image. The readout then proceeds from left-to-right on the first line, and then continues from left-to-right on subsequent lines. This is called **row major order**, and is almost always followed when a two-dimensional array (image) is converted to sequential data.

Television Video Signals

Although over 50 years old, the standard television signal is still one of the most common way to transmit an image. Figure 23-9 shows how the television signal appears on an oscilloscope. This is called **composite video**, meaning that there are vertical and horizontal synchronization (sync) pulses mixed with the actual picture information. These pulses are used in the television receiver to synchronize the vertical and horizontal deflection circuits to match the video being displayed. Each second of standard video contains 30 complete images, commonly called **frames**. A video engineer would say that each frame contains 525 **lines**, the television jargon for what programmers call *rows*. This number is a little deceptive because only 480 to 486 of these lines contain video information; the remaining 39 to 45 lines are reserved for sync pulses to keep the television's circuits synchronized with the video signal.

Standard television uses an **interlaced** format to reduce *flicker* in the displayed image. This means that all the odd lines of each frame are transmitted first, followed by the even lines. The group of odd lines is called the **odd field**, and the group of even lines is called the **even field**.

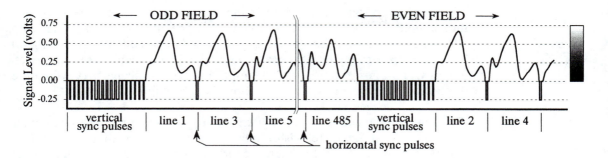

FIGURE 23-9

Composite video. The NTSC video signal consists of 30 complete frames (images) per second, with each frame containing 480 to 486 lines of video. Each frame is broken into two fields, one containing the odd lines and the other containing the even lines. Each field starts with a group of vertical sync pulses, followed by successive lines of video information separated by horizontal sync pulses. (The horizontal axis of this figure is not drawn to scale).

Since each frame consists of two fields, the video signal transmits 60 fields per second. Each field starts with a complex series of vertical sync pulses lasting 1.3 milliseconds. This is followed by either the even or odd lines of video. Each line lasts for 63.5 microseconds, including a 10.2 microsecond horizontal sync pulse, separating one line from the next. Within each line, the analog voltage corresponds to the grayscale of the image, with brighter values being in the direction *away* from the sync pulses. This places the sync pulses beyond the black range. In video jargon, the sync pulses are said to be *blacker than black*.

The hardware used for analog-to-digital conversion of video signals is called a **frame grabber**. This is usually in the form of an electronics card that plugs into a computer, and connects to a camera through a coaxial cable. Upon command from software, the frame grabber waits for the beginning of the next frame, as indicated by the vertical sync pulses. During the following two fields, each line of video is sampled many times, typically 512, 640 or 720 samples per line, at 8 bits per sample. These samples are stored in memory as one row of the digital image.

This way of acquiring a digital image results in an important difference between the vertical and horizontal directions. Each row in the digital image corresponds to one line in the video signal, and therefore to one row of wells in the CCD. Unfortunately, the columns are not so straight-forward. In the CCD, each row contains between about 400 and 800 wells (columns), depending on the particular device used. When a row of wells is read from the CCD, the resulting line of video is filtered into a smooth analog signal, such as in Fig. 23-9. In other words, the video signal does not depend on how many columns are present in the CCD. The resolution in the horizontal direction is limited by how rapidly the analog signal is allowed to change. This is usually set at 3.2 MHz for color television, resulting in a risetime of about 100 nanoseconds, i.e., about 1/500th of the 53.2 microsecond video line.

When the video signal is digitized in the frame grabber, it is converted back into columns. However, these columns in the digitized image have *no relation* to the columns in the CCD. The number of columns in the digital image depends solely on how many times the frame grabber samples each line of video. For example, a CCD might have 800 wells per row, while the digitized image might only have 512 pixels (i.e., columns) per row.

The number of columns in the digitized image is also important for another reason. The standard television image has an **aspect ratio** of 4 to 3, i.e., it is slightly wider than it is high. Motion pictures have the wider aspect ratio of 25 to 9. CCDs used for scientific applications often have an aspect ratio of 1 to 1, i.e., a perfect square. In any event, the aspect ratio of a CCD is fixed by the placement of the electrodes, and cannot be altered. However, the aspect ratio of the digitized image depends on the number of samples per line. This becomes a problem when the image is displayed, either on a video monitor or in a hardcopy. If the aspect ratio isn't properly reproduced, the image looks squashed horizontally or vertically.

The 525 line video signal described here is called **NTSC** (National Television Systems Committee), a standard defined way back in 1954. This is the system used in the United States and Japan. In Europe there are two similar standards called **PAL** (Phase Alternation by Line) and **SECAM** (Sequential Chrominance And Memory). The basic concepts are the same, just the numbers are different. Both PAL and SECAM operate with 25 interlaced frames per second, with 625 lines per frame. Just as with NTSC, some of these lines occur during the vertical sync, resulting in about 576 lines that carry picture information. Other more subtle differences relate to how color and sound are added to the signal.

The most straightforward way of transmitting color television would be to have three separate analog signals, one for each of the three colors the human eye can detect: red, green and blue. Unfortunately, the historical development of television did not allow such a simple scheme. The color television signal was developed to allow existing black and white television sets to remain in use without modification. This was done by retaining the same signal for brightness information, but adding a separate signal for color information. In video jargon, the brightness is called the *luminance signal*, while the color is the *chrominance signal*. The chrominance signal is contained on a 3.58 MHz carrier wave added to the black and white video signal. Sound is added in this same way, on a 4.5 MHz carrier wave. The television receiver separates these three signals, processes them individually, and recombines them in the final display.

Other Image Acquisition and Display

Not all images are acquired an entire frame at a time. Another very common way is by **line scanning**. This involves using a detector containing a one-dimensional array of pixels, say, 2048 pixels long by 1 pixel wide. As an object is moved past the detector, the image is acquired line-by-line.

Line scanning is used by fax machines and airport x-ray baggage scanners. As a variation, the object can be kept stationary while the detector is moved. This is very convenient when the detector is already mounted on a moving object, such as an aircraft taking images of the ground beneath it. The advantage of line scanning is that *speed* is traded for detector *simplicity*. For example, a fax machine may take several seconds to scan an entire page of text, but the resulting image contains thousands of rows and columns.

An even more simplified approach is to acquire the image **point-by-point**. For example, the microwave image of Venus was acquired one pixel at a time. Another example is the *scanning probe microscope*, capable of imaging individual atoms. A small probe, often consisting of only a single atom at its tip, is brought exceedingly close to the sample being imaged. Quantum mechanical effects can be detected between the probe and the sample, allowing the probe to be stopped an exact distance from the sample's surface. The probe is then moved over the surface of the sample, keeping a constant distance, tracing out the peaks and valleys. In the final image, each pixel's value represents the elevation of the corresponding location on the sample's surface.

Printed images are divided into two categories: **grayscale** and **halftone**. Each pixel in a grayscale image is a shade of gray between black and white, such as in a photograph. In comparison, each pixel in a halftone image is formed from many individual *dots*, with each dot being completely black or completely white. Shades of gray are produced by alternating various numbers of these black and white dots. For example, imagine a laser printer with a resolution of 600 dots-per-inch. To reproduce 256 levels of brightness between black and white, each pixel would correspond to an array of 16 by 16 printable dots. Black pixels are formed by making all of these 256 dots black. Likewise, white pixels are formed making all of these 256 dots white. Mid-gray has one-half of the dots white and one-half black. Since the individual dots are too small to be seen when viewed at a normal distance, the eye is fooled into thinking a grayscale has been formed.

Halftone images are easier for printers to handle, including photocopy machines. The disadvantage is that the image quality is often worse than grayscale pictures.

Brightness and Contrast Adjustments

An image must have the proper **brightness** and **contrast** for easy viewing. Brightness refers to the overall lightness or darkness of the image. Contrast is the *difference* in brightness between objects or regions. For example, a white rabbit running across a snowy field has *poor* contrast, while a black dog against the same white background has *good* contrast. Figure 23-10 shows four possible ways that brightness and contrast can be misadjusted. When the brightness is too high, as in (a), the whitest pixels are saturated, destroying the detail in these areas. The reverse is shown in (b), where the brightness is set too low, saturating the blackest pixels. Figure (c) shows

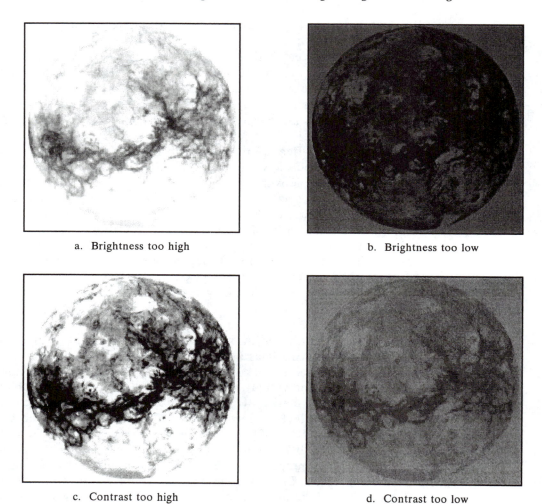

a. Brightness too high

b. Brightness too low

c. Contrast too high

d. Contrast too low

FIGURE 23-10
Brightness and contrast adjustments. Increasing the *brightness* makes every pixel in the image becomes lighter. In comparison, increasing the *contrast* makes the light areas become lighter, and the dark areas become darker. These images show the effect of misadjusting the brightness and contrast.

the contrast set to high, resulting in the blacks being too black, and the whites being too white. Lastly, (d) has the contrast set too low; all of the pixels are a mid-shade of gray making the objects fade into each other.

Figures 23-11 and 23-12 illustrate *brightness* and *contrast* in more detail. A test image is displayed in Fig. 23-12, using six different brightness and contrast levels. Figure 23-11 shows the construction of the test image, an array of 80×32 pixels, with each pixel having a value between 0 and 255. The backgound of the test image is filled with random noise, uniformly distributed between 0 and 255. The three square boxes have pixel values of 75, 150 and 225, from left-to-right. Each square contains two triangles with pixel values only slightly different from their surroundings. In other

FIGURE 23-11
Brightness and contrast test image. This is the structure of the digital image used in Fig. 23-12. The three squares form dark, medium, and bright objects, each containing two low contrast triangles. This figure indicates the digital numbers (DN) of the pixels in each region.

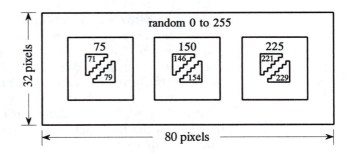

words, there is a dark region in the image with faint detail, this is a medium region in the image with faint detail, and there is a bright region in the image with faint detail.

Figure 23-12 shows how adjustment of the contrast and brightness allows different features in the image to be visualized. In (a), the brightness and contrast are set at the *normal* level, as indicated by the **B and C slide bars** at the left side of the image. Now turn your attention to the graph shown with each image, called an **output transform**, an **output look-up table**, or a **gamma curve**. This controls the hardware that displays the image. The value of each pixel in the stored image, a number between 0 and 255, is passed through this look-up table to produces another number between 0 and 255. This new digital number drives the video intensity circuit, with 0 through 255 being transformed into black through white, respectively. That is, the look-up table maps the stored numbers into the displayed brightness.

Figure (a) shows how the image appears when the output transform is set to do *nothing*, i.e., the digital output is identical to the digital input. Each pixel in the noisy background is a random shade of gray, equally distributed between black and white. The three boxes are displayed as dark, medium and light, clearly distinct from each other. The problem is, the triangles inside each square cannot be easily seen; the contrast is too low for the eye to distinguished these regions from their surroundings.

Figures (b) & (c) shows the effect of changing the brightness. Increasing the brightness shifts the output transform to the *left*, while decreasing the brightness shifts it to the *right*. Increasing the brightness makes *every* pixel in the image appear lighter. Conversely, decreasing the brightness makes *every* pixel in the image appear darker. These changes can improve the viewability of excessively dark or light areas in the image, but will **saturate** the image if taken too far. For example, all of the pixels in the far right square in (b) are displayed with full intensity, i.e., 255. The opposite effect is shown in (c), where all of the pixels in the far left square are displayed as blackest black, or digital number 0. Since all the pixels in these regions have the same value, the triangles are completely wiped out. Also notice that *none* of the triangles in (b) and (c) are easier to see than in (a). Changing the brightness provides little (if any) help in distinguishing low contrast objects from their surroundings.

Figure (d) shows the display optimized to view pixel values around digital number 75. This is done by turning up the *contrast*, resulting in the output transform increasing in *slope*. For example, the stored pixel values of 71 and 75 become 100 and 116 in the display, making the contrast a factor of four greater. Pixel values between 46 and 109 are displayed as the blackest black, to the whitest white. The price for this increased contrast is that pixel values 0 to 45 are saturated at black, and pixel values 110 to 255 are saturated at white. As shown in (d), the increased contrast allows the triangles in the left square to be seen, at the cost of saturating the middle and right squares.

Figure (e) shows the effect of increasing the contrast even further, resulting in only 16 of the possible 256 stored levels being displayed as nonsaturated. The brightness has also been decreased so that the 16 usable levels are centered on digital number 175. The details in the center square are now very visible; however, almost everything else in the image is saturated. For example, look at the noise around the border of the image. There are very few pixels with an intermediate gray shade; almost every pixel is either pure black or pure white. This technique of using high contrast to view only a few levels is sometimes called a **grayscale stretch**.

The contrast adjustment is a way of *zooming in* on a smaller range of pixel values. The brightness control *centers* the zoomed section on the pixel values of interest. Most digital imaging systems allow the brightness and contrast to be adjusted in just this manner, and often provide a graphical display of the output transform (as in Fig. 23-12). In comparison, the brightness and contrast controls on television and video monitors are *analog circuits*, and may operate differently. For example, the contrast control of a monitor may adjust the gain of the analog signal, while the brightness might add or subtract a DC offset. The moral is, don't be surprised if these analog controls don't respond in the way you think they should.

Grayscale Transforms

The last image, Fig. 23-12f, is different from the rest. Rather than having a slope in the curve over *one* range of input values, it has a slope in the curve over *two* ranges. This allows the display to simultaneously show the triangles in both the left and the right squares. Of course, this results in saturation of the pixel values that are *not* near these digital numbers. Notice that the slide bars for contrast and brightness are not shown in (f); this display is beyond what brightness and contrast adjustments can provide.

Taking this approach further results in a powerful technique for improving the appearance of images: the **grayscale transform**. The idea is to increase the contrast at pixel values of interest, at the expense of the pixel values we don't care about. This is done by defining the relative importance of each of the 0 to 255 possible pixel values. The more important the value, the greater its contrast is made in the displayed image. An example will show a systematic way of implementing this procedure.

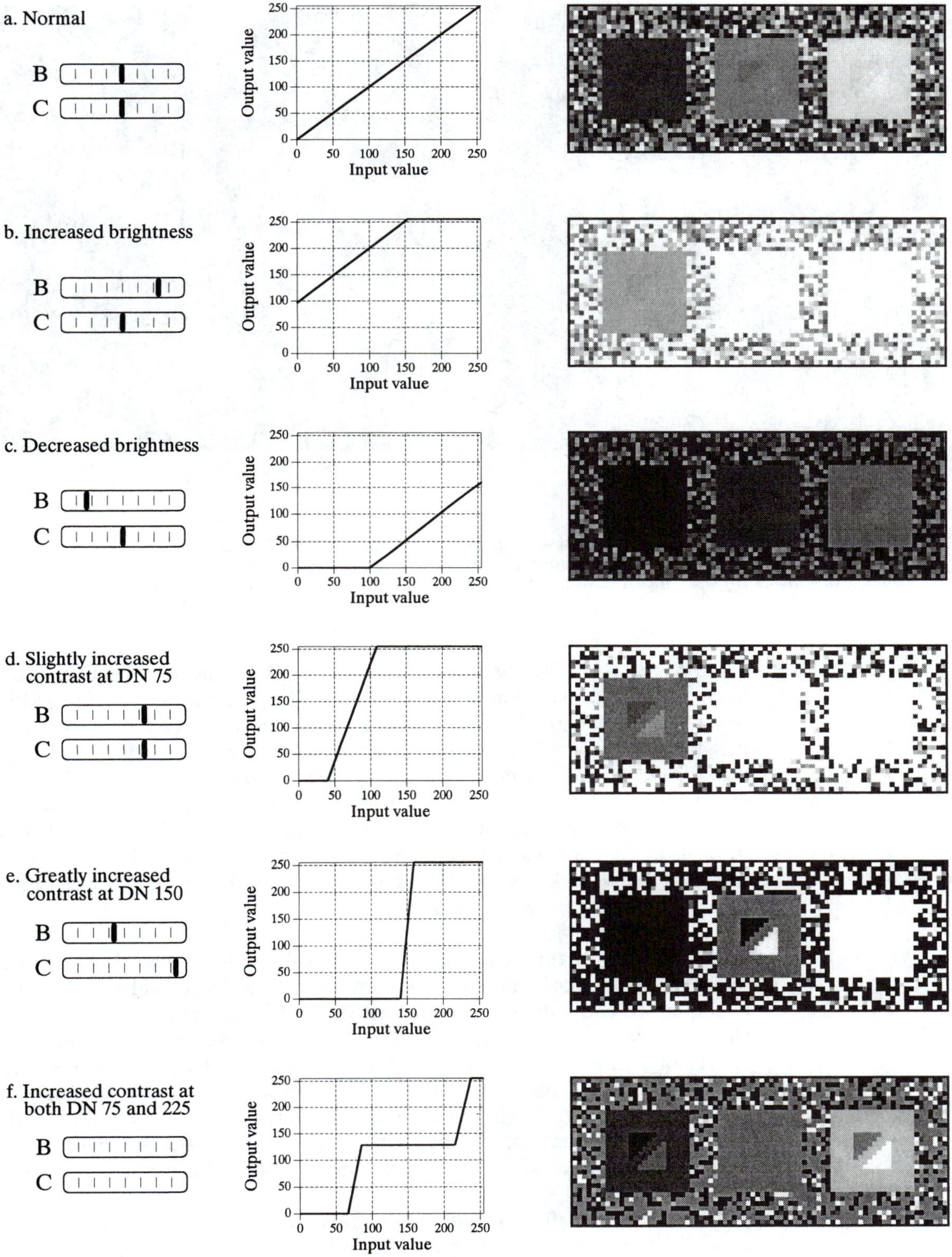
a. Normal
B
C
Output value
Input value
b. Increased brightness
B
C
Output value
Input value
c. Decreased brightness
B
C
Output value
Input value
d. Slightly increased contrast at DN 75
B
C
Output value
Input value
e. Greatly increased contrast at DN 150
B
C
Output value
Input value
f. Increased contrast at both DN 75 and 225
B
C
Output value
Input value

FIGURE 23-12

a. Original IR image   b. With grayscale transform

FIGURE 23-13
Grayscale processing. Image (a) was acquired with an infrared camera in total darkness. Brightness in the image is related to the temperature, accounting for the appearance of the warm human body and the hot truck grill. Image (b) was processed with the manual grayscale transform shown in Fig. 23-14c.

The image in Fig. 23-13a was acquired in total darkness by using a CCD camera that is sensitive in the far infrared. The parameter being imaged is *temperature*: the hotter the object, the more infrared energy it emits and the brighter it appears in the image. This accounts for the background being very black (cold), the body being gray (warm), and the truck grill being white (hot). These systems are great for the military and police; you can see the other guy when he can't even see himself! The image in (a) is difficult to view because of the uneven distribution of pixel values. Most of the image is so dark that details cannot be seen in the scene. On the other end, the grill is near white saturation.

The histogram of this image is displayed in Fig. 23-14a, showing that the background, human, and grill have reasonably separate values. In this example, we will increase the contrast in the background and the grill, at the expense of everything else, including the human body. Figure (b) represents this strategy. We declare that the lowest pixel values, the background, will have a relative contrast of twelve. Likewise, the highest pixel values, the grill, will have a relative contrast of six. The body will have a relative contrast of one, with a staircase transition between the regions. All these values are determined by trial and error.

The grayscale transform resulting from this strategy is shown in (c), labeled *manual*. It is found by taking the running sum (i.e., the discrete integral) of the curve in (b), and then normalizing so that it has a value of 255 at the

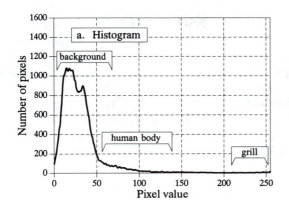

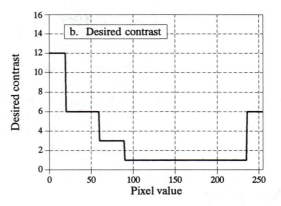

FIGURE 23-14
Developing a grayscale transform. Figure (a) is the histogram of the raw image in Fig. 23-13a. In (b), a curve is manually generated indicating the desired contrast at each pixel value. The LUT for the output transform is then found by integration and normalization of (b), resulting in the curve labeled *manual* in (c). In histogram equalization, the histogram of the raw image, shown in (a), is integrated and normalized to find the LUT, shown in (c).

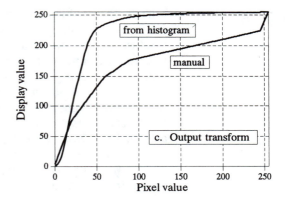

right side. Why take the *integral* to find the required curve? Think of it this way: The contrast at a particular pixel value is equal to the slope of the output transform. That is, we want (b) to be the derivative (slope) of (c). This means that (c) must be the integral of (b).

Passing the image in Fig. 23-13a through this manually determined grayscale transform produces the image in (b). The background has been made *lighter*, the grill has been made *darker*, and both have better contrast. These improvements are at the expense of the body's contrast, producing a less detailed image of the intruder (although it can't get much worse than in the original image).

Grayscale transforms can significantly improve the viewability of an image. The problem is, they can require a great deal of trial and error. **Histogram equalization** is a way to automate the procedure. Notice that the histogram in (a) and the contrast weighting curve in (b) have the same general shape. Histogram equalization blindly uses the histogram as the contrast weighing curve, eliminating the need for human judgement. That is, the output transform is found by integration and normalization of the *histogram*, rather than a manually generated curve. This results in the greatest contrast being given to those values that have the greatest number of pixels.

Histogram equalization is an interesting mathematical procedure because it maximizes the *entropy* of the image, a measure of how much information is transmitted by a fixed number of bits. The fault with histogram equalization is that it mistakes the shear *number* of pixels at a certain value with the *importance* of the pixels at that value. For example, the truck grill and human intruder are the most prominent features in Fig. 23-13. In spite of this, histogram equalization would almost completely ignore these objects because they contain relatively few pixels. Histogram equalization is quick and easy. Just remember, if it doesn't work well, a manually generated curve will probably do much better.

Warping

One of the problems in photographing a planet's surface is the distortion from the curvature of the spherical shape. For example, suppose you use a telescope to photograph a square region near the center of a planet, as illustrated in Fig. 23-15a. After a few hours, the planet will have rotated on its axis, appearing as in (b). The previously photographed region appears highly distorted because it is curved near the horizon of the planet. Each of the two images contain complete information about the region, just from a different perspective. It is quite common to acquire a photograph such as (a), but really want the image to look like (b), or vice versa. For example, a satellite mapping the surface of a planet may take thousands of images from straight above, as in (a). To make a natural looking picture of the entire planet, such as the image of Venus in Fig. 23-1, each image must be distorted and placed in the proper position. On the other hand, consider a weather satellite looking at a hurricane that is not directly below it. There is no choice but to acquire the image obliquely, as in (b). The image is then converted into how it would appear from above, as in (a).

These spatial transformations are called **warping**. Space photography is the most common use for warping, but there are others. For example, many vacuum tube imaging detectors have various amounts of spatial distortion. This includes night vision cameras used by the military and x-ray detectors used in the medical field. Digital warping (or *dewarping* if you prefer) can be used to correct the inherent distortion in these devices. Special effects artists for motion pictures love to warp images. For example, a technique called **morphing** gradually warps one object into another over a series of frames. This can produces illusions such as a child turning into an adult, or a man turning into a werewolf.

Warping takes the *original image* (a two-dimensional array) and generates a *warped image* (another two-dimensional array). This is done by looping through each pixel in the warped image and asking: What is the proper pixel value that should be placed here? Given the particular row and column being calculated in the warped image, there is a corresponding row and column in the original image. The pixel value from the original image is transferred to the warped image to carry out the algorithm. In the jargon of image processing, the row and column that the pixel *comes from* in the

a. Normal View

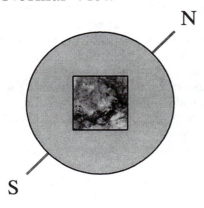

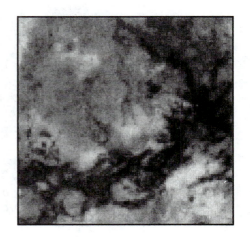

b. Oblique View

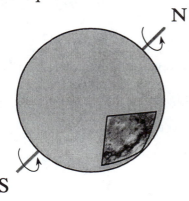

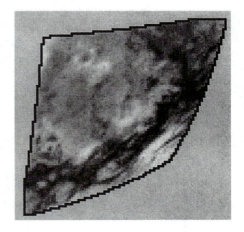

FIGURE 23-15
Image warping. As shown in (a), a normal view of a small section of a planet appears relatively distortion free. In comparison, an oblique view presents significant spatial distortion. *Warping* is the technique of changing one of these images into the other.

original image is called the **comes-from address**. Transferring each pixel from the original to the warped image is the easy part. The hard part is calculating the *comes-from address* associated with each pixel in the warped image. This is usually a pure math problem, and can become quite involved. Simply stretching the image in the horizontal or vertical direction is easier, involving only a multiplication of the row and/or column number to find the comes-from address.

One of the techniques used in warping is **subpixel interpolation**. For example, suppose you have developed a set of equations that turns a row and column address in the warped image into the comes-from address in the original image. Consider what might happen when you try to find the value of the pixel at row 10 and column 20 in the warped image. You pass the information: *row = 10, column = 20,* into your equations, and out pops: *comes-from row = 20.2, comes-from column = 14.5.* The point being, your calculations will likely use floating point, and therefore the comes-from addresses will not be integers. The easiest method to use is the **nearest neighbor** algorithm, that is, simply round the addresses to the nearest integer. This is simple, but can produce a very grainy appearance at the edges of objects where pixels may appear to be slightly misplaced.

Bilinear interpolation requires a little more effort, but provides significantly better images. Figure 23-16 shows how it works. You know the value of the four pixels *around* the fractional address, i.e., the value of the pixels at row 20 & 21, and column 14 and 15. In this example we will assume the pixels values are 91, 210, 162 and 95. The problem is to interpolate between these four values. This is done in two steps. First, interpolate in the *horizontal* direction between column 14 and 15. This produces two intermediate values, 150.5 on line 20, and 128.5 on line 21. Second, interpolate between these intermediate values in the vertical direction. This produces the bilinear interpolated pixel value of 139.5, which is then transferred to the warped image. Why interpolate in the horizontal direction *and then* the vertical direction instead of the reverse? It doesn't matter; the final answer is the same regardless of which order is used.

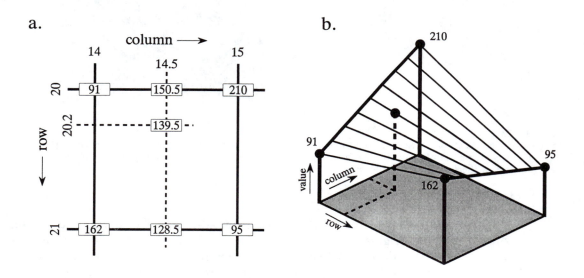

FIGURE 23-16
Subpixel interpolation. Subpixel interpolation for image warping is usually accomplished with bilinear interpolation. As shown in (a), two intermediate values are calculated by linear interpolation in the horizontal direction. The final value is then found by using linear interpolation in the vertical direction between the intermediate values. As shown by the three-dimensional illustration in (b), this procedure uniquely defines all values between the four known pixels at each of the corners.

Linear Image Processing

Linear image processing is based on the same two techniques as conventional DSP: *convolution* and *Fourier analysis*. Convolution is the more important of these two, since images have their information encoded in the spatial domain rather than the frequency domain. Linear filtering can improve images in many ways: sharpening the edges of objects, reducing random noise, correcting for unequal illumination, deconvolution to correct for blur and motion, etc. These procedures are carried out by convolving the original image with an appropriate filter kernel, producing the filtered image. A serious problem with image convolution is the enormous number of calculations that need to be performed, often resulting in unacceptably long execution times. This chapter presents strategies for designing filter kernels for various image processing tasks. Two important techniques for reducing the execution time are also described: *convolution by separability* and *FFT convolution*.

Convolution

Image convolution works in the same way as one-dimensional convolution. For instance, images can be viewed as a summation of *impulses*, i.e., scaled and shifted delta functions. Likewise, *linear systems* are characterized by how they respond to impulses; that is, by their *impulse responses*. As you should expect, the output image from a system is equal to the input image *convolved* with the system's impulse response.

The two-dimensional delta function is an image composed of all zeros, except for a single pixel at: *row* = 0, *column* = 0, which has a value of *one*. For now, assume that the row and column indexes can have both positive and negative values, such that the *one* is centered in a vast sea of zeros. When the delta function is passed through a linear system, the single nonzero point will be changed into some other two-dimensional pattern. Since the only thing that can happen to a point is that it *spreads out*, the impulse response is often called the **point spread function** (**PSF**) in image processing jargon.

The human eye provides an excellent example of these concepts. As described in the last chapter, the first layer of the retina transforms an image represented as a pattern of light into an image represented as a pattern of nerve impulses. The second layer of the retina *processes* this neural image and passes it to the third layer, the fibers forming the optic nerve. Imagine that the image being projected onto the retina is a very small spot of light in the center of a dark background. That is, an *impulse* is fed into the eye. Assuming that the system is linear, the image processing taking place in the retina can be determined by inspecting the image appearing at the optic nerve. In other words, we want to find the *point spread function* of the processing. We will revisit the assumption about linearity of the eye later in this chapter.

Figure 24-1 outlines this experiment. Figure (a) illustrates the impulse striking the retina while (b) shows the image appearing at the optic nerve. The middle layer of the eye passes the bright spike, but produces a circular region of increased *darkness*. The eye accomplishes this by a process known as *lateral inhibition*. If a nerve cell in the middle layer is activated, it decreases the ability of its nearby neighbors to become active. When a complete image is viewed by the eye, each point in the image contributes a scaled and shifted version of this impulse response to the image appearing at the optic nerve. In other words, the visual image is *convolved* with this PSF to produce the neural image transmitted to the brain. The obvious question is: how does convolving a viewed image with this PSF improve the ability of the eye to understand the world?

a. Image at first layer   b. Image at third layer

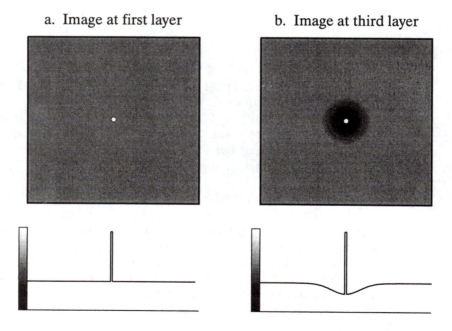

FIGURE 24-1
The PSF of the eye. The middle layer of the retina changes an impulse, shown in (a), into an impulse surrounded by a dark area, shown in (b). This point spread function enhances the edges of objects.

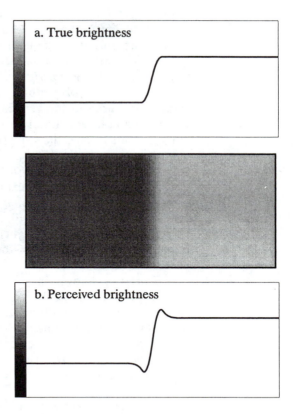

FIGURE 24-2
Mach bands. Image processing in the retina results in a slowly changing edge, as in (a), being sharpened, as in (b). This makes it easier to separate objects in the image, but produces an optical illusion called *Mach bands*. Near the edge, the overshoot makes the dark region look darker, and the light region look lighter. This produces dark and light bands that run parallel to the edge.

Humans and other animals use vision to identify nearby objects, such as enemies, food, and mates. This is done by distinguishing one region in the image from another, based on differences in brightness and color. In other words, the first step in recognizing an object is to identify its *edges*, the discontinuity that separates an object from its background. The middle layer of the retina helps this task by sharpening the edges in the viewed image. As an illustration of how this works, Fig. 24-2 shows an image that slowly changes from dark to light, producing a blurry and poorly defined edge. Figure (a) shows the intensity profile of this image, the pattern of brightness entering the eye. Figure (b) shows the brightness profile appearing on the optic nerve, the image transmitted to the brain. The processing in the retina makes the edge between the light and dark areas appear more abrupt, reinforcing that the two regions are different.

The overshoot in the edge response creates an interesting optical illusion. Next to the edge, the dark region appears to be unusually dark, and the light region appears to be unusually light. The resulting light and dark strips are called **Mach bands**, after Ernst Mach (1838-1916), an Austrian physicist who first described them.

As with one-dimensional signals, image convolution can be viewed in two ways: from the input, and from the output. From the input side, each pixel in the input image contributes a scaled and shifted version of the point spread function to the output image.   As viewed from the output side, each pixel in the output image is influenced by a group of pixels from the input signal. For one-dimensional signals, this region of influence is the impulse response flipped *left-for-right*. For image signals, it is the PSF flipped *left-for-right* and *top-for-bottom*. Since most of the PSFs used in DSP are symmetrical around the vertical and horizonal axes, these flips do nothing and can be ignored. Later in this chapter we will look at nonsymmetrical PSFs that must have the flips taken into account.

Figure 24-3 shows several common PSFs. In (a), the **pillbox** has a circular top and straight sides. For example, if the lens of a camera is not properly focused, each point in the image will be projected to a circular spot on the image sensor (look back at Fig. 23-2 and consider the effect of moving the projection screen left or right). In other words, the pillbox is the point spread function of an out-of-focus lens.

The **Gaussian**, shown in (b), is the PSF of imaging systems limited by *random* imperfections. For instance, the image from a telescope is blurred by atmospheric turbulence, causing each point of light to become a Gaussian in the final image. Image sensors, such as the CCD and retina, are often limited by the scattering of light and/or electrons. The Central Limit Theorem dictates that a Gaussian blur results from these types of random processes.

The pillbox and Gaussian are used in image processing the same as the *moving average filter* is used with one-dimensional signals. An image convolved with these PSFs will appear blurry and have less defined edges, but will be lower in random noise. These are called **smoothing filters**, for their action in the time domain, or **low-pass filters**, for how they treat the frequency domain.   The **square** PSF, shown in (c), can also be used as a smoothing filter, but it is not circularly symmetric. This results in the blurring being different in the diagonal directions compared to the vertical and horizontal. This may or may not be important, depending on the use.

The opposite of a smoothing filter is an **edge enhancement** or **high-pass filter**. The spectral inversion technique, discussed in Chapter 14, is used to change between the two. As illustrated in (d), an edge enhancement filter kernel is formed by taking the *negative* of a smoothing filter, and adding a delta function in the center. The image processing which occurs in the retina is an example of this type of filter.

Figure (e) shows the two-dimensional sinc function. One-dimensional signal processing uses the windowed-sinc to separate frequency bands. Since images do not have their information encoded in the frequency domain, the sinc function is seldom used as an imaging filter kernel, although it does find use in some theoretical problems. The sinc function can be hard to use because its tails decrease very slowly in amplitude ($1/x$), meaning it must be treated as *infinitely* wide. In comparison, the Gaussian's tails decrease very rapidly (e^{-x^2}) and can eventually be truncated with no ill effect.

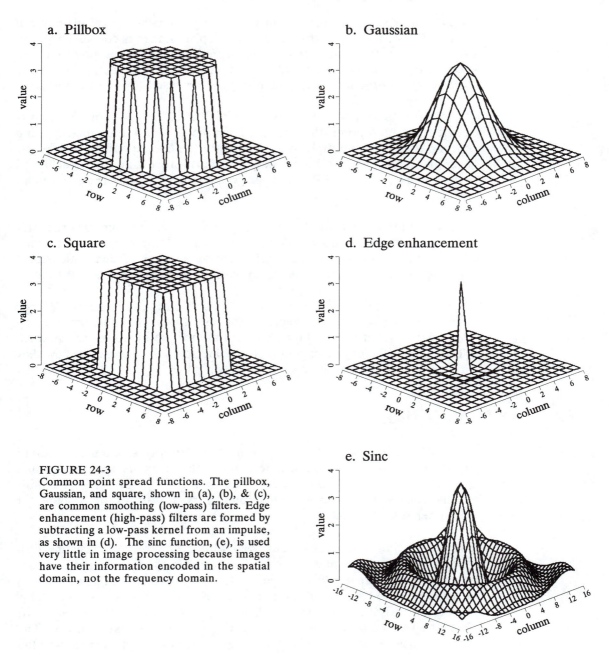

FIGURE 24-3
Common point spread functions. The pillbox, Gaussian, and square, shown in (a), (b), & (c), are common smoothing (low-pass) filters. Edge enhancement (high-pass) filters are formed by subtracting a low-pass kernel from an impulse, as shown in (d). The sinc function, (e), is used very little in image processing because images have their information encoded in the spatial domain, not the frequency domain.

All these filter kernels use *negative* indexes in the rows and columns, allowing the PSF to be centered at *row* = 0 and *column* = 0. Negative indexes are often eliminated in one-dimensional DSP by shifting the filter kernel to the right until all the nonzero samples have a positive index. This shift moves the output signal by an equal amount, which is usually of no concern. In comparison, a shift between the input and output images is generally not acceptable. Correspondingly, negative indexes are the norm for filter kernels in image processing.

A problem with image convolution is that a large number of calculations are involved. For instance, when a 512 by 512 pixel image is convolved with a 64 by 64 pixel PSF, more than a *billion* multiplications and additions are needed (i.e., $64 \times 64 \times 512 \times 512$). The long execution times can make the techniques impractical. Three approaches are used to speed things up.

The first strategy is to use a very small PSF, often only 3×3 pixels. This is carried out by looping through each sample in the output image, using optimized code to multiply and accumulate the corresponding nine pixels from the input image. A surprising amount of processing can be achieved with a mere 3×3 PSF, because it is large enough to affect the *edges* in an image.

The second strategy is used when a large PSF is needed, but its shape isn't critical. This calls for a filter kernel that is *separable*, a property that allows the image convolution to be carried out as a series of one-dimensional operations. This can improve the execution speed by *hundreds* of times.

The third strategy is FFT convolution, used when the filter kernel is large and has a specific shape. Even with the speed improvements provided by the highly efficient FFT, the execution time will be hideous. Let's take a closer look at the details of these three strategies, and examples of how they are used in image processing.

3×3 Edge Modification

Figure 24-4 shows several 3×3 operations. Figure (a) is an image acquired by an airport x-ray baggage scanner. When this image is convolved with a 3×3 delta function (a *one* surrounded by 8 zeros), the image remains unchanged. While this is not interesting by itself, it forms the baseline for the other filter kernels.

Figure (b) shows the image convolved with a 3×3 kernel consisting of a one, a negative one, and 7 zeros. This is called the **shift and subtract** operation, because a *shifted* version of the image (corresponding to the -1) is *subtracted* from the original image (corresponding to the 1). This processing produces the optical illusion that some objects are closer or farther away than the background, making a 3D or embossed effect. The brain interprets images as if the lighting is from *above*, the normal way the world presents itself. If the edges of an object are bright on the top and dark on the bottom, the object is perceived to be poking out from the background. To see another interesting effect, turn the picture upside down, and the objects will be pushed *into* the background.

Figure (c) shows an **edge detection** PSF, and the resulting image. Every edge in the original image is transformed into narrow dark and light bands that run parallel to the original edge. Thresholding this image can isolate either the dark or light band, providing a simple algorithm for detecting the edges in an image.

a. Delta function

0	0	0
0	1	0
0	0	0

b. Shift and subtract

0	0	0
0	1	0
0	0	-1

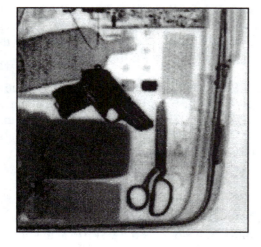

c. Edge detection

-1/8	-1/8	-1/8
-1/8	1	-1/8
-1/8	-1/8	-1/8

d. Edge enhancement

-k/8	-k/8	-k/8
-k/8	k+1	-k/8
-k/8	-k/8	-k/8

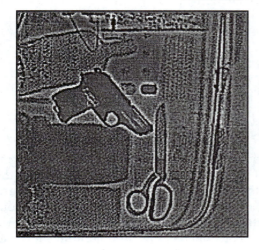

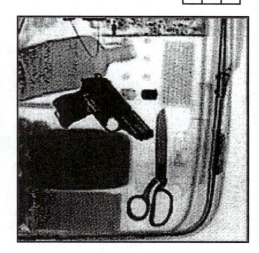

FIGURE 24-4

3×3 edge modification. The original image, (a), was acquired on an airport x-ray baggage scanner. The shift and subtract operation, shown in (b), results in a pseudo three-dimensional effect. The edge detection operator in (c) removes all contrast, leaving only the edge information. The edge enhancement filter, (d), adds various ratios of images (a) and (c), determined by the parameter, k. A value of $k = 2$ was used to create this image.

A common image processing technique is shown in (d): **edge enhancement**. This is sometimes called a *sharpening* operation. In (a), the objects have good contrast (an appropriate level of darkness and lightness) but very blurry edges. In (c), the objects have absolutely no contrast, but very sharp edges.   The strategy is to multiply the image with *good edges* by a constant, k, and add it to the image with *good contrast*. This is equivalent to convolving the original image with the 3×3 PSF shown in (d). If k is set to 0, the PSF becomes a delta function, and the image is left unchanged. As k is made larger, the image shows better edge definition. For the image in (d), a value of $k = 2$ was used: two parts of image (c) to one part of image (a).   This operation mimics the eye's ability to sharpen edges, allowing objects to be more easily separated from the background.

Convolution with any of these PSFs can result in negative pixel values appearing in the final image. Even if the program can handle negative values for pixels, the image display cannot. The most common way around this is to add an offset to each of the calculated pixels, as is done in these images. An alternative is to truncate out-of-range values.

Convolution by Separability

This is a technique for fast convolution, as long as the PSF is **separable**. A PSF is said to be *separable* if it can be broken into two one-dimensional signals: a vertical and a horizontal projection. Figure 24-5 shows an example of a separable image, the square PSF. Specifically, the value of each pixel in the image is equal to the corresponding point in the horizontal projection multiplied by the corresponding point in the vertical projection. In mathematical form:

EQUATION 24-1
Image separation. An image is referred to as *separable* if it can be decomposed into horizontal and vertical projections.

$$x[r,c] = vert[r] \times horz[c]$$

where $x[r,c]$ is the two-dimensional image, and *vert* $[r]$ & *horz* $[c]$ are the one-dimensional projections. Obviously, most images do not satisfy this requirement. For example, the pillbox is not separable. There are, however, an *infinite* number of separable images. This can be understood by generating arbitrary horizontal and vertical projections, and finding the image that corresponds to them. For example, Fig. 24-6 illustrates this with profiles that are double-sided exponentials. The image that corresponds to these profiles is then found from Eq. 24-1. When displayed, the image appears as a diamond shape that exponentially decays to zero as the distance from the origin increases.

In most image processing tasks, the ideal PSF is *circularly symmetric*, such as the pillbox. Even though digitized images are usually stored and processed in the rectangular format of rows and columns, it is desired to modify the image the same in all directions. This raises the question: is there a PSF that is circularly symmetric *and* separable? The answer is, yes,

FIGURE 24-5
Separation of the rectangular PSF. A PSF is said to be *separable* if it can be decomposed into horizontal and vertical profiles. Separable PSFs are important because they can be rapidly convolved.

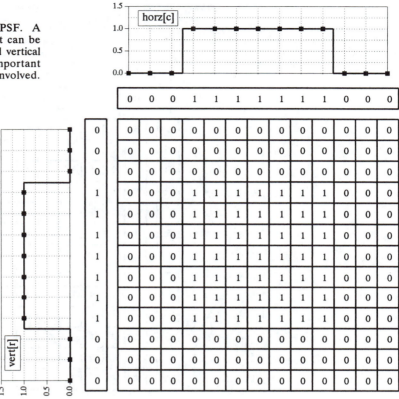

FIGURE 24-6
Creation of a separable PSF. An infinite number of separable PSFs can be generated by defining arbitrary projections, and then calculating the two-dimensional function that corresponds to them. In this example, the profiles are chosen to be double-sided exponentials, resulting in a diamond shaped PSF.

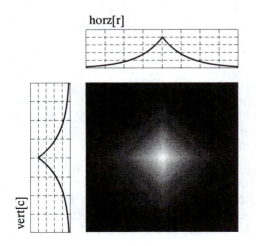
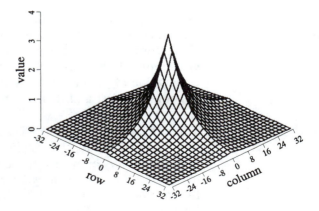

FIGURE 24-7
Separation of the Gaussian. The Gaussian is the only PSF that is circularly symmetric *and* separable. This makes it a common filter kernel in image processing.

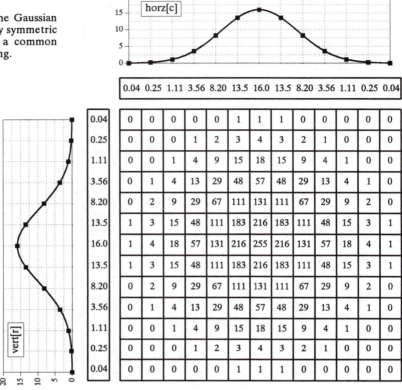

	0.04	0.25	1.11	3.56	8.20	13.5	16.0	13.5	8.20	3.56	1.11	0.25	0.04
0.04	0	0	0	0	0	1	1	1	0	0	0	0	0
0.25	0	0	0	1	2	3	4	3	2	1	0	0	0
1.11	0	0	1	4	9	15	18	15	9	4	1	0	0
3.56	0	1	4	13	29	48	57	48	29	13	4	1	0
8.20	0	2	9	29	67	111	131	111	67	29	9	2	0
13.5	1	3	15	48	111	183	216	183	111	48	15	3	1
16.0	1	4	18	57	131	216	255	216	131	57	18	4	1
13.5	1	3	15	48	111	183	216	183	111	48	15	3	1
8.20	0	2	9	29	67	111	131	111	67	29	9	2	0
3.56	0	1	4	13	29	48	57	48	29	13	4	1	0
1.11	0	0	1	4	9	15	18	15	9	4	1	0	0
0.25	0	0	0	1	2	3	4	3	2	1	0	0	0
0.04	0	0	0	0	0	1	1	1	0	0	0	0	0

but there is only one, the *Gaussian*. As is shown in Fig. 24-7, a two-dimensional Gaussian image has projections that are also Gaussians. The image and projection Gaussians have the *same* standard deviation.

To convolve an image with a separable filter kernel, convolve each *row* in the image with the *horizontal projection*, resulting in an intermediate image. Next, convolve each *column* of this intermediate image with the *vertical projection* of the PSF. The resulting image is identical to the direct convolution of the original image and the filter kernel. If you like, convolve the columns first and then the rows; the result is the same.

The convolution of an $N \times N$ image with an $M \times M$ filter kernel requires a time proportional to $N^2 M^2$. In other words, each pixel in the output image depends on *all* the pixels in the filter kernel. In comparison, convolution by separability only requires a time proportional to $N^2 M$. For filter kernels that are hundreds of pixels wide, this technique will reduce the execution time by a factor of *hundreds*.

Things can get even better. If you are willing to use a *rectangular* PSF (Fig. 24-5) or a *double-sided exponential* PSF (Fig. 24-6), the calculations are even more efficient. This is because the one-dimensional convolutions are the *moving average* filter (Chapter 15) and the *bidirectional single pole* filter

(Chapter 19), respectively. Both of these one-dimensional filters can be rapidly carried out by recursion. This results in an image convolution time proportional to only N^2, completely independent of the size of the PSF. In other words, an image can be convolved with as large a PSF as needed, with only a few integer operations per pixel. For example, the convolution of a 512×512 image requires only a few hundred milliseconds on a personal computer. That's fast! Don't like the shape of these two filter kernels? Convolve the image with one of them *several times* to approximate a Gaussian PSF (guaranteed by the Central Limit Theorem, Chapter 7). These are great algorithms, capable of snatching success from the jaws of failure. They are well worth remembering.

Example of a Large PSF: Illumination Flattening

A common application requiring a large PSF is the enhancement of images with unequal illumination. Convolution by separability is an ideal algorithm to carry out this processing. With only a few exceptions, the images seen by the eye are formed from *reflected* light. This means that a viewed image is equal to the reflectance of the objects multiplied by the ambient illumination. Figure 24-8 shows how this works. Figure (a) represents the *reflectance* of a scene being viewed, in this case, a series of light and dark bands. Figure (b) illustrates an example illumination signal, the pattern of light falling on (a). As in the real world, the illumination slowly varies over the imaging area. Figure (c) is the image seen by the eye, equal to the reflectance image, (a), multiplied by the illumination image, (b). The regions of poor illumination are difficult to view in (c) for two reasons: they are too dark and their contrast is too low (the difference between the peaks and the valleys).

To understand how this relates to the problem of every day vision, imagine you are looking at two identically dressed men. One of them is standing in the bright sunlight, while the other is standing in the shade of a nearby tree. The percent of the incident light reflected from both men is the same. For instance, their faces might reflect 80% of the incident light, their gray shirts 40% and their dark pants 5%. The problem is, the illumination of the two might be, say, ten times different. This makes the image of the man in the shade ten times darker than the person in the sunlight, and the contrast (between the face, shirt, and pants) ten times less.

The goal of the image processing is to *flatten* the illumination component in the acquired image. In other words, we want the final image to be representative of the objects' reflectance, not the lighting conditions. In terms of Fig. 24-8, given (c), find (a). This is a nonlinear filtering problem, since the component images were combined by multiplication, not addition. While this separation cannot be performed perfectly, the improvement can be dramatic.

To start, we will convolve image (c) with a large PSF, one-fifth the size of the entire image. The goal is to eliminate the sharp features in (c), resulting

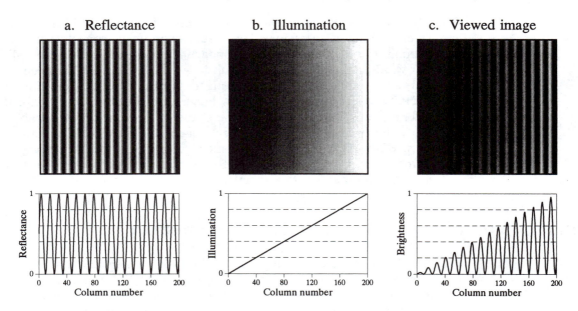

FIGURE 24-8
Model of image formation. A viewed image, (c), results from the multiplication of an illumination pattern, (b), by a reflectance pattern, (a). The goal of the image processing is to modify (c) to make it look more like (a). This is performed in Figs. (d), (e) and (f) on the opposite page.

in an approximation to the original illumination signal, (b). This is where convolution by separability is used. The exact shape of the PSF is not important, only that it is much wider than the features in the reflectance image. Figure (d) is the result, using a Gaussian filter kernel.

Since a smoothing filter provides an estimate of the illumination image, we will use an edge enhancement filter to find the reflectance image. That is, image (c) will be convolved with a filter kernel consisting of a delta function minus a Gaussian. To reduce execution time, this is done by subtracting the smoothed image in (d) from the original image in (c). Figure (e) shows the result. It doesn't work! While the dark areas have been properly lightened, the contrast in these areas is still terrible.

Linear filtering performs poorly in this application because the reflectance and illumination signals were original combined by multiplication, not addition. Linear filtering cannot correctly separate signals combined by a nonlinear operation. To separate these signals, they must be *unmultiplied*. In other words, the original image should be *divided* by the smoothed image, as is shown in (f). This corrects the brightness and restores the contrast to the proper level.

This procedure of dividing the images is closely related to **homomorphic processing**, previously described in Chapter 22. Homomorphic processing is a way of handling signals combined through a nonlinear operation. The strategy is to change the nonlinear problem into a linear one, through an appropriate mathematical operation. When two signals are combined by

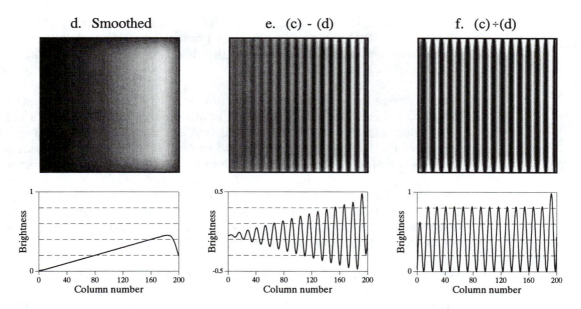

FIGURE 24-8 (continued)
Figure (d) is a smoothed version of (c), used as an approximation to the illumination signal. Figure (e) shows an approximation to the reflectance image, created by *subtracting* the smoothed image from the viewed image. A better approximation is shown in (f), obtained by the nonlinear process of *dividing* the two images.

multiplication, homomorphic processing starts by taking the *logarithm* of the acquired signal. With the identity: $\log(a \times b) = \log(a) + \log(b)$, the problem of separating *multiplied* signals is converted into the problem of separating *added* signals. For example, after taking the logarithm of the image in (c), a linear high-pass filter can be used to isolate the logarithm of the reflectance image. As before, the quickest way to carry out the high-pass filter is to subtract a smoothed version of the image. The antilogarithm (exponent) is then used to undo the logarithm, resulting in the desired approximation to the reflectance image.

Which is better, dividing or going along the homomorphic path? They are nearly the same, since taking the logarithm and subtracting is *equal* to dividing. The only difference is the approximation used for the illumination image. One method uses a smoothed version of the acquired image, while the other uses a smoothed version of the logarithm of the acquired image.

This technique of flattening the illumination signal is so useful it has been incorporated into the neural structure of the eye. The processing in the middle layer of the retina was previously described as an edge enhancement or high-pass filter. While this is true, it doesn't tell the whole story. The first layer of the eye is nonlinear, approximately taking the *logarithm* of the incoming image. This makes the eye a homomorphic processor. Just as described above, the logarithm followed by a linear edge enhancement filter flattens the illumination component, allowing the eye to see under poor lighting conditions. Another interesting use of homomorphic processing occurs in photography. The density (darkness) of a negative is equal to the logarithm of the brightness in the final photograph. This means that any manipulation of the negative during the development stage is a type of homomorphic processing.

Before leaving this example, there is a nuisance that needs to be mentioned. As discussed in Chapter 6, when an N point signal is convolved with an M point filter kernel, the resulting signal is $N+M$-1 points long. Likewise, when an $M \times M$ image is convolved with an $N \times N$ filter kernel, the result is an $M+N$-1 $\times M+N$-1 image. The problem is, it is often difficult to manage a changing image size. For instance, the allocated memory must change, the video display must be adjusted, the array indexing may need altering, etc. The common way around this is to *ignore* it; if we start with a 512×512 image, we want to end up with a 512×512 image. The pixels that do not fit within the original boundaries are discarded.

While this keeps the image size the same, it doesn't solve the whole problem; these is still the *boundary condition* for convolution. For example, imagine trying to calculate the pixel at the upper-right corner of (d). This is done by centering the Gaussian PSF on the upper-right corner of (c). Each pixel in (c) is then multiplied by the corresponding pixel in the overlaying PSF, and the products are added. The problem is, three-quarters of the PSF lies outside the defined image. The easiest approach is to assign the undefined pixels a value of zero. This is how (d) was created, accounting for the dark band around the perimeter of the image. That is, the brightness smoothly decreases to the pixel value of zero, exterior to the defined image.

Fortunately, this dark region around the boarder can be corrected (although it hasn't been in this example). This is done by dividing each pixel in (d) by a correction factor. The correction factor is the fraction of the PSF that was immersed in the input image when the pixel was calculated. That is, to correct an individual pixel in (d), imagine that the PSF is centered on the corresponding pixel in (c). For example, the upper-right pixel in (c) results from only 25% of the PSF overlapping the input image. Therefore, correct this pixel in (d) by dividing it by a factor of 0.25. This means that the pixels in the center of (d) will not be changed, but the dark pixels around the perimeter will be brightened. To find the correction factors, imagine convolving the filter kernel with an image having all the pixel values equal to *one*. The pixels in the resulting image are the correction factors needed to eliminate the edge effect.

Fourier Image Analysis

Fourier analysis is used in image processing in much the same way as with one-dimensional signals. However, images do not have their information encoded in the frequency domain, making the techniques much less useful. For example, when the Fourier transform is taken of an *audio* signal, the confusing time domain waveform is converted into an easy to understand frequency spectrum. In comparison, taking the Fourier transform of an image converts the straightforward information in the spatial domain into a scrambled form in the frequency domain. In short, don't expect the Fourier transform to help you understand the information encoded in images.

Likewise, don't look to the frequency domain for filter design. The basic feature in images is the edge, the line separating one *object* or *region* from another *object* or *region*. Since an edge is composed of a wide range of frequency components, trying to modify an image by manipulating the frequency spectrum is generally not productive. Image filters are normally designed in the spatial domain, where the information is encoded in its simplest form. Think in terms of *smoothing* and *edge enhancement* operations (the spatial domain) rather than *high-pass* and *low-pass* filters (the frequency domain).

In spite of this, Fourier image analysis does have several useful properties. For instance, *convolution* in the spatial domain corresponds to *multiplication* in the frequency domain. This is important because multiplication is a simpler mathematical operation than convolution. As with one-dimensional signals, this property enables FFT convolution and various deconvolution techniques. Another useful property of the frequency domain is the *Fourier Slice Theorem*, the relationship between an image and its projections (the image viewed from its sides). This is the basis of *computed tomography*, an x-ray imaging technique widely used medicine and industry.

The frequency spectrum of an image can be calculated in several ways, but the FFT method presented here is the only one that is practical. The original image must be composed of N rows by N columns, where N is a power of two, i.e., 256, 512, 1024, etc. If the size of the original image is not a power of two, pixels with a value of zero are added to make it the correct size. We will call the two-dimensional array that holds the image the **real array**. In addition, another array of the same size is needed, which we will call the **imaginary array**.

The recipe for calculating the Fourier transform of an image is quite simple: take the one-dimensional FFT of each of the rows, followed by the one-dimensional FFT of each of the columns. Specifically, start by taking the FFT of the N pixel values in row 0 of the real array. The real part of the FFT's output is placed back into row 0 of the real array, while the imaginary part of the FFT's output is placed into row 0 of the imaginary array. After repeating this procedure on rows 1 through N-1, both the real and imaginary arrays contain an intermediate image. Next, the procedure is repeated on each of the *columns* of the intermediate data. Take the N pixel values from column 0 of the real array, *and* the N pixel values from column 0 of the imaginary array, and calculate the FFT. The real part of the FFT's output is placed back into column 0 of the real array, while the imaginary part of the FFT's output is placed back into column 0 of the imaginary array. After this is repeated on columns 1 through N-1, both arrays have been overwritten with the image's frequency spectrum.

Since the vertical and horizontal directions are equivalent in an image, this algorithm can also be carried out by transforming the columns first and then transforming the rows. Regardless of the order used, the result is the same. From the way that the FFT keeps track of the data, the amplitudes of the low frequency components end up being at the corners of the two-dimensional spectrum, while the high frequencies are at the center. The inverse Fourier transform of an image is calculated by taking the inverse FFT of each row, followed by the inverse FFT of each column (or vice versa).

Figure 24-9 shows an example Fourier transform of an image. Figure (a) is the original image, a microscopic view of the input stage of a 741 op amp integrated circuit. Figure (b) shows the real and imaginary parts of the frequency spectrum of this image. Since the frequency domain can contain negative pixel values, the grayscale values of these images are offset such that negative values are dark, zero is gray, and positive values are light. The low-frequency components in an image are normally much larger in amplitude than the high-frequency components. This accounts for the very bright and dark pixels at the four corners of (b). Other than this, the spectra of *typical* images have no discernable order, appearing random. Of course, images can be *contrived* to have any spectrum you desire.

As shown in (c), the polar form of an image spectrum is only slightly easier to understand. The low-frequencies in the magnitude have large positive values (the white corners), while the high-frequencies have small positive values (the black center). The phase looks the same at low and high-frequencies, appearing to run randomly between $-\pi$ and π radians.

Figure (d) shows an alternative way of displaying an image spectrum. Since the spatial domain contains a *discrete* signal, the frequency domain is *periodic*. In other words, the frequency domain arrays are duplicated an infinite number of times to the left, right, top and bottom. For instance, imagine a tile wall, with each tile being the $N \times N$ magnitude shown in (c). Figure (d) is also an $N \times N$ section of this tile wall, but it straddles four tiles; the center of the image being where the four tiles touch. In other words, (c) is the same image as (d), except it has been shifted $N/2$ pixels horizontally (either left or right) and $N/2$ pixels vertically (either up or down) in the periodic frequency spectrum. This brings the bright pixels at the four corners of (c) together in the center of (d).

Figure 24-10 illustrates how the two-dimensional frequency domain is organized (with the low-frequencies placed at the corners). Row $N/2$ and column $N/2$ break the frequency spectrum into four quadrants. For the real part and the magnitude, the upper-right quadrant is a mirror image of the lower-left, while the upper-left is a mirror image of the lower-right. This symmetry also holds for the imaginary part and the phase, except that the values of the mirrored pixels are opposite in sign. In other words, every point in the frequency spectrum has a matching point placed symmetrically on the other side of the center of the image (row $N/2$ and column $N/2$). One of the points is the *positive* frequency, while the other is the matching

FIGURE 24-9
Frequency spectrum of an image. The example image, shown in (a), is a microscopic photograph of the silicon surface of an integrated circuit. The frequency spectrum can be displayed as the real and imaginary parts, shown in (b), or as the magnitude and phase, shown in (c). Figures (b) & (c) are displayed with the low-frequencies at the corners and the high-frequencies at the center. Since the frequency domain is periodic, the display can be rearranged to reverse these positions. This is shown in (d), where the magnitude and phase are displayed with the low-frequencies located at the center and the high-frequencies at the corners.

a. Image

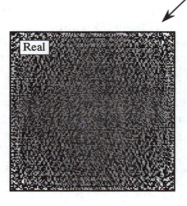

b. Frequency spectrum displayed in rectangular form (as the real and imaginary parts).

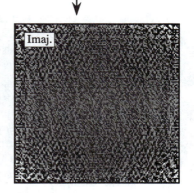

Real Imaj.

c. Frequency spectrum displayed in polar form (as the magnitude and phase).

Mag. Phase

d. Frequency spectrum displayed in polar form, with the spectrum shifted to place zero frequency at the center.

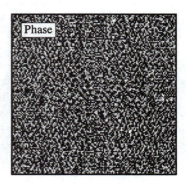

Mag. Phase

negative frequency, as discussed in Chapter 10 for one-dimensional signals. In equation form, this symmetry is expressed as:

EQUATION 24-2
Symmetry of the two-dimensional frequency domain. These equations can be used in both formats, when the low-frequencies are displayed at the corners, or when shifting places them at the center. In polar form, the magnitude has the same symmetry as the real part, while the phase has the same symmetry as the imaginary part.

$$Re\, X[r,c] = Re\, X[N-r, N-c]$$

$$Im\, X[r,c] = -Im\, X[N-r, N-c]$$

These equations take into account that the frequency spectrum is periodic, repeating itself every N samples with indexes running from 0 to $N-1$. In other words, $X[r,N]$ should be interpreted as $X[r,0]$, $X[N,c]$ as $X[0,c]$, and $X[N,N]$ as $X[0,0]$. This symmetry makes four points in the spectrum match with *themselves*. These points are located at: $[0,0]$, $[0,N/2]$, $[N/2,0]$ and $[N/2,N/2]$.

Each pair of points in the frequency domain corresponds to a sinusoid in the spatial domain. As shown in (a), the value at $[0,0]$ corresponds to the *zero frequency* sinusoid in the spatial domain, i.e., the DC component of the image. There is only one point shown in this figure, because this is one of the points that is its own match. As shown in (b), (c), and (d), other pairs of points correspond to two-dimensional sinusoids that look like waves on the ocean. One-dimensional sinusoids have a *frequency*, *phase*, and *amplitude*. Two dimensional sinusoids also have a *direction*.

The frequency and direction of each sinusoid is determined by the location of the pair of points in the frequency domain. As shown, draw a line from each point to the *zero frequency* location at the outside corner of the quadrant that the point is in, i.e., $[0,0]$, $[0,N/2]$, $[N/2,0]$, or $[N/2,N/2]$ (as indicated by the circles in this figure). The direction of this line determines the direction of the spatial sinusoid, while its length is proportional to the frequency of the wave. This results in the low frequencies being positioned near the corners, and the high frequencies near the center.

When the spectrum is displayed with zero frequency at the center (Fig. 24-9d), the line from each pair of points is drawn to the DC value at the *center* of the image, i.e., $[N/2,N/2]$. This organization is simpler to understand and work with, since all the lines are drawn to the same point. Another advantage of placing zero at the center is that it matches the frequency spectra of *continuous* images. When the spatial domain is continuous, the frequency domain is *aperiodic*. This places zero frequency at the center, with the frequency becoming higher in all directions out to infinity. In general, the frequency spectra of discrete images are displayed with zero frequency at the center whenever people will view the data, in textbooks, journal articles, and algorithm documentation. However, most calculations are carried out with the computer arrays storing data in the other format (low-frequencies at the corners). This is because the FFT has this format.

Spatial Domain

Frequency Domain

a.
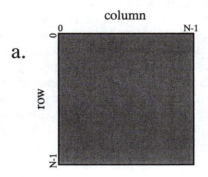
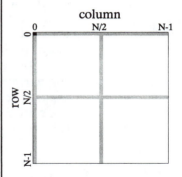

FIGURE 24-10
Two-dimensional sinusoids. Image sine and cosine waves have both a *frequency* and a *direction*. Four examples are shown here. These spectra are displayed with the low-frequencies at the corners. The circles in these spectra show the location of zero frequency.

b.
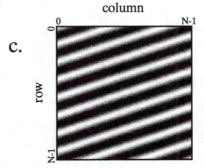
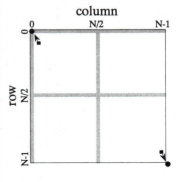

c.
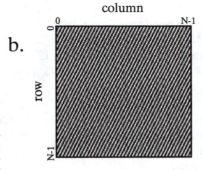
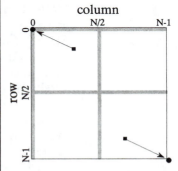

d.
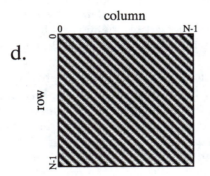
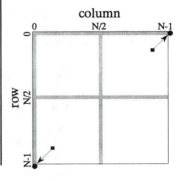

Even with the FFT, the time required to calculate the Fourier transform is a tremendous bottleneck in image processing. For example, the Fourier transform of a 512×512 image requires several minutes on a personal computer. This is roughly 10,000 times slower than needed for real time image processing, 30 frames per second. This long execution time results from the massive amount of information contained in images. For comparison, there are about the same number of *pixels* in a typical image, as there are *words* in this book. Image processing via the frequency domain will become more popular as computers become faster. This is a twenty-first century technology; watch it emerge!

FFT Convolution

Even though the Fourier transform is slow, it is still the fastest way to convolve an image with a large filter kernel. For example, convolving a 512×512 image with a 50×50 PSF is about 20 times faster using the FFT compared with conventional convolution. Chapter 18 discusses how FFT convolution works for one-dimensional signals. The two-dimensional version is a simple extension.

We will demonstrate FFT convolution with an example, an algorithm to locate a predetermined pattern in an image. Suppose we build a system for inspecting one-dollar bills, such as might be used for printing quality control, counterfeiting detection, or payment verification in a vending machine. As shown in Fig. 24-11, a 100×100 pixel image is acquired of the bill, centered on the portrait of George Washington. The goal is to search this image for a known pattern, in this example, the 29×29 pixel image of the face. The problem is this: given an acquired image and a known pattern, what is the most effective way to locate where (or if) the pattern appears in the image? If you paid attention in Chapter 6, you know that the solution to this problem is *correlation* (a matched filter) and that it can be implemented by using *convolution*.

Before performing the actual convolution, there are two modifications that need to be made to turn the target image into a PSF. These are illustrated in Fig. 24-12. Figure (a) shows the target signal, the pattern we are trying to detect. In (b), the image has been rotated by 180°, the same as being flipped left-for-right and then flipped top-for-bottom. As discussed in Chapter 7, when performing *correlation* by using *convolution*, the target signal needs to be reversed to counteract the reversal that occurs during convolution. We will return to this issue shortly.

The second modification is a trick for improving the effectiveness of the algorithm. Rather than trying to detect the face in the original image, it is more effective to detect the *edges of the face* in the *edges of the original image*. This is because the edges are sharper than the original features, making the correlation have a sharper peak. This step isn't required, but it makes the results significantly better. In the simplest form, a 3×3 edge detection filter is applied to both the original image and the target signal

a. Image to be searched

b. Target

FIGURE 24-11
Target detection example. The problem is to search the 100×100 pixel image of George Washington, (a), for the target pattern, (b), the 29×29 pixel face. The optimal solution is *correlation*, which can be carried out by *convolution*.

before the correlation is performed. From the associative property of convolution, this is the same as applying the edge detection filter to the target signal *twice*, and leaving the original image alone. In actual practice, applying the edge detection 3×3 kernel only once is generally sufficient. This is how (b) is changed into (c) in Fig. 24-12. This makes (c) the PSF to be used in the convolution

Figure 24-13 illustrates the details of FFT convolution. In this example, we will convolve image (a) with image (b) to produce image (c). The fact that these images have been chosen and preprocessed to implement *correlation* is irrelevant; this is a flow diagram of *convolution*. The first step is to pad both signals being convolved with enough zeros to make them a power of two in size, and big enough to hold the final image. That is, when images of 100×100 and 29×29 pixels are convolved, the resulting image will be 128×128 pixels. Therefore, enough zeros must be added to (a) and (b) to make them each 128×128 pixels in size. If this isn't done, circular

a. Original   b. Rotated   c. Edge detection

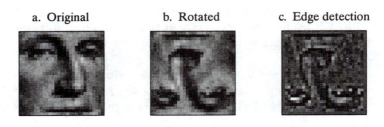

FIGURE 24-12
Development of a correlation filter kernel. The target signal is shown in (a). In (b) it is rotated by 180° to undo the rotation inherent in convolution, allowing correlation to be performed. Applying an edge detection filter results in (c), the filter kernel used for this example.

convolution takes place and the final image will be distorted. If you are having trouble understanding these concepts, go back and review Chapter 18, where the one-dimensional case is discussed in more detail.

The FFT algorithm is used to transform (a) and (b) into the frequency domain. This results in *four* 128×128 arrays, the real and imaginary parts of the two images being convolved. Multiplying the real and imaginary parts of (a) with the real and imaginary parts of (b), generates the real and imaginary parts of (c). (If you need to be reminded how this is done, see Eq. 9-1). Taking the Inverse FFT completes the algorithm by producing the final convolved image.

The value of each pixel in a correlation image is a measure of how well the target image matches the searched image *at that point*. In this particular example, the correlation image in (c) is composed of noise plus a single bright peak, indicating a good match to the target signal. Simply locating the brightest pixel in this image would specify the detected coordinates of the face. If we had not used the edge detection modification on the target signal, the peak would still be present, but much less distinct.

While correlation is a powerful tool in image processing, it suffers from a significant limitation: the target image must be exactly the same *size* and *rotational orientation* as the corresponding area in the searched image. Noise and other variations in the amplitude of each pixel are relatively unimportant, but an exact spatial match is critical. For example, this makes the method almost useless for finding enemy tanks in military reconnaissance photos, tumors in medical images, and handguns in airport baggage scans. One approach is to correlate the image *multiple times* with a variety of shapes and rotations of the target image. This works in principle, but the execution time will make you loose interest in a hurry.

A Closer Look at Image Convolution

Let's use this last example to explore two-dimensional convolution in more detail. Just as with one dimensional signals, image convolution can be viewed from either the *input side* or the *output side*. As you recall from Chapter 6, the input viewpoint is the best description of how convolution works, while the output viewpoint is how most of the mathematics and algorithms are written. You need to become comfortable with both these ways of looking at the operation.

Figure 24-14 shows the input side description of image convolution. Every pixel in the input image results in a scaled and shifted PSF being added to the output image. The output image is then calculated as the sum of all the contributing PSFs. This illustration show the contribution to the output image from the point at location [r,c] in the input image. The PSF is shifted such that pixel [0,0] in the PSF aligns with pixel [r,c] in the output image. If the PSF is defined with only positive indexes, such as in this example, the shifted PSF will be entirely to the lower-right of [r,c]. Don't

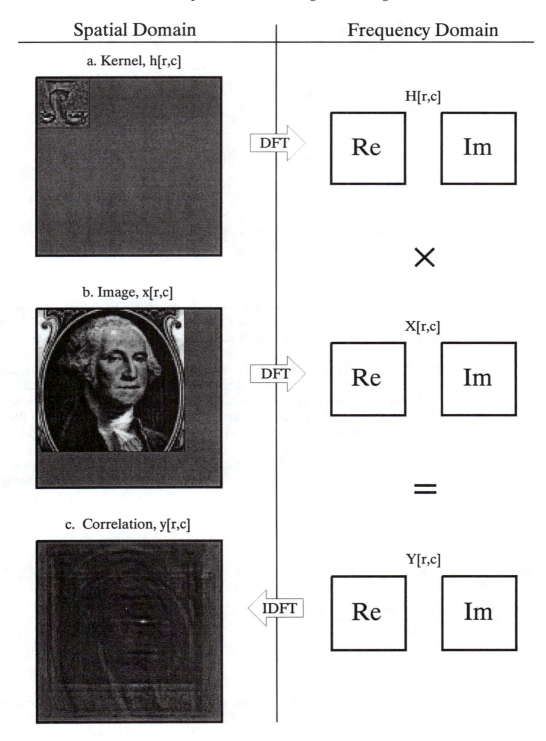

Spatial Domain

a. Kernel, h[r,c]

b. Image, x[r,c]

c. Correlation, y[r,c]

Frequency Domain

H[r,c]

Re   Im

×

X[r,c]

Re   Im

=

Y[r,c]

Re   Im

DFT

DFT

IDFT

FIGURE 24-13
Flow diagram of FFT image convolution. The images in (a) and (b) are transformed into the frequency domain by using the FFT. These two frequency spectra are multiplied, and the Inverse FFT is used to move back into the spatial domain. In this example, the original images have been chosen and preprocessed to implement *correlation* through the action of convolution.

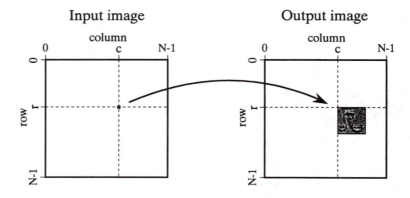

FIGURE 24-14
Image convolution viewed from the input side. Each pixel in the input image contributes a scaled and shifted PSF to the output image. The output image is the sum of these contributions. The face is inverted in this illustration because this is the PSF we are using.

be confused by the face appearing upside down in this figure; this upside down face *is* the PSF we are using in this example (Fig. 24-13a). In the input side view, there is no rotation of the PSF, it is simply shifted.

Image convolution viewed from the output is illustrated in Fig. 24-15. Each pixel in the output image, such as shown by the sample at [r,c], receives a contribution from many pixels in the input image. The PSF is rotated by 180° around pixel [0,0], and then shifted such that pixel [0,0] in the PSF is aligned with pixel [r,c] in the input image. If the PSF only uses positive indexes, it will be to the upper-left of pixel [r,c] in the input image. The value of the pixel at [r,c] in the output image is found by multiplying the pixels in the rotated PSF with the corresponding pixels in the input image, and summing the products. This procedure is given by Eq. 24-3, and in the program of Table 24-1.

EQUATION 24-3
Image convolution. The images $x[\ ,\ ]$ and $h[\ ,\ ]$ are convolved to produce image, $y[\ ,\ ]$. The size of $h[\ ,\ ]$ is $M \times M$ pixels, with the indexes running from 0 to $M-1$. In this equation, an individual pixel in the output image, $y[r,c]$, is calculated according to the output side view. The indexes j and k are used to loop through the rows and columns of $h[\ ,\ ]$ to calculate the sum-of-products.

$$y[r,c] = \sum_{k=0}^{M-1} \sum_{j=0}^{M-1} h[k,j]\, x[r-k, c-j]$$

Notice that the PSF rotation resulting from the convolution has undone the rotation made in the design of the PSF. This makes the face appear upright in Fig. 24-15, allowing it to be in the same orientation as the pattern being detected in the input image. That is, we have successfully used *convolution* to implement *correlation*. Compare Fig. 24-13c with Fig. 24-15 to see how the bright spot in the correlation image signifies that the target has been detected.

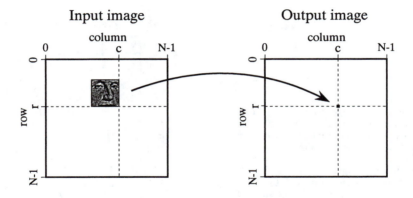

FIGURE 24-15
Image convolution viewed from the output side. Each pixel in the output signal is equal to the sum of the pixels in the rotated PSF multiplied by the corresponding pixels in the input image.

FFT convolution provides the same output image as the conventional convolution program of Table 24-1. Is the reduced execution time provided by FFT convolution really worth the additional program complexity? Let's take a closer look. Figure 24-16 shows an execution time comparison between conventional convolution using *floating point* (labeled FP), conventional convolution using *integers* (labeled INT), and FFT convolution using floating point (labeled FFT). Data for two different image sizes are presented, 512×512 and 128×128.

First, notice that the execution time required for FFT convolution does not depend on the size of the kernel, resulting in flat lines in this graph. On a 100 MHz Pentium personal computer, a 128×128 image can be convolved

```
100 CONVENTIONAL IMAGE CONVOLUTION
110 '
120 DIM X[99,99]              'holds the input image,   100×100 pixels
130 DIM H[28,28]              'holds the filter kernel,   29×29 pixels
140 DIM Y[127,127]            'holds the output image, 128×128 pixels
150 '
160 FOR R% = 0 TO 127         'loop through each row and column in the output
170 FOR C% = 0 TO 127         'image calculating the pixel value via Eq. 24-3
180 '
190   Y[R%,C%] = 0            'zero the pixel so it can be used as an accumulator
200 '
210   FOR J% = 0 TO 28        'multiply each pixel in the kernel by the corresponding
220   FOR K% = 0 TO 28        'pixel in the input image, and add to the accumulator
230     Y[R%,C%] = Y[R%,C%] + H[J%,K%] * X[R%-J%,C%-J%]
240   NEXT K%
250   NEXT J%
260 '
270 NEXT C%
280 NEXT R%
290 '
300 END
```

TABLE 24-1

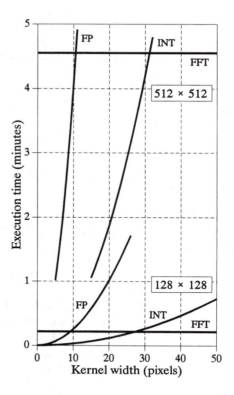

FIGURE 24-16
Execution time for image convolution. This graph shows the execution time on a 100 MHz Pentium processor for three image convolution methods: conventional convolution carried out with floating point math (FP), conventional convolution using integers (INT), and FFT convolution using floating point (FFT). The two sets of curves are for input image sizes of 512×512 and 128×128 pixels. Using FFT convolution, the time depends only on the image size, and not the size of the kernel. In contrast, conventional convolution depends on both the image and the kernel size.

in about 15 seconds using FFT convolution, while a 512×512 image requires more than 4 minutes. Adding up the number of calculations shows that the execution time for FFT convolution is proportional to $N^2 Log_2(N)$, for an $N{\times}N$ image. That is, a 512×512 image requires about 20 times as long as a 128×128 image.

Conventional convolution has an execution time proportional to $N^2 M^2$ for an $N{\times}N$ image convolved with an $M{\times}M$ kernel. This can be understood by examining the program in Table 24-1. In other words, the execution time for conventional convolution depends *very strongly* on the size of the kernel used. As shown in the graph, FFT convolution is faster than conventional convolution using floating point if the kernel is larger than about 10×10 pixels. In most cases, integers can be used for conventional convolution, increasing the break-even point to about 30×30 pixels. These break-even points depend slightly on the size of the image being convolved, as shown in the graph. The concept to remember is that FFT convolution is only useful for *large* filter kernels.

Special Imaging Techniques

This chapter presents four specific aspects of image processing. First, ways to characterize the *spatial resolution* are discussed. This describes the minimum size an object must be to be seen in an image. Second, the *signal-to-noise ratio* is examined, explaining how faint an object can be and still be detected. Third, *morphological* techniques are introduced. These are nonlinear operations used to manipulate binary images (where each pixel is either black or white). Fourth, the remarkable technique of *computed tomography* is described. This has revolutionized medical diagnosis by providing detailed images of the interior of the human body.

Spatial Resolution

Suppose we want to compare two imaging systems, with the goal of determining which has the best spatial resolution. In other words, we want to know which system can detect the smallest object. To simplify things, we would like the answer to be a *single number* for each system. This allows a direct comparison upon which to base design decisions. Unfortunately, a single parameter is not always sufficient to characterize all the subtle aspects of imaging. This is complicated by the fact that spatial resolution is limited by two distinct but interrelated effects: *sample spacing* and *sampling aperture size*. This section contains two main topics: (1) how a single parameter can best be used to characterize spatial resolution, and (2) the relationship between sample spacing and sampling aperture size.

Figure 25-1a shows profiles from three circularly symmetric PSFs: the pillbox, the Gaussian, and the exponential. These are representative of the PSFs commonly found in imaging systems. As described in the last chapter, the pillbox can result from an improperly focused lens system. Likewise, the Gaussian is formed when random errors are combined, such as viewing stars through a turbulent atmosphere. An exponential PSF is generated when electrons or x-rays strike a phosphor layer and are converted into

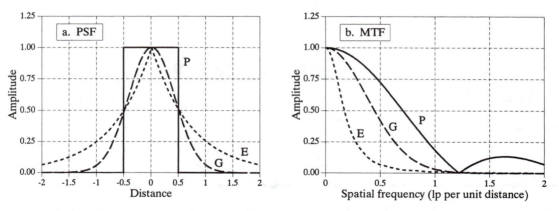

FIGURE 25-1

FWHM versus MTF. Figure (a) shows profiles of three PSFs commonly found in imaging systems: (P) pillbox, (G) Gaussian, and (E) exponential. Each of these has a FWHM of one unit. The corresponding MTFs are shown in (b). Unfortunately, similar values of FWHM do not correspond to similar MTF curves.

light. This is used in radiation detectors, night vision light amplifiers, and CRT displays. The exact shape of these three PSFs is not important for this discussion, only that they broadly represent the PSFs seen in real world applications.

The PSF contains complete information about the spatial resolution. To express the spatial resolution by a single number, we can ignore the *shape* of the PSF and simply measure its *width*. The most common way to specify this is by the Full-Width-at-Half-Maximum (FWHM) value. For example, all the PSFs in (a) have an FWHM of 1 unit.

Unfortunately, this method has two significant drawbacks. First, it does not match other measures of spatial resolution, including the subjective judgement of observers viewing the images. Second, it is usually very difficult to directly measure the PSF. Imagine feeding an impulse into an imaging system; that is, taking an image of a very small white dot on a black background. By definition, the acquired image will be the PSF of the system. The problem is, the measured PSF will only contain a few pixels, and its contrast will be low. Unless you are very careful, random noise will swamp the measurement. For instance, imagine that the impulse image is a 512×512 array of all zeros except for a single pixel having a value of 255. Now compare this to a normal image where all of the 512×512 pixels have an average value of about 128. In loose terms, the signal in the impulse image is about 100,000 times weaker than a normal image. No wonder the signal-to-noise ratio will be bad; there's hardly any signal!

A basic theme throughout this book is that signals should be understood in the domain where the information is encoded. For instance, audio signals should be dealt with in the frequency domain, while image signals should be handled in the spatial domain. In spite of this, one way to measure image resolution is by looking at the *frequency response*. This goes against the fundamental philosophy of this book; however, it is a common method and you need to become familiar with it.

Taking the two-dimensional Fourier transform of the PSF provides the two-dimensional frequency response. If the PSF is circularly symmetric, its frequency response will also be circularly symmetric. In this case, complete information about the frequency response is contained in its profile. That is, after calculating the frequency domain via the FFT method, columns 0 to $N/2$ in row 0 are all that is needed. In imaging jargon, this display of the frequency response is called the **Modulation Transfer Function (MTF)**. Figure 25-1b shows the MTFs for the three PSFs in (a). In cases where the PSF is not circularly symmetric, the entire two-dimensional frequency response contains information. However, it is usually sufficient to know the MTF curves in the vertical and horizontal directions (i.e., columns 0 to $N/2$ in row 0, and rows 0 to $N/2$ in column 0). Take note: this procedure of extracting a row or column from the two-dimensional frequency spectrum is *not* equivalent to taking the one-dimensional FFT of the profiles shown in (a). We will come back to this issue shortly. As shown in Fig. 25-1, similar values of FWHM do not correspond to similar MTF curves.

Figure 25-2 shows a **line pair gauge,** a device used to measure image resolution via the MTF. Line pair gauges come in different forms depending on the particular application. For example, the black and white pattern shown in this figure could be directly used to test video cameras. For an x-ray imaging system, the ribs might be made from lead, with an x-ray transparent material between. The key feature is that the black and white lines have a closer spacing toward one end. When an image is taken of a line pair gauge, the lines at the closely spaced end will be blurred together, while at the other end they will be distinct. Somewhere in the middle the lines will be just barely separable. An observer looks at the image, identifies this location, and reads the corresponding resolution on the calibrated scale.

FIGURE 25-2
Line pair gauge. The line pair gauge is a tool used to measure the resolution of imaging systems. A series of black and white ribs move together, creating a continuum of spatial frequencies. The resolution of a system is taken as the frequency where the eye can no longer distinguish the individual ribs. This example line pair gauge is shown several times larger than the calibrated scale indicates.

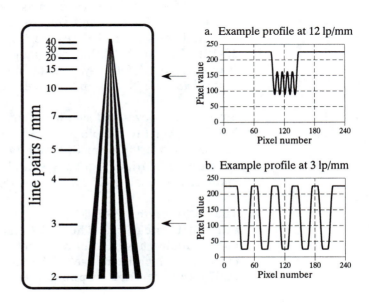

The *way* that the ribs blur together is important in understanding the limitations of this measurement. Imagine acquiring an image of the line pair gauge in Fig. 25-2. Figures (a) and (b) show examples of the profiles at low and high spatial frequencies. At the low frequency, shown in (b), the curve is flat on the top and bottom, but the edges are blurred, At the higher spatial frequency, (a), the *amplitude* of the modulation has been reduced. This is exactly what the MTF curve in Fig. 25-1b describes: higher spatial frequencies are reduced in amplitude. The individual ribs will be distinguishable in the image as long as the amplitude is greater than about 3% to 10% of the original height. This is related to the eye's ability to distinguish the low contrast difference between the peaks and valleys in the presence of image noise.

A strong advantage of the line pair gauge measurement is that it is simple and fast. The strongest disadvantage is that it relies on the human eye, and therefore has a certain subjective component. Even if the entire MTF curve is measured, the most common way to express the system resolution is to quote the frequency where the MTF is reduced to either 3%, 5% or 10%. Unfortunately, you will not always be told which of these values is being used; product data sheets frequently use vague terms such as "limiting resolution." Since manufacturers like their specifications to be as good as possible (regardless of what the device actually does), be safe and interpret these ambiguous terms to mean 3% on the MTF curve.

A subtle point to notice is that the MTF is defined in terms of *sine* waves, while the line pair gauge uses *square* waves. That is, the ribs are uniformly dark regions separated by uniformly light regions. This is done for manufacturing convenience; it is very difficult to make lines that have a sinusoidally varying darkness. What are the consequences of using a square wave to measure the MTF? At high spatial frequencies, all frequency components but the fundamental of the square wave have been removed. This causes the modulation to appear sinusoidal, such as is shown in Fig. 25-2a. At low frequencies, such as shown in Fig. 25-2b, the wave appears square. The fundamental sine wave contained in a square wave has an amplitude of $4/\pi = 1.27$ times the amplitude of the square wave (see Table 13-10). The result: the line pair gauge provides a slight overestimate of the true resolution of the system, by starting with an effective amplitude of more than pure black to pure white. Interesting, but almost always ignored.

Since square waves and sine waves are used interchangeably to measure the MTF, a special terminology has arisen. Instead of the word "cycle," those in imaging use the term **line pair** (a dark line next to a light line). For example, a spatial frequency would be referred to as *25 line pairs per millimeter*, instead of *25 cycles per millimeter*.

The width of the PSF doesn't track well with human perception and is difficult to measure. The MTF methods are in the *wrong domain* for understanding how resolution affects the encoded information. Is there a more favorable alternative? The answer is yes, the **line spread function (LSF)** and the **edge response**. As shown in Fig. 25-3, the line spread

a. Line Spread Function (LSF)   b. Edge Response

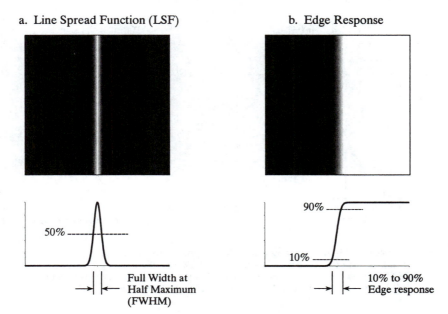

Full Width at Half Maximum (FWHM)   10% to 90% Edge response

FIGURE 25-3
Line spread function and edge response. The line spread function (LSF) is the derivative of the edge response. The width of the LSF is usually expressed as the Full-Width-at-Half-Maximum (FWHM). The width of the edge response is usually quoted by the 10% to 90% distance.

function is the response of the system to a thin line across the image. Similarly, the edge response is how the system responds to a sharp straight discontinuity (an edge). Since a line is the derivative (or first difference) of an edge, the LSF is the derivative (or first difference) of the edge response. The single parameter measurement used here is the distance required for the edge response to rise from 10% to 90%.

There are many advantages to using the edge response for measuring resolution. First, the measurement is in the same form as the image information is encoded. In fact, the main reason for wanting to know the resolution of a system is to understand how the edges in an image are *blurred*. The second advantage is that the edge response is simple to measure because edges are easy to generate in images. If needed, the LSF can easily be found by taking the first difference of the edge response.

The third advantage is that all common edges responses have a similar shape, even though they may originate from drastically different PSFs. This is shown in Fig. 25-4a, where the edge responses of the pillbox, Gaussian, and exponential PSFs are displayed. Since the shapes are similar, the 10%-90% distance is an excellent single parameter measure of resolution. The fourth advantage is that the MTF can be directly found by taking the one-dimensional FFT of the LSF (unlike the PSF to MTF calculation that must use a two-dimensional Fourier transform). Figure 25-4b shows the MTFs corresponding to the edge responses of (a). In other words, the curves in (a) are converted into the curves in (b) by taking the first difference (to find the LSF), and then taking the FFT.

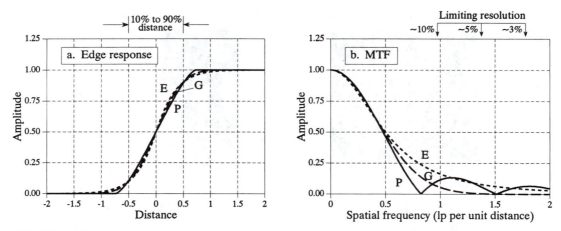

FIGURE 25-4
Edge response and MTF. Figure (a) shows the edge responses of three PSFs: (P) pillbox, (G) Gaussian, and (E) exponential. Each edge response has a 10% to 90% rise distance of 1 unit. Figure (b) shows the corresponding MTF curves, which are similar above the 10% level. *Limiting resolution* is a vague term indicating the frequency where the MTF has an amplitude of 3% to 10%.

The fifth advantage is that similar edge responses have similar MTF curves, as shown in Figs. 25-4 (a) and (b). This allows us to easily convert between the two measurements. In particular, a system that has a 10%-90% edge response of x distance, has a limiting resolution (10% contrast) of about 1 line pair per x distance. The units of the "distance" will depend on the type of system being dealt with. For example, consider three different imaging systems that have 10%-90% edge responses of 0.05 mm, 0.2 milliradian and 3.3 pixels. The 10% contrast level on the corresponding MTF curves will occur at about: 20 lp/mm, 5 lp/milliradian and 0.33 lp/pixel, respectively.

Figure 25-5 illustrates the mathematical relationship between the PSF and the LSF. Figure (a) shows a pillbox PSF, a circular area of value 1, displayed as white, surrounded by a region of all zeros, displayed as gray. A profile of the PSF (i.e., the pixel values along a line drawn across the center of the image) will be a rectangular pulse. Figure (b) shows the corresponding LSF. As shown, the LSF is mathematically equal to the *integrated profile* of the PSF. This is found by sweeping across the image in some direction, as illustrated by the rays (arrows). Each value in the integrated profile is the *sum* of the pixel values along the corresponding ray. In this example where the rays are vertical, each point in the integrated profile is found by adding all the pixel values in each column. This corresponds to the LSF of a line that is *vertical* in the image. The LSF of a line that is *horizontal* in the image is found by summing all of the pixel values in each *row*. For continuous images these concepts are the same, but the summations are replaced by integrals.

As shown in this example, the LSF can be directly calculated from the PSF. However, the PSF cannot always be calculated from the LSF. This is because the PSF contains information about the spatial resolution in *all directions*, while the LSF is limited to only one specific direction. A system

FIGURE 25-5
Relationship between the PSF and LSF. A pillbox PSF is shown in (a). Any row or column through the white center will be a rectangular pulse. Figure (b) shows the corresponding LSF, equivalent to an *integrated profile* of the PSF. That is, the LSF is found by sweeping across the image in some direction and adding (integrating) the pixel values along each ray. In the direction shown, this is done by adding all the pixels in each column.

a. Point Spread Function

b. "Integrated" profile of the PSF (the LSF)

has only one PSF, but an infinite number of LSFs, one for each angle. For example, imagine a system that has an oblong PSF. This makes the spatial resolution different in the vertical and horizontal directions, resulting in the LSF being different in these directions. Measuring the LSF at a single angle does not provide enough information to calculate the complete PSF except in the special instance where the PSF is circularly symmetric. Multiple LSF measurements at various angles make it possible to calculate a non-circular PSF; however, the mathematics is quite involved and usually not worth the effort. In fact, the problem of calculating the PSF from a number of LSF measurements is exactly the same problem faced in *computed tomography*, discussed later in this chapter.

As a practical matter, the LSF and the PSF are not dramatically different for most imaging systems, and it is very common to see one used as an approximation for the other. This is even more justifiable considering that there are two common cases where they are identical: the rectangular PSF has a rectangular LSF (with the same widths), and the Gaussian PSF has a Gaussian LSF (with the same standard deviations).

These concepts can be summarized into two skills: how to *evaluate* a resolution specification presented to you, and how to *measure* a resolution specification of your own. Suppose you come across an advertisement stating: "This system will resolve 40 line pairs per millimeter." You should interpret this to mean: "A sinusoid of 40 lp/mm will have its amplitude reduced to 3%-10% of its true value, and will be just barely visible in the image." You should also do the mental calculation that 40 lp/mm @ 10% contrast is equal to a 10%-90% edge response of 1/(40 lp/mm) = 0.025 mm. If the MTF specification is for a 3% contrast level, the edge response will be about 1.5 to 2 times wider.

When you measure the spatial resolution of an imaging system, the steps are carried out in reverse. Place a sharp edge in the image, and measure the resulting edge response. The 10%-90% distance of this curve is the best single parameter measurement of the system's resolution. To keep your boss and the marketing people happy, take the first difference of the edge response to find the LSF, and then use the FFT to find the MTF.

Sample Spacing and Sampling Aperture

Figure 25-6 shows two extreme examples of sampling, which we will call a **perfect detector** and a **blurry detector**. Imagine (a) being the surface of an imaging detector, such as a CCD. Light striking anywhere inside one of the square pixels will contribute *only* to that pixel value, and no others. This is shown in the figure by the black sampling aperture exactly filling one of the square pixels. This is an optimal situation for an image detector, because *all* of the light is detected, and there is *no* overlap or crosstalk between adjacent pixels. In other words, the sampling aperture is exactly equal to the sample spacing.

The alternative example is portrayed in (e). The sampling aperture is considerably larger than the sample spacing, and it follows a Gaussian distribution. In other words, each pixel in the detector receives a contribution from light striking the detector in a region *around* the pixel. This should sound familiar, because it is the output side viewpoint of convolution. From the corresponding input side viewpoint, a narrow beam of light striking the detector would contribute to the value of several neighboring pixels, also according to the Gaussian distribution.

Now turn your attention to the edge responses of the two examples. The markers in each graph indicate the actual pixel values you would find in an image, while the connecting lines show the *underlying curve* that is being sampled. An important concept is that the shape of this underlying curve is determined *only* by the *sampling aperture*. This means that the resolution in the final image can be limited in two ways. First, the underlying curve may have poor resolution, resulting from the sampling aperture being too large. Second, the sample spacing may be too large, resulting in small details being lost between the samples. Two edge response curves are presented for each example, illustrating that the actual samples can fall anywhere along the underlying curve. In other words, the edge being imaged may be sitting exactly upon a pixel, or be straddling two pixels. Notice that the perfect detector has *zero* or *one* sample on the rising part of the edge. Likewise, the blurry detector has *three* to *four* samples on the rising part of the edge.

What is limiting the resolution in these two systems? The answer is provided by the *sampling theorem*. As discussed in Chapter 3, sampling captures all frequency components below one-half of the sampling rate, while higher frequencies are lost due to aliasing. Now look at the MTF curve in (h). The sampling aperture of the blurry detector has removed all frequencies greater than one-half the sampling rate; therefore, *nothing* is lost during sampling. This means that the resolution of this system is

Example 1: Perfect detector

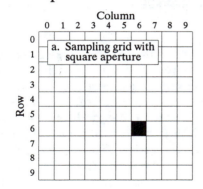

a. Sampling grid with square aperture

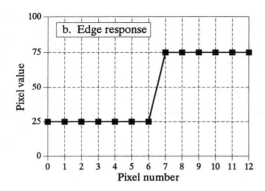

b. Edge response

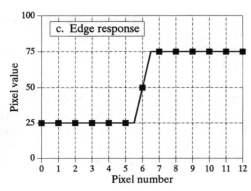

c. Edge response

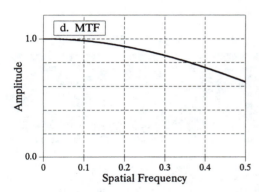

d. MTF

Example 2: Blurry detector

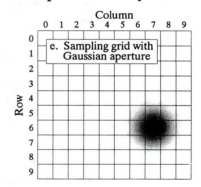

e. Sampling grid with Gaussian aperture

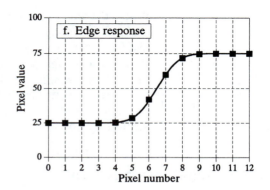

f. Edge response

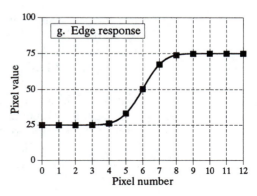

g. Edge response

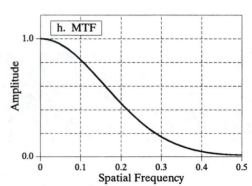

h. MTF

FIGURE 25-6

completely limited by the sampling aperture, and not the sample spacing. Put another way, the sampling aperture has acted as an antialias filter, allowing lossless sampling to take place.

In comparison, the MTF curve in (d) shows that *both* processes are limiting the resolution of this system. The high-frequency fall-off of the MTF curve represents information lost due to the *sampling aperture*. Since the MTF curve has not dropped to zero before a frequency of 0.5, there is also information lost during sampling, a result of the finite *sample spacing*. Which is limiting the resolution more? It is difficult to answer this question with a number, since they degrade the image in different ways. Suffice it to say that the resolution in the perfect detector (example 1) is mostly limited by the sample spacing.

While these concepts may seem difficult, they reduce to a very simple rule for practical usage. Consider a system with some 10%-90% edge response distance, for example 1 mm. If the sample spacing is greater than 1 mm (there is less than one sample along the edge), the system will be limited by the *sample spacing*. If the sample spacing is less than 0.33 mm (there are more than 3 samples along the edge), the resolution will be limited by the *sampling aperture*. When a system has 1-3 samples per edge, it will be limited by both factors.

Signal-to-Noise Ratio

An object is visible in an image because it has a different brightness than its surroundings. That is, the contrast of the object (i.e., the signal) must overcome the image noise. This can be broken into two classes: limitations of the *eye*, and limitations of the *data*.

Figure 25-7 illustrates an experiment to measure the eye's ability to detect weak signals. Depending on the observation conditions, the human eye can detect a minimum contrast of 0.5% to 5%. In other words, humans can distinguish about 20 to 200 shades of gray between the blackest black and the whitest white. The exact number depends on a variety of factors, such

Contrast

50%   40%   30%   20%   10%   8%   5%   3%   1%

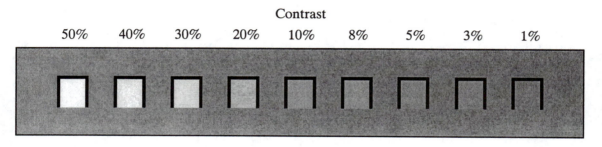

FIGURE 25-7
Contrast detection. The human eye can detect a minimum contrast of about 0.5 to 5%, depending on the observation conditions. 100% contrast is the difference between pure black and pure white.

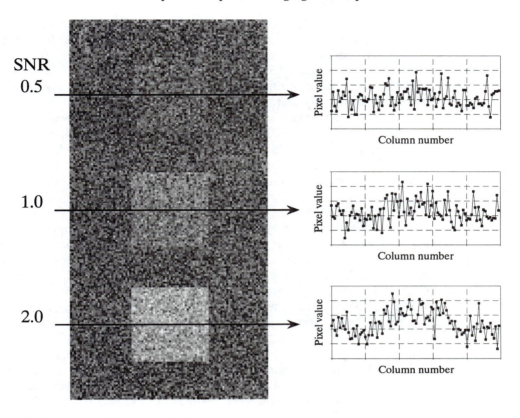

FIGURE 25-8
Minimum detectable SNR. An object is visible in an image only if its contrast is large enough to overcome the random image noise. In this example, the three squares have SNRs of 2.0, 1.0 and 0.5 (where the SNR is defined as the contrast of the object divided by the standard deviation of the noise).

as the brightness of the ambient lightning, the distance between the two regions being compared, and how the grayscale image is formed (video monitor, photograph, halftone, etc.).

The grayscale transform of Chapter 23 can be used to boost the contrast of a selected range of pixel values, providing a valuable tool in overcoming the limitations of the human eye. The contrast at one brightness level is increased, at the cost of reducing the contrast at another brightness level. However, this only works when the contrast of the object is not lost in random image noise. This is a more serious situation; the *signal* does not contain enough information to reveal the object, regardless of the performance of the eye.

Figure 25-8 shows an image with three squares having contrasts of 5%, 10%, and 20%. The background contains normally distributed random noise with a standard deviation of about 10% contrast. The SNR is defined as the contrast divided by the standard deviation of the noise, resulting in the three squares having SNRs of 0.5, 1.0 and 2.0. In general, trouble begins when the SNR falls below about 1.0.

The exact value for the minimum detectable SNR depends on the *size* of the object; the larger the object, the easier it is to detect. To understand this, imagine smoothing the image in Fig. 25-8 with a 3×3 square filter kernel. This leaves the contrast the same, but reduces the noise by a factor of three (i.e., the square root of the number of pixels in the kernel). Since the SNR is tripled, lower contrast objects can be seen. To see fainter objects, the filter kernel can be made even larger. For example, a 5×5 kernel improves the SNR by a factor of $\sqrt{25} = 5$. This strategy can be continued until the filter kernel is equal to the size of the object being detected. This means the ability to detect an object is proportional to the *square-root* of its *area*. If an object's diameter is doubled, it can be detected in twice as much noise.

Visual processing in the brain behaves in much the same way, smoothing the viewed image with various size filter kernels in an attempt to recognize low contrast objects. The three profiles in Fig. 25-8 illustrate just how good humans are at detecting objects in noisy environments. Even though the objects can hardly be identified in the profiles, they are obvious in the image. To really appreciate the capabilities of the human visual system, try writing algorithms that operate in this low SNR environment. You'll be humbled by what your brain can do, but your code can't!

Random image noise comes in two common forms. The first type, shown in Fig. 25-9a, has a constant amplitude. In other words, dark and light regions in the image are equally noisy. In comparison, (b) illustrates noise that *increases with the signal level*, resulting in the bright areas being more noisy than the dark ones. Both sources of noise are present in most images, but one or the other is usually dominant. For example, it is common for the noise to decrease as the signal level is decreased, until a plateau of constant amplitude noise is reached.

A common source of constant amplitude noise is the video *preamplifier*. All analog electronic circuits produce noise. However, it does the most harm where the signal being amplified is at its smallest, right at the CCD or other imaging sensor. Preamplifier noise originates from the random motion of electrons in the transistors. This makes the noise level depend on how the electronics are designed, but not on the level of the signal being amplified. For example, a typical CCD camera will have an SNR of about 300 to 1000 (40 to 60 dB), defined as the full scale signal level divided by the standard deviation of the constant amplitude noise.

Noise that increases with the signal level results when the image has been represented by a small number of individual particles. For example, this might be the *x-rays* passing through a patient, the *light photons* entering a camera, or the *electrons* in the well of a CCD. The mathematics governing these variations are called **counting statistics** or **Poisson statistics**. Suppose that the face of a CCD is uniformly illuminated such that an average of 10,000 electrons are generated in each well. By sheer chance, some wells will have more electrons, while some will have less. To be more exact, the number of electrons will be normally distributed with a mean of 10,000, with some standard deviation that describes how much variation there is from

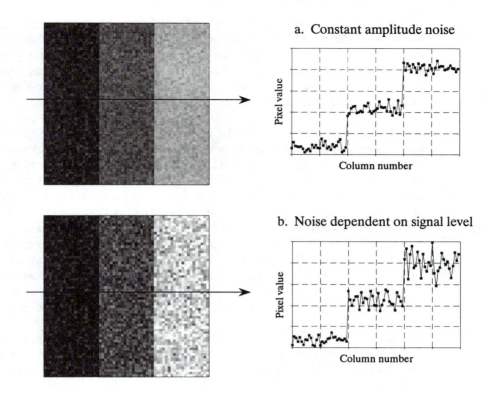

a. Constant amplitude noise

b. Noise dependent on signal level

FIGURE 25-9
Image noise. Random noise in images takes two general forms. In (a), the amplitude of the noise remains constant as the signal level changes. This is typical of electronic noise. In (b), the amplitude of the noise increases as the square-root of the signal level. This type of noise originates from the detection of a small number of particles, such as light photons, electrons, or x-rays.

well-to-well. A key feature of Poisson statistics is that the standard deviation is equal to the square-root of the number of individual particles. That is, if there are N particles in each pixel, the mean is equal to N and the standard deviation is equal to $\sqrt{N}$. This makes the signal-to-noise ratio equal to $N/\sqrt{N}$, or simply, $\sqrt{N}$. In equation form:

Equation 25-1
Poisson statistics. In a Poisson distributed signal, the mean, μ, is the average number of individual particles, N. The standard deviation, σ, is equal to the square-root of the average number of individual particles. The signal-to-noise ratio (SNR) is the mean divided by the standard deviation.

$$\mu = N$$

$$\sigma = \sqrt{N}$$

$$SNR = \sqrt{N}$$

In the CCD example, the standard deviation is $\sqrt{10,000} = 100$. Likewise the signal-to-noise ratio is also $\sqrt{10,000} = 100$. If the average number of electrons per well is increased to one million, both the standard deviation and the SNR increase to 1,000. That is, the noise becomes larger as the signal becomes larger, as shown in Fig. 25-9b. However, the signal is becoming larger *faster* than the noise, resulting in an overall improvement in the SNR. Don't be confused into thinking that a lower signal will provide less noise and therefore better information. Remember, your goal is *not* to reduce the noise, but to extract a signal *from* the noise. This makes the SNR the key parameter.

Many imaging systems operate by converting one particle type to another. For example, consider what happens in a medical x-ray imaging system. Within an x-ray tube, *electrons* strike a metal target, producing *x-rays*. After passing through the patient, the x-rays strike a vacuum tube detector known as an image intensifier. Here the x-rays are subsequently converted into *light photons*, then *electrons*, and then back to *light photons*. These light photons enter the camera where they are converted into *electrons* in the well of a CCD. In each of these intermediate forms, the image is represented by a finite number of particles, resulting in added noise as dictated by Eq. 25-1. The final SNR reflects the combined noise of *all* stages; however, one stage is usually dominant. This is the stage with the *worst* SNR because it has the *fewest* particles. This limiting stage is called the **quantum sink**.

In night vision systems, the quantum sink is the number of light photons that can be captured by the camera. The darker the night, the noisier the final image. Medical x-ray imaging is a similar example; the quantum sink is the number of x-rays striking the detector. Higher radiation levels provide less noisy images at the expense of more radiation to the patient.

When is the noise from Poisson statistics the primary noise in an image? It is dominant whenever the noise resulting from the quantum sink is greater than the other sources of noise in the system, such as from the electronics. For example, consider a typical CCD camera with an SNR of 300. That is, the noise from the CCD preamplifier is 1/300th of the full scale signal. An equivalent noise would be produced if the quantum sink of the system contains 90,000 particles per pixel. If the quantum sink has a smaller number of particles, Poisson noise will dominate the system. If the quantum sink has a larger number of particles, the preamplifier noise will be predominant. Accordingly, most CCD's are designed with a full well capacity of 100,000 to 1,000,000 electrons, minimizing the Poisson noise.

Morphological Image Processing

The identification of objects within an image can be a very difficult task. One way to simplify the problem is to change the grayscale image into a **binary image**, in which each pixel is restricted to a value of either 0 or 1. The techniques used on these binary images go by such names as: **blob analysis**, **connectivity analysis**, and **morphological image processing** (from the Greek word *morphē*, meaning shape or form). The foundation of morphological processing is in the mathematically rigorous field of *set theory*; however, this level of sophistication is seldom needed. Most morphological algorithms are simple logic operations and very *ad hoc*. In

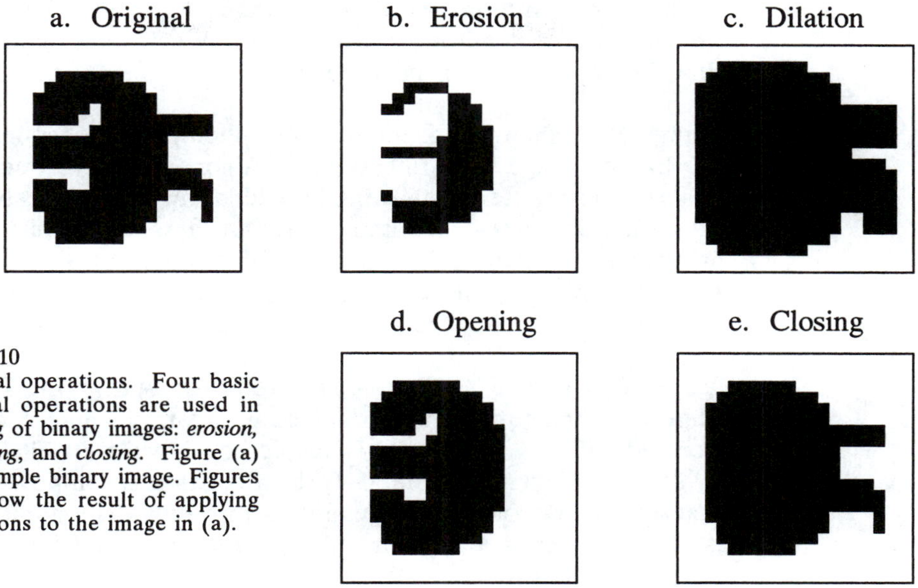

a. Original

b. Erosion

c. Dilation

d. Opening

e. Closing

FIGURE 25-10
Morphological operations. Four basic morphological operations are used in the processing of binary images: *erosion, dilation, opening,* and *closing*. Figure (a) shows an example binary image. Figures (b) to (e) show the result of applying these operations to the image in (a).

other words, each application requires a custom solution developed by trial-and-error. This is usually more of an art than a science. A bag of tricks is used rather than standard algorithms and formal mathematical properties. Here are some examples.

Figure 25-10a shows an example binary image. This might represent an enemy tank in an infrared image, an asteroid in a space photograph, or a suspected tumor in a medical x-ray. Each pixel in the background is displayed as white, while each pixel in the object is displayed as black. Frequently, binary images are formed by thresholding a grayscale image; pixels with a value greater than a threshold are set to 1, while pixels with a value below the threshold are set to 0. It is common for the grayscale image to be processed with linear techniques before the thresholding. For instance, *illumination flattening* (described in Chapter 24) can often improve the quality of the initial binary image.

Figures (b) and (c) show how the image is changed by the two most common morphological operations, **erosion** and **dilation**. In erosion, every object pixel that is touching a background pixel is changed into a background pixel. In dilation, every background pixel that is touching an object pixel is changed into an object pixel. Erosion makes the objects smaller, and can break a single object into multiple objects. Dilation makes the objects larger, and can merge multiple objects into one.

As shown in (d), **opening** is defined as an erosion followed by a dilation. Figure (e) shows the opposite operation of **closing**, defined as a dilation followed by an erosion. As illustrated by these examples, *opening* removes small islands and thin filaments of *object pixels*. Likewise, *closing* removes islands and thin filaments of *background pixels*. These techniques are useful for handling noisy images where some pixels have the wrong binary value. For instance, it might be known that an object cannot contain a "hole", or that the object's border must be smooth.

Figure 25-11 shows an example of morphological processing. Figure (a) is the binary image of a fingerprint. Algorithms have been developed to analyze these patterns, allowing individual fingerprints to be matched with those in a database. A common step in these algorithms is shown in (b), an operation called **skeletonization**. This simplifies the image by *removing* redundant pixels; that is, changing appropriate pixels from black to white. This results in each ridge being turned into a line only a single pixel wide.

Tables 25-1 and 25-2 show the skeletonization program. Even though the fingerprint image is binary, it is held in an array where each pixel can run from 0 to 255. A black pixel is denoted by 0, while a white pixel is denoted by 255. As shown in Table 25-1, the algorithm is composed of 6 iterations that gradually erode the ridges into a thin line. The number of iterations is chosen by trial and error. An alternative would be to stop when an iteration makes no changes.

During an iteration, each pixel in the image is evaluated for being *removable*; the pixel meets a set of criteria for being changed from black to white. Lines 200-240 loop through each pixel in the image, while the subroutine in Table 25-2 makes the evaluation. If the pixel under consideration is not removable, the subroutine does nothing. If the pixel is removable, the subroutine changes its value from 0 to 1. This indicates that the pixel is still black, but will be changed to white at the end of the iteration. After all the pixels have been evaluated, lines 260-300 change the value of the marked pixels from 1 to 255. This two-stage process results in the thick ridges being eroded equally from all directions, rather than a pattern based on how the rows and columns are scanned.

a. Original fingerprint   b. Skeletonized fingerprint

FIGURE 25-11
Binary skeletonization. The binary image of a fingerprint, (a), contains ridges that are many pixels wide. The skeletonized version, (b), contains ridges only a single pixel wide.

```
100 'SKELETONIZATION PROGRAM
110 'Object pixels have a value of 0 (displayed as black)
120 'Background pixels have a value of 255 (displayed as white)
130 '
140 DIM X%[149,149]              'X%[ , ] holds the image being processed
150 '
160 GOSUB XXXX                   'Mythical subroutine to load X%[ , ]
170 '
180 FOR ITER% = 0 TO 5          'Run through six iteration loops
190 '
200   FOR R% = 1 TO 148          'Loop through each pixel in the image.
210   FOR C% = 1 TO 148          'Subroutine 5000 (Table 25-2) indicates which
220     GOSUB 5000               'pixels can be changed from black to white,
230   NEXT C%                    'by marking the pixels with a value of 1.
240   NEXT R%
250 '
260   FOR R% = 0 TO 149          'Loop through each pixel in the image changing
270   FOR C% = 0 TO 149          'the marked pixels from black to white.
280    IF X%(R%,C%) = 1 THEN X%(R%,C%) = 255
290   NEXT C%
300   NEXT R%
310 '
320 NEXT ITER%
330 '
340 END
```

TABLE 25-1

The decision to remove a pixel is based on four rules, as contained in the subroutine shown in Table 25-2. *All* of these rules must be satisfied for a pixel to be changed from black to white. The first three rules are rather simple, while the fourth is quite complicated. As shown in Fig. 25-12a, a pixel at location [R,C] has eight neighbors. The four neighbors in the horizontal and vertical directions (labeled 2,4,6,8) are frequently called the **close neighbors**. The diagonal pixels (labeled 1,3,5,7) are correspondingly called the **distant neighbors**. The four rules are as follows:

Rule one: The pixel under consideration must presently be black. If the pixel is already white, no action needs to be taken.

Rule two: At least one of the pixel's close neighbors must be white. This insures that the erosion of the thick ridges takes place from the outside. In other words, if a pixel is black, and it is completely surrounded by black pixels, it is to be left alone on this iteration. Why use only the *close neighbors*, rather than *all* of the neighbors? The answer is simple: running the algorithm both ways shows that it works better. Remember, this is very common in morphological image processing; trial and error is used to find if one technique performs better than another.

Rule three: The pixel must have more than one black neighbor. If it has only one, it must be the end of a line, and therefore shouldn't be removed.

Rule four: A pixel cannot be removed if it results in its neighbors being *disconnected*. This is so each ridge is changed into a continuous line, not a group of interrupted segments. As shown by the examples in Fig. 25-12,

a. Pixel numbering

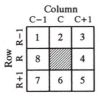

FIGURE 25-12
Neighboring pixels. A pixel at row and column [R,C] has eight neighbors, referred to by the numbers in (a). Figures (b) and (c) show examples where the neighboring pixels are *connected* and *unconnected*, respectively. This definition is used by rule number four of the skeletonization algorithm.

b. Connected neighbors

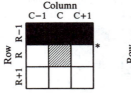

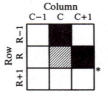

c. Unconnected neighbors

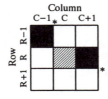
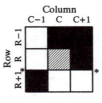
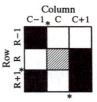

connected means that all of the black neighbors touch each other. Likewise, *unconnected* means that the black neighbors form two or more groups.

The algorithm for determining if the neighbors are connected or unconnected is based on counting the black-to-white transitions between adjacent neighboring pixels, in a *clockwise* direction. For example, if pixel 1 is black and pixel 2 is white, it is considered a black-to-white transition. Likewise, if pixel 2 is black and both pixel 3 and 4 are white, this is also a black-to-white transition. In total, there are eight locations where a black-to-white transition may occur. To illustrate this definition further, the examples in (b) and (c) have an asterisk placed by each black-to-white transition. The key to this algorithm is that there will be exactly *one* black-to-white transition if the neighbors are *connected*. More than one such transition indicates that the neighbors are *unconnected*.

As additional examples of binary image processing, consider the types of algorithms that might be useful after the fingerprint is skeletonized. A disadvantage of this particular skeletonization algorithm is that it leaves a considerable amount of *fuzz*, short offshoots that stick out from the sides of longer segments. There are several different approaches for eliminating these artifacts. For example, a program might loop through the image removing the pixel at the end of every line. These pixels are identified

```
5000 ' Subroutine to determine if the pixel at X%[R%,C%] can be removed.
5010 ' If all four of the rules are satisfied, then X%(R%,C%], is set to a value of 1,
5020 ' indicating it should be removed at the end of the iteration.
5030 '
5040 'RULE #1:  Do nothing if the pixel already white
5050 IF X%(R%,C%) = 255 THEN RETURN
5060 '
5070 '
5080 'RULE #2:  Do nothing if all of the close neighbors are black
5090 IF  X%[R% -1,C%  ] <> 255 AND X%[R%  ,C%+1] <> 255 AND
         X%[R%+1,C%  ] <> 255 AND X%[R%  ,C% -1] <> 255  THEN RETURN
5100 '
5110 '
5120 'RULE #3:  Do nothing if only a single neighbor pixel is black
5130 COUNT% = 0
5140 IF X%[R% -1,C% -1]  = 0 THEN COUNT% = COUNT% + 1
5150 IF X%[R% -1,C%   ]  = 0 THEN COUNT% = COUNT% + 1
5160 IF X%[R% -1,C%+1]  = 0 THEN COUNT% = COUNT% + 1
5170 IF X%[R%   ,C%+1]  = 0 THEN COUNT% = COUNT% + 1
5180 IF X%[R%+1,C%+1]  = 0 THEN COUNT% = COUNT% + 1
5190 IF X%[R%+1,C%   ]  = 0 THEN COUNT% = COUNT% + 1
5200 IF X%[R%+1,C% -1]  = 0 THEN COUNT% = COUNT% + 1
5210 IF X%[R%   ,C% -1]  = 0 THEN COUNT% = COUNT% + 1
5220 IF COUNT% = 1 THEN RETURN
5230 '
5240 '
5250 'RULE 4:  Do nothing if the neighbors are unconnected.
5260 'Determine this by counting the black-to-white transitions
5270 'while moving clockwise through the 8 neighboring pixels.
5280 COUNT% = 0
5290 IF X%[R% -1,C% -1]  = 0 AND X%[R% -1,C%   ] > 0  THEN COUNT% = COUNT% + 1
5300 IF X%[R% -1,C%   ]  = 0 AND X%[R% -1,C%+1] > 0  AND X%[R%  ,C%+1] > 0
         THEN COUNT% = COUNT% + 1
5310 IF X%[R% -1,C%+1]  = 0 AND X%[R%   ,C%+1] > 0  THEN COUNT% = COUNT% + 1
5320 IF X%[R%   ,C%+1]  = 0 AND X%[R%+1,C%+1] > 0  AND X%[R%+1,C%  ] > 0
         THEN COUNT% = COUNT% + 1
5330 IF X%[R%+1,C%+1]  = 0 AND X%[R%+1,C%   ] > 0 THEN COUNT% = COUNT% + 1
5340 IF X%[R%+1,C%   ]  = 0 AND X%[R%+1,C% -1] > 0  AND X%[R%  ,C%-1] > 0
         THEN COUNT% = COUNT% + 1
5350 IF X%[R%+1,C% -1]  = 0 AND X%[R%   ,C% -1] > 0  THEN COUNT% = COUNT% + 1
5360 IF X%[R%   ,C% -1]  = 0 AND X%[R% -1,C% -1] > 0  AND X%[R%-1,C%  ] > 0
         THEN COUNT% = COUNT% + 1
5370 IF COUNT% > 1 THEN RETURN
5380 '
5390 '
5400 'If all rules are satisfied, mark the pixel to be set to white at the end of the iteration
5410 X%(R%,C%) = 1
5420 '
5430 RETURN
```

TABLE 25-2

by having only one black neighbor. Do this several times and the fuzz is removed at the expense of making each of the correct lines shorter. A better method would loop through the image identifying *branch pixels* (pixels that have more than two neighbors). Starting with each branch pixel, count the number of pixels in each offshoot. If the number of pixels in an offshoot is less than some value (say, 5), declare it to be fuzz, and change the pixels in the branch from black to white.

Another algorithm might change the data from a *bitmap* to a *vector mapped* format. This involves creating a list of the ridges contained in the image and the pixels contained in each ridge. In the vector mapped form, each ridge in the fingerprint has an individual identity, as opposed to an image composed of many unrelated pixels. This can be accomplished by looping through the image looking for the endpoints of each line, the pixels that have only one black neighbor. Starting from the endpoint, each line is traced from pixel to connecting pixel. After the opposite end of the line is reached, all the traced pixels are declared to be a single *object*, and treated accordingly in future algorithms.

Computed Tomography

A basic problem in imaging with x-rays (or other penetrating radiation) is that a *two-dimensional image* is obtained of a *three-dimensional object*. This means that structures can *overlap* in the final image, even though they are completely separate in the object. This is particularly troublesome in medical diagnosis where there are many anatomic structures that can interfere with what the physician is trying to see. During the 1930's, this problem was attacked by moving the x-ray source and detector in a coordinated motion during image formation. From the geometry of this motion, a single *plane* within the patient remains in focus, while structures outside this plane become blurred. This is analogous to a camera being focused on an object at 5 feet, while objects at a distance of 1 and 50 feet are blurry. These related techniques based on motion blurring are now collectively called **classical tomography.** The word *tomography* means "*a picture of a plane.*"

In spite of being well developed for more than 50 years, classical tomography is rarely used. This is because it has a significant limitation: the interfering objects are not *removed* from the image, only *blurred*. The resulting image quality is usually too poor to be of practical use. The long sought solution was a system that could create an image representing a 2D slice through a 3D object with *no* interference from other structures in the 3D object.

This problem was solved in the early 1970s with the introduction of a technique called **computed tomography (CT).** CT revolutionized the medical x-ray field with its unprecedented ability to visualize the anatomic structure of the body. Figure 25-13 shows a typical medical CT image. Computed tomography was originally introduced to the marketplace under the names *Computed Axial Tomography* and *CAT scanner*. These terms are now frowned upon in the medical field, although you hear them used frequently by the general public.

Figure 25-14 illustrates a simple geometry for acquiring a CT slice through the center of the head. A narrow pencil beam of x-rays is passed from the x-ray source to the x-ray detector. This means that the measured value at the detector is related to the total amount of material placed *anywhere*

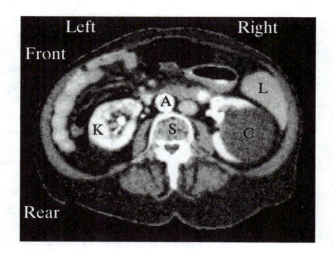

FIGURE 25-13
Computed tomography image. This CT slice is of a human abdomen, at the level of the navel. Many organs are visible, such as the (L) Liver, (K) Kidney, (A) Aorta, (S) Spine, and (C) Cyst covering the right kidney. CT can visualize internal anatomy far better than conventional medical x-rays.

along the beam's path. Materials such as bone and teeth block more of the x-rays, resulting in a lower signal compared to soft tissue and fat. As shown in the illustration, the source and detector assemblies are translated to acquire a **view** (CT jargon) at this particular angle. While this figure shows only a single view being acquired, a complete CT scan requires 300 to 1000 views taken at rotational increments of about 0.3° to 1.0°. This is accomplished by mounting the x-ray source and detector on a rotating gantry that surrounds the patient. A key feature of CT data acquisition is that x-rays pass *only* through the slice of the body being examined. This is unlike classical tomography where x-rays are passing through structures that you try to suppress in the final image. Computed tomography doesn't allow information from irrelevant locations to even enter the acquired data.

FIGURE 25-14
CT data acquisition. A simple CT system passes a narrow beam of x-rays through the body from source to detector. The source and detector are then translated to obtain a complete view. The remaining views are obtained by rotating the source and detector in about 1° increments, and repeating the translation process.

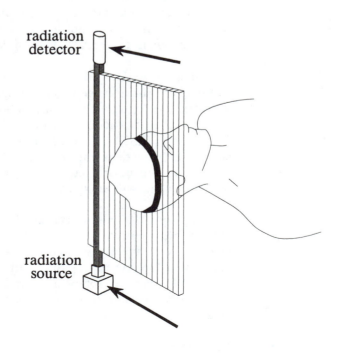

Several preprocessing steps are usually needed before the image reconstruction can take place. For instance, the logarithm must be taken of each x-ray measurement. This is because x-rays decrease in intensity exponentially as they pass through material. Taking the logarithm provides a signal that is linearly related to the characteristics of the material being measured. Other preprocessing steps are used to compensate for the use of *polychromatic* (more than one energy) x-rays, and *multielement* detectors (as opposed to the single element shown in Fig. 25-14). While these are a key step in the overall technique, they are not related to the reconstruction algorithms and we won't discuss them further.

Figure 25-15 illustrates the relationship between the measured views and the corresponding image. Each sample acquired in a CT system is equal to the sum of the image values along a ray pointing to that sample. For example, view 1 is found by adding all the pixels in each row. Likewise, view 3 is found by adding all the pixels in each column. The other views, such as view 2, sum the pixels along rays that are at an angle.

There are four main approaches to calculating the slice image given the set of its views. These are called **CT reconstruction algorithms**. The first method is totally impractical, but provides a better understanding of the problem. It is based on solving many **simultaneous linear equations**. One equation can be written for each measurement. That is, a particular sample in a particular profile is the sum of a particular group of pixels in the image. To calculate N^2 unknown variables (i.e., the image pixel values), there must be N^2 independent equations, and therefore N^2 measurements. Most CT scanners acquire about 50% more samples than rigidly required by this analysis. For example, to reconstruct a 512×512 image, a system might take 700 views with 600 samples in each view. By making the problem *overdetermined* in this manner, the final image has reduced noise and artifacts. The problem with this first method of CT reconstruction is computation time. Solving several hundred thousand simultaneous linear equations is an daunting task.

The second method of CT reconstruction uses **iterative** techniques to calculate the final image in small steps. There are several variations of this method: the Algebraic Reconstruction Technique (ART), Simultaneous Iterative Reconstruction Technique (SIRT), and Iterative Least Squares Technique (ILST). The difference between these methods is how the successive corrections are made: ray-by-ray, pixel-by-pixel, or simultaneously correcting the entire data set, respectively. As an example of these techniques, we will look at ART.

To start the ART algorithm, all the pixels in the image array are set to some arbitrary value. An iterative procedure is then used to gradually change the image array to correspond to the profiles. An iteration cycle consists of looping through each of the measured data points. For each measured value, the following question is asked: *how can the pixel values in the array be changed to make them consistent with this particular measurement?* In other words, the measured sample is compared with the

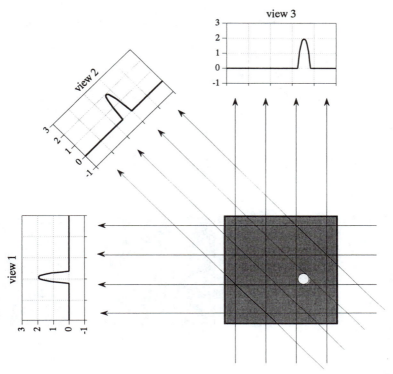

FIGURE 25-15
CT views. Computed tomography acquires a set of views and then reconstructs the corresponding image. Each sample in a view is equal to the sum of the image values along the ray that points to that sample. In this example, the image is a small pillbox surrounded by zeros. While only three views are shown here, a typical CT scan uses hundreds of views at slightly different angles.

sum of the image pixels along the ray pointing to the sample. If the ray sum is lower than the measured sample, all the pixels along the ray are increased in value. Likewise, if the ray sum is higher than the measured sample, all of the pixel values along the ray are decreased. After the first complete iteration cycle, there will still be an error between the ray sums and the measured values. This is because the changes made for any one measurement disrupts all the previous corrections made. The idea is that the errors become smaller with repeated iterations until the image converges to the proper solution.

Iterative techniques are generally slow, but they are useful when better algorithms are not available. In fact, ART was used in the first commercial medical CT scanner released in 1972, the EMI Mark I. We will revisit iterative techniques in the next chapter on neural networks. Development of the third and forth methods have almost entirely replaced iterative techniques in commercial CT products.

The last two reconstruction algorithms are based on formal mathematical solutions to the problem. These are elegant examples of DSP. The third method is called **filtered backprojection**. It is a modification of an older

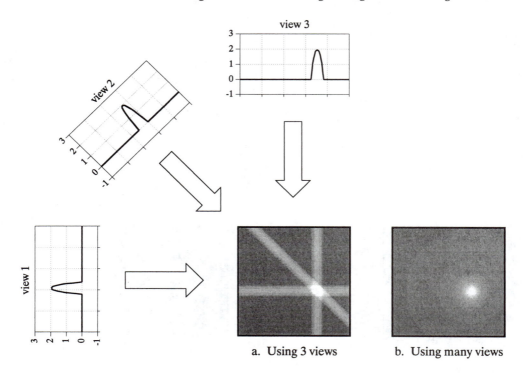

FIGURE 25-16
Backprojection. Backprojection reconstructs an image by taking each view and *smearing* it along
the path it was originally acquired. The resulting image is a blurry version of the correct image.

technique, called **backprojection** or **simple backprojection.** Figure 25-16
shows that simple backprojection is a common sense approach, but very
unsophisticated.   An individual sample is backprojected by setting all the
image pixels along the ray pointing to the sample to the same value.   In
less technical terms, a backprojection is formed by *smearing* each view back
through the image in the direction it was originally acquired.   The final
backprojected image is then taken as the sum of all the backprojected views.

While backprojection is conceptually simple, it does not correctly solve the
problem. As shown in (b), a backprojected image is very *blurry*. A single
point in the *true* image is reconstructed as a circular region that decreases
in intensity away from the center.   In more formal terms, the point spread
function of backprojection is circularly symmetric, and decreases as the
reciprocal of its radius.

Filtered backprojection is a technique to correct the blurring encountered
in simple backprojection. As illustrated in Fig. 25-17, each view is *filtered*
before the backprojection to counteract the blurring PSF. That is, each of
the one-dimensional views is convolved with a one-dimensional filter kernel
to create a set of *filtered views*. These filtered views are then backprojected
to provide the reconstructed image, a close approximation to the "correct"
image. In fact, the image produced by filtered backprojection is *identical*

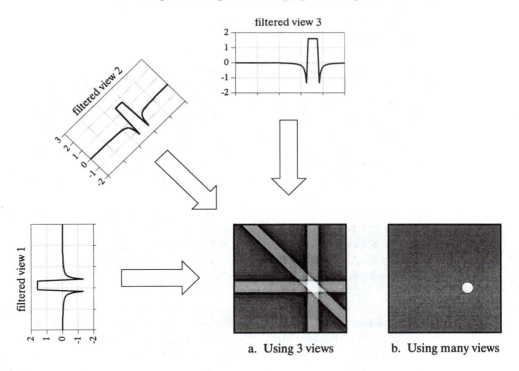

a. Using 3 views   b. Using many views

FIGURE 25-17
Filtered backprojection. Filtered backprojection reconstructs an image by filtering each view before backprojection. This removes the blurring seen in simple backprojection, and results in a mathematically exact reconstruction of the image. Filtered backprojection is the most commonly used algorithm for computed tomography systems.

to the "correct" image when there are an *infinite* number of views and an *infinite* number of points per view.

The filter kernel used in this technique will be discussed shortly. For now, notice how the profiles have been changed by the filter. The image in this example is a uniform white circle surrounded by a black background (a pillbox). Each of the acquired views has a flat background with a rounded region representing the white circle. Filtering changes the views in two significant ways. First, the top of the pulse is made flat, resulting in the final backprojection creating a *uniform* signal level within the circle. Second, negative spikes have been introduced at the sides of the pulse. When backprojected, these negative regions counteract the blur.

The fourth method is called **Fourier reconstruction**. In the spatial domain, CT reconstruction involves the relationship between a two-dimensional image and its set of one-dimensional views. By taking the two-dimensional Fourier transform of the image and the one-dimensional Fourier transform of each of its views, the problem can be examined in the frequency domain. As it turns out, the relationship between an image and its views is far simpler in the frequency domain than in the spatial domain. The frequency domain analysis of this problem is a milestone in CT technology called the **Fourier slice theorem**.

Figure 25-18 shows how the problem looks in both the spatial and the frequency domains. In the spatial domain, each view is found by integrating the image along rays at a particular angle. In the frequency domain, the image spectrum is represented in this illustration by a two-dimensional grid. The spectrum of each view (a one-dimensional signal) is represented by a dark line superimposed on the grid. As shown by the positioning of the lines on the grid, the Fourier slice theorem states that the spectrum of a view is identical to the values along a line (slice) through the image spectrum. For instance, the spectrum of view 1 is the same as the center column of the image spectrum, and the spectrum of view 3 is the same as the center row of the image spectrum. Notice that the spectrum of each view is positioned on the grid at the same angle that the view was originally acquired. All these frequency spectra include the negative frequencies and are displayed with zero frequency at the center.

Fourier reconstruction of a CT image requires three steps. First, the one-dimensional FFT is taken of each view. Second, these view spectra are used to calculate the two-dimensional frequency spectrum of the image, as outlined by the Fourier slice theorem. Since the view spectra are arranged *radially*, and the correct image spectrum is arranged *rectangularly*, an interpolation routine is needed to make the conversion. Third, the inverse FFT is taken of the image spectrum to obtain the reconstructed image.

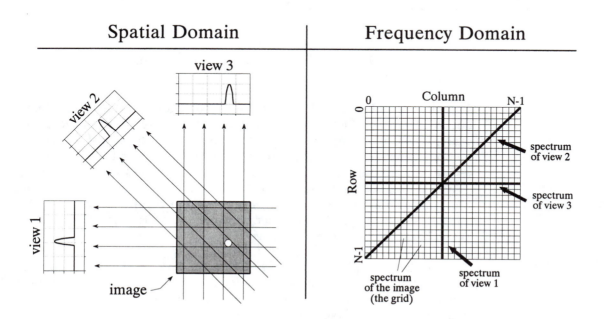

FIGURE 25-18

The Fourier Slice Theorem. The Fourier Slice Theorem describes the relationship between an image and its views in the frequency domain. In the spatial domain, each view is found by integrating the image along rays at a particular angle. In the frequency domain, the spectrum of each view is a one-dimensional "slice" of the two-dimensional image spectrum.

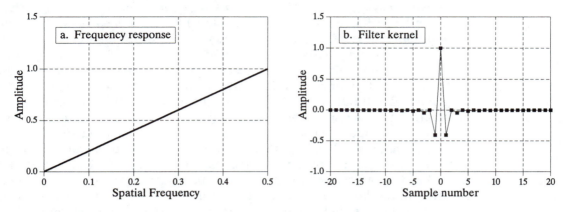

FIGURE 25-20
Backprojection filter. The frequency response of the backprojection filter is shown in (a), and the corresponding filter kernel is shown in (b). Equation 25-2 provides the values for the filter kernel.

This "radial to rectangular" conversion is also the key for understanding filtered backprojection. The radial arrangement is the spectrum of the backprojected image, while the rectangular grid is the spectrum of the correct image. If we compare one small region of the radial spectrum with the corresponding region of the rectangular grid, we find that the sample values are identical. However, they have a different *sample density*. The correct spectrum has uniformly spaced points throughout, as shown by the even spacing of the rectangular grid. In comparison, the backprojected spectrum has a higher sample density near the center because of its radial arrangement. In other words, the spokes of a wheel are closer together near the hub. This issue does not affect Fourier reconstruction because the interpolation is from the *values* of the nearest neighbors, not their *density*.

The filter in filtered backprojection cancels this unequal sample density. In particular, the frequency response of the filter must be the *inverse* of the sample density. Since the backprojected spectrum has a density of $1/f$, the appropriate filter has a frequency response of $H[f] = f$. This frequency response is shown in Fig. 25-19a. The filter kernel is then found by taking the inverse Fourier transform, as shown in (b). Mathematically, the filter kernel is given by:

EQUATION 25-2
The filter kernel for filtered backprojection. Figure 25-19b shows a graph of this kernel.

$$h[0] = 1$$

$$h[k] = 0 \qquad \text{for even values of } k$$

$$h[k] = \frac{4/\pi^2}{k^2} \qquad \text{for odd values of } k$$

Before leaving the topic of computed tomography, it should be mentioned that there are several similar imaging techniques in the medical field. All use extensive amounts of DSP. **Positron emission tomography (PET)** involves injecting the patient with a mildly radioactive compound that emits *positrons*. Immediately after emission, the positron annihilates with an electron, creating two gamma rays that exit the body in exactly opposite directions. Radiation detectors placed around the patient look for these back-to-back gamma rays, identifying the location of the *line* that the gamma rays traveled along. Since the point where the gamma rays were created must be somewhere along this line, a reconstruction algorithm similar to computed tomography can be used. This results in an image that looks similar to CT, except that *brightness* is related to the amount of the radioactive material present at each location. A unique advantage of PET is that the radioactive compounds can be attached to various substances used by the body in some manner, such as glucose. The reconstructed image is then related to the concentration of this biological substance. This allows the imaging of the body's *physiology* rather than simple *anatomy*. For example, images can be produced showing which portions of the human brain are involved in various mental tasks.

A more direct competitor to computed tomography is **magnetic resonance imaging (MRI)**, which is now found in most major hospitals. This technique was originally developed under the name **nuclear magnetic resonance (NMR)**. The name change was for public relations when local governments protested the use of anything *nuclear* in their communities. It was often an impossible task to educate the public that the term *nuclear* simply referred to the fact that all atoms contain a *nucleus*. An MRI scan is conducted by placing the patient in the center of a powerful magnet. Radio waves in conjunction with the magnetic field cause selected nuclei in the body to resonate, resulting in the emission of secondary radio waves. These secondary radio waves are digitized and form the data set used in the MRI reconstruction algorithms. The result is a set of images that appear very similar to computed tomography. The advantages of MRI are numerous: good soft tissue discrimination, flexible slice selection, and not using potentially dangerous x-ray radiation. On the negative side, MRI is a more expensive technique than CT, and poor for imaging bones and other hard tissues. CT and MRI will be the mainstays of medical imaging for many years to come.

Neural Networks (and more!)

Traditional DSP is based on *algorithms*, changing data from one form to another through step-by-step procedures. Most of these techniques also need *parameters* to operate. For example: recursive filters use recursion *coefficients*, feature detection can be implemented by correlation and *thresholds*, an image display depends on the *brightness* and *contrast* settings, etc. Algorithms describe what is to be done, while parameters provide a benchmark to judge the data. The proper selection of parameters is often more important than the algorithm itself. Neural networks take this idea to the extreme by using very simple algorithms, but many highly optimized parameters. This is a revolutionary departure from the traditional mainstays of science and engineering: mathematical logic and theorizing followed by experimentation. Neural networks replace these problem solving strategies with trial & error, pragmatic solutions, and a "this works better than that" methodology. This chapter presents a variety of issues regarding parameter selection in both neural networks and more traditional DSP algorithms.

Target Detection

Scientists & engineers often need to know if a particular object or condition is present. For instance, geophysicists explore the earth for oil, physicians examine patients for disease, astronomers search the universe for extra-terrestrial intelligence, etc. These problems usually involve comparing the acquired data against a threshold. If the threshold is exceeded, the **target** (the object or condition being sought) is deemed present.

For example, suppose you invent a device for detecting cancer in humans. The apparatus is waved over a patient, and a number between 0 and 30 pops up on the video screen. Low numbers correspond to healthy subjects, while high numbers indicate that cancerous tissue is present. You find that the device works quite well, but isn't perfect and occasionally makes an error. The question is: how do you use this system to the benefit of the patient being examined?

Figure 26-1 illustrates a systematic way of analyzing this situation. Suppose the device is tested on two groups: several hundred volunteers known to be healthy (nontarget), and several hundred volunteers known to have cancer (target). Figures (a) & (b) show these test results displayed as histograms. The healthy subjects generally produce a lower number than those that have cancer (good), but there is some overlap between the two distributions (bad).

As discussed in Chapter 2, the histogram can be used as an estimate of the **probability distribution function (pdf)**, as shown in (c). For instance, imagine that the device is used on a randomly chosen healthy subject. From (c), there is about an 8% chance that the test result will be 3, about a 1% chance that it will be 18, etc. (This example does not specify if the output is a real number, requiring a *pdf*, or an integer, requiring a *pmf*. Don't worry about it here; it isn't important).

Now, think about what happens when the device is used on a patient of unknown health. For example, if a person we have never seen before receives a value of 15, what can we conclude? Do they have cancer or not? We know that the probability of a healthy person generating a 15 is 2.1%. Likewise, there is a 0.7% chance that a person with cancer will produce a 15. If no other information is available, we would conclude that the subject is three times as likely not to have cancer, as to have cancer. That is, the test result of 15 implies a 25% probability that the subject is from the target group. This method can be generalized to form the curve in (d), the probability of the subject having cancer based only on the number produced by the device [mathematically, $pdf_t / (pdf_t + pdf_{nt})$].

If we stopped the analysis at this point, we would be making one of the most common (and serious) errors in target detection. Another source of information must usually be taken into account to make the curve in (d) meaningful. This is the relative number of targets versus nontargets in the population to be tested. For instance, we may find that only one in one-thousand people have the cancer we are trying to detect. To include this in the analysis, the amplitude of the nontarget pdf in (c) is adjusted so that the area under the curve is 0.999. Likewise, the amplitude of the target pdf is adjusted to make the area under the curve be 0.001. Figure (d) is then calculated as before to give the probability that a patient has cancer.

Neglecting this information is a serious error because it greatly affects how the test results are interpreted. In other words, the curve in figure (d) is drastically altered when the prevalence information is included. For instance, if the fraction of the population having cancer is 0.001, a test result of 15 corresponds to only a 0.025% probability that this patient has cancer. This is very different from the 25% probability found by relying on the output of the machine alone.

This method of converting the output value into a probability can be useful for understanding the problem, but it is not the main way that target detection is accomplished. Most applications require a yes/no decision on

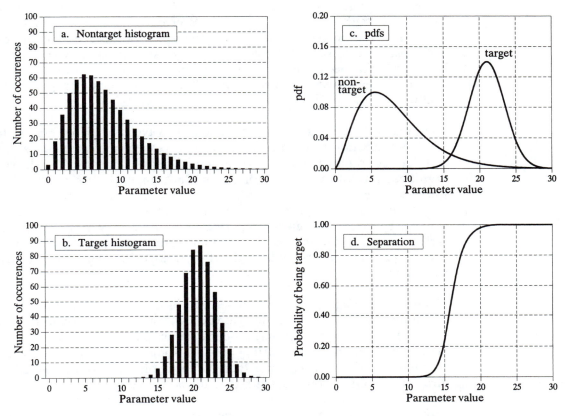

FIGURE 26-1
Probability of target detection. Figures (a) and (b) shows histograms of target and nontarget groups with respect to some parameter value. From these histograms, the probability distribution functions of the two groups can be estimated, as shown in (c). Using only this information, the curve in (d) can be calculated, giving the probability that a target has been found, based on a specific value of the parameter.

the presence of a target, since *yes* will result in one action and *no* will result in another. This is done by comparing the output value of the test to a **threshold**. If the output is above the threshold, the test is said to be **positive**, indicating that the target is present. If the output is below the threshold, the test is said to be **negative**, indicating that the target is not present. In our cancer example, a negative test result means that the patient is told they are healthy, and sent home. When the test result is positive, additional tests will be performed, such as obtaining a sample of the tissue by insertion of a biopsy needle.

Since the target and nontarget distributions overlap, some test results will not be correct. That is, some patients sent home will actually have cancer, and some patients sent for additional tests will be healthy. In the jargon of target detection, a *correct* classification is called **true**, while an *incorrect* classification is called **false**. For example, if a patient has cancer, and the test properly detects the condition, it is said to be a **true-positive**. Likewise, if a patient does not have cancer, and the test indicates that cancer is not present, it is said to be a **true-negative**. A **false-positive** occurs when the patient does not have cancer, but the test erroneously indicates that they do. This results in needless worry, and the pain and expense of additional tests. An even worse scenario occurs with the **false-negative**, where cancer is present, but the test indicates the patient is healthy. As we all know, untreated cancer can cause many health problems, including premature death.

The human suffering resulting from these two types of errors makes the threshold selection a delicate balancing act. How many *false-positives* can be tolerated to reduce the number of *false-negatives*? Figure 26-2 shows a graphical way of evaluating this problem, the **ROC curve** (short for Receiver Operating Characteristic). The ROC curve plots the percent of target signals reported as positive (higher is better), against the percent of nontarget signals erroneously reported as positive (lower is better), for various values of the threshold. In other words, each point on the ROC curve represents one possible tradeoff of true-positive and false-positive performance.

Figures (a) through (d) show four settings of the threshold in our cancer detection example. For instance, look at (b) where the threshold is set at 17. Remember, every test that produces an output value *greater* than the threshold is reported as a *positive* result. About 13% of the area of the nontarget distribution is greater than the threshold (i.e., to the *right* of the threshold). Of all the patients that do not have cancer, 87% will be reported as negative (i.e., a true-negative), while 13% will be reported as positive (i.e., a false-positive). In comparison, about 80% of the area of the target distribution is greater than the threshold. This means that 80% of those that have cancer will generate a positive test result (i.e., a true-positive). The other 20% that have cancer will be incorrectly reported as a negative (i.e., a false-negative). As shown in the ROC curve in (b), this threshold results in a point on the curve at: *% nontargets positive = 13%*, and *% targets positive = 80%*.

The more efficient the detection process, the more the ROC curve will bend toward the upper-left corner of the graph. Pure guessing results in a straight line at a 45° diagonal. Setting the threshold relatively low, as shown in (a), results in nearly all the target signals being detected. This comes at the price of many false alarms (false-positives). As illustrated in (d), setting the threshold relatively high provides the reverse situation: few false alarms, but many missed targets.

These analysis techniques are useful in understanding the consequences of threshold selection, but the final decision is based on what some *human* will accept. Suppose you initially set the threshold of the cancer detection apparatus to some value you feel is appropriate. After many patients have been screened with the system, you speak with a dozen or so patients that have been subjected to false-positives. Hearing how *your* system has unnecessarily disrupted the lives of these people affects you deeply, motivating you to increase the threshold. Eventually you encounter a

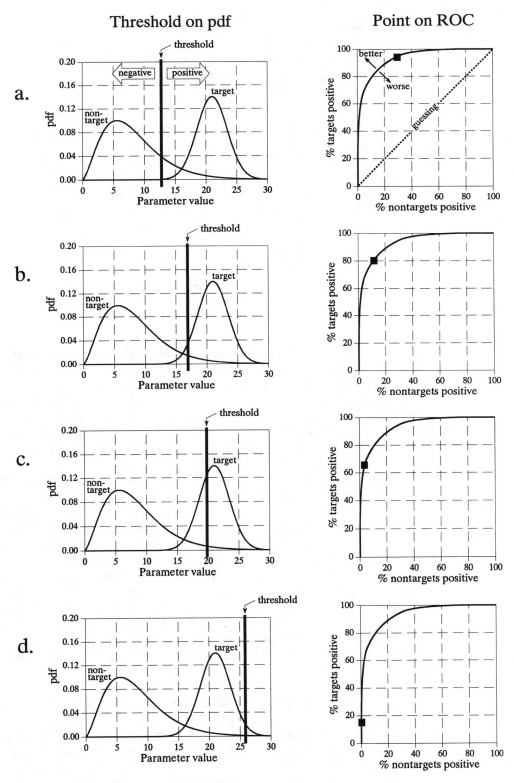

FIGURE 26-2
Relationship between ROC curves and pdfs.

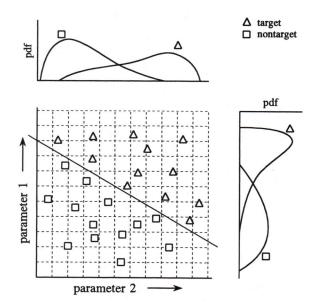

FIGURE 26-3
Example of a two-parameter space. The target ($\triangle$) and nontarget ($\square$) groups are completely separate in two-dimensions; however, they overlap in each individual parameter. This overlap is shown by the one-dimensional pdfs along each of the parameter axes.

situation that makes you feel even worse: you speak with a patient who is terminally ill with a cancer that *your* system failed to detect. You respond to this difficult experience by *greatly* lowering the threshold. As time goes on and these events are repeated many times, the threshold gradually moves to an *equilibrium* value. That is, the false-positive rate multiplied by a significance factor (lowering the threshold) is balanced by the false-negative rate multiplied by another significance factor (raising the threshold).

This analysis can be extended to devices that provide more than one output. For example, suppose that a cancer detection system operates by taking an x-ray image of the subject, followed by automated image analysis algorithms to identify tumors. The algorithms identify suspicious regions, and then measure key characteristics to aid in the evaluation. For instance, suppose we measure the *diameter* of the suspect region (parameter 1) and its *brightness* in the image (parameter 2). Further suppose that our research indicates that tumors are generally larger and brighter than normal tissue. As a first try, we could go through the previously presented ROC analysis for each parameter, and find an acceptable threshold for each. We could then classify a test as positive only if it met both criteria: parameter 1 greater than some threshold *and* parameter 2 greater than another threshold.

This technique of thresholding the parameters separately and then invoking logic functions (AND, OR, etc.) is very common. Nevertheless, it is very inefficient, and much better methods are available. Figure 26-3 shows why this is the case. In this figure, each triangle represents a single occurrence of a target (a patient with cancer), plotted at a location that corresponds to the value of its two parameters. Likewise, each square represents a single occurrence of a nontarget (a patient without cancer). As shown in the pdf

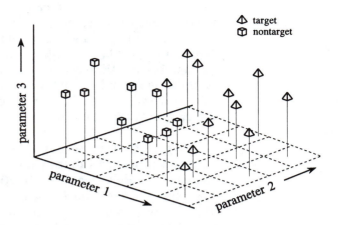

FIGURE 26-4
Example of a three-parameter space. Just as a two-parameter space forms a plane surface, a three parameter space can be graphically represented using the conventional x, y, and z axes. Separation of a three-parameter space into regions requires a dividing *plane*, or a curved *surface*.

graph on the side of each axis, both parameters have a large overlap between the target and nontarget distributions. In other words, each parameter, taken individually, is a poor predictor of cancer. Combining the two parameters with simple logic functions would only provide a small improvement. This is especially interesting since the two parameters contain information to *perfectly* separate the targets from the nontargets. This is done by drawing a diagonal line between the two groups, as shown in the figure.

In the jargon of the field, this type of coordinate system is called a **parameter space**. For example, the two-dimensional plane in this example could be called a diameter-brightness space. The idea is that targets will occupy one region of the parameter space, while nontargets will occupy another. Separation between the two regions may be as simple as a straight line, or as complicated as closed regions with irregular borders. Figure 26-4 shows the next level of complexity, a three-parameter space being represented on the x, y and z axes. For example, this might correspond to a cancer detection system that measures *diameter*, *brightness*, and some third parameter, say, *edge sharpness*. Just as in the two-dimensional case, the important idea is that the members of the target and nontarget groups will (hopefully) occupy different regions of the space, allowing the two to be separated. In three dimensions, regions are separated by planes and curved surfaces. The term **hyperspace** (over, above, or beyond normal space) is often used to describe parameter spaces with more than three dimensions. Mathematically, hyperspaces are no different from one, two and three-dimensional spaces; however, they have the practical problem of not being able to be displayed in a graphical form in our three-dimensional universe.

The threshold selected for a single parameter problem cannot (usually) be classified as right or wrong. This is because each threshold value results in a unique combination of false-positives and false-negatives, i.e., some point along the ROC curve. This is trading one goal for another, and has no absolutely correct answer. On the other hand, parameter spaces with two or more parameters can definitely have wrong divisions between regions. For instance, imagine increasing the number of data points in Fig. 26-3, revealing a small overlap between the target and nontarget groups. It would be possible to move the threshold line between the groups to trade the number of false-positives against the number of false-negatives. That is, the diagonal line would be moved toward the top-right, or the bottom-left. However, it would be wrong to rotate the line, because it would increase *both* types of errors.

As suggested by these examples, the conventional approach to target detection (sometimes called pattern recognition) is a two step process. The first step is called **feature extraction**. This uses algorithms to reduce the raw data to a few parameters, such as diameter, brightness, edge sharpness, etc. These parameters are often called **features** or **classifiers**. Feature extraction is needed to reduce the amount of data. For example, a medical x-ray image may contain more than a million pixels. The goal of feature extraction is to distill the information into a more concentrated and manageable form. This type of algorithm development is more of an art than a science. It takes a great deal of experience and skill to look at a problem and say: *"These are the classifiers that best capture the information."* Trial-and-error plays a significant role.

In the second step, an evaluation is made of the classifiers to determine if the target is present or not. In other words, some method is used to divide the parameter space into a region that corresponds to the targets, and a region that corresponds to the nontargets. This is quite straightforward for one and two-parameter spaces; the known data points are plotted on a graph (such as Fig. 26-3), and the regions separated by eye. The division is then written into a computer program as an equation, or some other way of defining one region from another. In principle, this same technique can be applied to a three-dimensional parameter space. The problem is, three-dimensional graphs are very difficult for humans to understand and visualize (such as Fig. 26-4). Caution: Don't try this in hyperspace; your brain will explode!

In short, we need a machine that can carry out a multi-parameter space division, according to examples of target and nontarget signals. This ideal target detection system is remarkably close to the main topic of this chapter, the *neural network*.

Neural Network Architecture

Humans and other animals process information with *neural networks*. These are formed from *trillions* of **neurons** (nerve cells) exchanging brief electrical pulses called **action potentials**. Computer algorithms that mimic these biological structures are formally called **artificial neural networks** to distinguish them from the squishy things inside of animals. However, most scientists and engineers are not this formal and use the term *neural network* to include both biological and nonbiological systems.

Neural network research is motivated by two desires: to obtain a better understanding of the human brain, and to develop computers that can deal with abstract and poorly defined problems. For example, conventional computers have trouble understanding speech and recognizing people's faces. In comparison, humans do extremely well at these tasks.

Many different neural network structures have been tried, some based on imitating what a biologist sees under the microscope, some based on a more mathematical analysis of the problem. The most commonly used structure is shown in Fig. 26-5. This neural network is formed in three layers, called the **input layer, hidden layer, and output layer**. Each layer consists of one or more **nodes**, represented in this diagram by the small circles. The lines between the nodes indicate the flow of information from one node to the next. In this particular type of neural network, the information flows only from the input to the output (that is, from left-to-right). Other types of neural networks have more intricate connections, such as feedback paths.

The nodes of the input layer are **passive**, meaning they do not modify the data. They receive a single value on their input, and duplicate the value to

FIGURE 26-5
Neural network architecture. This is the most common structure for neural networks: three layers with full inter-connection. The input layer nodes are passive, doing nothing but relaying the values from their single input to their multiple outputs. In comparison, the nodes of the hidden and output layers are active, modifying the signals in accordance with Fig. 26-6. The action of this neural network is determined by the weights applied in the hidden and output nodes.

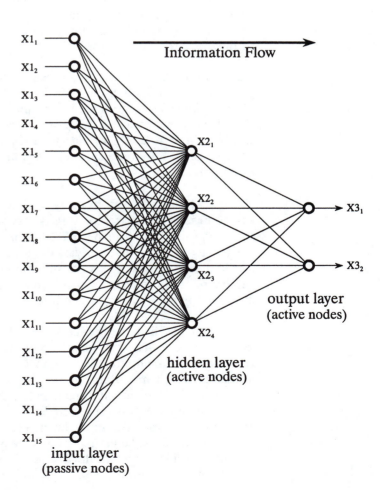

their multiple outputs.   In comparison, the nodes of the hidden and output layer are **active**. This means they modify the data as shown in Fig. 26-6. The variables: $X1_1, X1_2 \cdots X1_{15}$ hold the data to be evaluated (see Fig. 26-5). For example, they may be pixel values from an image, samples from an audio signal, stock market prices on successive days, etc. They may also be the output of some other algorithm, such as the classifiers in our cancer detection example: diameter, brightness, edge sharpness, etc.

Each value from the input layer is duplicated and sent to *all* of the hidden nodes. This is called a **fully interconnected** structure.   As shown in Fig. 26-6, the values entering a hidden node are multiplied by **weights**, a set of predetermined numbers stored in the program. The weighted inputs are then added to produce a single number. This is shown in the diagram by the symbol, Σ. Before leaving the node, this number is passed through a nonlinear mathematical function called a *sigmoid*. This is an "s" shaped curve that limits the node's output. That is, the input to the sigmoid is a value between $-\infty$ and $+\infty$, while its output can only be between 0 and 1.

The outputs from the hidden layer are represented in the flow diagram (Fig 26-5) by the variables: $X2_1, X2_2, X2_3$ and $X2_4$. Just as before, each of these values is duplicated and applied to the next layer.   The active nodes of the output layer combine and modify the data to produce the two output values of this network, $X3_1$ and $X3_2$.

Neural networks can have any number of layers, and any number of nodes per layer. Most applications use the three layer structure with a maximum of a few hundred input nodes. The hidden layer is usually about 10% the size of the input layer. In the case of target detection, the output layer only needs a single node. The output of this node is thresholded to provide a positive or negative indication of the target's presence or absence in the input data.

Table 26-1 is a program to carry out the flow diagram of Fig. 26-5.   The key point is that this architecture is very simple and very generalized. This same flow diagram can be used for many problems, regardless of their particular quirks. The ability of the neural network to provide useful data manipulation lies in the proper selection of the *weights*. This is a dramatic departure from conventional information processing where solutions are described in step-by-step procedures.

As an example, imagine a neural network for recognizing objects in a sonar signal. Suppose that 1000 samples from the signal are stored in a computer. How does the computer determine if these data represent a submarine, whale, undersea mountain, or nothing at all? Conventional DSP would approach this problem with mathematics and algorithms, such as correlation and frequency spectrum analysis. With a neural network, the 1000 samples are simply fed into the input layer, resulting in values popping from the output layer. By selecting the proper weights, the output can be configured to report a wide range of information. For instance, there might be outputs for: submarine (yes/no), whale (yes/no), undersea mountain (yes/no), etc.

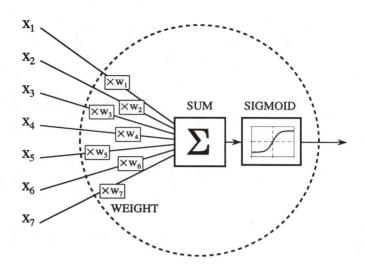

FIGURE 26-6
Neural network active node. This is a flow diagram of the active nodes used in the hidden and output layers of the neural network. Each input is multiplied by a weight (the w_N values), and then summed. This produces a single value that is passed through an "s" shaped nonlinear function called a *sigmoid*. The sigmoid function is shown in more detail in Fig. 26-7.

With other weights, the outputs might classify the objects as: metal or non-metal, biological or nonbiological, enemy or ally, etc. No algorithms, no rules, no procedures; only a relationship between the input and output dictated by the values of the weights selected.

```
100 'NEURAL NETWORK (FOR THE FLOW DIAGRAM IN FIG. 26-5)
110 '
120 DIM X1[15]                    'holds the input values
130 DIM X2[4]                     'holds the values exiting the hidden layer
140 DIM X3[2]                     'holds the values exiting the output layer
150 DIM WH[4,15]                  'holds the hidden layer weights
160 DIM WO[2,4]                   'holds the output layer weights
170 '
180 GOSUB XXXX                    'mythical subroutine to load X1[ ] with the input data
190 GOSUB XXXX                    'mythical subroutine to load the weights, WH[ , ] & W0[ , ]
200 '
210 '                            'FIND THE HIDDEN NODE VALUES, X2[ ]
220 FOR J% = 1 TO 4               'loop for each hidden layer node
230   ACC = 0                     'clear the accumulator variable, ACC
240   FOR I% = 1 TO 15            'weight and sum each input node
250     ACC = ACC + X1[I%] * WH[J%,I%]
260   NEXT I%
270   X2[J%] = 1 / (1 + EXP(-ACC) )    'pass summed value through the sigmoid
280 NEXT J%
290 '
300 '                            'FIND THE OUTPUT NODE VALUES, X3[ ]
310 FOR J% = 1 TO 2               'loop for each output layer node
320   ACC = 0                     'clear the accumulator variable, ACC
330   FOR I% = 1 TO 4             'weight and sum each hidden node
340     ACC = ACC + X2[I%] * WO[J%,I%]
350   NEXT I%
360   X3[J%] = 1 / (1 + EXP(-ACC) )    'pass summed value through the sigmoid
370 NEXT J%
380 '
390 END
```

TABLE 26-1

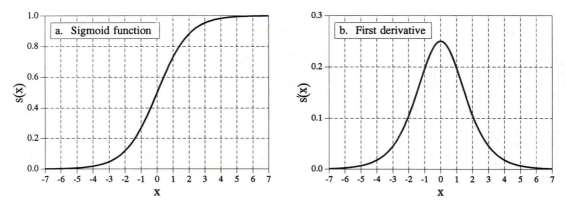

FIGURE 26-7
The sigmoid function and its derivative. Equations 26-1 and 26-2 generate these curves.

Figure 26-7a shows a closer look at the sigmoid function, mathematically described by the equation:

EQUATION 26-1
The sigmoid function. This is used in neural networks as a smooth threshold. This function is graphed in Fig. 26-7a.

$$s(x) = \frac{1}{1+e^{-x}}$$

The exact shape of the sigmoid is not important, only that it is a **smooth threshold**. For comparison, a **simple threshold** produces a value of *one* when $x > 0$, and a value of *zero* when $x < 0$. The sigmoid performs this same basic thresholding function, but is also *differentiable*, as shown in Fig. 26-7b. While the derivative is not used in the flow diagram (Fig. 25-5), it is a critical part of finding the proper weights to use. More about this shortly. An advantage of the sigmoid is that there is a shortcut to calculating the value of its derivative:

EQUATION 26-2
First derivative of the sigmoid function. This is calculated by using the value of the sigmoid function itself.

$$s'(x) = s(x)\big(1 - s(x)\big)$$

For example, if $x = 0$, then $s(x) = 0.5$ (by Eq. 26-1), and the first derivative is calculated: $s'(x) = 0.5(1 - 0.5) = 0.25$. This isn't a critical concept, just a trick to make the algebra shorter.

Wouldn't the neural network be more flexible if the sigmoid could be adjusted left-or-right, making it centered on some other value than $x = 0$? The answer is yes, and most neural networks allow for this. It is very simple to implement; an additional node is added to the input layer, with its input always having a value of *one*. When this is multiplied by the weights of the hidden layer, it provides a *bias* (DC offset) to each sigmoid. This addition is called a **bias node**. It is treated the same as the other nodes, except for the constant input.

Can neural networks be made without a sigmoid or similar nonlinearity? To answer this, look at the three-layer network of Fig. 26-5. If the sigmoids were not present, the three layers would *collapse* into only two layers. In other words, the summations and weights of the hidden and output layers could be combined into a single layer, resulting in only a two-layer network.

Why Does It Work?

The weights required to make a neural network carry out a particular task are found by a **learning algorithm**, together with **examples** of how the system *should* operate. For instance, the examples in the sonar problem would be a database of several hundred (or more) of the 1000 sample segments. Some of the example segments would correspond to submarines, others to whales, others to random noise, etc. The learning algorithm uses these examples to calculate a set of weights appropriate for the task at hand. The term **learning** is widely used in the neural network field to describe this process; however, a better description might be: *determining an optimized set of weights based on the statistics of the examples*. Regardless of what the method is called, the resulting weights are virtually impossible for humans to understand. Patterns may be observable in some rare cases, but generally they appear to be random numbers. A neural network using these weights can be observed to have the proper input/output relationship, but *why* these particular weights work is quite baffling. This mystic quality of neural networks has caused many scientists and engineers to shy away from them. Remember all those science fiction movies of renegade computers taking over the earth?

In spite of this, it is common to hear neural network advocates make statements such as: "neural networks are well understood." To explore this claim, we will first show that it is possible to pick neural network weights through traditional DSP methods. Next, we will demonstrate that the learning algorithms provide *better* solutions than the traditional techniques. While this doesn't explain *why* a particular set of weights works, it does provide confidence in the method.

In the most sophisticated view, the neural network is a method of labeling the various regions in *parameter space*. For example, consider the sonar system neural network with 1000 inputs and a single output. With proper weight selection, the output will be near *one* if the input signal is an echo from a submarine, and near *zero* if the input is only noise. This forms a parameter hyperspace of 1000 dimensions. The neural network is a method of assigning a value to each location in this hyperspace. That is, the 1000 input values define a *location* in the hyperspace, while the output of the neural network provides the *value* at that location. A look-up table could perform this task perfectly, having an output value stored for each possible input address. The difference is that the neural network *calculates* the value at each location (address), rather than the impossibly large task of *storing* each value. In fact, neural network architectures are often evaluated by how well they separate the hyperspace for a given number of weights.

This approach also provides a clue to the number of nodes required in the hidden layer. A parameter space of N dimensions requires N numbers to specify a location. Identifying a *region* in the hyperspace requires $2N$ values (i.e., a minimum and maximum value along each axis defines a hyperspace rectangular solid). For instance, these simple calculations would indicate that a neural network with 1000 inputs needs 2000 weights to identify one region of the hyperspace from another. In a fully interconnected network, this would require two hidden nodes. The number of regions needed depends on the particular problem, but can be expected to be far less than the number of dimensions in the parameter space. While this is only a crude approximation, it generally explains why most neural networks can operate with a hidden layer of 2% to 30% the size of the input layer.

A completely different way of understanding neural networks uses the DSP concept of *correlation*. As discussed in Chapter 7, correlation is the optimal way of detecting if a known pattern is contained within a signal. It is carried out by multiplying the signal with the pattern being looked for, and adding the products. The higher the sum, the more the signal resembles the pattern. Now, examine Fig. 26-5 and think of each hidden node as looking for a specific pattern in the input data. That is, each of the hidden nodes *correlates* the input data with the set of weights associated with that hidden node. If the pattern is present, the sum passed to the sigmoid will be large, otherwise it will be small.

The action of the sigmoid is quite interesting in this viewpoint. Look back at Fig. 26-1d and notice that the probability curve separating two bell shaped distributions resembles a sigmoid. If we were manually designing a neural network, we could make the output of each hidden node be the *fractional probability* that a specific pattern is present in the input data. The output layer repeats this operation, making the entire three-layer structure a correlation of correlations, a network that looks for *patterns of patterns*.

Conventional DSP is based on two techniques, *convolution* and *Fourier analysis*. It is reassuring that neural networks can carry out both these operations, plus *much more*. Imagine an N sample signal being filtered to produce another N sample signal. According to the output side view of convolution, each sample in the output signal is a weighted sum of samples from the input. Now, imagine a two-layer neural network with N nodes in each layer. The value produced by each output layer node is also a weighted sum of the input values. If each output layer node uses the same weights as all the other output nodes, the network will implement linear convolution. Likewise, the DFT can be calculated with a two layer neural network with N nodes in each layer. Each output layer node finds the amplitude of one frequency component. This is done by making the weights of each output layer node the same as the sinusoid being looked for. The resulting network correlates the input signal with each of the basis function sinusoids, thus calculating the DFT. Of course, a two-layer neural network is much less powerful than the standard three layer architecture. This means neural networks can carry out *nonlinear* as well as *linear* processing.

Suppose that one of these conventional DSP strategies is used to design the weights of a neural network. Can it be claimed that the network is *optimal*? Traditional DSP algorithms are usually based on assumptions about the characteristics of the input signal. For instance, Wiener filtering is optimal for maximizing the signal-to-noise ratio *assuming* the signal and noise spectra are both known; correlation is optimal for detecting targets *assuming* the noise is white; deconvolution counteracts an undesired convolution *assuming* the deconvolution kernel is the inverse of the original convolution kernel, etc. The problem is, scientist and engineer's seldom have a perfect knowledge of the input signals that will be encountered. While the underlying mathematics may be elegant, the overall performance is limited by how well the data are understood.

For instance, imagine testing a traditional DSP algorithm with actual input signals. Next, repeat the test with the algorithm changed slightly, say, by increasing one of the parameters by one percent. If the second test result is better than the first, the original algorithm is not optimized for the task at hand. Nearly all conventional DSP algorithms can be significantly improved by a trial-and-error evaluation of small changes to the algorithm's parameters and procedures. This is the strategy of the neural network.

Training the Neural Network

Neural network design can best be explained with an example. Figure 26-8 shows the problem we will attack, identifying individual letters in an image of text. This pattern recognition task has received much attention. It is easy enough that many approaches achieve partial success, but difficult enough that there are no perfect solutions. Many successful commercial products have been based on this problem, such as: reading the addresses on letters for postal routing, document entry into word processors, etc.

The first step in developing a neural network is to create a database of examples. For the text recognition problem, this is accomplished by printing the 26 capital letters: A,B,C,D ⋯ Y,Z, 50 times on a sheet of paper. Next, these 1300 letters are converted into a digital image by using one of the many scanning devices available for personal computers. This large digital image is then divided into small images of 10×10 pixels, each containing a single letter. This information is stored as a 1.3 Megabyte database: 1300 images; 100 pixels per image; 8 bits per pixel. We will use the first 260 images in this database to *train* the neural network (i.e., determine the weights), and the remainder to *test* its performance. The database must also contain a way of identifying the letter contained in each image. For instance, an additional byte could be added to each 10×10 image, containing the letter's ASCII code. In another scheme, the position

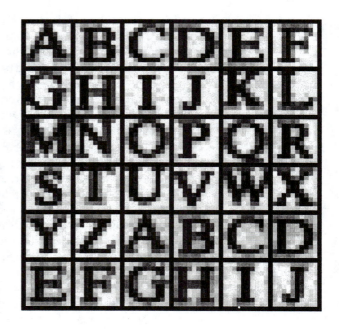

FIGURE 26-8
Example image of text. Identifying letters in images of text is one of the classic pattern recognition problems. In this example, each letter is contained in a 10×10 pixel image, with 256 gray levels per pixel. The database used to train and test the example neural network consists of 50 sets of the 26 capital letters, for a total of 1300 images. The images shown here are a portion of this database.

of each 10×10 image in the database could indicate what the letter is. For example, images 0 to 49 might all be an "A", images 50-99 might all be a "B", etc.

For this demonstration, the neural network will be designed for an arbitrary task: determine which of the 10×10 images contains a *vowel*, i.e., A, E, I, O, or U. This may not have any practical application, but it does illustrate the ability of the neural network to learn very abstract pattern recognition problems. By including ten examples of each letter in the training set, the network will (hopefully) learn the key features that distinguish the target from the nontarget images.

The neural network used in this example is the traditional three-layer, fully interconnected architecture, as shown in Figs. 26-5 and 26-6. There are 101 nodes in the input layer (100 pixel values plus a bias node), 10 nodes in the hidden layer, and 1 node in the output layer. When a 100 pixel image is applied to the input of the network, we want the output value to be close to *one* if a vowel is present, and near *zero* if a vowel is not present. Don't be worried that the input signal was acquired as a two-dimensional array (10×10), while the input to the neural network is a one-dimensional array. This is *your* understanding of how the pixel values are interrelated; the *neural network* will find relationships of its own.

Table 26-2 shows the main program for calculating the neural network weights, with Table 26-3 containing three subroutines called from the main program. The array elements: X1[1] through X1[100], hold the input layer values. In addition, X1[101] always holds a value of 1, providing the input to the bias node. The output values from the hidden nodes are contained

```
100 'NEURAL NETWORK TRAINING (Determination of weights)
110 '
120                                     'INITIALIZE
130 MU = .000005                        'iteration step size
140 DIM X1[101]                         'holds the input layer signal + bias term
150 DIM X2[10]                          'holds the hidden layer signal
160 DIM WH[10,101]                      'holds hidden layer weights
170 DIM WO[10]                          'holds output layer weights
180 '
190 FOR H% = 1 TO 10                    'SET WEIGHTS TO RANDOM VALUES
200   WO[H%] = (RND-0.5)                'output layer weights: -0.5 to 0.5
210   FOR I% = 1 TO 101                 'hidden layer weights: -0.0005 to 0.0005
220     WH[H%,I%] = (RND-0.5)/1000
230   NEXT I%
240 NEXT H%
250 '
260 '                                   'ITERATION LOOP
270 FOR ITER% = 1 TO 800                'loop for 800 iterations
280 '
290   ESUM = 0                          'clear the error accumulator, ESUM
300   '
310   FOR LETTER% = 1 TO 260            'loop for each letter in the training set
320     GOSUB 1000                      'load X1[ ] with training set
330     GOSUB 2000                      'find the error for this letter, ELET
340     ESUM = ESUM + ELET^2            'accumulate error for this iteration
350     GOSUB 3000                      'find the new weights
360   NEXT LETTER%
370   '
380   PRINT ITER% ESUM                  'print the progress to the video screen
390 '
400 NEXT ITER%
410 '
420 GOSUB XXXX                          'mythical subroutine to save the weights
430 END
```

TABLE 26-2

in the array elements: X2[1] through X2[10]. The variable, X3, contains the network's output value. The weights of the hidden layer are contained in the array, WH[,], where the first index identifies the hidden node (1 to 10), and the second index is the input layer node (1 to 101). The weights of the output layer are held in WO[1] to WO[10]. This makes a total of 1020 weight values that define how the network will operate.

The first action of the program is to set each weight to an arbitrary initial value by using a random number generator. As shown in lines 190 to 240, the hidden layer weights are assigned initial values between -0.0005 and 0.0005, while the output layer weights are between -0.5 and 0.5. These ranges are chosen to be the same order of magnitude that the final weights *must* be. This is based on: (1) the range of values in the input signal, (2) the number of inputs summed at each node, and (3) the range of values over which the sigmoid is active, an input of about $-5 < x < 5$, and an output of 0 to 1. For instance, when 101 inputs with a typical value of 100 are multiplied by the typical weight value of 0.0002, the sum of the products is about 2, which is in the active range of the sigmoid's input.

If we evaluated the performance of the neural network using these random weights, we would expect it to be the same as random guessing. The learning algorithm improves the performance of the network by gradually changing each weight in the proper direction. This is called an **iterative** procedure, and is controlled in the program by the FOR-NEXT loop in lines 270-400. Each iteration makes the weights slightly more efficient at separating the target from the nontarget examples. The iteration loop is usually carried out until no further improvement is being made. In typical neural networks, this may be anywhere from ten to ten-thousand iterations, but a few hundred is common. This example carries out 800 iterations.

In order for this iterative strategy to work, there must be a *single* parameter that describes how well the system is currently performing. The variable ESUM (for error sum) serves this function in the program. The first action inside the iteration loop is to set ESUM to zero (line 290) so that it can be used as an accumulator. At the end of each iteration, the value of ESUM is printed to the video screen (line 380), so that the operator can insure that progress is being made. The value of ESUM will start high, and gradually decrease as the neural network is trained to recognize the targets. Figure 26-9 shows examples of how ESUM decreases as the iterations proceed.

All 260 images in the training set are evaluated during each iteration, as controlled by the FOR-NEXT loop in lines 310-360. Subroutine 1000 is used to retrieve images from the database of examples. Since this is not something of particular interest here, we will only describe the parameters passed to and from this subroutine. Subroutine 1000 is entered with the parameter, LETTER%, being between 1 and 260. Upon return, the input node values, X1[1] to X1[100], contain the pixel values for the image in the database corresponding to LETTER%. The bias node value, X1[101], is always returned with a constant value of *one*. Subroutine 1000 also returns another parameter, CORRECT. This contains the desired output value of the network for this particular letter. That is, if the letter in the image is a vowel, CORRECT will be returned with a value of *one*. If the letter in the image is not a vowel, CORRECT will be returned with a value of *zero*.

After the image being worked on is loaded into X1[1] through X1[100], subroutine 2000 passes the data through the current neural network to produce the output node value, X3. In other words, subroutine 2000 is the same as the program in Table 26-1, except for a different number of nodes in each layer. This subroutine also calculates how well the current network identifies the letter as a target or a nontarget. In line 2210, the variable ELET (for error-letter) is calculated as the difference between the output value actually generated, X3, and the desired value, CORRECT. This makes ELET a value between -1 and 1. All of the 260 values for ELET are combined (line 340) to form ESUM, the total squared error of the network for the entire training set.

Line 2220 shows an option that is often included when calculating the error: assigning a different *importance* to the errors for targets and nontargets. For example, recall the cancer example presented earlier in this chapter,

```
1000 'SUBROUTINE TO LOAD X1[ ] WITH IMAGES FROM THE DATABASE
1010 'Variables entering routine: LETTER%
1020 'Variables exiting routine: X1[1] to X1[100], X1[101] = 1, CORRECT
1030 '
1040 'The variable, LETTER%, between 1 and 260, indicates which image in the database is to be
1050 'returned in X1[1] to X1[100].  The bias node, X1[101], always has a value of one.  The variable, 1060
'CORRECT, has a value of one if the image being returned is a vowel, and zero otherwise.
1070 '(The details of this subroutine are unimportant, and not listed here).
1900 RETURN


2000 'SUBROUTINE TO CALCULATE THE ERROR WITH THE CURRENT WEIGHTS
2010 'Variables entering routine:  X1[ ], X2[ ], WI[ , ], WH[ ], CORRECT
2020 'Variables exiting routine:   ELET
2030 '
2040 '                                       'FIND THE HIDDEN NODE VALUES, X2[ ]
2050 FOR H% = 1 TO 10                        'loop for each hidden nodes
2060   ACC = 0                               'clear the accumulator
2070   FOR I% = 1 TO 101                     'weight and sum each input node
2080     ACC = ACC + X1[I%] * WH[H%,I%]
2090   NEXT I%
2100   X2[H%] = 1 / (1 + EXP(-ACC) )         'pass summed value through sigmoid
2110 NEXT H%
2120 '
2130 '                                       'FIND THE OUTPUT VALUE: X3
2140 ACC = 0                                 'clear the accumulator
2150 FOR H% = 1 TO 10                        'weight and sum each hidden node
2160   ACC = ACC + X2[H%] * WO[H%]
2170 NEXT H%
2180 X3 = 1 / (1 + EXP(-ACC) )               'pass summed value through sigmoid
2190 '
2200 '                                       'FIND ERROR FOR THIS LETTER, ELET
2210 ELET = (CORRECT - X3)                   'find the error
2220 IF CORRECT = 1 THEN ELET = ELET*5       'give extra weight to targets
2230 '
2240 RETURN


3000 'SUBROUTINE TO FIND NEW WEIGHTS
3010 'Variables entering routine:  X1[ ], X2[ ], X3, WI[ , ], WH[ ], ELET, MU
3020 'Variables exiting routine:   WI[ , ], WH[ ]
3030 '
3040 '                                       'FIND NEW WEIGHTS FOR INPUT NODES
3050 FOR H% = 1 TO 10
3060   FOR I% = 1 TO 101
3070     SLOPEO = X3 * (1 - X3)
3080     SLOPEH = X2(H%) * (1 - X2[H%])
3090     DX3DW = X1[I%] * SLOPEH * WO[H%] * SLOPEO
3100     WH[H%,I%] = WH[H%,I%] + DX3DW * ELET * MU
3110   NEXT I%
3120 NEXT H%
3130 '
3140 '                                       'FIND NEW WEIGHTS FOR HIDDEN NODES
3150 FOR H% = 1 TO 10
3160   SLOPEO = X3 * (1 - X3)
3170   DX3DW = X2[H%] * SLOPEO
3180   WO[H%] = WO[H%] + DX3DW * ELET * MU
3190 NEXT H%
3200 '
3210 RETURN
```

TABLE 26-3

and the consequences of making a false-positive error versus a false-negative error. In the present example, we will arbitrarily declare that the error in detecting a target is *five* times as bad as the error in detecting a nontarget. In effect, this tells the network to do a better job with the targets, even if it hurts the performance of the nontargets.

Subroutine 3000 is the heart of the neural network strategy, the algorithm for changing the weights on each iteration. We will use an analogy to explain the underlying mathematics. Consider the predicament of a military paratrooper dropped behind enemy lines. He parachutes to the ground in unfamiliar territory, only to find it is so dark he can't see more than a few feet away. His orders are to proceed to the bottom of the nearest valley to begin the remainder of his mission. The problem is, without being able to see more than a few feet, how does he make his way to the valley floor? Put another way, he needs an algorithm to adjust his x and y position on the earth's surface in order to *minimize* his elevation. This is analogous to the problem of adjusting the neural network weights, such that the network's error, ESUM, is minimized.

We will look at two algorithms to solve this problem: **evolution** and **steepest descent**. In evolution, the paratrooper takes a flying jump in some random direction. If the new elevation is *higher* than the previous, he curses and returns to his starting location, where he tries again. If the new elevation is *lower*, he feels a measure of success, and repeats the process from the new location. Eventually he will reach the bottom of the valley, although in a very inefficient and haphazard path. This method is called *evolution* because it is the same type of algorithm employed by nature in biological evolution. Each new generation of a species has random variations from the previous. If these differences are of benefit to the species, they are more likely to be retained and passed to the *next* generation. This is a result of the improvement allowing the animal to receive more food, escape its enemies, produce more offspring, etc. If the new trait is detrimental, the disadvantaged animal becomes lunch for some predator, and the variation is discarded. In this sense, each new generation is an iteration of the evolutionary optimization procedure.

When evolution is used as the training algorithm, each weight in the neural network is slightly changed by adding the value from a random number generator. If the modified weights make a better network (i.e., a lower value for ESUM), the changes are retained, otherwise they are discarded. While this works, it is very slow in **converging**. This is the jargon used to describe that continual improvement is being made toward an optimal solution (the bottom of the valley). In simpler terms, the program is going to need days to reach a solution, rather than minutes or hours.

Fortunately, the *steepest descent* algorithm is much faster. This is how the paratrooper would naturally respond: evaluate which way is *downhill*, and move in that direction. Think about the situation this way. The paratrooper can move one step to the north, and record the change in elevation. After returning to his original position, he can take one step to the east, and

FIGURE 26-9
Neural network convergence. This graph shows how the neural network error (the value of ESUM) decreases as the iterations proceed. Three separate trials are shown, each starting with different initial weights.

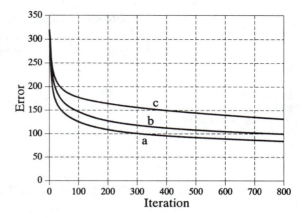

record that elevation change. Using these two values, he can determine which direction is downhill. Suppose the paratrooper drops 10 cm when he moves one step in the northern direction, and drops 20 cm when he moves one step in the eastern direction. To travel directly downhill, he needs to move along each axis an amount proportional to the slope along that axis. In this case, he might move north by 10 steps and east by 20 steps. This moves him down the steepest part of the slope a distance of $\sqrt{10^2+20^2}$ = 22.4 steps. Alternatively, he could move in a straight line to the new location, 22.4 steps along the diagonal. The key point is: *the steepest descent is achieved by moving along each axis a distance proportional to the slope along that axis*.

Subroutine 3000 implements this same steepest decent algorithm for the network weights. Before entering subroutine 3000, one of the example images has been applied to the input layer, and the information propagated to the output. This means that the values for: X1[ ], X2[ ] and X3 are all specified, as well as the current weight values: WH[,] and WO[ ]. In addition, we know the error the network produces for this particular image, ELET. The hidden layer weights are updated in lines 3050 to 3120, while the output layer weights are modified in lines 3150 to 3190. *This is done by calculating the slope for each weight, and then changing each weight by an amount proportional to that slope*. In the paratrooper case, the slope along an axis is found by moving a small distance along the axis (say, Δx), measuring the change in elevation (say, ΔE), and then dividing the two ($\Delta E/\Delta x$). The slope of a neural network weight can be found in this same way: add a small increment to the weight value (Δw), find the resulting change in the output signal ($\Delta X3$), and divide the two ($\Delta X3/\Delta w$). Later in this chapter we will look at an example that calculates the slope this way. However, in the present example we will use a more efficient method.

Earlier we said that the nonlinearity (the sigmoid) needs to be *differentiable*. Here is where we will use this property. If we know the slope at each point on the nonlinearity, we can directly write an equation for the slope of each weight ($\Delta X3/\Delta w$) without actually having to perturb it. Consider a specific weight, for example, WO[1], corresponding to the first input of the output node. Look at the structure in Figs. 26-5 and 26-6, and ask: how will the output (*X3*) be affected if this particular weight (*w*) is changed slightly, but everything else is kept the same? The answer is:

EQUATION 26-3
Slope of the output layer weights. This equation is written for the weight, WO[1].

$$\frac{\Delta X3}{\Delta w} = X2[1] \ \text{SLOPE}_O$$

where SLOPE$_O$ is the first derivative of the output layer sigmoid, evaluated where we are operating on its curve. In other words, SLOPE$_O$ describes how much the *output* of the sigmoid changes in response to a change in the *input* to the sigmoid. From Eq. 26-2, SLOPE$_O$ can be calculated from the current output value of the sigmoid, X3. This calculation is shown in line 3160. In line 3170, the slope for this weight is calculated via Eq. 26-3, and stored in the variable DX3DW (i.e., $\Delta X3/\Delta w$).

Using a similar analysis, the slope for a weight on the hidden layer, such as WH[1,1], can be found by:

EQUATION 26-4
Slope of the hidden layer weights. This equation is written for the weight, WH[1,1].

$$\frac{\Delta X3}{\Delta w} = X1[1] \ \text{SLOPE}_{H1} \ WO[1] \ \text{SLOPE}_O$$

SLOPE$_{H1}$ is the first derivative of the hidden layer sigmoid, evaluated where we are operating on its curve. The other values, X1[1] and WO[1], are simply constants that the weight change sees as it makes its way to the output. In lines 3070 and 3080, the slopes of the sigmoids are calculated using Eq. 26-2. The slope of the hidden layer weight, DX3DW is calculated in line 3090 via Eq. 26-4.

Now that we know the *slope* of each of the weights, we can look at how each weight is changed for the next iteration. The new value for each weight is found by taking the current weight, and adding an amount that is proportional to the slope:

EQUATION 26-5
Updating the weights. Each of the weights is adjusted by adding an amount proportional to the slope of the weight.

$$w_{new} = w_{old} + \frac{\Delta X3}{\Delta w} \ \text{ELET} \ \text{MU}$$

This calculation is carried out in line 3100 for the hidden layer, and line 3180 for the output layer. The proportionality constant consists of two factors, ELET, the error of the network for this particular input, and MU, a constant set at the beginning of the program. To understand the need for ELET in this calculation, imagine that an image placed on the input produces a *small* error in the output signal. Next, imagine that another image applied to the input produces a *large* output error. When adjusting the weights, we want to nudge the network more for the second image than the first. If something is working poorly, we want to change it; if it is working well, we want to leave it alone. This is accomplished by changing each weight in proportion to the current error, ELET.

To understand how MU affects the system, recall the example of the paratrooper. Once he determines the downhill direction, he must decide how far to proceed before reevaluating the slope of the terrain. By making this distance short, one meter for example, he will be able to precisely follow the contours of the terrain and always be moving in an optimal direction. The problem is that he spends most of his time evaluating the slope, rather than actually moving down the hill. In comparison, he could choose the distance to be large, say 1000 meters. While this would allow the paratrooper to move rapidly along the terrain, he might overshoot the downhill path. Too large of a distance makes him jump all over the country-side without making the desired progress.

In the neural network, MU controls how much the weights are changed on each iteration. The value to use depends on the particular problem, being as low as 10^{-6}, or as high as 0.1. From the analogy of the paratrooper, it can be expected that too small of a value will cause the network to converge too slowly. In comparison, too large of a value will cause the convergence to be erratic, and will exhibit chaotic oscillation around the final solution. Unfortunately, the way neural networks react to various values of MU can be difficult to understand or predict. This makes it critical that the network error (i.e., ESUM) be monitored during the training, such as printing it to the video screen at the end of each iteration. If the system isn't converging properly, stop the program and try another value for MU.

Evaluating the Results

So, how does it work? The training program for vowel recognition was run three times using different random values for the initial weights. About one hour is required to complete the 800 iterations on a 100 MHz Pentium personnel computer. Figure 26-9 shows how the error of the network, ESUM, changes over this period. The gradual decline indicates that the network is learning the task, and that the weights reach a near optimal value after several hundred iterations. Each trial produces a different solution to the problem, with a different final performance. This is analogous to the paratrooper starting at different locations, and thereby ending up at the bottom of different valleys. Just as some valleys are deeper than others, some neural network solutions are better than others. This means that the learning algorithm should be run several times, with the best of the group taken as the final solution.

trial (a)

trial (b)

trial (c)

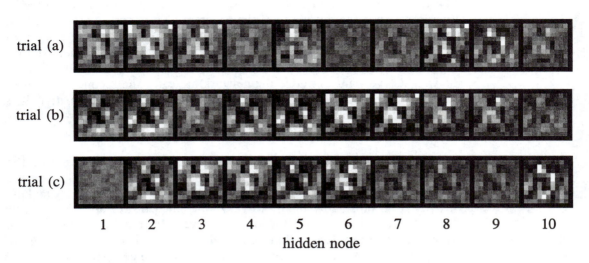

1   2   3   4   5   6   7   8   9   10

hidden node

FIGURE 26-10
Example of neural network weights. In this figure, the hidden layer weights for the three solutions are displayed as images. All three of these solutions appear random to the human eye.

In Fig. 26-10, the hidden layer weights of the three solutions are displayed as images. This means the first action taken by the neural network is to correlate (multiply and sum) these images with the input signal. They look like random noise! These weights values can be *shown* to work, but *why* they work is something of a mystery. Here is something else to ponder. The human brain is composed of about 100 *trillion* neurons, each with an average of 10,000 interconnections. If we can't understand the simple neural network in this example, how can we study something that is at least 100,000,000,000,000 times more complex? This is 21st century research.

Figure 26-11a shows a histogram of the neural network's output for the 260 letters in the training set. Remember, the weights were selected to make the output near *one* for vowel images, and near *zero* otherwise. Separation has been perfectly achieved, with no overlap between the two distributions. Also notice that the vowel distribution is narrower than the nonvowel distribution. This is because we declared the target error to be five times more important than the nontarget error (see line 2220).

In comparison, Fig. 26-11b shows the histogram for images 261 through 1300 in the database. While the target and nontarget distributions are reasonably distinct, they are not completely separated. Why does the neural network perform better on the first 260 letters than the last 1040? Figure (a) is cheating! It's easy to take a test if you have already seen the answers. In other words, the neural network is recognizing specific images in the training set, not the general patterns identifying vowels from nonvowels.

Figure 26-12 shows the performance of the three solutions, displayed as ROC curves. Trial (b) provides a significantly better network than the

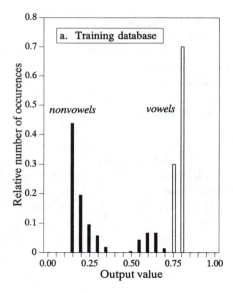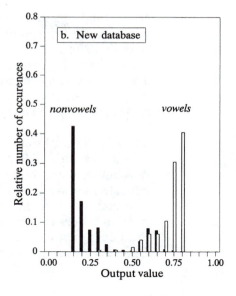

FIGURE 26-11
Neural network performance. These are histograms of the neural network's output values, (a) for the training data, and (b) for the remaining images. The neural network performs better with the training data because it has already seen the answers to the test.

other two. This is a matter of random chance depending on the initial weights used. At one threshold setting, the neural network designed in trial "b" can detect 24 out of 25 targets (i.e., 96% of the vowel images), with a false alarm rate of only 1 in 25 nontargets (i.e., 4% of the nonvowel images). Not bad considering the abstract nature of this problem, and the very general solution applied.

FIGURE 26-12
ROC analysis of neural network examples. These curves compare three neural networks designed to detect images of vowels. Trial (b) is the best solution, shown by its curve being closer to the upper-left corner of the graph. This network can correctly detect 24 out of 25 targets, while providing only 1 false alarm for each 25 nontargets. That is, there is a the point on the ROC curve at $x = 4\%$ and $y = 96\%$

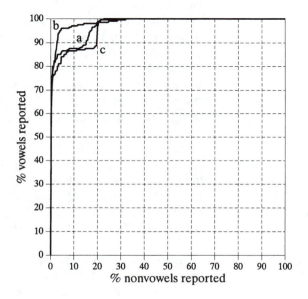

Some final comments on neural networks. Getting a neural network to converge during training can be tricky.   If the network error (ESUM) doesn't steadily decrease, the program must be terminated, changed, and then restarted. This may take several attempts before success is reached. Three things can be changed to affect the convergence: (1) MU, (2) the magnitude of the initial random weights, and (3) the number of hidden nodes (in the order they should be changed).

The most critical item in neural network development is the *validity* of the training examples. For instance, when new commercial products are being developed, the only test data available are from prototypes, simulations, educated guesses, etc. If a neural network is trained on this preliminary information, it might not operate properly in the final application. Any difference between the training database and the eventual data will degrade the neural network's performance (Murphy's law for neural networks). Don't try to second guess the neural network on this issue; you can't!

Recursive Filter Design

Chapters 19 and 20 show how to design recursive filters with the standard frequency responses: high-pass, low-pass, band-pass, etc. What if you need something custom? The answer is to design a recursive filter just as you would a neural network: start with a generic set of recursion coefficients, and use iteration to slowly mold them into what you want. This technique is important for two reasons. First, it allows custom recursive filters to be designed without having to hassle with the mathematics of the z-transform. Second, it shows that the ideas from conventional DSP and neural networks can be combined to form superb algorithms.

The main program for this method is shown in Table 26-4, with two subroutines in Table 26-5. The array, T[ ], holds the desired frequency response, some kind of curve that we have manually designed. Since this program is based around the FFT, the lengths of the signals must be a power of two. As written, this program uses an FFT length of 256, as defined by the variable, N%, in line 130. This means that T[0] to T[128] correspond to the frequencies between 0 and 0.5 of the sampling rate. Only the magnitude is contained in this array; the phase is not controlled in this design, and becomes whatever it becomes.

The recursion coefficients are set to their initial values in lines 270-310, typically selected to be the *identity* system. Don't use random numbers here, or the initial filter will be unstable. The recursion coefficients are held in the arrays, A[ ] and B[ ]. The variable, NP%, sets the number of poles in the designed filter. For example, if NP% is 5, the "a" coefficients run from A[0] to A[5], while the "b" coefficients run from B[1] to B[5].

As previously mentioned, the iterative procedure requires a *single* value that describes how well the current system is functioning. This is provided by the variable, ER (for error), and is calculated in subroutine 3000.   Lines

```
100 'ITERATIVE DESIGN OF RECURSIVE FILTER
110 '
120                                      'INITIALIZE
130 N%  = 256                           'number of points in FFT
140 NP% = 8                             'number of poles in filter
150 DELTA = .00001                      'perturbation increment
160 MU = .2                             'iteration step size
170 DIM REX[255]                        'real part of signal during FFT
180 DIM IMX[255]                        'imaginary part of signal during FFT
190 DIM T[128]                          'desired frequency response (mag only)
200 DIM A[8]                            'the "a" recursion coefficients
210 DIM B[8]                            'the "b" recursion coefficients
220 DIM SA[8]                           'slope for "a" coefficients
230 DIM SB[8]                           'slope for "b" coefficients
240 '
250 GOSUB XXXX                          'mythical subroutine to load T[ ]
260 '
270 FOR P% = 0 TO NP%                   'initialize coefficients to the identity system
280   A[P%] = 0
290   B[P%] = 0
300 NEXT P%
310 A[0] = 1
320 '
330 '                                   'ITERATION LOOP
340 FOR ITER% = 1 TO 100                'loop for desired number of iterations
350   GOSUB 2000                        'calculate new coefficients
360   PRINT ITER% ENEW  MU              'print current status to video screen
370   IF ENEW > EOLD THEN MU = MU/2     'adjust the value of MU
380 NEXT ITER%
390 '
400 '
410 FOR P% = 0 TO NP%                   'PRINT OUT THE COEFFICIENTS
420   PRINT A[P%]    B[P%]
430 NEXT P%
440 '
450 END
```

TABLE 26-4

3040 to 3080 load an impulse in the array, IMX[ ]. Next, lines 3100-3150 use this impulse as an input signal to the recursive filter defined by the current values of A[ ] and B[ ]. The output of this filter is thus the *impulse response* of the current system, and is stored in the array, REX[ ]. The system's frequency response is then found by taking the FFT of the impulse response, as shown in line 3170. Subroutine 1000 is the FFT program listed in Table 12-4 in Chapter 12. This FFT subroutine returns the frequency response in rectangular form, overwriting the arrays REX[ ] and IMX[ ].

Lines 3200-3250 then calculate ER, the *mean squared error* between the magnitude of the current frequency response, and the desired frequency response. Pay particular attention to how this error is found. The iterative action of this program optimizes this error, making the way it is defined very important. The FOR-NEXT loop runs through each frequency in the frequency response. For each frequency, line 3220 calculates the magnitude of the current frequency response from the rectangular data. In line 3230, the error at this frequency is found by subtracting the desired magnitude, T[ ], from the current magnitude, MAG. This error is then squared, and added to the accumulator variable, ER. After looping through each frequency, line 3250 completes the calculation to make ER the mean squared error of the entire frequency response.

Lines 340 to 380 control the iteration loop of the program. Subroutine 2000 is where the changes to the recursion coefficients are made. The first action in this subroutine is to determine the current value of ER, and store it in another variable, EOLD (lines 2040 & 2050). After the subroutine updates the coefficients, the value of ER is again determined, and assigned to the variable, ENEW (lines 2270 and 2280).

The variable, MU, controls the iteration step size, just as in the previous neural network program. An advanced feature is used in this program: an *automated* adjustment to the value of MU. This is the reason for having the two variables, EOLD and ENEW. When the program starts, MU is set to the relatively high value of 0.2 (line 160). This allows the convergence to proceed rapidly, but will limit how close the filter can come to an optimal solution. As the iterations proceed, points will be reached where no progress is being made, identified by ENEW being *higher* than EOLD. Each time this occurs, line 370 reduces the value of MU.

Subroutine 2000 updates the recursion coefficients according to the steepest decent method: calculate the slope for each coefficient, and then change the coefficient an amount proportional to its slope. Lines 2080-2130 calculate the slopes for the "a" coefficients, storing them in the array, SA[ ]. Likewise, lines 2150-2200 calculate the slopes for the "b" coefficients, storing them in the array, SB[ ]. Lines 2220-2250 then modify each of the recursion coefficients by an amount proportional to these slopes. In this program, the proportionality constant is simply the step size, MU. No error term is required in the proportionality constant because there is only *one* example to be matched: the desired frequency response.

The last issue is how the program calculates the slopes of the recursion coefficients. In the neural network example, an *equation* for the slope was derived. This procedure cannot be used here because it would require taking the derivative *across* the DFT. Instead, a brute force method is applied: actually change the recursion coefficient by a small increment, and then directly calculate the new value of ER. The slope is then found as the change in ER divided by the amount of the increment. Specifically, the current value of ER is found in lines 2040-2050, and stored in the variable, EOLD. The loop in lines 2080-2130 runs through each of the "a" coefficients. The first action inside this loop is to add a small increment, DELTA, to the recursion coefficient being worked on (line 2090). Subroutine 3000 is invoked in line 2100 to find the value of ER with the modified coefficient. Line 2110 then calculates the slope of this coefficient as: $(ER - EOLD)/DELTA$. Line 2120 then restores the modified coefficient by subtracting the value of DELTA.

Figure 26-13 shows several examples of filters designed using this program. The dotted line is the desired frequency response, while the solid line is the

```
2000 'SUBROUTINE TO CALCULATE THE NEW RECURSION COEFFICIENTS
2010 'Variables entering routine:   A[ ], B[ ], DELTA, MU
2020 'Variables exiting routine:    A[ ], B[ ], EOLD, ENEW
2030 '
2040 GOSUB 3000                              'FIND THE CURRENT ERROR
2050 EOLD = ER                               'store current error in variable, EOLD
2060 '
2070                                         'FIND THE ERROR SLOPES
2080 FOR P% = 0 TO NP%                       'loop through each "a" coefficient
2090   A[P%] = A[P%] + DELTA                 'add a small increment to the coefficient
2100   GOSUB 3000                            'find the error with the change
2110   SA[P%] = (ER-EOLD)/DELTA              'calculate the error slope, store in SA[ ]
2120   A[P%] = A[P%] - DELTA                 'return coefficient to original value
2130 NEXT P%
2140 '
2150 FOR P% = 1 TO NP%                       'repeat process for each "b" coefficient
2160   B[P%] = B[P%] + DELTA
2170   GOSUB 3000
2180   SB[P%] = (ER-EOLD)/DELTA              'calculate the error slope, store in SB[ ]
2190   B[P%] = B[P%] - DELTA
2200 NEXT P%
2210 '                                       'CALCULATE NEW COEFFICIENTS
2220 FOR P% = 0 TO NP%                       'loop through each coefficient
2230   A[P%] = A[P%] - SA[P%] * MU           'adjust coefficients to move "downhill"
2240   B[P%] = B[P%] - SB[P%] * MU
2250 NEXT P%
2260 '
2270 GOSUB 3000                              'FIND THE NEW ERROR
2280 ENEW = ER                               'store new error in variable, ENEW
2290 '
2300 RETURN


3000 'SUBROUTINE TO CALCULATE THE FREQUENCY DOMAIN ERROR
3010 'Variables entering routine:   A[ ], B[ ], T[ ]
3020 'Variables exiting routine:    ER
3030 '
3040 FOR I% = 0 TO N%-1                      'LOAD SHIFTED  IMPULSE INTO IMX[ ]
3050   REX[I%] = 0
3060   IMX[I%] = 0
3070 NEXT I%
3080 IMX[12] = 1
3090 '                                       'CALCULATE IMPULSE RESPONSE
3100 FOR I% = 12 TO N%-1
3110   FOR J% = 0 TO NP%
3120     REX[I%] = REX[I%]  +  A[J%] * IMX[I%-J%]  +  B[J%] * REX[I%-J%]
3130   NEXT J%
3140 NEXT I%
3150 IMX[12] = 0
3160 '                                       'CALCULATE THE FFT
3170 GOSUB 1000                              'Table 12-4, uses REX[ ], IMX[ ], N%
3180 '
3190                                         'FIND FREQUENCY DOMAIN ERROR
3200 ER = 0                                  'zero ER, to use as an accumulator
3210 FOR I% = 0 TO N%/2                      'loop through each positive frequency
3220   MAG = SQR(REX[I%]^2 + IMX[I%]^2)      'rectangular --> polar conversion
3230   ER = ER + ( MAG - T[I%] )^2           'calculate and accumulate squared error
3240 NEXT I%
3250 ER = SQR( ER/(N%/2+1) )                 'finish calculation of error, ER
3260 '
3270 RETURN
```

TABLE 26-5

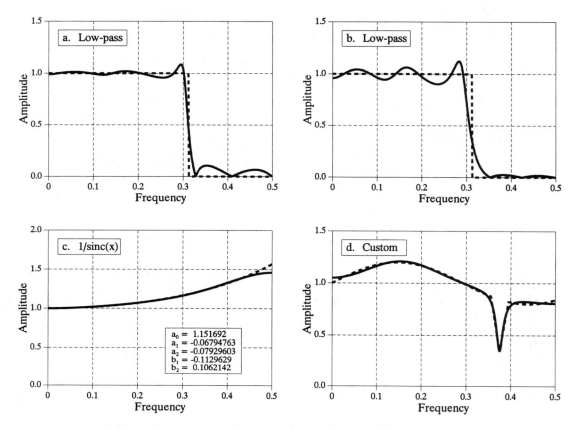

FIGURE 26-13
Iterative design of recursive filters. Figure (a) shows an 8 pole low-pass filter with the error equally distributed between 0 and 0.5. In (b), the error has been weighted to force better performance in the stopband, at the expense of error in the passband. Figure (c) shows a 2 pole filter used for the 1/sinc(x) correction in digital-to-analog conversion. The frequency response in (d) is completely custom. In each figure, the desired frequency response is shown by the dotted line, and the actual frequency response by the solid curve.

frequency response of the designed filter. Each of these filters requires several minutes to converge on a 100 MHz Pentium. Figure (a) is an 8 pole low-pass filter, where the error is equally weighted over the entire frequency spectrum (the program as written). Figure (b) is the same filter, except the error in the stopband is multiplied by *eight* when ER is being calculated. This forces the filter to have less stopband ripple, at the expense of greater ripple in the passband.

Figure (c) shows a 2 pole filter for: $1/sinc(x)$. As discussed in Chapter 3, this can be used to counteract the zeroth-order hold during digital-to-analog conversion (see Fig. 3-6). The error in this filter was only summed between 0 and 0.45, resulting in a better match over this range, at the expense of a worse match between 0.45 and 0.5. Lastly, (d) is a very irregular 6 pole frequency response that includes a sharp dip. To achieve convergence, the recursion coefficients were initially set to those of a notch filter.

CHAPTER

27

Data Compression

Data transmission and storage cost money. The more information being dealt with, the more it costs. In spite of this, most digital data are not stored in the most compact form. Rather, they are stored in whatever way makes them easiest to use, such as: ASCII text from word processors, binary code that can be executed on a computer, individual samples from a data acquisition system, etc. Typically, these easy-to-use encoding methods require data files about twice as large as actually needed to represent the information. Data compression is the general term for the various algorithms and programs developed to address this problem. A *compression program* is used to convert data from an easy-to-use format to one optimized for compactness. Likewise, an *uncompression program* returns the information to its original form. We examine five techniques for data compression in this chapter. The first three are simple encoding techniques, called: run-length, Huffman, and delta encoding. The last two are elaborate procedures that have established themselves as industry standards: LZW and JPEG.

Data Compression Strategies

Table 27-1 shows two different ways that data compression algorithms can be categorized. In (a), the methods have been classified as either **lossless** or **lossy**. A lossless technique means that the restored data file is *identical* to the original. This is absolutely necessary for many types of data, for example: executable code, word processing files, tabulated numbers, etc. You cannot afford to misplace even a single bit of this type of information. In comparison, data files that represent images and other acquired signals do not have to be keep in perfect condition for storage or transmission. All real world measurements inherently contain a certain amount of *noise*. If the changes made to these signals resemble a small amount of additional noise, no harm is done. Compression techniques that allow this type of degradation are called **lossy**. This distinction is important because lossy techniques are much more effective at compression than lossless methods. The higher the compression ratio, the more noise added to the data.

Lossless	Lossy		Method	Group size: input	output
run-length	CS&Q		CS&Q	*fixed*	*fixed*
Huffman	JPEG		Huffman	*fixed*	*variable*
delta	MPEG		Arithmetic	*variable*	*variable*
LZW			run-length, LZW	*variable*	*fixed*

a. Lossless or Lossy   b. Fixed or variable group size

TABLE 27-1

Compression classifications. Data compression methods can be divided in two ways. In (a), the techniques are classified as *lossless* or *lossy*. Lossless methods restore the compressed data to exactly the same form as the original, while lossy methods only generate an approximation. In (b), the methods are classified according to a *fixed* or *variable* size of group taken from the original file and written to the compressed file.

Images transmitted over the world wide web are an excellent example of why data compression is important. Suppose we need to download a digitized color photograph over a computer's 33.6 kbps modem. If the image is not compressed (a *TIFF* file, for example), it will contain about 600 kbytes of data. If it has been compressed using a *lossless* technique (such as used in the *GIF* format), it will be about one-half this size, or 300 kbytes. If *lossy* compression has been used (a JPEG file), it will be about 50 kbytes. The point is, the download times for these three equivalent files are 142 seconds, 71 seconds, and 12 seconds, respectively. That's a big difference! JPEG is the best choice for digitized photographs, while GIF is used with *drawn* images, such as company logos that have large areas of a single color.

Our second way of classifying data compression methods is shown in Table 27-1b. Most data compression programs operate by taking a group of data from the original file, compressing it in some way, and then writing the compressed group to the output file. For instance, one of the techniques in this table is **CS&Q**, short for **coarser sampling and/or quantization**. Suppose we are compressing a digitized waveform, such as an audio signal that has been digitized to 12 bits. We might read two adjacent samples from the original file (24 bits), discard one of the sample completely, discard the least significant 4 bits from the other sample, and then write the remaining 8 bits to the output file. With 24 bits in and 8 bits out, we have implemented a 3:1 compression ratio using a lossy algorithm. While this is rather crude in itself, it is very effective when used with a technique called *transform compression*. As we will discuss later, this is the basis of JPEG.

Table 27-1b shows CS&Q to be a fixed-input fixed-output scheme. That is, a fixed number of bits are read from the input file and a smaller fixed number of bits are written to the output file. Other compression methods allow a variable number of bits to be read or written. As you go through the description of each of these compression methods, refer back to this table to understand how it fits into this classification scheme. Why are JPEG and MPEG not listed in this table? These are composite algorithms that combine many of the other techniques. They are too sophisticated to be classified into these simple categories.

Run-Length Encoding

Data files frequently contain the same character repeated many times in a row. For example, text files use multiple spaces to separate sentences, indent paragraphs, format tables & charts, etc. Digitized *signals* can also have runs of the same value, indicating that the signal is not changing. For instance, an image of the nighttime sky would contain long runs of the character or characters representing the black background. Likewise, digitized music might have a long run of zeros between songs. Run-length encoding is a simple method of compressing these types of files.

Figure 27-1 illustrates run-length encoding for a data sequence having frequent runs of *zeros*. Each time a zero is encountered in the input data, *two* values are written to the output file. The first of these values is a zero, a flag to indicate that run-length compression is beginning. The second value is the number of zeros in the run. If the average run-length is longer than two, compression will take place. On the other hand, many single zeros in the data can make the encoded file larger than the original.

Many different run-length schemes have been developed. For example, the input data can be treated as individual bytes, or groups of bytes that represent something more elaborate, such as floating point numbers.   Run-length encoding can be used on only *one* of the characters (as with the *zero* above), *several* of the characters, or *all* of the characters.

A good example of a generalized run-length scheme is **PackBits**, created for Macintosh users. Each byte (eight bits) from the input file is replaced by nine bits in the compressed file.   The added ninth bit is interpreted as the *sign* of the number. That is, each character read from the input file is between 0 to 255, while each character written to the encoded file is between -255 and 255. To understand how this is used, consider the input file: 1,2,3,4,2,2,2,2,4, and the compressed file generated by the PackBits algorithm: 1,2,3,4,2,-3,4. The compression program simply transfers each number from the input file to the compressed file, with the exception of the run: 2,2,2,2. This is represented in the compressed file by the two numbers: 2,-3. The first number ("2") indicates what character the run consists of. The second number ("-3") indicates the number of characters in the run, found by taking the absolute value and adding one. For instance, 4,-2 means 4,4,4; 21,-4 means 21,21,21,21,21, etc.

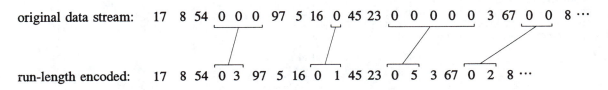

FIGURE 27-1
Example of run-length encoding. Each run of zeros is replaced by two characters in the compressed file: a zero to indicate that compression is occurring, followed by the number of zeros in the run.

An inconvenience with PackBits is that the nine bits must be reformatted into the standard eight bit bytes used in computer storage and transmission. A useful modification to this scheme can be made when the input is restricted to be ASCII text. As shown in Table 27-2, each ASCII character is usually stored as a full byte (eight bits), but really only uses *seven* of the bits to identify the character. In other words, the values 127 through 255 are not defined with any standardized meaning, and do not need to be stored or transmitted. This allows the eighth bit to indicate if run-length encoding is in progress.

Huffman Encoding

This method is named after D.A. Huffman, who developed the procedure in the 1950s. Figure 27-2 shows a histogram of the byte values from a large ASCII file. More than 96% of this file consists of only 31 characters: the lower case letters, the space, the comma, the period, and the carriage return. This observation can be used to make an appropriate compression scheme for this file. To start, we will assign each of these 31 common characters a five bit binary code: 00000 = "a", 00001 = "b", 00010 = "c", etc. This allows 96% of the file to be reduced in size by 5/8. The last of the five bit codes, 11111, will be a flag indicating that the character being transmitted is not one of the 31 common characters. The next eight bits in the file indicate what the character is, according to the standard ASCII assignment. This results in 4% of the characters in the input file requiring 5+8=13 bits. The idea is to assign frequently used characters fewer bits,

TABLE 27-2
ASCII codes. This is a long established standard for allowing letters and numbers to be represented in digital form. Each printable character is assigned a number between 32 and 127, while the numbers between 0 and 31 are used for various control actions. Even though only 128 codes are defined, ASCII characters are usually stored as a full byte (8 bits). The undefined values (128 to 255) are often used for Greek letters, math symbols, and various geometric patterns; however, this is not standardized. Many of the control characters (0 to 31) are based on older communications networks, and are not applicable to computer technology.

#		#		#		#	
0	null	32	space	64	@	96	`
1	start heading	33	!	65	A	97	a
2	start of text	34	"	66	B	98	b
3	end of text	35	#	67	C	99	c
4	end of xmit	36	$	68	D	100	d
5	enquiry	37	%	69	E	101	e
6	acknowledge	38	&	70	F	102	f
7	bell, beep	39	'	71	G	103	g
8	backspace	40	(	72	H	104	h
9	horz. tab	41	)	73	I	105	i
10	line feed	42	*	74	J	106	j
11	vert. tab, home	43	+	75	K	107	k
12	form feed, cls	44	,	76	L	108	l
13	carriage return	45	-	77	M	109	m
14	shift out	46	.	78	N	110	n
15	shift in	47	/	79	O	111	o
16	data line esc	48	0	80	P	112	p
17	device control 1	49	1	81	Q	113	q
18	device control 2	50	2	82	R	114	r
19	device control 3	51	3	83	S	115	s
20	device control 4	52	4	84	T	116	t
21	negative ack.	53	5	85	U	117	r
22	synch. idle	54	6	86	V	118	v
23	end xmit block	55	7	87	W	119	w
24	cancel	56	8	88	X	120	x
25	end of medium	57	9	89	Y	121	y
26	substitute	58	:	90	Z	122	z
27	escape	59	;	91	[	123	{
28	file separator	60	<	92	\	124	|
29	group separator	61	=	93	]	125	}
30	record separator	62	>	94	^	126	~
31	unit separator	63	?	95	_	127	del

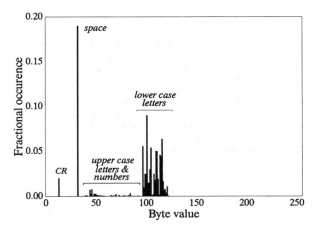

FIGURE 27-2
Histogram of text. This is a histogram of the ASCII values from a chapter in this book. The most common characters are the lower case letters, the space and the carriage return.

and seldom used characters more bits. In this example, the *average* number of bits required per original character is: $0.96 \times 5 + 0.04 \times 13 = 5.32$. In other words, an overall compression ratio of: 8 bits/5.32 bits, or about 1.5:1.

Huffman encoding takes this idea to the extreme. Characters that occur most often, such the space and period, may be assigned as few as one or two bits. Infrequently used characters, such as: !, @, #, $ and %, may require a dozen or more bits. In mathematical terms, the optimal situation is reached when the number of bits used for each character is proportional to the logarithm of the character's probability of occurrence.

A clever feature of Huffman encoding is how the variable length codes can be packed together. Imagine receiving a serial data stream of ones and zeros. If each character is represented by eight bits, you can directly separate one character from the next by breaking off 8 bit chunks. Now consider a Huffman encoded data stream, where each character can have a variable number of bits. How do you separate one character from the next? The answer lies in the proper selection of the Huffman codes that enable the correct separation. An example will illustrate how this works.

Figure 27-3 shows a simplified Huffman encoding scheme. The characters *A* through *G* occur in the original data stream with the probabilities shown. Since the character *A* is the most common, we will represent it with a single bit, the code: 1. The next most common character, *B*, receives two bits, the code: 01. This continues to the least frequent character, *G*, being assigned six bits, 000011. As shown in this illustration, the variable length codes are resorted into eight bit groups, the standard for computer use.

When uncompression occurs, all the eight bit groups are placed end-to-end to form a long serial string of ones and zeros. Look closely at the encoding table of Fig. 27-3, and notice how each code consists of two parts: a number of zeros before a *one*, and an optional binary code after the *one*. This allows the binary data stream to be separated into codes without the need for delimiters or other marker between the codes. The uncompression program

FIGURE 27-3
Huffman encoding. The encoding table assigns each of the seven letters used in this example a variable length binary code, based on its probability of occurrence. The original data stream composed of these 7 characters is translated by this table into the Huffman encoded data. Since each of the Huffman codes is a different length, the binary data need to be regrouped into standard 8 bit bytes for storage and transmission.

Example Encoding Table

letter	probability	Huffman code
A	. 54	
B	. 0	0
C	.0 2	00 0
D	.063	00
E	.059	000
F	.0 5	0000 0
G	.0	0000

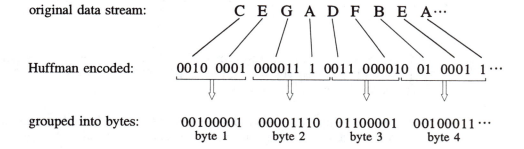

original data stream: C E G A D F B E A ···

Huffman encoded: 0010 0001 000011 1 0011 000010 01 0001 1 ···

grouped into bytes:

00100001	00001110	01100001	00100011 ···
byte 1	byte 2	byte 3	byte 4

looks at the stream of ones and zeros until a valid code is formed, and then starting over looking for the next character. The way that the codes are formed insures that no ambiguity exists in the separation.

A more sophisticated version of the Huffman approach is called **arithmetic encoding**. In this scheme, *sequences* of characters are represented by individual codes, according to their probability of occurrence. This has the advantage of better data compression, say 5-10%. Run-length encoding followed by either Huffman or arithmetic encoding is also a common strategy. As you might expect, these types of algorithms are very complicated, and usually left to data compression specialists.

To implement Huffman or arithmetic encoding, the compression and un-compression algorithms must agree on the binary codes used to represent each character (or groups of characters). This can be handled in one of two ways. The simplest is to use a predefined encoding table that is always the same, regardless of the information being compressed. More complex schemes use encoding optimized for the particular data being used. This requires that the encoding table be included in the compressed file for use by the uncompression program. Both methods are common.

Delta Encoding

In science, engineering, and mathematics, the Greek letter *delta* (Δ) is used to denote the *change* in a variable. The term *delta encoding*, refers to original data stream: 17 19 24 24 24 21 15 10 89 95 96 96 96 95 94 94 95 93 90 87 86 86 ⋯

move *delta* *delta* *delta*   ⋯

delta encoded: 17 2 5 0 0 -3 -6 -5 79 6 1 0 0 -1 -1 0 1 -2 -3 -3 -1 0 ⋯

FIGURE 27-4
Example of delta encoding. The first value in the encoded file is the same as the first value in the original
file. Thereafter, each sample in the encoded file is the difference between the current and last sample in
the original file.

several techniques that store data as the *difference* between successive
samples (or characters), rather than directly storing the samples themselves.
Figure 27-4 shows an example of how this is done. The first value in the
delta encoded file is the same as the first value in the original data. All the
following values in the encoded file are equal to the difference (delta)
between the corresponding value in the input file, and the *previous* value in
the input file.

Delta encoding can be used for data compression when the values in the
original data are *smooth*, that is, there is typically only a small change
between adjacent values. This is not the case for ASCII text and executable
code; however, it is very common when the file represents a *signal*. For
instance, Fig. 27-5a shows a segment of an audio signal, digitized to 8 bits,
with each sample between -127 and 127. Figure 27-5b shows the delta
encoded version of this signal. The key feature is that the delta encoded
signal has a *lower amplitude* than the original signal. In other words, delta
encoding has increased the probability that each sample's value will be near
zero, and decreased the probability that it will be far from zero. This
uneven probability is just the thing that Huffman encoding needs to
operate. If the original signal is not changing, or is changing in a straight
line, delta encoding will result in runs of samples having the same value.

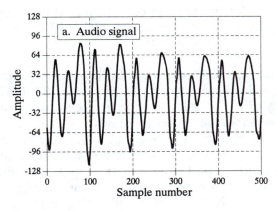
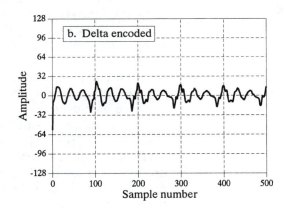

FIGURE 27-5
Example of delta encoding. Figure (a) is an audio signal digitized to 8 bits. Figure (b) shows the delta
encoded version of this signal. Delta encoding is useful for data compression if the signal being encoded
varies slowly from sample-to-sample.

This is what run-length encoding requires. Correspondingly, delta encoding followed by Huffman and/or run-length encoding is a common strategy for compressing signals.

The idea used in delta encoding can be expanded into a more complicated technique called **Linear Predictive Coding**, or **LPC**. To understand LPC, imagine that the first 99 samples from the input signal have been encoded, and we are about to work on sample number 100. We then ask ourselves: based on the first 99 samples, what is the most likely value for sample 100? In delta encoding, the answer is that the most likely value for sample 100 is the same as the previous value, sample 99. This expected value is used as a reference to encode sample 100. That is, the *difference* between the sample and the expectation is placed in the encoded file. LPC expands on this by making a better guess at what the most probable value is. This is done by looking at the last several samples, rather than just the last sample. The algorithms used by LPC are similar to recursive filters, making use of the z-transform and other intensively mathematical techniques.

LZW Compression

LZW compression is named after its developers, A. Lempel and J. Ziv, with later modifications by Terry A. Welch. It is the foremost technique for general purpose data compression due to its simplicity and versatility. Typically, you can expect LZW to compress text, executable code, and similar data files to about one-half their original size. LZW also performs well when presented with extremely redundant data files, such as tabulated numbers, computer source code, and acquired signals. Compression ratios of 5:1 are common for these cases. LZW is the basis of several personal computer utilities that claim to *"double the capacity of your hard drive."*

LZW compression is always used in GIF image files, and offered as an option in TIFF and PostScript. LZW compression is protected under U.S. patent number 4,558,302, granted December 10, 1985 to Sperry Corporation (now the Unisys Corporation). For information on commercial licensing, contact: Welch Licensing Department, Law Department, M/SC2SW1, Unisys Corporation, Blue Bell, Pennsylvania, 19424-0001.

LZW compression uses a **code table**, as illustrated in Fig. 27-6. A common choice is to provide 4096 entries in the table. In this case, the LZW encoded data consists entirely of 12 bit codes, each referring to one of the entries in the code table. Uncompression is achieved by taking each code from the compressed file, and translating it through the code table to find what character or characters it represents. Codes 0-255 in the code table are always assigned to represent single bytes from the input file. For example, if only these first 256 codes were used, each byte in the original file would be converted into 12 bits in the LZW encoded file, resulting in a 50% larger file size. During uncompression, each 12 bit code would be translated via the code table back into the single bytes. Of course, this wouldn't be a useful situation.

Example Code Table

code number	translation
0000	0
0001	1
⋮	⋮
0254	254
0255	255
0256	145 201 4
0257	243 245
⋮	⋮
4095	xxx xxx xxx

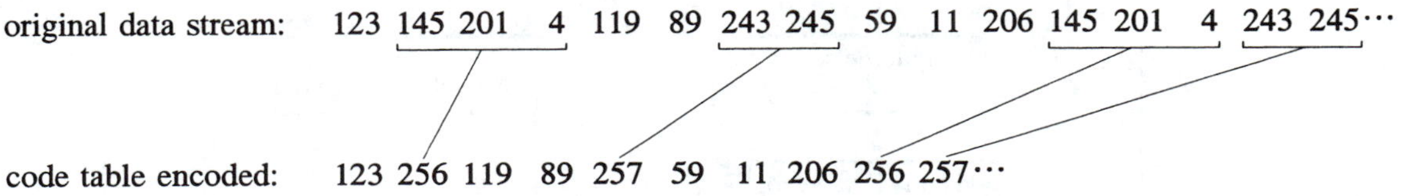

identical code (brace over 0000–0255)
unique code (brace over 0256–4095)

FIGURE 27-6

Example of code table compression. This is the basis of the popular LZW compression method. Encoding occurs by identifying sequences of bytes in the original file that exist in the code table. The 12 bit code representing the sequence is placed in the compressed file instead of the sequence. The first 256 entries in the table correspond to the single byte values, 0 to 255, while the remaining entries correspond to *sequences* of bytes. The LZW algorithm is an efficient way of generating the code table based on the particular data being compressed. (The code table in this figure is a simplified example, not one actually generated by the LZW algorithm).

original data stream: 123 145 201 4 119 89 243 245 59 11 206 145 201 4 243 245···

code table encoded: 123 256 119 89 257 59 11 206 256 257···

The LZW method achieves compression by using codes 256 through 4095 to represent *sequences* of bytes. For example, code 523 may represent the sequence of three bytes: 231 124 234. Each time the compression algorithm encounters this sequence in the input file, code 523 is placed in the encoded file. During uncompression, code 523 is translated via the code table to recreate the true 3 byte sequence. The longer the sequence assigned to a single code, and the more often the sequence is repeated, the higher the compression achieved.

Although this is a simple approach, there are two major obstacles that need to be overcome: (1) how to determine what sequences should be in the code table, and (2) how to provide the uncompression program the same code table used by the compression program. The LZW algorithm exquisitely solves both these problems.

When the LZW program starts to encode a file, the code table contains only the first 256 entries, with the remainder of the table being blank. This means that the first codes going into the compressed file are simply the single bytes from the input file being converted to 12 bits. As the encoding continues, the LZW algorithm identifies repeated sequences in the data, and adds them to the code table. Compression starts the second time a sequence is encountered. The key point is that a sequence from the input file is not added to the code table until it has already been placed in the compressed file as individual characters (codes 0 to 255). This is important because it allows the uncompression program to *reconstruct* the code table directly from the compressed data, without having to transmit the code table separately.

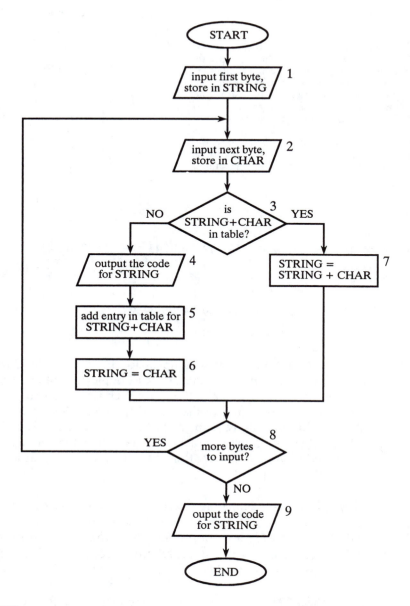

FIGURE 27-7
LZW compression flowchart. The variable, *CHAR*, is a single byte. The variable, *STRING*, is a variable length sequence of bytes. Data are read from the input file (box 1 & 2) as single bytes, and written to the compressed file (box 4) as 12 bit codes. Table 27-3 shows an example of this algorithm.

Figure 27-7 shows a flowchart for LZW compression. Table 27-3 provides the step-by-step details for an example input file consisting of 45 bytes, the ASCII text string: *the/rain/in/Spain/falls/mainly/on/the/plain*. When we say that the LZW algorithm reads the character "a" from the input file, we mean it reads the value: 01100001 (97 expressed in 8 bits), where 97 is "a" in ASCII. When we say it writes the character "a" to the encoded file, we mean it writes: 000001100001 (97 expressed in 12 bits).

	CHAR	STRING + CHAR	In Table?	Output	Add to Table	New STRING	Comments
1	t	t				t	first character- no action
2	h	th	no	t	256 = th	h	
3	e	he	no	h	257 = he	e	
4	/	e/	no	e	258 = e/	/	
5	r	/r	no	/	259 = /r	r	
6	a	ra	no	r	260 = ra	a	
7	i	ai	no	a	261 = ai	i	
8	n	in	no	i	262 = in	n	
9	/	n/	no	n	263 = n/	/	
10	i	/i	no	/	264 = /i	i	
11	n	in	yes (262)			in	first match found
12	/	in/	no	262	265 = in/	/	
13	S	/S	no	/	266 = /S	S	
14	p	Sp	no	S	267 = Sp	p	
15	a	pa	no	p	268 = pa	a	
16	i	ai	yes (261)			ai	matches *ai*, *ain* not in table yet
17	n	ain	no	261	269 = ain	n	*ain* added to table
18	/	n/	yes (263)			n/	
19	f	n/f	no	263	270 = n/f	f	
20	a	fa	no	f	271 = fa	a	
21	l	al	no	a	272 = al	l	
22	l	ll	no	l	273 = ll	l	
23	s	ls	no	l	274 = ls	s	
24	/	s/	no	s	275 = s/	/	
25	m	/m	no	/	276 = /m	m	
26	a	ma	no	m	277 = ma	a	
27	i	ai	yes (261)			ai	matches *ai*
28	n	ain	yes (269)			ain	matches longer string, *ain*
29	l	ainl	no	269	278 = ainl	l	
30	y	ly	no	l	279 = ly	y	
31	/	y/	no	y	280 = y/	/	
32	o	/o	no	/	281 = /o	o	
33	n	on	no	o	282 = on	n	
34	/	n/	yes (263)			n/	
35	t	n/t	no	263	283 = n/t	t	
36	h	th	yes (256)			th	matches *th*, *the* not in table yet
37	e	the	no	256	284 = the	e	*the* added to table
38	/	e/	yes			e/	
39	p	e/p	no	258	285 = e/p	p	
40	l	pl	no	p	286 = pl	l	
41	a	la	no	l	287 = la	a	
42	i	ai	yes (261)			ai	matches *ai*
43	n	ain	yes (269)			ain	matches longer string *ain*
44	/	ain/	no	269	288 = ain/	/	
45	EOF	/		/			end of file, output *STRING*

TABLE 27-3
LZW example. This shows the compression of the phrase: *the/rain/in/Spain/falls/mainly/on/the/plain/*.

The compression algorithm uses two variables: *CHAR* and *STRING*. The variable, *CHAR*, holds a single character, i.e., a single byte value between 0 and 255. The variable, *STRING*, is a variable length string, i.e., a group of one or more characters, with each character being a single byte. In box 1 of Fig. 27-7, the program starts by taking the first byte from the input file, and placing it in the variable, *STRING*. Table 27-3 shows this action in line 1. This is followed by the algorithm looping for each additional byte in the input file, controlled in the flow diagram by box 8. Each time a byte is read from the input file (box 2), it is stored in the variable, *CHAR*. The data table is then searched to determine if the concatenation of the two variables, *STRING+CHAR*, has already been assigned a code (box 3).

If a match in the code table is *not* found, three actions are taken, as shown in boxes 4, 5 & 6. In box 4, the 12 bit code corresponding to the contents of the variable, *STRING*, is written to the compressed file. In box 5, a new code is created in the table for the concatenation of *STRING+CHAR*. In box 6, the variable, *STRING*, takes the value of the variable, *CHAR*. An example of these actions is shown in lines 2 through 10 in Table 27-3, for the first 10 bytes of the example file.

When a match in the code table *is* found (box 3), the concatenation of *STRING+CHAR* is stored in the variable, *STRING*, without any other action taking place (box 7). That is, if a matching sequence is found in the table, no action should be taken before determining if there is a *longer* matching sequence also in the table. An example of this is shown in line 11, where the sequence: *STRING+CHAR = in*, is identified as already having a code in the table. In line 12, the next character from the input file, /, is added to the sequence, and the code table is searched for: *in/*. Since this longer sequence is not in the table, the program *adds* it to the table, outputs the code for the shorter sequence that *is* in the table (code 262), and starts over searching for sequences beginning with the character, /. This flow of events is continued until there are no more characters in the input file. The program is wrapped up with the code corresponding to the current value of *STRING* being written to the compressed file (as illustrated in box 9 of Fig. 27-7 and line 45 of Table 27-3).

A flowchart of the LZW uncompression algorithm is shown in Fig. 27-8. Each code is read from the compressed file and compared to the code table to provide the translation. As each code is processed in this manner, the code table is updated so that it continually matches the one used during the compression. However, there is a small complication in the uncompression routine. There are certain combinations of data that result in the uncompression algorithm receiving a code that does not yet exist in its code table. This contingency is handled in boxes 4,5 & 6.

Only a few dozen lines of code are required for the most elementary LZW programs. The real difficulty lies in the efficient management of the code table. The brute force approach results in large memory requirements and a slow program execution. Several tricks are used in commercial LZW programs to improve their performance. For instance, the memory problem

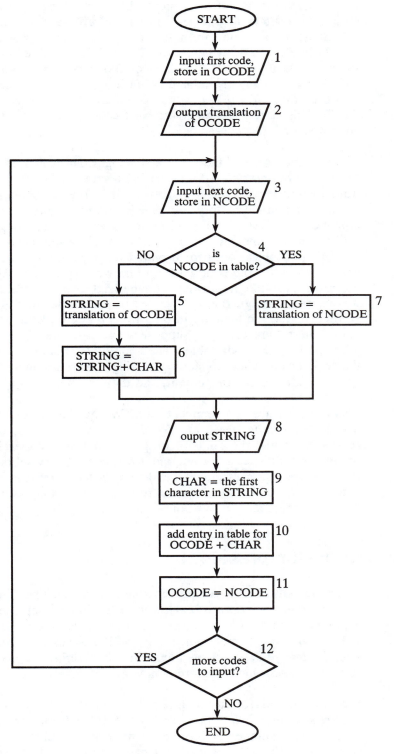

FIGURE 27-8
LZW uncompression flowchart. The variables, *OCODE* and *NCODE* (oldcode and newcode), hold the
12 bit codes from the compressed file, *CHAR* holds a single byte, STRING holds a string of bytes.

arises because it is not know beforehand how long each of the character strings for each code will be. Most LZW programs handle this by taking advantage of the redundant nature of the code table. For example, look at line 29 in Table 27-3, where code 278 is defined to be *ainl*.   Rather than storing these four bytes, code 278 could be stored as: *code 269 + l*, where code 269 was previously defined as *ain* in line 17. Likewise, code 269 would be stored as: *code 261 + n*, where code 261 was previously defined as *ai* in line 7. This pattern always holds: every code can be expressed as a previous code plus one new character.

The execution time of the compression algorithm is limited by searching the code table to determine if a match is present. As an analogy, imagine you want to find if a friend's name is listed in the telephone directory. The catch is, the only directory you have is arranged by telephone number, not alphabetical order. This requires you to search page after page trying to find the name you want. This inefficient situation is exactly the same as searching all 4096 codes for a match to a specific character string. The answer: organize the code table so that what you are looking for tells you where to look (like a partially alphabetized telephone directory). In other words, don't assign the 4096 codes to sequential locations in memory. Rather, divide the memory into sections based on what sequences will be stored there. For example, suppose we want to find if the sequence: *code 329 + x,* is in the code table. The code table should be organized so that the "*x*" indicates where to starting looking. There are many schemes for this type of code table management, and they can become quite complicated.

This brings up the last comment on LZW and similar compression schemes: *it is a very competitive field*. While the basics of data compression are relatively simple, the kinds of programs sold as commercial products are extremely sophisticated. Companies make money by selling you programs that perform compression, and jealously protect their trade-secrets through patents and the like. Don't expect to achieve the same level of performance as these programs in a few hours work.

JPEG (Transform Compression)

Many methods of lossy compression have been developed; however, a family of techniques called *transform compression* has proven the most valuable. The best example of transform compression is embodied in the popular JPEG standard of image encoding. JPEG is named after its origin, the *Joint Photographers Experts Group*. We will describe the operation of JPEG to illustrate how lossy compression works.

We have already discussed a simple method of lossy data compression, *coarser sampling and/or quantization* (CS&Q in Table 27-1). This involves reducing the number of bits per sample or entirely discard some of the samples. Both these procedures have the desired effect: the data file becomes smaller at the expense of signal quality. As you might expect, these simple methods do not work very well.

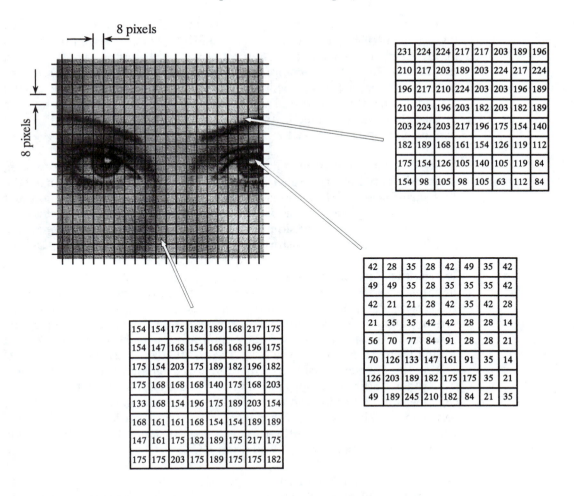

FIGURE 27-9
JPEG image division. JPEG transform compression starts by breaking the image into 8×8 groups, each containing 64 pixels. Three of these 8×8 groups are enlarged in this figure, showing the values of the individual pixels, a single byte value between 0 and 255.

Transform compression is based on a simple premise: when the signal is passed through the Fourier (or other) transform, the resulting data values will no longer be equal in their information carrying roles. In particular, the low frequency components of a signal are more important than the high frequency components. Removing 50% of the bits from the high frequency components might remove, say, only 5% of the encoded information.

As shown in Fig. 27-9, JPEG compression starts by breaking the image into 8×8 pixel groups. The full JPEG algorithm can accept a wide range of bits per pixel, including the use of color information. In this example, each pixel is a single byte, a grayscale value between 0 and 255. These 8×8 pixel groups are treated independently during compression. That is, each group is initially represented by 64 bytes. After transforming and removing data, each group is represented by, say, 2 to 20 bytes. During uncompression, the inverse transform is taken of the 2 to 20 bytes to create an approximation of the original 8×8 group. These approximated groups are then fitted together to form the uncompressed image. Why use 8×8 pixel groups instead of, for instance, 16×16? The 8×8 grouping was based on the maximum size that integrated circuit technology could handle at the time the standard was developed. In any event, the 8×8 size works well, and it may or may not be changed in the future.

Many different transforms have been investigated for data compression, some of them invented specifically for this purpose. For instance, the *Karhunen-Loeve* transform provides the best possible compression ratio, but is difficult to implement. The *Fourier transform* is easy to use, but does not provide adequate compression. After much competition, the winner is a relative of the Fourier transform, the **Discrete Cosine Transform (DCT)**.

Just as the Fourier transform uses sine and cosine waves to represent a signal, the DCT only uses cosine waves. There are several versions of the DCT, with slight differences in their mathematics. As an example of one version, imagine a 129 point signal, running from sample 0 to sample 128. Now, make this a 256 point signal by duplicating samples 1 through 127 and adding them as samples 255 to 130. That is: 0, 1, 2, ⋯, 127, 128, 127, ⋯, 2, 1. Taking the Fourier transform of this 256 point signal results in a frequency spectrum of 129 points, spread between 0 and 128. Since the time domain signal was forced to be symmetrical, the spectrum's imaginary part will be composed of all zeros. In other words, we started with a 129 point time domain signal, and ended with a frequency spectrum of 129 points, each the amplitude of a cosine wave. Voila, the DCT!

When the DCT is taken of an 8×8 group, it results in an 8×8 spectrum. In other words, 64 numbers are changed into 64 other numbers. All these values are *real*; there is no complex mathematics here. Just as in Fourier analysis, each value in the spectrum is the amplitude of a **basis function**. Figure 27-10 shows 6 of the 64 basis functions used in an 8×8 DCT, according to where the amplitude sits in the spectrum. The 8×8 DCT basis functions are given by:

EQUATION 27-1
DCT basis functions. The variables *x* & *y* are the indexes in the spatial domain, and *u* & *v* are the indexes in the frequency spectrum. This is for an 8×8 DCT, making all the indexes run from 0 to 7.

$$b[x,y] = \cos\left[\frac{(2x+1)\,u\pi}{16}\right] \cos\left[\frac{(2y+1)\,v\pi}{16}\right]$$

The low frequencies reside in the upper-left corner of the spectrum, while the high frequencies are in the lower-right. The DC component is at [0,0], the upper-left most value. The basis function for [0,1] is one-half cycle of a cosine wave in one direction, and a constant value in the other. The basis function for [1,0] is similar, just rotated by 90°.

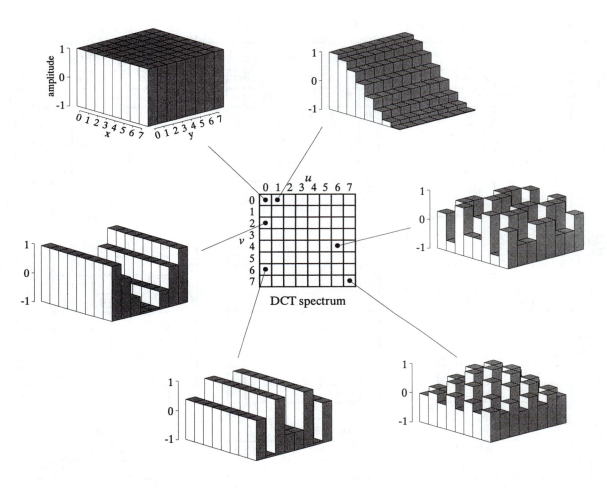

FIGURE 27-10
The DCT basis functions. The DCT spectrum consists of an 8×8 array, with each element in the array being an amplitude of one of the 64 basis functions. Six of these basis functions are shown here, referenced to where the corresponding amplitude resides.

The DCT calculates the spectrum by *correlating* the 8×8 pixel group with each of the basis functions. That is, each spectral value is found by multiplying the appropriate basis function by the 8×8 pixel group, and then summing the products. Two adjustments are then needed to finish the DCT calculation (just as with the Fourier transform). First, divide the 15 spectral values in row 0 and column 0 by *two*. Second, divide all 64 values in the spectrum by 16. The inverse DCT is calculated by assigning each of the amplitudes in the spectrum to the proper basis function, and summing to recreate the spatial domain. No extra steps are required. These are exactly the same concepts as in Fourier analysis, just with different basis functions.

Figure 27-11 illustrates JPEG encoding for the three 8×8 groups identified in Fig. 27-9. The left column, Figs. a, b & c, show the original pixel values. The center column, Figs. d, e & f, show the DCT spectra of these groups.

Original Group

a. Eyebrow

231	224	224	217	217	203	189	196
210	217	203	189	203	224	217	224
196	217	210	224	203	203	196	189
210	203	196	203	182	203	182	189
203	224	203	217	196	175	154	140
182	189	168	161	154	126	119	112
175	154	126	105	140	105	119	84
154	98	105	98	105	63	112	84

b. Eye

42	28	35	28	42	49	35	42
49	49	35	28	35	35	35	42
42	21	21	28	42	35	42	28
21	35	35	42	42	28	28	14
56	70	77	84	91	28	28	21
70	126	133	147	161	91	35	14
126	203	189	182	175	175	35	21
49	189	245	210	182	84	21	35

c. Nose

154	154	175	182	189	168	217	175
154	147	168	154	168	168	196	175
175	154	203	175	189	182	196	182
175	168	168	168	140	175	168	203
133	168	154	196	175	189	203	154
168	161	161	168	154	154	189	189
147	161	175	182	189	175	217	175
175	175	203	175	189	175	175	182

DCT Spectrum

d. Eyebrow spectrum

174	19	0	3	1	0	-3	1
52	-13	-3	-4	-4	-4	5	-8
-18	-4	8	3	3	2	0	9
5	12	-4	0	0	-5	-1	0
1	2	-2	-1	4	4	2	0
-1	2	1	3	0	0	1	1
-2	5	-5	-5	3	2	-1	-1
3	5	-7	0	0	0	-4	0

e. Eye spectrum

70	24	-28	-4	-2	-10	-1	0
-53	-35	43	13	7	13	1	3
23	9	-10	-8	-7	-6	5	-3
6	2	-2	8	2	-1	0	-1
-10	-2	-1	-12	2	1	-1	4
3	0	0	11	-4	-1	5	6
-3	-5	-5	-4	3	2	-3	5
3	0	4	5	1	2	1	0

f. Nose spectrum

174	-11	-2	-3	-3	6	-3	4
-2	-3	1	2	0	3	1	2
3	0	-4	0	0	0	-1	9
-4	-6	-2	1	-1	4	-10	-3
1	2	-2	0	0	-2	0	-5
3	-1	3	-2	2	1	1	0
3	5	2	-2	3	0	4	3
4	-3	-13	3	-4	3	-5	3

Quantization Error

g. Using 10 bits

0	0	0	0	-1	0	0	0
-1	0	0	0	0	0	0	-1
0	0	0	0	0	0	0	0
0	0	0	0	0	0	0	0
0	0	0	0	0	0	0	0
0	0	1	0	0	0	-1	0
0	0	0	0	0	0	0	0
0	0	0	0	0	0	0	0

h. Using 8 bits

0	-3	-1	-1	1	0	0	-1
1	0	-1	-1	0	0	0	-1
-1	-2	1	0	-2	0	-2	-2
-1	-2	-1	2	0	2	0	1
0	-2	1	0	0	1	0	0
0	-4	-1	0	1	0	0	0
0	-2	0	1	-1	-1	1	-1
-1	-3	1	1	1	-3	-2	-1

i. Using 5 bits

-13	-7	1	4	0	0	10	-2
-22	6	-13	5	-5	2	-2	-13
-9	-15	0	-17	-8	8	12	25
-9	16	1	9	1	-5	-5	13
-20	-3	-13	-16	-19	-1	-4	-22
-11	6	-8	16	-9	-3	-7	6
-14	10	-9	4	-15	3	3	-4
-13	19	12	9	18	5	-5	10

FIGURE 27-11

Example of JPEG encoding. The left column shows three 8×8 pixel groups, the same ones shown in Fig. 27-9. The center column shows the DCT spectra of these three groups. The third column shows the error in the uncompressed pixel values resulting from using a finite number of bits to represent the spectrum.

The right column, Figs. g, h & i, shows the effect of reducing the number of bits used to represent each component in the frequency spectrum. For instance, (g) is formed by truncating each of the samples in (d) to ten bits, taking the inverse DCT, and then subtracting the reconstructed image from the original. Likewise, (h) and (i) are formed by truncating each sample in the spectrum to eight and five bits, respectively. As expected, the error in the reconstruction increases as fewer bits are used to represent the data. As an example of this bit truncation, the spectra shown in the center column are represented with 8 bits per spectral value, arranged as 0 to 255 for the DC component, and -127 to 127 for the other values.

The second method of compressing the frequency domain is to discard some of the 64 spectral values. As shown by the spectra in Fig. 27-11, nearly all of the signal is contained in the low frequency components. This means the highest frequency components can be eliminated, while only degrading the signal a small amount. Figure 27-12 shows an example of the image distortion that occurs when various numbers of the high frequency components are deleted. The 8×8 group used in this example is the *eye* image of Fig. 27-10. Figure (d) shows the correct reconstruction using all 64 spectral values. The remaining figures show the reconstruction using the indicated number of lowest frequency coefficients. As illustrated in (c), even removing three-fourths of the highest frequency components produces little error in the reconstruction. Even better, the error that does occur looks very much like random noise.

JPEG is good example of how several data compression schemes can be combined for greater effectiveness. The entire JPEG procedure is outlined in the following steps. First, the image is broken into the 8×8 groups. Second, the DCT is taken of each group. Third, each 8×8 spectrum is compressed by the above methods: reducing the number of bits and eliminating some of the components. This takes place in a single step, controlled by a **quantization table**. Two examples of quantization tables are shown in Fig. 27-13. Each value in the spectrum is divided by the matching value in the quantization table, and the result rounded to the nearest integer. For instance, the upper-left value of the quantization table is *one*,

a. 3 coefficients

b. 6 coefficients

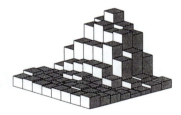

c. 15 coefficients

FIGURE 27-12
Example of JPEG reconstruction. The 8×8 pixel group used in this example is the *eye* in Fig. 27-9. As shown, less than 1/4 of the 64 values are needed to achieve a good approximation to the correct image.

d. 64 coefficients
(correct image)

a. Low compression

1	1	1	1	1	2	2	4
1	1	1	1	1	2	2	4
1	1	1	1	2	2	2	4
1	1	1	1	2	2	4	8
1	1	2	2	2	2	4	8
2	2	2	2	2	4	8	8
2	2	2	4	4	8	8	16
4	4	4	4	8	8	16	16

b. High compression

1	2	4	8	16	32	64	128
2	4	4	8	16	32	64	128
4	4	8	16	32	64	128	128
8	8	16	32	64	128	128	256
16	16	32	64	128	128	256	256
32	32	64	128	128	256	256	256
64	64	128	128	256	256	256	256
128	128	128	256	256	256	256	256

FIGURE 27-13
JPEG quantization tables. These are two example quantization tables that might be used during compression. Each value in the DCT spectrum is divided by the corresponding value in the quantization table, and the result rounded to the nearest integer.

resulting in the DC value being left unchanged. In comparison, the lower-right entry in (a) is 16, meaning that the original range of -127 to 127 is reduced to only -7 to 7. In other words, the value has been reduced in precision from eight bits to four bits. In a more extreme case, the lower-right entry in (b) is 256, completely eliminating the spectral value.

In the fourth step of JPEG encoding, the modified spectrum is converted from an 8×8 array into a linear sequence. The serpentine pattern shown in Figure 27-14 is used for this step, placing all of the high frequency components together at the end of the linear sequence. This groups the *zeros* from the eliminated components into long runs. The fifth step compresses these runs of zeros by run-length encoding. In the sixth step, the sequence is encoded by either Huffman or arithmetic encoding to form the final compressed file.

The amount of compression, and the resulting loss of image quality, can be selected when the JPEG compression program is run. Figure 27-15 shows the type of image distortion resulting from high compression ratios. With the 45:1 compression ratio shown, each of the 8×8 groups is represented by only about 12 bits. Close inspection of this image shows that six of the lowest frequency basis functions are represented to some degree.

FIGURE 27-14
JPEG serial conversion. A serpentine pattern used to convert the 8×8 DCT spectrum into a linear sequence of 64 values. This places all of the high frequency components together, where the large number of zeros can be efficiently compressed with run-length encoding.

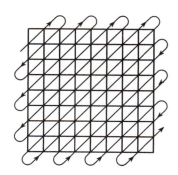

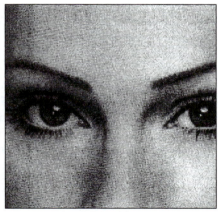

a. Original image

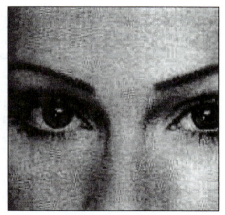

b. With 10:1 compression

FIGURE 27-15
Example of JPEG distortion. Figure (a) shows the original image, while (b) and (c) shows restored images using compression ratios of 10:1 and 45:1, respectively. The high compression ratio used in (c) results in each 8×8 pixel group being represented by less than 12 bits.

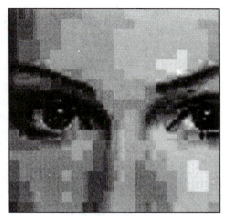

c. With 45:1 compression

Why is the DCT better than the Fourier transform for image compression? The main reason is that the DCT has one-half cycle basis functions, i.e., S[0,1] and S[1,0]. As shown in Fig. 27-10, these gently slope from one side of the array to the other. In comparison, the lowest frequencies in the Fourier transform form *one complete cycle*. Images nearly always contain regions where the brightness is gradually changing over a region. Using a basis function that matches this basic pattern allows for better compression.

MPEG

MPEG is a compression standard for digital video sequences, such as used in computer video and digital television networks. In addition, MPEG also provides for the compression of the sound track associated with the video. The name comes from its originating organization, the *Moving Pictures Experts Group*. If you think JPEG is complicated, MPEG is a nightmare! MPEG is something you buy, not try to write yourself. The future of this technology is to encode the compression and uncompression algorithms directly into integrated circuits. The potential of MPEG is vast. Think of thousands of video channels being carried on a single optical fiber running into your home. This is a key technology of the 21st century.

In addition to reducing the data rate, MPEG has several important features. The movie can be played *forward* or in *reverse*, and at either *normal* or *fast* speed. The encoded information is *random access*, that is, any individual frame in the sequence can be easily displayed as a still picture. This goes along with making the movie *editable*, meaning that short segments from the movie can be encoded only with reference to themselves, not the entire sequence. MPEG is designed to be robust to errors. The last thing you want is for a single bit error to cause a disruption of the movie.

The approach used by MPEG can be divided into two types of compression: *within-the-frame* and *between-frame*. Within-the-frame compression means that individual frames making up the video sequence are encoded as if they were ordinary still images. This compression is preformed using the JPEG standard, with just a few variations. In MPEG terminology, a frame that has been encoded in this way is called an intra-coded or **I-picture**.

Most of the pixels in a video sequence change very little from one frame to the next. Unless the camera is moving, most of the image is composed of a background that remains constant over dozens of frames. MPEG takes advantage of this with a sophisticated form of *delta encoding* to compress the redundant information *between frames*. After compressing one of the frames as an I-picture, MPEG encodes successive frames as predictive-coded or **P-pictures**. That is, only the pixels that have changed since the I-picture are included in the P-picture.

While these two compression schemes form the backbone of MPEG, the actual implementation is immensely more sophisticated than described here. For example, a P-picture can be referenced to an I-picture that has been *shifted*, accounting for motion of objects in the image sequence. There are also bidirectional predictive-coded or **B-pictures**. These are referenced to both a previous and a future I-picture. This handles regions in the image that gradually change over many of frames. The individual frames can also be stored out-of-order in the compressed data to facilitate the proper sequencing of the I, P, and B-pictures. The addition of color and sound makes this all the more complicated.

The main distortion associated with MPEG occurs when large sections of the image change quickly. In effect, a burst of information is needed to keep up with the rapidly changing scenes. If the data rate is fixed, the viewer notices "blocky" patterns when changing from one scene to the next. This can be minimized in networks that transmit multiple video channels simultaneously, such as cable television. The sudden burst of information needed to support a rapidly changing scene in one video channel, is averaged with the modest requirements of the relatively static scenes in the other channels.

CHAPTER

28

Complex Numbers

Complex numbers are an extension of the ordinary numbers used in everyday math. They have the unique property of representing and manipulating *two* variables as a *single* quantity. This fits very naturally with Fourier analysis, where the frequency domain is composed of two signals, the real and the imaginary parts. Complex numbers shorten the equations used in DSP, and enable techniques that are difficult or impossible with real numbers alone. For instance, the Fast Fourier Transform is based on complex numbers. Unfortunately, complex techniques are very mathematical, and it requires a great deal of study and practice to use them effectively. Many scientists and engineers regard complex techniques as the dividing line between DSP as a *tool*, and DSP as a *career*. In this chapter, we look at the mathematics of complex numbers, and elementary ways of using them in science and engineering. The following three chapters discuss important techniques based on complex numbers: the *complex Fourier transform*, the *Laplace transform*, and the *z-transform*. These complex transforms are the heart of theoretical DSP. Get ready, here comes the math!

The Complex Number System

To illustrate complex numbers, consider a child throwing a ball into the air. For example, assume that the ball is thrown straight up, with an initial velocity of 9.8 meters per second. One second after it leaves the child's hand, the ball has reached a height of 4.9 meters, and the acceleration of gravity (9.8 meters per second^2) has reduced its velocity to zero. The ball then accelerates toward the ground, being caught by the child two seconds after it was thrown. From basic physics equations, the height of the ball at any instant of time is given by:

$$h = \frac{-gt^2}{2} + vt$$

where h is the height above the ground (in meters), g is the acceleration of gravity (9.8 meters per second^2), v is the initial velocity (9.8 meters per second), and t is the time (in seconds).

Now, suppose we want to know *when* the ball passes a certain height. Plugging in the known values and solving for t:

$$t = 1 \pm \sqrt{1 - h/4.9}$$

For instance, the ball is at a height of 3 meters *twice*: $t = 0.38$ (going up) and $t = 1.62$ seconds (going down).

As long as we ask reasonable questions, these equations give reasonable answers. But what happens when we ask unreasonable questions? For example: At what time does the ball reach a height of 10 meters? This question has no answer in reality because the ball *never* reaches this height. Nevertheless, plugging the value of $h = 10$ into the above equation gives two answers: $t = 1 + \sqrt{-1.041}$ and $t = 1 - \sqrt{-1.041}$. Both these answers contain the square-root of a negative number, something that does not exist in the world as we know it. This unusual property of polynomial equations was first used by the Italian mathematician Girolamo Cardano (1501-1576). Two centuries later, the great German mathematician Carl Friedrich Gauss (1777-1855) coined the term **complex numbers**, and paved the way for the modern understanding of the field.

Every complex number is the sum of two components: a **real part** and an **imaginary part**. The real part is a **real number**, one of the ordinary numbers we all learned in childhood. The imaginary part is an **imaginary number**, that is, the *square-root of a negative number*. To keep things standardized, the imaginary part is usually reduced to an ordinary number multiplied by the square-root of negative one. As an example, the complex number: $t = 1 + \sqrt{-1.041}$, is first reduced to: $t = 1 + \sqrt{1.041}\sqrt{-1}$, and then to the final form: $t = 1 + 1.02\sqrt{-1}$. The real part of this complex number is 1, while the imaginary part is $1.02\sqrt{-1}$. This notation allows the abstract term, $\sqrt{-1}$, to be given a special symbol. Mathematicians have long used i to denote $\sqrt{-1}$. In comparison, electrical engineers use the symbol, j, because i is used to represent electrical current. Both symbols are common in DSP. In this book the electrical engineering convention, j, will be used.

For example, all the following are valid complex numbers: $1 + 2j$, $1 - 2j$, $-1 + 2j$, $3.14159 + 2.7183j$, $(4/3) + (19/2)j$, etc. All ordinary numbers, such as: 2, 6.34, and -1.414, can be viewed as a complex number with *zero* for the imaginary part, i.e., $2 + 0j$, $6.34 + 0j$, and $-1.414 + 0j$.

Just as real numbers are described as having positions along a number line, complex numbers are represented by locations in a two-dimensional display called the **complex plane**. As shown in Fig. 28-1, the horizontal axis of the

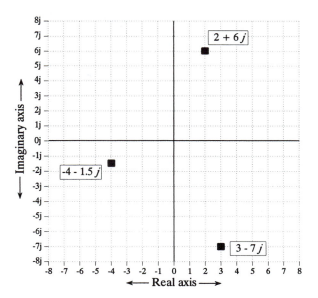

FIGURE 28-1
The complex plane. Every complex number
has a unique location in the complex plane,
as illustrated by the three examples shown
here. The horizontal axis represents the real
part, while the vertical axis represents the
imaginary part.

complex plane is the real part of the complex number, while the vertical axis
is the imaginary part. Since real numbers are those complex numbers that
have an imaginary part equal to zero, the *real number line* is the same as the
x-axis of the complex plane.

In mathematical equations, a complex number is represented by a single
variable, even though it is composed of two parts. For example, the three
complex variables in Fig. 28-1 could be written:

$$A = 2 + 6j$$
$$B = -4 - 1.5j$$
$$C = 3 - 7j$$

where A, B, & C are complex variables. This illustrates a strong *advantage*
and a strong *disadvantage* of using complex numbers. The advantage is the
inherent shorthand of representing two things by a single symbol. The disadvantage is having to remember which variables are complex and which
variables are ordinary numbers.

The mathematical notation for separating a complex number into its real
and imaginary parts uses the operators: *Re* ( ) and *Im* ( ). For example,
using the above complex numbers:

$$Re\ A = 2 \qquad\qquad Im\ A = 6$$
$$Re\ B = -4 \qquad\qquad Im\ B = -1.5$$
$$Re\ C = 3 \qquad\qquad Im\ C = -7$$

Notice that the value returned by the mathematical operator, *Im* ( ), does not include the *j*. For example, $Im(3+4j)$ is equal to 4, not $4j$.

Complex numbers follow the same algebra as ordinary numbers, treating the quantity, *j*, as a constant. For instance, addition, subtraction, multiplication and division are given by:

EQUATION 28-1
Addition of complex numbers.

$$(a + bj) + (c + dj) = (a + c) + j(b + d)$$

EQUATION 28-2
Subtraction of complex numbers.

$$(a + bj) - (c + dj) = (a - c) + j(b - d)$$

EQUATION 28-3
Multiplication of complex numbers.

$$(a + bj)(c + dj) = (ac - bd) + j(bc + ad)$$

EQUATION 28-4
Division of complex numbers.

$$\frac{(a + bj)}{(c + dj)} = \left(\frac{ac + bd}{c^2 + d^2}\right) + j\left(\frac{bc - ad}{c^2 + d^2}\right)$$

Two tricks are used when manipulating equations such as these. First, whenever a j^2 term is encountered, it is replaced by -1. This follows from the definition of *j*, that is: $j^2 = (\sqrt{-1})^2 = -1$. The second trick is a way to eliminate the *j* term from the denominator of a fraction. For instance, the left side of Eq. 28-4 has a denominator of $c + dj$. This is handled by multiplying the numerator and denominator by the term $c - jd$, cancelling all the imaginary terms from the denominator. In the jargon of the field, switching the sign of the imaginary part of a complex number is called taking the **complex conjugate**. This is denoted by a star at the upper right corner of the variable. For example, if $Z = a + bj$, then $Z^* = a - bj$. In other words, Eq. 28-4 is derived by multiplying both the numerator and denominator by the complex conjugate of the denominator.

The following properties hold even when the variables A, B, and C are complex. These relations can be proven by breaking each variable into its real and imaginary parts and working out the algebra.

EQUATION 28-5
Commutative property.

$$AB = BA$$

EQUATION 28-6
Associative property.

$$(A + B) + C = A + (B + C)$$

EQUATION 28-7
Distributive property.

$$A(B + C) = AB + AC$$

Polar Notation

Complex numbers can also be expressed in *polar notation*, besides the *rectangular notation* just described. For example, Fig. 28-2 shows three complex numbers in polar form, the same ones previously presented in Fig. 28-1. The **magnitude** is the length of the vector starting at the origin and ending at the complex point, while the **phase angle** is measured between this vector and the positive x-axis. Complex numbers can be converted between rectangular and polar notation by the following equations (paying attention to the polar notation *nuisances* discussed in Chapter 8):

EQUATION 28-8
Rectangular-to-polar conversion. The complex variable, A, can be changed from rectangular form: *Re* A & *Im* A, to polar form: *M* & *θ*.

$$M = \sqrt{(Re\,A)^2 + (Im\,A)^2}$$

$$\theta = \arctan\left[\frac{Im\,A}{Re\,A}\right]$$

EQUATION 28-9
Polar-to-rectangular conversion. This is changing the complex number from *M* & *θ* to *Re* A & *Im* A.

$$Re\,A = M\cos(\theta)$$

$$Im\,A = M\sin(\theta)$$

This brings up a giant leap in the mathematics. (Yes, this means you should pay extra attention). A complex number written in rectangular notation

FIGURE 28-2
Complex numbers in polar form. Three example points in the complex plane are shown in polar coordinates. Figure 28-1 shows these same points in rectangular form.

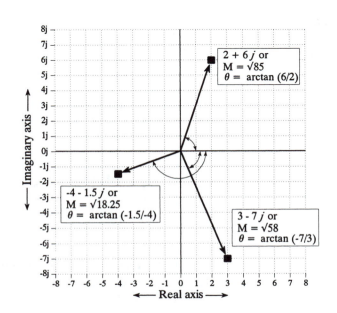

is in the form: $a + bj$. The information is carried in the variables: a & b, but the proper complex number is the entire expression: $a + bj$. In polar form, the key information is contained in M & θ, but what is the full expression for the proper complex number?

The key to this is Eq. 28-9, the polar-to-rectangular conversion. If we start with the proper complex number, $a + bj$, and apply Eq. 28-9, we obtain:

EQUATION 28-10
Rectangular and polar complex numbers.
The left side is the rectangular form of a
complex number, while the expression on
the right is the polar representation. The
conversion between: M & θ and a & b, is
given by Eqs. 28-8 and 28-9.

$$a + jb = M(\cos\theta + j\sin\theta)$$

The expression on the left is the proper *rectangular* description of a complex number, while the expression on the right is the proper *polar* description.

Before continuing with the next step, let's review how we arrived at this point. First, we gave the rectangular form of a complex number a graphical representation, that is, a location in a two-dimensional plane. Second, we defined the terms M & θ to be consistent with our previous experience about the relationship between polar and rectangular coordinates (Eq. 28-8 and 28-9). Third, we followed the mathematical consequences of these actions, arriving at what the correct polar form of a complex number must be, i.e., $M(\cos\theta + j\sin\theta)$. Even though this logic is straightforward, the result is difficult to see with "intuition." Unfortunately, it gets worse.

One of the most important equations in complex mathematics is **Euler's relation,** named for the clever and very prolific Swiss mathematician, Leonhard Euler (1707-1783):

EQUATION 28-11
Euler's relation. This is a key equation
for using complex numbers in science
and engineering.

$$e^{jx} = \cos x + j\sin x$$

If you like such things, this relation can be proven by expanding the exponential term into a Taylor series:

$$e^{jx} = \sum_{n=0}^{\infty} \frac{(jx)^n}{n!} = \left[\sum_{k=0}^{\infty} (-1)^k \frac{x^{2k}}{(2k)!}\right] + j\left[\sum_{k=0}^{\infty} (-1)^k \frac{x^{2k+1}}{(2k+1)!}\right]$$

The two bracketed terms on the right of this expression are the Taylor series for $\cos(x)$ and $\sin(x)$. Don't spend too much time on this proof; we aren't going to use it for anything.

Rewriting Eq. 28-10 using Euler's relation results in the most common way of expressing a complex number in polar notation, a **complex exponential**:

EQUATION 28-12
Exponential form of complex numbers. The rectangular form, on the left, is equal to the exponential polar form, on the right.

$$a + jb = M e^{j\theta}$$

Complex numbers in this exponential form are the backbone of DSP mathematics. Start your understanding by memorizing Eqs. 28-8 through 28-12. A strong advantage of using this exponential polar form is that it is very simple to multiply and divide complex numbers:

EQUATION 28-13
Multiplication of complex numbers.

$$\left[M_1 e^{j\theta_1} \right]\left[M_2 e^{j\theta_2} \right] = M_1 M_2 e^{j(\theta_1 + \theta_2)}$$

EQUATION 28-14
Division of complex numbers.

$$\frac{M_1 e^{j\theta_1}}{M_2 e^{j\theta_2}} = \left[\frac{M_1}{M_2}\right] e^{j(\theta_1 - \theta_2)}$$

That is, complex numbers in polar form are multiplied by multiplying their magnitudes and adding their angles. The easiest way to perform addition and subtraction in polar form is to convert the numbers to rectangular form, perform the operation, and reconvert back into polar. Complex numbers are usually expressed in rectangular form in computer routines, but in polar form when writing and manipulating equations. Just as *Re* ( ) and *Im* ( ) are used to extract the rectangular components from a complex number, the operators *Mag* ( ) and *Phase* ( ) are used to extract the polar components. For example, if $A = 5e^{j\pi/7}$, then $Mag(A) = 5$ and $Phase(A) = \pi/7$.

Using Complex Numbers by Substitution

Let's summarize where we are at. Solutions to common algebraic equations often contain the square-root of a negative number. These are called *complex numbers*, and represent solutions that cannot exist in the world as we know it. Complex numbers are expressed in one of two forms: $a + bj$ (rectangular), or $M e^{j\theta}$ (polar), where j is a symbol representing $\sqrt{-1}$. using either notation, a single complex number contains two separate pieces of information, either $a \& b$ or $M \& \theta$. In spite of their elusive nature, complex numbers follow mathematical laws that are similar (or identical) to those governing ordinary numbers.

This describes what complex numbers are and how they fit into the world of pure mathematics. Our next task is to describe ways they are useful in science and engineering problems. How is it possible to use a mathematics that has no connection with our everyday experience? The answer: *If the tool we have is a hammer, make the problem look like a nail.* In other words, we *change* the physical problem into a complex number form, manipulate the complex numbers, and then *change* back into a physical answer.

There are two ways that physical problems can be represented using complex numbers: a simple method of **substitution**, and a more elegant method we will call **mathematical equivalence**. Mathematical equivalence will be discussed in the next chapter on the *complex Fourier transform*. The remainder of this chapter is devoted to substitution.

Substitution takes two real physical parameters and places one in the real part of the complex number and one in the imaginary part. This allows the two values to be manipulated as a single entity, i.e., a single complex number. After the desired mathematical operations, the complex number is separated into its real and imaginary parts, which again correspond to the physical parameters we are concerned with.

A simple example will show how this works. As you recall from elementary physics, *vectors* can represent such things as: force, velocity, acceleration, etc. For example, imagine a sailboat being pushed in one direction by the wind, and in another direction by the ocean current. The resulting force on the boat is the vector sum of the two individual force vectors. This example is shown in Fig. 28-3, where two vectors, A and B, are added through the parallelogram law, resulting in C.

We can represent this problem with complex numbers by placing the east/west coordinate into the real part of a complex number, and the north/south coordinate into the imaginary part. This allows us to treat each vector as a single complex number, even though it is composed of two parts. For instance, the force of the wind, vector A, might be in the direction of 2 parts to the east and 6 parts to the north, represented as the complex number: $2+6j$. Likewise, the force of the ocean current, vector B, might be in the direction of 4 parts to the east and 3 parts to the south, represented as the complex number: $4-3j$. These two vectors can be added via Eq. 28-1, resulting in the complex number representing vector C: $6+3j$. Converting this back into a physical meaning, the combined force on the sailboat is in the direction of 6 parts to the north and 3 parts to the east.

Could this problem be solved without complex numbers? Of course! The complex numbers merely provide a formalized way of keeping track of the *two* components that form a *single* vector. The idea to remember is that some physical problems can be converted into a complex form by simply adding a j to one of the components. Converting back to the physical problem is nothing more than dropping the j. This is the essence of the *substitution* method.

Here's the rub. How do we know that the rules and laws that apply to complex mathematics are the same rules and laws that apply to the original

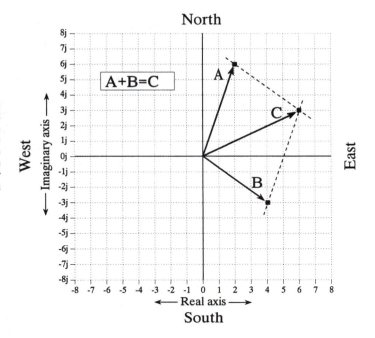

FIGURE 28-3
Adding vectors with complex numbers. The vectors A & B represent forces measured with respect to north/south and east/west. The east/west dimension is replaced by the real part of the complex number, while the north/south dimension is replaced by the imaginary part. This substitution allow complex mathematics to be used with an entirely *real* problem.

physical problem? For instance, we used Eq. 28-1 to add the force vectors in the sailboat problem. How do we know that the addition of complex numbers provides the same result as the addition of force vectors? In most cases, we know that complex mathematics can be used for a particular application because *someone else* said it does. Some brilliant and well respected mathematician or engineer worked out the details and published the results. The point to remember is that we cannot substitute just *any* problem into a complex form and expect the answer to make sense. We must stick to applications that have been shown to be applicable to complex analysis.

Let's look at an example where complex number substitution *does not* work. Imagine that you buy apples for $5 a box, and oranges for $10 a box. You represent this by the complex number: $5 + 10j$. During a particular week, you buy 6 boxes of apples and 2 boxes of oranges, which you represent by the complex number: $6 + 2j$. The total price you must pay for the goods is equal to number of items multiplied by the price of each item, that is, $(5 + 10j)(6 + 2j) = 10 + 70j$. In other words, the complex math indicates you must pay a total of $10 for the apples and $70 for the oranges. The problem is, the answer is completely wrong! The rules of complex mathematics *do not* follow the rules of this particular physical problem.

Complex Representation of Sinusoids

Complex numbers find a niche in electronics and signal processing because they are a compact way to represent and manipulate the most useful of all waveforms: sine and cosine waves. The conventional way to represent a sinusoid is: $M\cos(\omega t + \phi)$ or $A\cos(\omega t) + B\sin(\omega t)$, in polar and rectangular notation, respectively. Notice that we are representing frequency by ω, the *natural frequency* in *radians per second*. If it makes you more comfortable, you can replace each ω with $2\pi f$ to make the expressions in hertz. However, most DSP mathematics is written using the shorter notation, and you should get use to it. Since it requires two parameters to represent a single sinusoid (i.e., A & B or M & θ), the use of complex numbers to represent these important waveforms is a natural. Using substitution, the change from the conventional sinusoid representation to a complex number is straight-forward. In rectangular form:

$$A\cos(\omega t) + B\sin(\omega t) \quad \rightleftarrows \quad a + jb$$

(conventional representation)   *(complex number)*

where $A = a$, and $B = -b$. Put in words, the amplitude of the cosine wave becomes the real part of the complex number, while the *negative* of the sine wave's amplitude becomes the imaginary part. It is important to understand that this is *not* an equation, but merely a way of letting a complex number *represent* a sinusoid. This substitution also can be applied in polar form:

$$M\cos(\omega t + \phi) \quad \rightleftarrows \quad Me^{j\theta}$$

(conventional representation) *(complex number)*

where $M = M$, and $\theta = -\phi$. In words, the polar notation substitution leaves the magnitude the same, but changes the sign of the phase angle.

Why change the sign of the imaginary part & phase angle? This is to make the substitution appear in the same form as the *complex Fourier transform* described in the next chapter. The *substitution* techniques of this chapter gain nothing from this sign change, but it is almost always done to keep things consistent with the more advanced methods.

Using complex numbers to represent sine and cosine waves is a common technique in electrical circuit analysis and DSP. This is because many (but not all) of the rules and laws governing complex numbers are the same as those governing sinusoids. In other words, we can represent the sine and cosine waves with complex numbers, manipulate the numbers in various ways, and have the resulting answer match the way the sinusoids behave.

However, we must be careful to use only those mathematical operations that mimic the physical problem being represented (sinusoids in this case). For example, suppose we use the complex variables, A and B, to represent two sinusoids of the same frequency, but with different amplitudes and phase shifts. When the two complex numbers are *added*, a third complex number is produced. Likewise, a third sinusoid is created when the two sinusoids are added. As you would hope, the third complex number represents the third sinusoid. The complex addition matches the physical system.

Now, imagine multiplying the complex numbers A and B, resulting in another complex number. Does this match what happens when the two sinusoids are multiplied? No! Multiplying two sinusoids does *not* produce another sinusoid. Complex multiplication fails to match the physical system, and therefore cannot be used.

Fortunately, the valid operations are clearly defined. Two conditions must be satisfied. First, all of the sinusoids must be at the *same frequency*. For example, if the complex numbers: $1+1j$ and $2+2j$ represent sinusoids at the same frequency, then the sum of the two sinusoids is represented by the complex number: $3+3j$. However, if $1+1j$ and $2+2j$ represent sinusoids with different frequencies, there is nothing that can be done with the complex representation. In this case, the sum of the complex numbers, $3+3j$, is meaningless.

In spite of this, frequency can be left as a variable when using complex numbers, but it must be the *same* frequency everywhere. For instance, it is perfectly valid to add: $2\omega+3\omega j$ and $3\omega+1j$, to produce: $5\omega+(3\omega+1)j$. These represent sinusoids where the amplitude and phase vary as frequency changes. While we do not know *what* the particular frequency is, we do know that it is the *same* everywhere, i.e., ω.

The second requirement is that the operations being represented must be *linear*, as discussed in Chapter 5. For instance, sinusoids can be combined by addition and subtraction, but not by multiplication or division. Likewise, systems may be amplifiers, attenuators, high or low-pass filters, etc., but not such actions as: squaring, clipping and thresholding. Remember, even convolution and Fourier analysis are only valid for linear systems.

Complex Representation of Systems

Figure 28-4 shows an example of using complex numbers to represent a sinusoid passing through a linear system. We will use continuous signals for this example, although discrete signals are handled the same way. Since the input signal is a sinusoid, and the system is linear, the output will also be a sinusoid, and at the same frequency as the input. As shown, the example input signal has a conventional representation of: $3\cos(\omega t+\pi/4)$, or the equivalent expression: $2.1213\cos(\omega t)+2.1213\sin(\omega t)$. When represented by a complex number this becomes: $3e^{-j\pi/4}$ or $2.1213-j\,2.1213$. Likewise, the conventional representation of the output is: $1.5\cos(\omega t-\pi/8)$, or in the alternate form: $1.3858\cos(\omega t)-0.5740\sin(\omega t)$. This is represented by the complex number: $1.5e^{j\pi/8}$ or $1.3858+j\,0.5740$.

The system's characteristics can also be represented by a complex number. The magnitude of the complex number is the ratio between the magnitudes

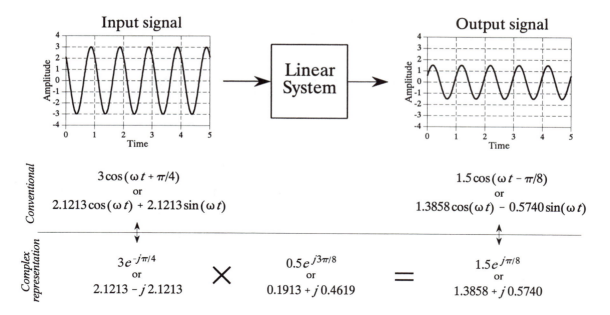

FIGURE 28-4
Sinusoids represented by complex numbers. Complex numbers are popular in DSP and electronics because they are a convenient way to represent and manipulate *sinusoids*. As shown in this example, sinusoidal input and output signals can be represented as complex numbers, expressed in either polar or rectangular form. In addition, the *change* that a linear system makes to a sinusoid can also be expressed as a complex number.

of the input and output (i.e., M_{out}/M_{in}). Likewise, the angle of the complex number is the *negative* of the difference between the input and output angles (i.e., $-[\phi_{out} - \phi_{in}]$). In the example used here, the system is described by the complex number, $0.5e^{j3\pi/8}$. In other words, the amplitude of the sinusoid is reduced by 0.5, while the phase angle is changed by $-3\pi/8$.

The complex number representing the system can be converted into rectangular form as: $0.1913 + j\,0.4619$, but we must be careful in interpreting what this means. It does *not* mean that a sine wave passing through the system is changed in amplitude by 0.1913, nor that a cosine wave is changed by 0.4619. In general, a pure sine or cosine wave entering a linear system is converted into a *mixture* of sine and cosine waves.

Fortunately, the complex math automatically keeps track of these cross-terms. When a sinusoid passes through a linear system, the complex numbers representing the input signal and the system are *multiplied*, producing the complex number representing the output. If any two of the complex numbers are known, the third can be found. The calculations can be carried out in either polar or rectangular form, as shown in Fig. 28-4.

In previous chapters we described how the Fourier transform decomposes a signal into cosine and sine waves. The amplitudes of the cosine waves are called the *real part*, while the amplitudes of the sine waves are called the *imaginary part*. We stressed that these amplitudes are ordinary numbers, and the terms *real* and *imaginary* are just names used to keep the two separate. Now that complex numbers have been introduced, it should be quite obvious were the names come from. For example, imagine a 1024 point signal being decomposed into 513 cosine waves and 513 sine waves. Using substitution, we can represent the spectrum by 513 complex numbers. However, don't be misled into thinking that this is the *complex Fourier transform*, the topic of Chapter 29. This is still the *real Fourier transform*; the spectrum has just been placed in a complex format by using substitution.

Electrical Circuit Analysis

This method of substituting complex numbers for cosine & sine waves is called the **Phasor transform**. It is the main tool used to analyze networks composed of resistors, capacitors and inductors. [Electrical engineers use a more formal definition of the phasor transform, based on multiplying by the complex term: $e^{j\omega t}$ and taking the real part. This allows the procedure to be written as an equation, making it easier to deal with in mathematical work. It provides the same result as substitution, but is more elegant].

The first step is to understand the relationship between the current and voltage for each of these devices. For the resistor, this is expressed in Ohm's law: $v = iR$, where i is the instantaneous current through the device, v is the instantaneous voltage across the device, and R is the resistance. In contrast, the capacitor and inductor are governed by the differential equations: $i = C \, dv/dt$, and $v = L \, di/dt$, where C is the capacitance and L is the inductance. In the most general method of circuit analysis, these nasty differential equations are combined as dictated by the circuit configuration, and then solved for the parameters of interest. While this will answer *everything* about the circuit, the math can become a real mess.

This can be greatly simplified by restricting the signals to be sinusoids. By representing these sinusoids with complex numbers, the difficult *differential* equations can be directly replaced with much simpler *algebraic* equations. Figure 28-5 illustrates how this works. We treat each of these three components (resistor, capacitor & inductor) as a *system*. The input to the system is the sinusoidal current through the device, while the output is the sinusoidal voltage across its two terminals. This means we can represent the input and output of the system by the two complex variables: I (for current) and V (for voltage), respectively. The relation between the input and output can also be expressed by a complex number. This complex number is called the **impedance**, and is given the symbol: Z. This means:

$$I \times Z = V$$

In words, the complex number representing the sinusoidal voltage is equal to the complex number representing the sinusoidal current multiplied by the impedance (another complex number). Given any two, the third can be

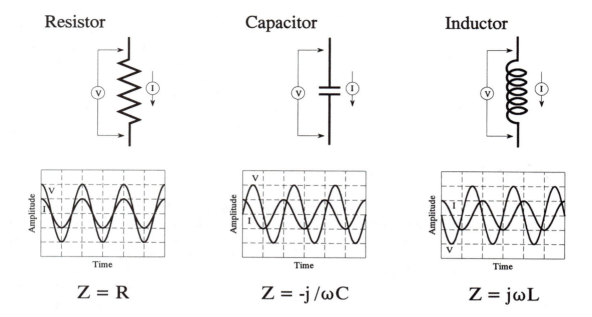

Resistor
$$Z = R$$

Capacitor
$$Z = -j/\omega C$$

Inductor
$$Z = j\omega L$$

FIGURE 28-5
Definition of impedance. When sinusoidal voltages and currents are represented by complex numbers, the ratio between the two is called the *impedance*, and is denoted by the complex variable, Z. Resistors, capacitors and inductors have impedances of R, $-j/\omega C$, and $j\omega L$, respectively.

found. In polar form, the magnitude of the impedance is the ratio between the amplitudes of V and I. Likewise, the phase of the impedance is the phase difference between V and I.

This relation can be thought of as *Ohm's law for sinusoids*. Ohms's law ($v = iR$) describes how the *resistance* relates the instantaneous current and voltage in a resistor. When the signals are sinusoids represented by complex numbers, the relation becomes: $V = IZ$. That is, the *impedance* relates the current and voltage. Resistance is an ordinary number, since it deals with two ordinary numbers. Impedance is a complex number, since it relates two complex numbers. Impedance contains more information than resistance, because it dictates both the amplitudes *and* the phase angles.

From the differential equations that govern their operation, it can be shown that the impedance of the resistor, capacitor, and inductor are: R, $-j/\omega C$, and $j\omega L$, respectively. As an example, imagine that the current in each of these components is a unity amplitude cosine wave, as shown in Fig. 28-5. Using substitution, this is represented by the complex number: $1 + 0j$. The voltage across the resistor will be: $V = IZ = (1+0j)R = R+0j$. In other words, a cosine wave of amplitude R. The voltage across the capacitor is found to be: $V = IZ = (1+0j)(-j/\omega C)$. This reduces to: $0-j/\omega C$, a sine wave of amplitude, $1/\omega C$. Likewise, the voltage across the inductor can be calculated: $V = IZ = (1+0j)(j\omega L)$. This reduces to: $0+j\omega L$, a negative sine wave of amplitude, ωL.

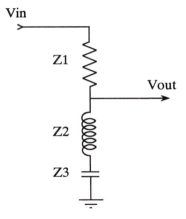

Vin

Z1

Vout

Z2

Z3

FIGURE 28-6
RLC notch filter. This circuit removes a narrow band of frequencies from a signal. The use of complex substitution greatly simplifies the analysis of this and similar circuits.

The beauty of this method is that *RLC* circuits can be analyzed without having to resort to differential equations. The *impedance* of the resistors, capacitors and inductors is treated the same as *resistance* in a DC circuit. This includes all of the basic combinations, such as: resistors in series, resistors in parallel, voltage dividers, etc.

As an example, Fig. 28-6 shows an *RLC* circuit called a **notch filter**, used to remove a narrow band of frequencies. For instance, it could eliminate 60 hertz interference in an audio or instrumentation signal. If this circuit were composed of three resistors (instead of the resistor, capacitor and inductor), the relationship between the input and output signals would be given by the voltage divider formula: $v_{out}/v_{in} = (R2+R3)/(R1+R2+R3)$. Since the circuit contains capacitors and inductors, the equation is rewritten with impedances:

$$\frac{Vout}{Vin} = \frac{Z2 + Z3}{Z1 + Z2 + Z3}$$

where: *Vout, Vin, Z1, Z2,* and *Z3* are all complex variables. Plugging in the impedance of each component:

$$\frac{Vout}{Vin} = \frac{j\omega L - \dfrac{j}{\omega C}}{R + j\omega L - \dfrac{j}{\omega C}}$$

Next, we crank through the algebra to separate everything containing a j, from everything that does not contain a j. In other words, we separate the equation into its real and imaginary parts. This algebra can be tiresome and long, but the alternative is to write and solve *differential equations*, an

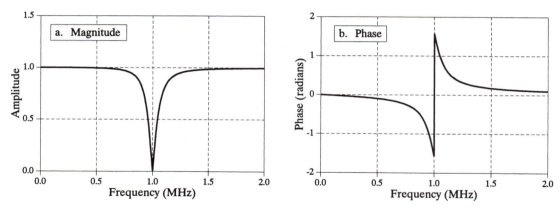

FIGURE 28-7
Notch filter frequency response. These curves are for the component values: R = 50Ω, C = 470ρF, and L = 54 μH.

even nastier task. When separated into the real and imaginary parts, the complex representation of the notch filter becomes:

$$\frac{Vout}{Vin} = \frac{k^2}{R^2 + k^2} + j\frac{Rk}{R^2 + k^2}$$

where: $k = \omega L - 1/\omega C$

Lastly, the relation is converted to polar notation, and graphed in Fig. 28-7:

$$Mag = \frac{\omega L - 1/\omega C}{\left(R^2 + \left[\omega L - 1/\omega C\right]^2\right)^{1/2}} \qquad Phase = \arctan\left[\frac{R}{\omega L - 1/\omega C}\right]$$

The key point to remember from these examples is how *substitution* allows complex numbers to represent real world problems. In the next chapter we will look at a more advanced way to use complex numbers in science and engineering, *mathematical equivalence*.

The Complex Fourier Transform

Although complex numbers are fundamentally disconnected from our reality, they can be used to solve science and engineering problems in two ways. First, the parameters from a real world problem can be substituted into a complex form, as presented in the last chapter. The second method is much more elegant and powerful, a way of making the complex numbers mathematically equivalent to the physical problem. This approach leads to the *complex Fourier transform*, a more sophisticated version of the *real Fourier transform* discussed in Chapter 8. The complex Fourier transform is important in itself, but also as a stepping stone to more powerful complex techniques, such as the *Laplace* and *z-transforms*. These complex transforms are the foundation of theoretical DSP.

The Real DFT

All four members of the Fourier transform family (DFT, DTFT, Fourier Transform & Fourier Series) can be carried out with either real numbers or complex numbers. Since DSP is mainly concerned with the DFT, we will use it as an example. Before jumping into the complex math, let's review the real DFT with a special emphasis on things that are awkward with the mathematics. In Eq. 8-4 we defined the *real* version of the Discrete Fourier Transform according to the equations:

EQUATION 29-1
The real DFT. This is the forward transform, calculating the frequency domain from the time domain. In spite of using the names: *real part* and *imaginary part*, these equations only involve ordinary numbers. The frequency index, k, runs from 0 to $N/2$. The $2/N$ term can be applied to the forward or inverse transform.

$$Re\,X[k] = \frac{2}{N}\sum_{n=0}^{N-1}x[n]\cos(2\pi kn/N)$$

$$Im\,X[k] = \frac{-2}{N}\sum_{n=0}^{N-1}x[n]\sin(2\pi kn/N)$$

In words, an N sample time domain signal, $x[n]$, is decomposed into a set of $N/2 + 1$ cosine waves, and $N/2 + 1$ sine waves, with frequencies given by the index, k. The amplitudes of the cosine waves are contained in $Re\,X[k]$, while the amplitudes of the sine waves are contained in $Im\,X[k]$. These equations operate by *correlating* the respective cosine or sine wave with the time domain signal. In spite of using the names: *real part* and *imaginary part*, there are no complex numbers in these equations. There isn't a j anywhere in sight! We have also included the normalization factor, $2/N$ in these equations. Remember, this can be placed in front of either the synthesis or analysis equation, or be handled as a separate step (as described by Eq. 8-3). These equations should be very familiar from previous chapters. If they aren't, go back and brush up on these concepts before continuing. If you don't understand the *real* DFT, you will never be able to understand the *complex* DFT.

Even though the real DFT uses only real numbers, *substitution* allows the frequency domain to be *represented* using complex numbers. As suggested by the names of the arrays, $Re\,X[k]$ becomes the real part of the complex frequency spectrum, and $Im\,X[k]$ becomes the imaginary part. In other words, we place a j with each value in the imaginary part, and add the result to the real part. However, do not make the mistake of thinking that this is the "complex DFT." This is nothing more than the *real DFT with complex substitution*.

While the real DFT is adequate for many applications in science and engineering, it is mathematically awkward in three respects. First, it can only take advantage of complex numbers through the use of **substitution**. This makes mathematicians uncomfortable; they want to say: "this *equals* that," not simply: "this *represents* that." For instance, imagine we are given the mathematical statement: A equals B. We immediately know countless consequences: $5A = 5B$, $1+A = 1+B$, $A/x = B/x$, etc. Now suppose we are given the statement: A represents B. Without additional information, we know absolutely nothing! When things are equal, we have access to four-thousand years of mathematics. When things only represent each other, we must start from scratch with new definitions. For example, when sinusoids are represented by complex numbers, we allow addition and subtraction, but prohibit multiplication and division.

The second thing handled poorly by the real Fourier transform is the **negative frequency** portion of the spectrum. As you recall from Chapter 10, sine and cosine waves can be described as having a *positive* frequency or a *negative* frequency. Since the two views are identical, the real Fourier transform ignores the negative frequencies. However, there are applications where the negative frequencies are important. This occurs when negative frequency components are forced to move into the positive frequency portion of the spectrum. The ghosts take human form, so to speak. For instance, this is what happens in aliasing, circular convolution, and amplitude modulation. Since the real Fourier transform doesn't use negative frequencies, its ability to deal with these situations is very limited.

Our third complaint is the **special handing of $Re\,X[0]$ and $Re\,X[N/2]$**, the first and last points in the frequency spectrum. Suppose we start with an

N point signal, $x[n]$. Taking the DFT provides the frequency spectrum contained in $Re\, X[k]$ and $Im\, X[k]$, where k runs from 0 to $N/2$. However, these are not the amplitudes needed to reconstruct the time domain waveform; samples $Re\, X[0]$ and $Re\, X[N/2]$ must first be divided by two. (See Eq. 8-3 to refresh your memory). This is easily carried out in computer programs, but inconvenient to deal with in equations.

The complex Fourier transform is an elegant solution to these problems. It is natural for complex numbers and negative frequencies to go hand-in-hand. Let's see how it works.

Mathematical Equivalence

Our first step is to show how sine and cosine waves can be written in an *equation* with complex numbers. The key to this is Euler's relation, presented in the last chapter:

EQUATION 29-2
Euler's relation.

$$e^{jx} = \cos(x) + j\sin(x)$$

At first glance, this doesn't appear to be much help; one complex expression is equal to another complex expression. Nevertheless, a little algebra can rearrange the relation into two other forms:

EQUATION 29-3
Euler's relation for
sine & cosine.

$$\cos(x) = \frac{e^{jx} - e^{-jx}}{2} \qquad \sin(x) = \frac{e^{jx} - e^{-jx}}{2j}$$

This result is extremely important, we have developed a way of writing *equations* between complex numbers and ordinary sinusoids. Although Eq. 29-3 is the standard form of the identity, it will be more useful for this discussion if we change a few terms around:

EQUATION 29-4
Sinusoids as complex numbers. Using
complex numbers, cosine and sine waves
can be written as the sum of a positive
and a negative frequency.

$$\cos(\omega t) = \frac{1}{2}e^{j(-\omega)t} + \frac{1}{2}e^{j\omega t}$$

$$\sin(\omega t) = \frac{1}{2}je^{j(-\omega)t} - \frac{1}{2}je^{j\omega t}$$

Each expression is the sum of two exponentials: one containing a *positive* frequency (ω), and the other containing a *negative* frequency ($-\omega$). In other words, when sine and cosine waves are written as complex numbers, the negative portion of the frequency spectrum is automatically included. The positive and negative frequencies are treated with an equal status; it required one-half of each to form a complete waveform.

The Complex DFT

The forward complex DFT, written in polar form, is given by:

Equation 29-5
The forward complex DFT. Both the time domain, $x[n]$, and the frequency domain, $X[k]$, are arrays of complex numbers, with k and n running from 0 to N-1. This equation is in polar form, the most common for DSP.

$$X[k] = \frac{1}{N} \sum_{n=0}^{N-1} x[n]\, e^{-j2\pi kn/N}$$

Alternatively, Euler's relation can be used to rewrite the forward transform in rectangular form:

EQUATION 29-6
The forward complex DFT (rectangular form).

$$X[k] = \frac{1}{N} \sum_{n=0}^{N-1} x[n]\left[\cos(2\pi kn/N) - j\sin(2\pi kn/N)\right]$$

To start, compare this equation of the *complex Fourier transform* with the equation of the *real Fourier transform*, Eq. 29-1. At first glance, they appear to be identical, with only small amount of algebra being required to turn Eq. 29-6 into Eq. 29-1. However, this is very misleading; the differences between these two equations are very subtle and easy to overlook, but tremendously important. Let's go through the differences in detail.

First, the real Fourier transform converts a real time domain signal, $x[n]$, into two real frequency domain signals, $Re\,X[k]$ & $Im\,X[k]$. By using complex substitution, the frequency domain can be *represented* by a single complex array, $X[k]$. In the complex Fourier transform, both $x[n]$ & $X[k]$ are arrays of complex numbers. A practical note: Even though the time domain is complex, there is nothing that *requires* us to use the imaginary part. Suppose we want to process a real signal, such as a series of voltage measurements taken over time. This group of data becomes the real part of the time domain signal, while the imaginary part is composed of zeros.

Second, the real Fourier transform only deals with *positive* frequencies. That is, the frequency domain index, k, only runs from 0 to $N/2$. In comparison, the complex Fourier transform includes both *positive* and *negative* frequencies. This means k runs from 0 to N-1. The frequencies between 0 and $N/2$ are positive, while the frequencies between $N/2$ and N-1 are negative. Remember, the frequency spectrum of a discrete signal is periodic, making the negative frequencies between $N/2$ and N-1 the same as between $-N/2$ and 0. The samples at 0 and $N/2$ straddle the line between positive and negative. If you need to refresh your memory on this, look back at Chapters 10 and 12.

Third, in the real Fourier transform with substitution, a j was added to the sine wave terms, allowing the frequency spectrum to be represented by complex numbers. To convert back to ordinary sine and cosine waves, we can simply drop the j. This is the sloppiness that comes when one thing only *represents* another thing. In comparison, the complex DFT, Eq. 29-5, is a formal mathematical equation with j being an integral part. In this view, we cannot arbitrary add or remove a j any more than we can add or remove any other variable in the equation.

Forth, the real Fourier transform has a scaling factor of *two* in front, while the complex Fourier transform does not. Say we take the real DFT of a cosine wave with an amplitude of *one*. The spectral value corresponding to the cosine wave is also *one*. Now, let's repeat the process using the complex DFT. In this case, the cosine wave corresponds to *two* spectral values, a positive and a negative frequency. Both these frequencies have a value of ½. In other words, a positive frequency with an amplitude of ½, combines with a negative frequency with an amplitude of ½, producing a cosine wave with an amplitude of *one*.

Fifth, the real Fourier transform requires special handling of two frequency domain samples: $Re\,X[0]$ & $Re\,X[N/2]$, but the complex Fourier transform does not. Suppose we start with a time domain signal, and take the DFT to find the frequency domain signal. To reverse the process, we take the Inverse DFT of the frequency domain signal, reconstructing the original time domain signal. However, there is scaling required to make the reconstructed signal be identical to the original signal. For the complex Fourier transform, a factor of $1/N$ must be introduced somewhere along the way. This can be tacked-on to the forward transform, the inverse transform, or kept as a separate step between the two. For the real Fourier transform, an additional factor of two is required ($2/N$), as described above. However, the real Fourier transform also requires an additional scaling step: $Re\,X[0]$ and $Re\,X[N/2]$ must be divided by *two* somewhere along the way. Put in other words, a scaling factor of $1/N$ is used with these two samples, while $2/N$ is used for the remainder of the spectrum. As previously stated, this awkward step is one of our complaints about the real Fourier transform.

Why are the real and complex DFTs different in how these two points are handled? To answer this, remember that a cosine (or sine) wave in the time domain becomes split between a positive and a negative frequency in the complex DFT's spectrum. However, there are two exceptions to this, the spectral values at 0 and $N/2$. These correspond to zero frequency (DC) and the Nyquist frequency (one-half the sampling rate). Since these points straddle the positive and negative portions of the spectrum, they do not have a matching point. Because they are not combined with another value, they inherently have only one-half the contribution to the time domain as the other frequencies.

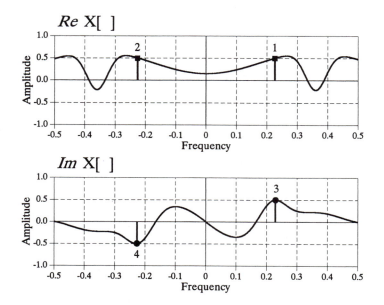

FIGURE 29-1
Complex frequency spectrum. These curves correspond to an entirely real time domain signal, because the real part of the spectrum has an *even* symmetry, and the imaginary part has an *odd* symmetry. The two square markers in the real part correspond to a cosine wave with an amplitude of one, and a frequency of 0.23. The two round markers in the imaginary part correspond to a sine wave with an amplitude of one, and a frequency of 0.23.

Figure 29-1 illustrates the complex DFT's frequency spectrum. This figure assumes the time domain is entirely *real*, that is, its imaginary part is zero. We will discuss the idea of imaginary time domain signals shortly. There are two common ways of displaying a complex frequency spectrum. As shown here, zero frequency can be placed in the center, with positive frequencies to the right and negative frequencies to the left. This is the best way to *think* about the complete spectrum, and is the *only* way that an aperiodic spectrum can be displayed.

The problem is that the spectrum of a discrete signal is *periodic* (such as with the DFT and the DTFT). This means that everything between -0.5 and 0.5 repeats itself an infinite number of times to the left and to the right. In this case, the spectrum between 0 and 1.0 contains the same information as from -0.5 to 0.5. When graphs are made, such as Fig. 29-1, the -0.5 to 0.5 convention is usually used. However, many equations and programs use the 0 to 1.0 form. For instance, in Eqs. 29-5 and 29-6 the frequency index, *k*, runs from 0 to *N*-1 (coinciding with 0 to 1.0). However, we could write it to run from -*N*/2 to *N*/2-1 (coincide with -0.5 to 0.5), if we desired.

Using the spectrum in Fig. 29-1 as a guide, we can examine how the inverse complex DFT reconstructs the time domain signal. The inverse complex DFT, written in polar form, is given by:

Equation 29-7
The inverse complex DFT. This is matching equation to the forward complex DFT in Eq. 29-5.

$$x[n] = \sum_{k=0}^{N-1} X[k]\, e^{j2\pi kn/N}$$

Using Euler's relation, this can be written in rectangular form as:

Equation 29-8
The inverse complex DFT.
This is Eq. 21-7 rewritten to
show how each value in the
frequency spectrum affects
the time domain.

$$x[n] = \sum_{k=0}^{N-1} Re X[k] \left(\cos(2\pi kn/N)] + j\sin(2\pi kn/N) \right)$$

$$- \sum_{k=0}^{N-1} Im X[k] \left(\sin(2\pi kn/N)] + j\cos(2\pi kn/N) \right)$$

The compact form of Eq. 29-7 is how the inverse DFT is usually written, although the expanded version in Eq. 29-9 can be easier to understand. In words, each value in the real part of the frequency domain contributes a real cosine wave *and* an imaginary sine wave to the time domain. Likewise, each value in the imaginary part of the frequency domain contributes a real sine wave *and* an imaginary cosine wave. The time domain is found by adding all these real and imaginary sinusoids. The important concept is that each value in the frequency domain produces both a *real* sinusoid and an *imaginary* sinusoid in the time domain.

For example, imagine we want to reconstruct a unity amplitude cosine wave at a frequency of $2\pi k/N$. This requires a positive frequency and a negative frequency, both from the real part of the frequency spectrum. The two square markers in Fig. 29-1 are an example of this, with the frequency set at: $k/N = 0.23$. The positive frequency at 0.23 (labeled 1 in Fig. 29-1) contributes a cosine wave and an imaginary sine wave to the time domain:

$$\tfrac{1}{2}\cos(2\pi\, 0.23\, n) + \tfrac{1}{2}j\sin(2\pi\, 0.23\, n)$$

Likewise, the negative frequency at -0.23 (labeled 2 in Fig. 29-1) also contributes a cosine and an imaginary sine wave to the time domain:

$$\tfrac{1}{2}\cos(2\pi\, (-0.23)\, n) + \tfrac{1}{2}j\sin(2\pi\, (-0.23)\, n)$$

The negative sign within the cosine and sine terms can be eliminated by the relations: $\cos(-x) = \cos(x)$ and $\sin(-x) = -\sin(x)$. This allows the negative frequency's contribution to be rewritten:

$$\tfrac{1}{2}\cos(2\pi\, 0.23\, n) - \tfrac{1}{2}j\sin(2\pi\, 0.23\, n)$$

Adding the contributions from the positive and the negative frequencies reconstructs the time domain signal:

contribution from positive frequency $\rightarrow$ $\qquad$ $\frac{1}{2}\cos(2\pi 0.23\,n) + \frac{1}{2}j\sin(2\pi 0.23\,n)$

contribution from negative frequency $\rightarrow$ $\qquad$ $\frac{1}{2}\cos(2\pi 0.23\,n) - \frac{1}{2}j\sin(2\pi 0.23\,n)$

resultant time domain signal $\rightarrow$ $\qquad$ $\cos(2\pi 0.23\,n)$

In this same way, we can synthesize a sine wave in the time domain. In this case, we need a positive and negative frequency from the imaginary part of the frequency spectrum. This is shown by the round markers in Fig. 29-1. From Eq. 29-8, these spectral values contribute a sine wave and an imaginary cosine wave to the time domain. The imaginary cosine waves cancel, while the real sine waves add:

contribution from positive frequency $\rightarrow$ $\qquad$ $-\frac{1}{2}\sin(2\pi 0.23\,n) - \frac{1}{2}j\cos(2\pi 0.23\,n)$

contribution from negative frequency $\rightarrow$ $\qquad$ $-\frac{1}{2}\sin(2\pi 0.23\,n) + \frac{1}{2}j\cos(2\pi 0.23\,n)$

resultant time domain signal $\rightarrow$ $\qquad$ $-\sin(2\pi 0.23\,n)$

Notice that a *negative* sine wave is generated, even though the positive frequency had a value that was *positive*. This sign inversion is an inherent part of the mathematics of the complex DFT. As you recall, this same sign inversion is commonly used in the real DFT. That is, a *positive* value in the imaginary part of the frequency spectrum corresponds to a *negative* sine wave. Most authors include this sign inversion in the definition of the real Fourier transform to make it consistent with its complex counterpart. The point is, this sign inversion *must* be used in the complex Fourier transform, but is merely an option in the real Fourier transform.

The *symmetry* of the complex Fourier transform is very important. As illustrated in Fig. 29-1, a real time domain signal corresponds to a frequency spectrum with an *even* real part, and an *odd* imaginary part. In other words, the negative and positive frequencies have the same sign in the real part (such as points 1 and 2 in Fig. 29-1), but opposite signs in the imaginary part (points 3 and 4).

This brings up another topic: the imaginary part of the time domain. Until now we have assumed that the time domain is completely real, that is, the imaginary part is zero. However, the complex Fourier transform does not require this. What is the physical meaning of an imaginary time domain signal? Usually, there is none. This is just something allowed by the complex mathematics, without a correspondence to the world we live in. However, there are applications where it can be used or manipulated for a mathematical purpose.

An example of this is presented in Chapter 12. The imaginary part of the time domain produces a frequency spectrum with an odd real part, and an even imaginary part. This is just the opposite of the spectrum produced by the real part of the time domain (Fig. 29-1). When the time domain contains both a real part and an imaginary part, the frequency spectrum is the sum of the two spectra, had they been calculated individually. Chapter 12 describes how this can be used to make the FFT algorithm calculate the frequency spectra of *two* real signals at once. One signal is placed in the real part of the time domain, while the other is place in the imaginary part. After the FFT calculation, the spectra of the two signals are separated by an even/odd decomposition.

The Family of Fourier Transforms

Just as the DFT has a real and complex version, so do the other members of the Fourier transform family. This produces the zoo of equations shown in Table 29-1. Rather than studying these equations individually, try to understand them as a well organized and symmetrical *group*. The following comments describe the organization of the Fourier transform family. It is detailed, repetitive, and boring. Nevertheless, this is the background needed to understand theoretical DSP. Study it well.

1. Four Fourier Transforms
A time domain signal can be either *continuous* or *discrete*, and it can be either *periodic* or *aperiodic*. This defines four types of Fourier transforms: the **Discrete Fourier Transform** (discrete, periodic), the **Discrete Time Fourier Transform** (discrete, aperiodic), the **Fourier Series** (continuous, periodic), and the **Fourier Transform** (continuous, aperiodic). Don't try to understand the reasoning behind these names, there isn't any.

If a signal is discrete in one domain, it will be periodic in the other. Likewise, if a signal is continuous in one domain, will be aperiodic in the other. Continuous signals are represented by parenthesis, ( ), while discrete signals are represented by brackets, [ ]. There is no notation to indicate if a signal is periodic or aperiodic.

2. Real versus Complex
Each of these four transforms has a complex version and a real version. The complex versions have a complex time domain signal and a complex frequency domain signal. The real versions have a real time domain signal and two real frequency domain signals. Both positive and negative frequencies are used in the complex cases, while only positive frequencies are used for the real transforms. The complex transforms are usually

written in an exponential form; however, Euler's relation can be used to change them into a cosine and sine form if needed.

3. Analysis and Synthesis
Each transform has an analysis equation (also called the forward transform) and a synthesis equation (also called the inverse transform). The analysis equations describe how to calculate each value in the frequency domain based on all of the values in the time domain. The synthesis equations describe how to calculate each value in the time domain based on all of the values in the frequency domain.

4. Time Domain Notation
Continuous time domain signals are called $x(t)$, while discrete time domain signals are called $x[n]$. For the complex transforms, these signals are complex. For the real transforms, these signals are real. All of the time domain signals extend from minus infinity to positive infinity. However, if the time domain is periodic, we are only concerned with a single cycle, because the rest is redundant. The variables, T and N, denote the periods of continuous and discrete signals in the time domain, respectively.

5. Frequency Domain Notation
Continuous frequency domain signals are called $X(\omega)$ if they are complex, and $Re\,X(\omega)$ & $Im\,X(\omega)$ if they are real. Discrete frequency domain signals are called $X[k]$ if they are complex, and $Re\,X[k]$ & $Im\,X[k]$ if they are real. The complex transforms have negative frequencies that extend from minus infinity to zero, and positive frequencies that extend from zero to positive infinity. The real transforms only use positive frequencies. If the frequency domain is periodic, we are only concerned with a single cycle, because the rest is redundant. For continuous frequency domains, the independent variable, ω, makes one complete period from $-\pi$ to π. In the discrete case, we use the period where k runs from 0 to N-1

6. The Analysis Equations
The analysis equations operate by *correlation*, i.e., multiplying the time domain signal by a sinusoid and integrating (continuous time domain) or summing (discrete time domain) over the appropriate time domain section. If the time domain signal is aperiodic, the appropriate section is from minus infinity to positive infinity. If the time domain signal is periodic, the appropriate section is over any one complete period. The equations shown here are written with the integration (or summation) over the period: 0 to T (or 0 to N-1). However, any other complete period would give identical results, i.e., $-T$ to 0, $-T/2$ to $T/2$, etc.

7. The Synthesis Equations
The synthesis equations describe how an individual value in the time domain is calculated from all the points in the frequency domain. This is done by multiplying the frequency domain by a sinusoid, and integrating (continuous frequency domain) or summing (discrete frequency domain) over the appropriate frequency domain section. If the frequency domain is complex and aperiodic, the appropriate section is negative infinity to

positive infinity. If the frequency domain is complex and periodic, the appropriate section is over one complete cycle, i.e., $-\pi$ to π (continuous frequency domain), or 0 to N-1 (discrete frequency domain). If the frequency domain is real and aperiodic, the appropriate section is zero to positive infinity, that is, only the positive frequencies. Lastly, if the frequency domain is real and periodic, the appropriate section is over the one-half cycle containing the positive frequencies, either 0 to π (continuous frequency domain) or 0 to $N/2$ (discrete frequency domain).

9. Scaling

To make the analysis and synthesis equations undo each other, a scaling factor must be placed on one or the other equation. In Table 29-1, we have placed the scaling factors with the analysis equations. In the complex case, these scaling factors are: $1/N$, $1/T$, or $1/2\pi$. Since the real transforms do not use negative frequencies, the scaling factors are twice as large: $2/N$, $2/T$, or $1/\pi$. The real transforms also include a negative sign in the calculation of the imaginary part of the frequency spectrum (an option used to make the real transforms more consistent with the complex transforms). Lastly, the synthesis equations for the real DFT and the real Fourier Series have special scaling instructions involving $Re\,X(0)$ and $Re\,X[N/2]$.

10. Variations

These equations may look different in other publications. Here are a few variations to watch out for:

- Using f instead of ω by the relation: $\omega = 2\pi f$
- Integrating over other periods, such as: $-T$ to 0, $-T/2$ to $T/2$, or 0 to T
- Moving all or part of the scaling factor to the synthesis equation
- Replacing the period with the fundamental frequency, $f_0 = 1/T$
- Using other variable names, for example, ω can become Ω in the DTFT, and $Re\,X[k]$ & $Im\,X[k]$ can become a_k & b_k in the Fourier Series

Why the Complex Fourier Transform is Used

It is painfully obvious from this chapter that the complex DFT is much more complicated than the real DFT. Are the benefits of the complex DFT really worth the effort to learn the intricate mathematics? The answer to this question depends on who you are, and what you plan on using DSP for. A basic premise of this book is that most practical DSP techniques can be understood and used without resorting to complex transforms. If you are learning DSP to assist in your non-DSP research or engineering, the complex DFT is probably overkill.

Nevertheless, complex mathematics is the primary language of those that specialize in DSP. If you do not understand this language, you cannot communicate with professionals in the field. This includes the ability to understand the DSP literature: books, papers, technical articles, etc. Why are complex techniques so popular with the professional DSP crowd?

Discrete Fourier Transform (DFT)

complex transform	real transform
synthesis $$x[n] = \sum_{k=0}^{N-1} X[k]\, e^{j2\pi kn/N}$$	*synthesis* $$x[n] = \sum_{k=0}^{N/2} Re\,X[k] \cos(2\pi kn/N)$$ $$- Im\,X[k] \sin(2\pi kn/N)$$
analysis $$X[k] = \frac{1}{N}\sum_{n=0}^{N-1} x[n]\, e^{-j2\pi kn/N}$$	*analysis* $$Re\,X[k] = \frac{2}{N}\sum_{n=0}^{N-1} x[n] \cos(2\pi kn/N)$$ $$Im\,X[k] = \frac{-2}{N}\sum_{n=0}^{N-1} x[n] \sin(2\pi kn/N)$$

Time domain:
   x[n] is complex, discrete and periodic
   n runs over one period, from 0 to N-1
 	
Frequency domain:
   X[k] is complex, discrete and periodic
   k runs over one period, from 0 to N-1
   k = 0 to N/2 are positive frequencies
   k = N/2 to N-1 are negative frequencies
	Time domain:
	   x[n] is real, discrete and periodic
	   n runs over one period, from 0 to N-1
	
	Frequency domain:
	   Re X[k] is real, discrete and periodic
	   Im X[k] is real, discrete and periodic
	   k runs over one-half period, from 0 to N/2
	
	Note: Before using the synthesis equation, the values for Re X[0] and Re X[N/2] must be divided by two.

Discrete Time Fourier Transform (DTFT)

complex transform	real transform
synthesis $$x[n] = \int_{0}^{2\pi} X(\omega)\, e^{j\omega n}\, d\omega$$	*synthesis* $$x[n] = \int_{0}^{\pi} Re\,X(\omega) \cos(\omega n)$$ $$- Im\,X(\omega) \sin(\omega n)\, d\omega$$
analysis $$X(\omega) = \frac{1}{2\pi}\sum_{n=-\infty}^{+\infty} x[n]\, e^{-j\omega n}$$	*analysis* $$Re\,X(\omega) = \frac{1}{\pi}\sum_{n=-\infty}^{+\infty} x[n] \cos(\omega n)$$ $$Im\,X(\omega) = \frac{-1}{\pi}\sum_{n=-\infty}^{+\infty} x[n] \sin(\omega n)$$

Time domain:
   x[n] is complex, discrete and aperiodic
   n runs from -∞ to +∞
 	
Frequency domain:
   X(ω) is complex, continuous, and periodic
   ω runs over a single period, from 0 to 2π
   ω = 0 to π are positive frequencies
   ω = π to 2π are negative frequencies
	Time domain:
	   x[n] is real, discrete and aperiodic
	   n runs from -∞ to +∞
	
	Frequency domain:
	   Re X(ω) is real, continuous and periodic
	   Im X(ω) is real, continuous and periodic
	   ω runs over one-half period, from 0 to π

TABLE 29-1   The Fourier Transforms.

Fourier Series

complex transform	real transform

synthesis

$$x(t) = \sum_{k=-\infty}^{+\infty} X[k]\, e^{j2\pi kt/T}$$

synthesis

$$x(t) = \sum_{k=0}^{+\infty} Re\,X[k]\cos(2\pi kt/T)$$
$$- Im\,X[k]\sin(2\pi kt/T)$$

analysis

$$X[k] = \frac{1}{T}\int_{0}^{T} x(t)\, e^{-j2\pi kt/T}dt$$

analysis

$$Re\,X[k] = \frac{2}{T}\int_{0}^{T} x(t)\cos(2\pi kt/T)\, dt$$

$$Im\,X[k] = \frac{-2}{T}\int_{0}^{T} x(t)\sin(2\pi kt/T)\, dt$$

Time domain:
   x(t) is complex, continuous and periodic
   t runs over one period, from 0 to T

Frequency domain:
   X[k] is complex, discrete, and aperiodic
   k runs from -∞ to +∞
   k > 0 are positive frequencies
   k < 0 are negative frequencies

Time domain:
   x(t) is real, continuous, and periodic
   t runs over one period, from 0 to T

Frequency domain:
   Re X[k] is real, discrete and aperiodic
   Im X[k] is real, discrete and aperiodic
   k runs from 0 to +∞

Note: Before using the synthesis equation, the value for Re X[0] must be divided by two.

Fourier Transform

complex transform	real transform

synthesis

$$x(t) = \int_{-\infty}^{+\infty} X(\omega)\, e^{j\omega t}\, d\omega$$

synthesis

$$x(t) = \int_{0}^{+\infty} Re\,X(\omega)\cos(\omega t)$$
$$- Im\,X(\omega)\sin(\omega t)\, dt$$

analysis

$$X(\omega) = \frac{1}{2\pi}\int_{-\infty}^{+\infty} x(t)\, e^{-j\omega t}\, dt$$

analysis

$$Re\,X(\omega) = \frac{1}{\pi}\int_{-\infty}^{+\infty} x(t)\cos(\omega t)\, dt$$

$$Im\,X(\omega) = \frac{-1}{\pi}\int_{-\infty}^{+\infty} x(t)\sin(\omega t)\, dt$$

Time domain:
   x(t) is complex, continious and aperiodic
   t runs from -∞ to +∞

Frequency domain:
   X(ω) is complex, continious, and aperiodic
   ω runs from -∞ to +∞
   ω > 0 are positive frequencies
   ω < 0 are negative frequencies

Time domain:
   x(t) is real, continuous, and aperiodic
   t runs from -∞ to +∞

Frequency domain:
   Re X[ω] is real, continuous and aperiodic
   Im X[ω] is real, continuous and aperiodic
   ω runs from 0 to +∞

TABLE 29-1 The Fourier Transforms.

There are several reasons we have already mentioned: compact equations, symmetry between the analysis and synthesis equations, symmetry between the time and frequency domains, inclusion of negative frequencies, a stepping stone to the Laplace and z-transforms, etc.

There is also a more philosophical reason we have not discussed, something called *truth*. We started this chapter by listing several ways that the real Fourier transform is awkward. When the complex Fourier transform was introduced, the problems vanished. Wonderful, we said, the complex Fourier transform has solved the difficulties.

While this is true, it does not give the complex Fourier transform its proper due. Look at this situation this way. In spite of its abstract nature, the complex Fourier transform *properly* describes how physical systems behave. When we restrict the mathematics to be real numbers, problems arise. In other words, these problems are not *solved* by the complex Fourier transform, they are *introduced* by the real Fourier transform. In the world of mathematics, the complex Fourier transform is a greater truth than the real Fourier transform. This holds great appeal to mathematicians and academicians, a group that strives to expand human knowledge, rather than simply solving a particular problem at hand.

The Laplace Transform

The two main techniques in signal processing, convolution and Fourier analysis, teach that a linear system can be completely understood from its impulse or frequency response. This is a very generalized approach, since the impulse and frequency responses can be of nearly any shape or form. In fact, it is *too* general for many applications in science and engineering. Many of the parameters in our universe interact through *differential equations*. For example, the voltage across an inductor is proportional to the derivative of the current through the device. Likewise, the force applied to a mass is proportional to the derivative of its velocity. Physics is filled with these kinds of relations. The frequency and impulse responses of these systems cannot be arbitrary, but must be consistent with the solution of these differential equations. This means that their impulse responses can only consist of *exponentials* and *sinusoids*. The Laplace transform is a technique for analyzing these special systems when the signals are *continuous*. The z-transform is a similar technique used in the *discrete* case.

The Nature of the s-Domain

The Laplace transform is a well established mathematical technique for solving differential equations. It is named in honor of the great French mathematician, Pierre Simon De Laplace (1749-1827). Like all transforms, the Laplace transform changes one signal into another according to some fixed set of rules or equations. As illustrated in Fig. 30-1, the Laplace transform changes a signal in the time domain into a signal in the **s-domain**, also called the **s-plane**. The time domain signal is continuous, extends to both positive and negative infinity, and may be either periodic or aperiodic. The Laplace transform allows the time domain to be *complex*; however, this is seldom needed in signal processing. In this discussion, and nearly all practical applications, the time domain signal is completely real.

As shown in Fig. 30-1, the s-domain is a complex plane, i.e., there are real numbers along the horizontal axis and imaginary numbers along the vertical axis. The distance along the real axis is expressed by the variable, σ, a lower case Greek sigma. Likewise, the imaginary axis uses the variable, ω, the natural frequency. This coordinate system allows the location of any point to be specified by providing values for σ and ω. Using complex notation, each location is represented by the complex variable, s, where: $s = σ + jω$. Just as with the Fourier transform, signals in the s-domain are represented by capital letters. For example, a time domain signal, $x(t)$, is transformed into an s-domain signal, $X(s)$, or alternatively, $X(σ,ω)$. The s-plane is continuous, and extends to infinity in all four directions.

In addition to having a *location* defined by a complex number, each point in the s-domain has a *value* that is a complex number. In other words, each location in the s-plane has a real part and an imaginary part. As with all complex numbers, the real & imaginary parts can alternatively be expressed as the magnitude & phase.

Just as the Fourier transform analyzes signals in terms of sinusoids, the Laplace transform analyzes signals in terms of sinusoids *and* exponentials. From a mathematical standpoint, this makes the Fourier transform a *subset* of the more elaborate Laplace transform. Figure 30-1 shows a graphical description of how the s-domain is related to the time domain. To find the values along a vertical line in the s-plane (the values at a particular σ), the time domain signal is first multiplied by the exponential curve: $e^{-σt}$. The left half of the s-plane multiplies the time domain with exponentials that *increase* with time (σ < 0), while in the right half the exponentials *decrease* with time (σ > 0). Next, take the complex Fourier transform of the exponentially weighted signal. The resulting spectrum is placed along a vertical line in the s-plane, with the top half of the s-plane containing the positive frequencies and the bottom half containing the negative frequencies. Take special note that the values on the y-axis of the s-plane (σ = 0) are exactly equal to the Fourier transform of the time domain signal.

As discussed in the last chapter, the complex Fourier Transform is given by:

$$X(ω) = \int_{-∞}^{∞} x(t) e^{-jωt} dt$$

This can be expanded into the Laplace transform by first multiplying the time domain signal by the exponential term:

$$X(σ,ω) = \int_{-∞}^{∞} [x(t) e^{-σt}] e^{-jωt} dt$$

While this is not the simplest form of the Laplace transform, it is probably the best description of the strategy and operation of the technique. To

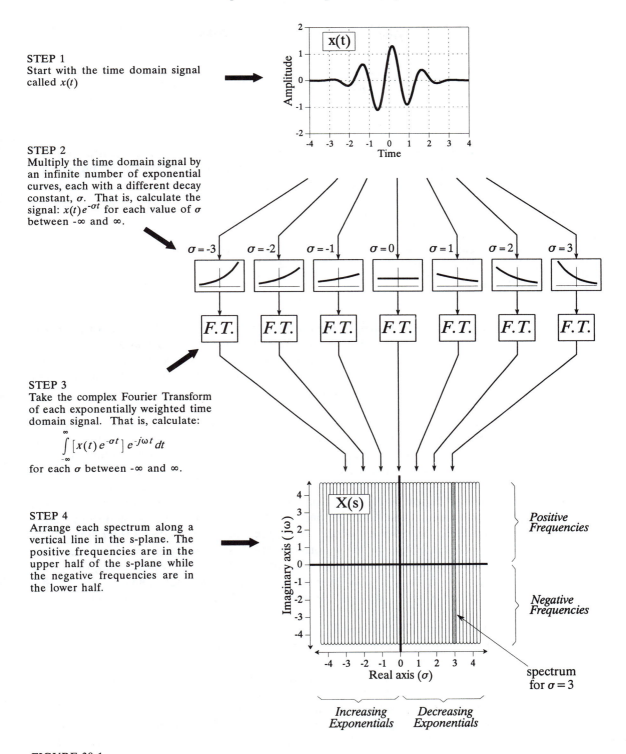

STEP 1
Start with the time domain signal
called $x(t)$

STEP 2
Multiply the time domain signal by
an infinite number of exponential
curves, each with a different decay
constant, σ. That is, calculate the
signal: $x(t)e^{-\sigma t}$ for each value of σ
between $-\infty$ and ∞.

STEP 3
Take the complex Fourier Transform
of each exponentially weighted time
domain signal. That is, calculate:

$$\int_{-\infty}^{\infty} \left[x(t)\, e^{-\sigma t} \right] e^{-j\omega t}\, dt$$

for each σ between $-\infty$ and ∞.

STEP 4
Arrange each spectrum along a
vertical line in the s-plane. The
positive frequencies are in the
upper half of the s-plane while
the negative frequencies are in
the lower half.

FIGURE 30-1
The Laplace transform. The Laplace transform converts a signal in the *time domain*, $x(t)$, into a signal in the *s-domain*, $X(s)$ or $X(\sigma,\omega)$. The values along each vertical line in the s-domain can be found by multiplying the time domain signal by an exponential curve with a decay constant σ, and taking the complex Fourier transform. When the time domain is entirely real, the upper half of the s-plane is a mirror image of the lower half.

place the equation in a shorter form, the two exponential terms can be combined:

$$X(\sigma,\omega) = \int_{-\infty}^{\infty} x(t)\, e^{-(\sigma+j\omega)t}\, dt$$

Finally, the *location* in the complex plane can be represented by the complex variable, s, where $s = \sigma + j\omega$. This allows the equation to be reduced to an even more compact expression:

EQUATION 30-1
The Laplace transform. This equation defines how a time domain signal, $x(t)$, is related to an s-domain signal, $X(s)$. The s-domain variables, s, and $X(\ )$, are complex. While the time domain *may* be complex, it is usually real.

$$X(s) = \int_{-\infty}^{\infty} x(t)\, e^{-st}\, dt$$

This is the final form of the Laplace transform, one of the most important equations in signal processing and electronics. Pay special attention to the term: e^{-st}, called a *complex exponential*. As shown by the above derivation, complex exponentials are a compact way of representing both sinusoids and exponentials in a single expression.

Although we have explained the Laplace transform as a two stage process (multiplication by an exponential curve followed by the Fourier transform), keep in mind that this is only a teaching aid, a way of breaking Eq. 30-1 into simpler components. The Laplace transform is a single equation relating $x(t)$ and $X(s)$, not a step-by-step procedure. Equation 30-1 describes how to calculate each *point* in the s-plane (identified by its values for σ and ω) based on the values of σ, ω, and the time domain signal, $x(t)$. Using the Fourier transform to *simultaneously* calculate all the points along a vertical line is merely a convenience, not a requirement. However, it is very important to remember that the values in the s-plane along the y-axis ($\sigma = 0$) are *exactly* equal to the Fourier transform. As explained later in this chapter, this is a key part of why the Laplace transform is useful.

To explore the nature of Eq. 30-1 further, let's look at several individual points in the s-domain and examine how the values at these locations are related to the time domain signal. To start, recall how individual points in the *frequency domain* are related to the time domain signal. Each point in the frequency domain, identified by a specific value of ω, corresponds to two sinusoids, $\cos(\omega t)$ and $\sin(\omega t)$. The real part is found by multiplying the time domain signal by the cosine wave, and then integrating from $-\infty$ to ∞. The imaginary part is found in the same way, except the sine wave is used. If we are dealing with the *complex* Fourier transform, the values at the corresponding negative frequency, $-\omega$, will be the complex conjugate (same real part, negative imaginary part) of the values at ω. The Laplace transform is just an extension of these same concepts.

s-Domain

Associated Waveforms

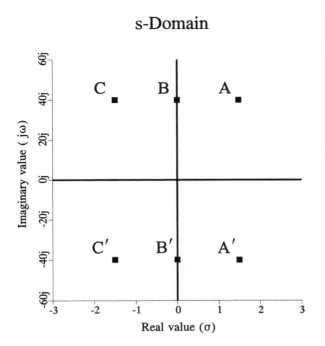

FIGURE 30-2

Waveforms associated with the s-domain. Each location in the s-domain is identified by two parameters: σ and ω. These parameters also define two waveforms associated with each location. If we only consider *pairs* of points (such as: A&A', B&B', and C&C'), the two waveforms associated with each location are sine and cosine waves of frequency ω, with an exponentially changing amplitude controlled by σ.

Figure 30-2 shows three *pairs* of points in the s-plane: A&A', B&B', and C&C'. Just as in the complex frequency spectrum, the points at A, B, & C (the positive frequencies) are the complex conjugates of the points at A', B', & C' (the negative frequencies). The top half of the s-plane is a mirror image of the lower half, and both halves are needed to correspond with a real time domain signal. In other words, treating these points in pairs bypasses the complex math, allowing us to operate in the time domain with only real numbers.

Since each of these pairs has specific values for σ and $\pm\omega$, there are two waveforms associated with each pair: $\cos(\omega t)e^{-\sigma t}$ and $\sin(\omega t)e^{-\sigma t}$. For instance, points A&A' are at a location of $\sigma = 1.5$ and $\omega = \pm 40$, and therefore correspond to the waveforms: $\cos(40t)e^{-1.5t}$ and $\sin(40t)e^{-1.5t}$. As shown in Fig. 30-2, these are sinusoids that exponentially *decreases* in amplitude as time progresses. In this same way, the sine and cosine waves associated with B&B' have a *constant* amplitude, resulting from the value of σ being zero. Likewise, the sine and cosine waves that are associated with locations C&C' exponentially *increases* in amplitude, since σ is negative.

The value at each location in the s-plane consists of a *real part* and an *imaginary part*. The real part is found by multiplying the time domain signal by the exponentially weighted cosine wave and then integrated from -∞ to ∞. The imaginary part is found in the same way, except the exponentially weighted sine wave is used instead. It looks like this in equation form, using the real part of A&A' as an example:

$$Re X(\sigma{=}1.5, \omega{=}\pm40) = \int_{-\infty}^{\infty} x(t)\cos(40t)e^{-1.5t}\,dt$$

Figure 30-3 shows an example of a time domain waveform, its frequency spectrum, and its s-domain representation. The example time domain signal is a rectangular pulse of width *two* and height *one*. As shown, the complex Fourier transform of this signal is a sinc function in the real part, and an entirely zero signal in the imaginary part. The s-domain is an undulating two-dimensional signal, displayed here as topographical surfaces of the real and imaginary parts. The mathematics works like this:

$$X(s) = \int_{-\infty}^{\infty} x(t)\,e^{-st}\,dt = \int_{-1}^{1} 1\,e^{-st}\,dt$$

In words, we start with the definition of the Laplace transform (Eq. 30-1), plug in the unity value for $x(t)$, and change the limits to match the length of the nonzero portion of the time domain signal. Evaluating this integral provides the s-domain signal, expressed in terms of the complex location, s, and the complex value, $X(s)$:

$$X(s) = \frac{e^s - e^{-s}}{s}$$

While this is the most compact form of the answer, the use of complex variables makes it difficult to understand, and impossible to generate a visual display, such as Fig. 30-3. The solution is to replace the complex variable, s, with $\sigma{+}j\omega$, and then separate the real and imaginary parts:

$$Re X(\sigma, \omega) = \frac{\sigma\cos(\omega)[e^{\sigma}-e^{-\sigma}] + \omega\sin(\omega)[e^{\sigma}+e^{-\sigma}]}{\sigma^2 + \omega^2}$$

$$Im X(\sigma, \omega) = \frac{\sigma\sin(\omega)[e^{\sigma}+e^{-\sigma}] - \omega\cos(\omega)[e^{\sigma}-e^{-\sigma}]}{\sigma^2 + \omega^2}$$

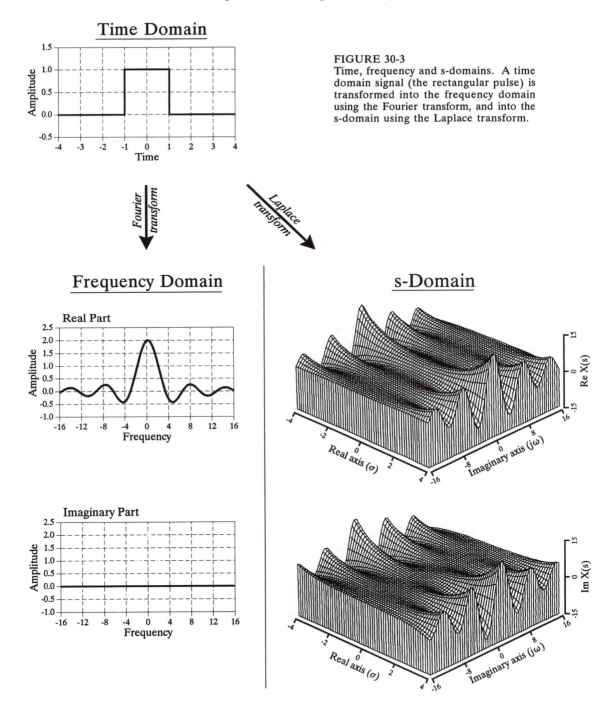

Time Domain

FIGURE 30-3
Time, frequency and s-domains. A time domain signal (the rectangular pulse) is transformed into the frequency domain using the Fourier transform, and into the s-domain using the Laplace transform.

Fourier transform

Laplace transform

Frequency Domain

Real Part

Imaginary Part

s-Domain

The topographical surfaces in Fig. 30-3 are graphs of these equations. These equations are quite long and the mathematics to derive them is very tedious. This brings up a practical issue: with algebra of this complexity, how do we know that we haven't made an error in the calculations? One check is to verify that these equations reduce to the *Fourier transform* along the y-axis. This is done by setting σ to zero in the equations, and simplifying:

$$Re\, X(\sigma, \omega) \bigg|_{\sigma=0} = \frac{2\sin(\omega)}{\omega} \qquad\qquad Im\, X(\sigma, \omega) \bigg|_{\sigma=0} = 0$$

As illustrated in Fig. 30-3, these are the correct frequency domain signals, the same as found by directly taking the Fourier transform of the time domain waveform.

Strategy of the Laplace Transform

An analogy will help in explaining how the Laplace transform is used in signal processing. Imagine you are traveling by train at night between two cities. Your map indicates that the path is very straight, but the night is so dark you cannot see any of the surrounding countryside. With nothing better to do, you notice an altimeter on the wall of the passenger car and decide to keep track of the elevation changes along the route.

Being bored after a few hours, you strike up a conversation with the conductor: "Interesting terrain," you say. "It seems we are generally increasing in elevation, but there are a few interesting irregularities that I have observed." Ignoring the conductor's obvious disinterest, you continue: "Near the start of our journey, we passed through some sort of abrupt rise, followed by an equally abrupt descent. Later we encountered a shallow depression." Thinking you might be dangerous or demented, the conductor decides to respond: "Yes, I guess that is true. Our destination is located at the base of a large mountain range, accounting for the general increase in elevation. However, along the way we pass on the outskirts of a large mountain and through the center of a valley."

Now, think about how you understand the relationship between elevation and distance along the train route, compared to that of the conductor. Since you have directly measured the elevation along the way, you can rightly claim that you know *everything* about the relationship. In comparison, the conductor knows this same complete information, but in a simpler and more intuitive form: the location of the hills and valleys that *cause* the dips and humps along the path. While your description of the signal might consist of thousands of individual measurements, the conductor's description of the signal will contain only a few parameters.

To show how this is analogous to signal processing, imagine we are trying to understand the characteristics of some electric circuit. To aid in our investigation, we carefully measure the impulse response and/or the frequency response. As discussed in previous chapters, the impulse and frequency responses contain *complete* information about this linear system.

However, this does not mean that you know the information in the *simplest* way. In particular, you understand the frequency response as a set of values that change with frequency. Just as in our train analogy, the frequency response can be more easily understood in terms of the terrain *surrounding* the frequency response. That is, by the characteristics of the s-plane.

With the train analogy in mind, look back at Fig. 30-3, and ask: how does the shape of this s-domain aid in understanding the frequency response? The answer is, it doesn't! The s-plane in this example makes a nice graph, but it provides no insight into why the frequency domain behaves as it does. This is because the Laplace transform is designed to analyze a specific class of time domain signals: *impulse responses that consist of sinusoids and exponentials*. If the Laplace transform is taken of some other waveform (such as the rectangular pulse in Fig. 30-3), the resulting s-domain is meaningless.

As mentioned in the introduction, systems that belong to this class are extremely common in science and engineering. This is because sinusoids and exponentials are solutions to *differential equations*, the mathematics that controls much of our physical world. For example, all of the following systems are governed by differential equations: electric circuits, wave propagation, linear and rotational motion, electric and magnetic fields, heat flow, etc.

Imagine we are trying to understand some linear system that is controlled by differential equations, such as an electric circuit. Solving the differential equations provides a mathematical way to find the impulse response. Alternatively, we could measure the impulse response using suitable pulse generators, oscilloscopes, data recorders, etc. Before we inspect the newly found impulse response, we ask ourselves what we *expect* to find. There are several characteristics of the waveform that we know without even looking. First, the impulse response must be *causal*. In other words, the impulse response must have a value of zero until the input becomes nonzero at $t = 0$. This is the cause and effect that our universe is based upon.

The second thing we know about the impulse response is that it will be composed of *sinusoids and exponentials*, because these are the solutions to the differential equations that govern the system. Try as we might, we will never find this type of system having an impulse response that is, for example, a square pulse or triangular waveform. Third, the impulse response will be *infinite* in length. That is, it has nonzero values that extend from $t = 0$ to $t = +\infty$. This is because sine and cosine waves have a constant amplitude, and exponentials decay toward zero without ever actually reaching it. If the system we are investigating is **stable**, the amplitude of the impulse response will become smaller as time increases, reaching a value of zero at $t = +\infty$. There is also the possibility that the system is **unstable**, for example, an amplifier that spontaneously oscillates due to an excessive amount of feedback. In this case, the impulse response will *increase* in amplitude as time increases, becoming infinitely large. Even the smallest disturbance to this system will produce an unbounded output.

The general mathematics of the Laplace transform is very similar to that of the Fourier transform. In both cases, predetermined waveforms are multiplied by the time domain signal, and the result integrated. At first glance, it would appear that the strategy of the Laplace transform is the same as the Fourier transform: correlate the time domain signal with a set of basis functions to decompose the waveform. Not true! Even though the mathematics is much the same, the rationale behind the two techniques is very different. The Laplace transform *probes* the time domain waveform to identify its key features: the *frequencies* of the sinusoids, and the *decay constants* of the exponentials. An example will show how this works.

The center column in Fig. 30-5 shows the impulse response of the *RLC* notch filter discussed in Chapter 28. It contains an impulse at $t = 0$, followed by an exponentially decaying sinusoid. As illustrated in (a) through (e), we will *probe* this impulse response with various exponentially decaying sinusoids. Each of these probing waveforms is characterized by two parameters: ω, that determines the sinusoidal frequency, and σ, that determines the decay rate. In other words, each probing waveform corresponds to a different location in the s-plane, as shown by the s-plane diagram in Fig. 30-4. The impulse response is probed by *multiplying* it with these waveforms, and then integrating the result from $t = -\infty$ to $+\infty$. This action is shown in the right column. Our goal is to find combinations of σ and ω that exactly *cancel* the impulse response being investigated. This cancellation can occur in two forms: the area under the curve can be either *zero*, or just *barely infinite*. All other results are uninteresting and can be ignored. Locations in the s-plane that produce a zero cancellation are called **zeros** of the system. Likewise, locations that produce the "just barely infinite" type of cancellation are called **poles**. Poles and zeros are analogous to the mountains and valleys in our train story, representing the terrain "around" the frequency response.

To start, consider what happens when the probing waveform decreases in amplitude as time advances, as shown in (a). This will occur whenever $\sigma > 0$ (the right half of the s-plane). Since both the impulse response and the probe becomes smaller with increasing time, the product of the two will also have this same characteristic. When the product of the two waveforms is integrated from negative to positive infinity, the result will be some number that is not especially interesting. In particular, a decreasing probe

s-plane diagram

FIGURE 30-4
Pole-zero example. The notch filter has two poles (represented by ×) and two zeros (represented by ○). This s-plane diagram shows the five locations we will "probe" in this example to analyze this system. (Figure 30-5 is a continuation of this example).

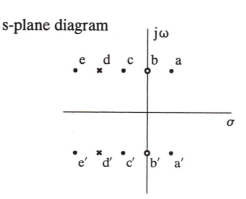

Probing waveform, $p(t)$ Impulse response, $h(t)$ Multiply: $p(t) \times h(t)$

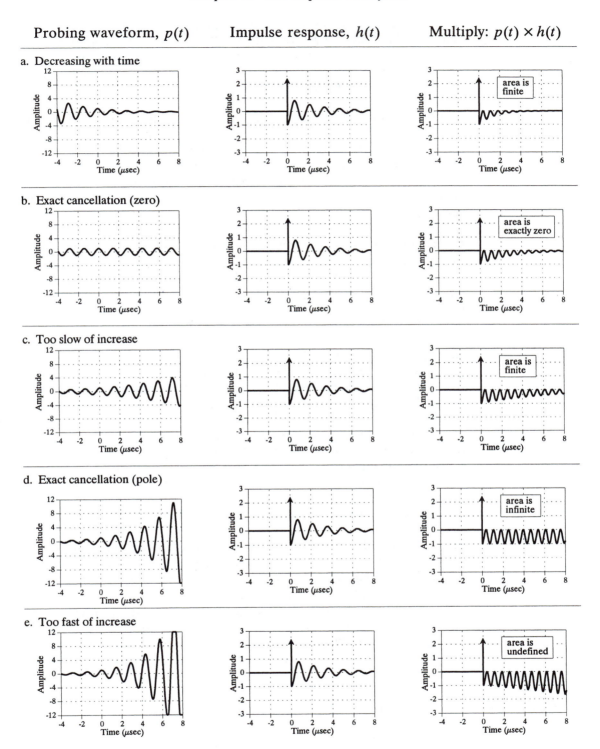

a. Decreasing with time

b. Exact cancellation (zero)

c. Too slow of increase

d. Exact cancellation (pole)

e. Too fast of increase

FIGURE 30-5
Probing the impulse response. The Laplace transform can be viewed as probing the system's impulse response with various exponentially decaying sinusoids. Probing waveforms that produce a cancellation are called *poles* and *zeros*. This illustration shows five probing waveforms (left column) being applied to the impulse response of a notch filter (center column). The locations in the s-plane that correspond to these five waveforms are shown in Fig. 30-4.

cannot cancel a decreasing impulse response. This means that a stable system will not have any poles with $\sigma > 0$. In other words, all of the poles in a stable system are confined to the left half of the s-plane. In fact, poles in the right half of the s-place show that the system is unstable (i.e., an impulse response that *increases* with time).

Figure (b) shows one of the special cases we have been looking for. When this waveform is multiplied by the impulse response, the resulting integral has a value of zero. This occurs because the area above the x-axis (from the delta function) is exactly equal to the area below (from the rectified sinusoid). The values for σ and ω that produce this type of cancellation are called a *zero* of the system. As shown in the s-plane diagram of Fig. 30-4, zeros are indicated by small circles (O).

Figure (c) shows the next probe we can try. Here we are using a sinusoid that exponentially *increases* with time, but at a rate *slower* than the impulse response is *decreasing* with time. This results in the product of the two waveforms also decreasing as time advances. As in (a), this makes the integral of the product some uninteresting real number. The important point being that no type of exact cancellation occurs.

Jumping out of order, look at (e), a probing waveform that increases at a *faster* rate than the impulse response decays. When multiplied, the resulting signal increases in amplitude as time advances. This means that the area under the curve becomes larger with increasing time, and the total area from $t = -\infty$ to $+\infty$ is not defined. In mathematical jargon, the *integral does not converge*. In other words, not all areas of the s-plane have a defined value. The portion of the s-plane where the integral is defined is called the **region-of-convergence**. In some mathematical techniques it is important to know what portions of the s-plane are within the region-of-convergence. However, this information is not needed for the applications in this book. Only the exact cancellations are of interest for this discussion.

In (d), the probing waveform increases at exactly the same rate that the impulse response decreases. This makes the product of the two waveforms have a constant amplitude. In other words, this is the dividing line between (c) and (e), resulting in a total area that is just *barely undefined* (if the mathematicians will forgive this loose description). In more exact terms, this point is on the borderline of the region of convergence. As mentioned, values for σ and ω that produce this type of exact cancellation are called *poles* of the system. Poles are indicated in the s-plane by crosses (×).

Analysis of Electric Circuits

We have introduced the Laplace transform in graphical terms, describing what the waveforms look like and how they are manipulated. This is the most intuitive way of understanding the approach, but is very different from how it is actually used. The Laplace transform is inherently a mathematical technique; it is used by writing and manipulating *equations*. The problem is, it is easy to become lost in the abstract nature of the complex algebra and loose all connection to the real world. Your task is to merge the two views together. The Laplace transform is the primary method for analyzing electric circuits. Keep in mind that *any* system governed by differential equations can be handled the same way; electric circuits are just an example we are using.

The brute force approach is to solve the differential equations controlling the system, providing the system's impulse response. The impulse response can then be converted into the s-domain via Eq. 30-1. Fortunately, there is a better way: transform each of the individual components into the s-domain, and *then* account for how they interact. This is very similar to the *phasor transform* presented in Chapter 28, where resistors, inductors and capacitors are represented by R, $j\omega L$, and $1/j\omega C$, respectively. In the Laplace transform, resistors, inductors and capacitors become the complex variables: R, sL, and $1/sC$. Notice that the phasor transform is a *subset* of the Laplace transform. That is, when σ is set to zero in $s = \sigma + j\omega$, R becomes R, sL becomes $j\omega L$, and $1/sC$ becomes $1/j\omega C$.

Just as in the last chapter, we will treat each of the three components as an individual system, with the current waveform being the input signal, and the voltage waveform being the output signal. When we say that resistors, inductors and capacitors become R, sL , and $1/sC$ in the s-domain, this refers to the output divided by the input. In other words, the Laplace transform of the *voltage waveform* divided by the Laplace transform of the *current waveform* is equal to these expressions.

As an example of this, imagine we force the current through an inductor to be a unity amplitude cosine wave with a frequency given by ω_0. The resulting *voltage* waveform across the inductor can be found by solving the differential equation that governs its operation:

$$v(t) \;=\; L\frac{d}{dt}i(t) \;=\; L\frac{d}{dt}\cos(\omega_0 t) \;=\; \omega_0 L\sin(\omega_0 t)$$

If we start the current waveform at $t = 0$, the voltage waveform will also start at this same time (i.e., $i(t) = 0$ and $v(t) = 0$ for $t < 0$). These voltage and current waveforms are converted into the s-domain by Eq. 30-1:

$$I(s) \;=\; \int_0^\infty \cos(\omega_0 t)e^{-st}dt \;=\; \frac{\omega_0}{\omega_0^2 + s^2}$$

$$V(s) \;=\; \int_0^\infty \omega_0 L\sin(\omega_0 t)e^{-st}dt \;=\; \frac{\omega_0 L s}{\omega_0^2 + s^2}$$

To complete this example, we will divide the s-domain voltage by the s-domain current, just as if we were using Ohm's law ($R = V/I$):

$$\frac{V(s)}{I(s)} = \frac{\dfrac{\omega_0 L s}{\omega_0^2 + s^2}}{\dfrac{\omega_0}{\omega_0^2 + s^2}} = sL$$

We find that the s-domain representation of the voltage across the inductor, divided by the s-domain representation of the current through the inductor, is equal to sL. This is *always* the case, regardless of the current waveform we start with. In a similar way, the ratio of s-domain voltage to s-domain current is always equal to R for resistors, and $1/sC$ for capacitors.

Figure 30-6 shows an example circuit we will analyze with the Laplace transform, the *RLC* notch filter discussed in Chapter 28. Since this analysis is the same for all electric circuits, we will outline it in steps.

Step 1. Transform each of the components into the s-domain. In other words, replace the value of each resistor with R, each inductor with sL, and each capacitor with $1/sC$. This is shown in Fig. 30-6.

Step 2: Find $H(s)$, the output divided by the input. As described in Chapter 28, this is done by treating each of the components as if they obey Ohm's law, with the "resistances" given by: R, sL, and $1/sC$. This allows us to use the standard equations for resistors in series, resistors in parallel, voltage dividers, etc. Treating the *RLC* circuit in this example as a voltage divider (just as in Chapter 28), $H(s)$ is found:

$$H(s) = \frac{V_{out}(s)}{V_{in}(s)} = \frac{sL + 1/sC}{R + sL + 1/sC} = \frac{sL + 1/sC}{R + sL + 1/sC}\left[\frac{s}{s}\right] = \frac{Ls^2 + 1/C}{Ls^2 + Rs + 1/C}$$

As you recall from *Fourier* analysis, the frequency spectrum of the output signal divided by the frequency spectrum of the input signal is equal to the system's *frequency response*, given the symbol, $H(\omega)$. The above equation is an extension of this into the s-domain. The signal, $H(s)$, is called the system's **transfer function**, and is equal to the s-domain representation of the output signal divided by the s-domain representation of the input signal. Further, $H(s)$ is equal to the *Laplace transform* of the impulse response, just the same as $H(\omega)$ is equal to the *Fourier transform* of the impulse response.

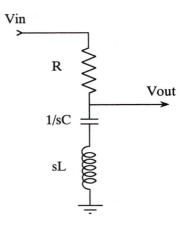

FIGURE 30-6
Notch filter analysis in the s-domain. The first step in this procedure is to replace the resistor, inductor & capacitor values with their s-domain equivalents.

So far, this is identical to the techniques of the last chapter, except for using s instead of $j\omega$. The difference between the two methods is what happens from this point on. This is as far as we can go with $j\omega$. We might graph the frequency response, or examining it in some other way; however, this is a mathematical dead end. In comparison, the interesting aspects of the Laplace transform have just begun. Finding $H(s)$ is the key to Laplace analysis; however, it must be expressed in a particular form to be useful. This requires the algebraic manipulation of the next two steps.

Step 3: Arrange $H(s)$ to be one polynomial over another. This makes the transfer function written as:

EQUATION 30-2
Transfer function in polynomial form.

$$H(s) = \frac{as^2 + bs + c}{as^2 + bs + c}$$

It is always possible to express the transfer function in this form *if* the system is controlled by differential equations. For example, the rectangular pulse shown in Fig. 30-3 is not the solution to a differential equation and its Laplace transform cannot be written in this way. In comparison, any electric circuit composed of resistors, capacitors, and inductors can be written in this form. For the *RLC* notch filter used in this example, the algebra shown in *step 2* has already placed the transfer function in the correct form, that is:

$$H(s) = \frac{as^2 + bs + c}{as^2 + bs + c} = \frac{Ls^2 + 1/C}{Ls^2 + Rs + 1/C}$$

where: $a = L$, $b = 0$, $c = 1/C$; and $a = L$, $b = R$, $c = 1/C$

Step 4: Factor the numerator and denominator polynomials. That is, break the numerator and denominator polynomials into components that each contain a single *s*. When the components are multiplied together, they must equal the original numerator and denominator. In other words, the equation is placed into the form:

Equation 30-3
The factored s-domain. This form allows the s-domain to be expressed as poles and zeros.

$$H(s) = \frac{(s-z_1)(s-z_2)(s-z_3)\cdots}{(s-p_1)(s-p_2)(s-p_3)\cdots}$$

The roots of the numerator, $z_1, z_2, z_3 \cdots$, are the **zeros** of the equation, while the roots of the denominator, $p_1, p_2, p_3 \cdots$, are the **poles**. These are the same zeros and poles we encountered earlier in this chapter, and we will discuss how they are used in the next section.

Factoring an s-domain expression is straightforward if the numerator and denominator are *second-order polynomials*, or less. In other words, we can easily handle the terms: s and s^2, but not: $s^3, s^4, s^5, \cdots$. This is because the roots of a second-order polynomial, $ax^2 + bx + c$, can be found by the quadratic equation: $x = -b \pm \sqrt{b^2 - 4ac}/2a$. Using this method, the transfer function of the example notch filter is factored into:

$$H(s) = \frac{(s-z_1)(s-z_2)}{(s-p_1)(s-p_2)}$$

where:

$$z_1 = j/\sqrt{LC} \qquad\qquad p_1 = \frac{-R + \sqrt{R^2 - 4L/C}}{2L}$$

$$z_2 = -j/\sqrt{LC} \qquad\qquad p_2 = \frac{-R - \sqrt{R^2 - 4L/C}}{2L}$$

As in this example, a second-order system has a maximum of two zeros and two poles. The number of poles in a system is equal to the number of independent energy storing components. For instance, inductors and capacitors store energy, while resistors do not. The number of zeros will be equal to, or less than, the number of poles.

Polynomials greater than second order cannot generally be factored using algebra, requiring more complicated numerical methods. As an alternative, circuits can be constructed as a *cascade* of *second-order* stages. A good example is the family of analog filters presented in Chapter 3. For instance, an eight pole filter is designed by cascading four stages of two poles each. The important point is that this multistage approach is used to overcome limitations in the *mathematics*, not limitations in the *electronics*.

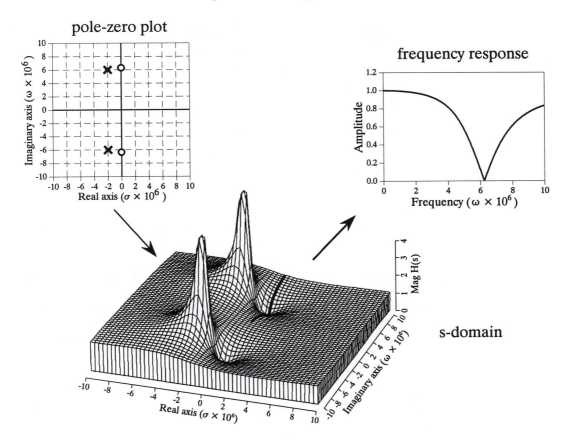

FIGURE 30-6
Poles and zeros in the s-domain. These illustrations show the relationship between the pole-zero plot, the s-domain, and the frequency response. The notch filter component values used in these graphs are: R=220 Ω, C=470 pF, and L = 54 μH. These values place the center of the notch at ω = 6.277 million, i.e., a frequency of approximately 1 MHz.

The Importance of Poles and Zeros

To make this less abstract, we will use actual component values for the notch filter we just analyzed: $R = 220\,\Omega$, $L = 54\,\mu H$, $C = 470\,pF$. Plugging these values into the above equations, places the poles and zeros at:

$$z_1 = 0 + j\,6.277 \times 10^6 \qquad p_1 = -2.037 \times 10^6 + j\,5.937 \times 10^6$$

$$z_2 = 0 - j\,6.277 \times 10^6 \qquad p_2 = -2.037 \times 10^6 - j\,5.937 \times 10^6$$

These pole and zero locations are shown in Fig. 30-7. Each zero is represented by a circle, while each pole is represented by a cross. This is called a **pole-zero diagram**, and is the most common way that s-domain data are displayed. Figure 30-7 also shows a topographical display of the s-plane. For simplicity, only the magnitude is shown, but don't forget that there is a corresponding phase. Just as mountains and valleys determine the shape of the surface of the earth, the poles and zeros determine the shape of the s-plane. Unlike mountains and valleys, every pole and zero is exactly the same shape and size as every other pole and zero. The only unique characteristic a pole or zero has is its *location*. Poles and zeros are important because they provide a concise representation of the value at *any point in the s-plane*. That is, we can completely describe the characteristics of the system using only a *few parameters*. In the case of the *RLC* notch filter, we only need to specify four complex parameters to represent the system: z_1, z_2, p_1, p_2 (each consisting of a real and an imaginary part).

To better understand poles and zeros, imagine an ant crawling around the s-plane. At any particular location the ant happens to be (i.e., some value of s), there is a corresponding value of the transfer function, $H(s)$. This value is a complex number that can be expressed as the magnitude & phase, or as the real & imaginary parts. Now, let the ant carry us to one of the zeros in the s-plane. The value we measure for the real and imaginary parts will be *zero* at this location. This can be understood by examining the mathematical equation for $H(s)$ in Eq. 30-3. If the location, s, is equal to any of the z's, one of the terms in the numerator will be zero. This makes the entire expression equal to zero, regardless of the other values.

Next, our ant journey takes us to one of the poles, where we again measure the value of the real and imaginary parts of $H(s)$. The measured value becomes larger and larger as we come close to the exact location of the pole (hence the name). This can also be understood from Eq. 30-3. If the location, s, is equal to any of the p's, the denominator will be equal to zero, and the division by zero makes the entire expression infinity large.

Having explored the unique locations, our ant journey now moves randomly throughout the s-plane. The value of $H(s)$ at each location depends entirely on the positioning of the poles and the zeros, because there are *no* other types of features allowed in this strange terrain. If we are near a pole, the value will be large; if we are near a zero, the value will be small.

Equation 30-3 also describes how *multiple* poles and zeros interact to form the s-domain signal. Remember, subtracting two complex numbers provides the *distance* between them in the complex plane. For example, $(s - z_1)$ is the distance between the arbitrary location, s, and the zero located at z_1. Therefore, Eq. 30-3 specifies that the value at each location, s, is equal to the distance to all of the zeros *multiplied*, divided by the distance to all of the poles *multiplied*.

This brings us to the heart of this chapter: how the location of the poles & zeros provides a deeper understanding of the system's *frequency response*. The frequency response is equal to the values of $H(s)$ along the imaginary axis, signified by the dark line in the topographical plot of Fig. 30-7. Imagine our ant starting at the origin and crawling along this path. Near the origin, the distance to the zeros is approximately equal to the distance to the poles. This makes the numerator and denominator in Eq. 30-3

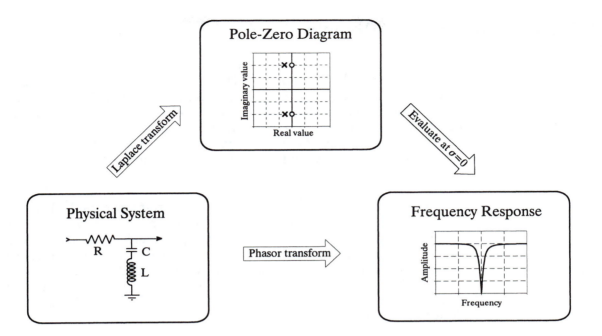

FIGURE 30-8
Strategy for using the Laplace transform. The phasor transform presented in the last chapter (the method using R, $j\omega L$, & $-j/\omega C$) allows the frequency response to be directly calculated from the parameters of the physical system. In comparison, the Laplace transform calculates an s-domain representation from the physical system, usually displayed in the form of a pole-zero diagram. In turn, the frequency response can be obtained from the s-domain by evaluating the transfer function along the imaginary axis. While both methods provide the same end result, the intermediate step of the s-domain provides insight into why the frequency response behaves as it does.

cancel, providing a unity frequency response at low frequencies. The situation doesn't change significantly until the ant moves near the pole and zero location. When approaching the zero, the value of $H(s)$ drops suddenly, becoming *zero* when the ant is upon the *zero*. As the ant moves past the pole and zero pair, the value of $H(s)$ again returns to unity. Using this type of visualization, it can be seen that the width of the notch depends on the distance between the pole and zero.

Figure 30-8 summarizes how the Laplace transform is used. We start with a physical system, such as an electric circuit. If we desire, the phasor transform can directly provide the frequency response of the system, as described in Chapter 28. An alternative is to take the Laplace transform using the four step method previously outlined. This results in a mathematical expression for the transfer function, $H(s)$, which can be represented in a pole-zero diagram. The frequency response can then be found by evaluating the transfer function along the imaginary axis, that is, by replacing each s with $j\omega$. While both methods provide the same result, the intermediate pole-zero diagram provides an understanding of *why* the system behaves as it does, and how it can be changed.

Filter Design in the s-Domain

The most powerful application of the Laplace transform is the design of systems *directly* in the s-domain. This involves two steps: First, the s-domain is designed by specifying the number and location of the poles and zeros. This is a pure mathematical problem, with the goal of obtaining the best frequency response. In the second step, an electronic circuit is derived that provides this s-domain representation. This is something of an art, since there are many circuit configurations that have a given pole-zero diagram.

As previously mentioned, *step 4* of the Laplace transform method is very difficult if the system contains more than two poles or two zeros. A common solution is to implement multiple poles and zeros in *successive stages*. For example, a 6 pole filter is implemented as three successive stages, with each stage containing up to two poles and two zeros. Since each of these stages can be represented in the s-domain by a quadratic numerator divided by a quadratic denominator, this approach is called designing with **biquads**.

Figure 30-7 shows a common biquad circuit, the one used in the filter design method of Chapter 3. This is called the **Sallen-Key** circuit, after R.P. Sallen and E.L. Key, authors of a paper that described this technique in the mid 1950s. While there are several variations, the most common circuit uses two resistors of equal value, two capacitors of equal value, and an amplifier with an amplification of between 1 and 3. Although not available to Sallen and Key, the amplifiers can now be made with low-cost op amps with appropriate feedback resistors. Going through the four step circuit analysis procedure, the location of this circuit's two poles can be related to the component values:

EQUATION 30-4
Sallen-Key pole locations. These equations relate the pole position, ω and σ, to the amplifier gain, A, the resistor, R, and capacitor, C.

$$\sigma = \frac{A-3}{2RC}$$

$$\omega = \frac{\pm\sqrt{-A^2+6A-5}}{2RC}$$

These equations show that the two poles always lie somewhere on a circle of radius: $1/RC$. The exact position along the circle depends on the gain of the amplifier. As shown in (a), an amplification of 1 places both of the poles on the real axis. The frequency response of this configuration is a low-pass filter with a relatively smooth transition between the passband and stopband. The -3dB (0.707) cutoff frequency of this circuit, denoted by ω_0, is where the circle intersects the imaginary axis, i.e., $\omega_0 = 1/RC$.

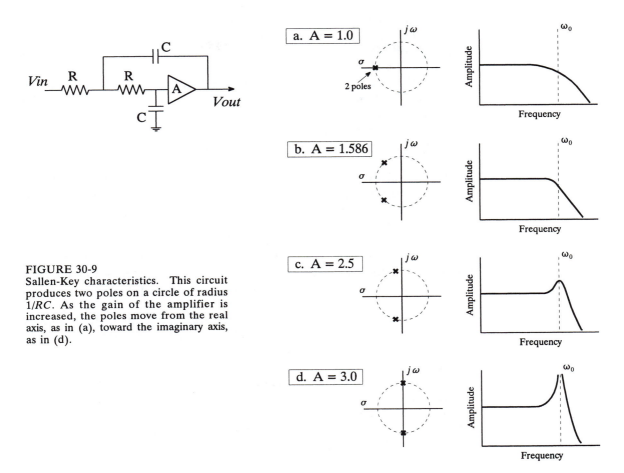

FIGURE 30-9
Sallen-Key characteristics. This circuit produces two poles on a circle of radius $1/RC$. As the gain of the amplifier is increased, the poles move from the real axis, as in (a), toward the imaginary axis, as in (d).

As the amplification is increased, the poles move along the circle, with a corresponding change in the frequency response. As shown in (b), an amplification of 1.586 places the poles at 45 degree angles, resulting in the frequency response having a sharper transition. Increasing the amplification further moves the poles even closer to the imaginary axis, resulting in the frequency response showing a peaked curve. This condition is illustrated in (c), where the amplification is set at 2.5. The amplitude of the peak continues to grow as the amplification is increased, until a gain of 3 is reached. As shown in (d), this is a special case that places the poles directly on the imaginary axis. The corresponding frequency response now has an infinity large value at the peak. In practical terms, this means the circuit has turned into an oscillator. Increasing the gain further pushes the poles deeper into the right half of the s-plane. As mentioned before, this correspond to the system being unstable (spontaneous oscillation).

Using the Sallen-Key circuit as a building block, a wide variety of filter types can be constructed. For example, a low-pass **Butterworth filter** is designed by placing a selected number of poles evenly around the left-half of the circle, as shown in Fig. 30-10. Each two poles in this configuration requires one Sallen-Key stage. As described in Chapter 3, the Butterworth filter is maximally flat, that is, it has the sharpest transition between the passband and stopband *without peaking* in the frequency response. The more poles used, the faster the transition. Since all the poles in the Butterworth filter lie on the same circle, all the cascaded stages use the same values for R and C. The only thing different between the stages is the amplification. Why does this circular pattern of poles provide the optimally flat response? Don't look for an obvious or intuitive answer to this question; it just falls out of the mathematics.

Figure 30-11 shows how the pole positions of the Butterworth filter can be modified to produce the **Chebyshev filter**. As discussed in Chapter 3, the Chebyshev filter achieves a sharper transition than the Butterworth at the expense of ripple being allowed into the passband. In the s-domain, this corresponds to the circle of poles being flattened into an *ellipse*. The more flattened the ellipse, the more ripple in the passband, and the sharper the transition. When formed from a cascade of Sallen-Key stages, this requires different values of resistors and capacitors in each stage.

Figure 30-11 also shows the next level of sophistication in filter design strategy: the **elliptic filter**. The elliptic filter achieves the sharpest possible transition by allowing ripple in both the passband and the stopband. In the s-domain, this corresponds to placing zeros directly on the real axis, with the first one near the cutoff frequency. Elliptic filters come in several varieties and are significantly more difficult to design than Butterworth and Chebyshev configurations. This is because the poles and zeros of the elliptic filter do not lie in a simple geometric pattern, but in a mathematical arrangement involving elliptic functions and integrals (hence the name).

FIGURE 30-10
The Butterworth s-plane. The low-pass Butterworth filter is created by placing poles equally around the left-half of a circle. The more poles used in the filter, the faster the roll-off.

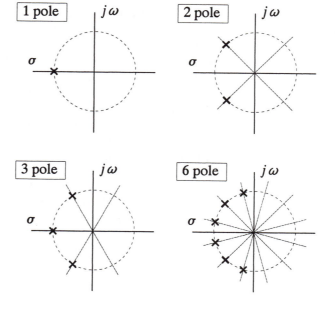

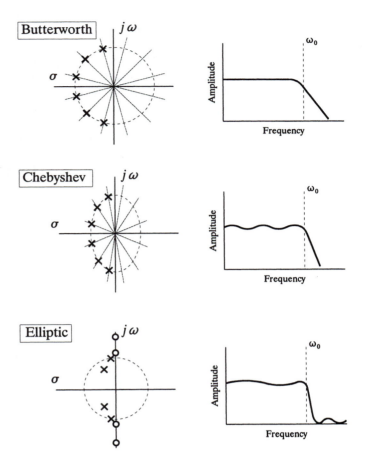

FIGURE 30-11
Classic pole-zero patterns. These are the three classic pole-zero patterns in filter design. Butterworth filters have poles equally spaced around a circle, resulting in a maximally flat response. Chebyshev filters have poles placed on an ellipse, providing a sharper transition, but at the cost of ripple in the passband. Elliptic filters add zeros to the stopband. This results in a faster transition, but with ripple in the passband *and* stopband.

Since each biquad produces two poles, **even order filters** (2 pole, 4 pole, 6 pole, etc.) can be constructed by cascading biquad stages. However, **odd order filters** (1 pole, 3 pole, 5 pole, etc.) require something that the biquad cannot provide: a single pole on the imaginary axis. This turns out to be nothing more than a simple RC circuit added to the cascade. For example, a 9 pole filter can be constructed from 5 stages: 4 Sallen-Key biquads, plus one stage consisting of a single capacitor and resistor.

These classic pole-zero patterns are for low-pass filters; however, they can be modified for other frequency responses. This is done by designing a low-pass filter, and then performing a mathematical transformation in the s-domain. We start by calculating the low-pass filter pole locations, and then writing the transfer function, $H(s)$, in the form of Eq. 30-3. The transfer function of the corresponding high-pass filter is found by replacing each "s" with "$1/s$", and then rearranging the expression to again be in the pole-zero form of Eq. 30-3. This defines new pole and zero locations that implement the high-pass filter. More complicated s-domain transforms can create band-pass and band-reject filters from an initial low-pass design. This type of mathematical manipulation in the s-domain is the central theme of filter design, and entire books are devoted to the subject. Analog filter design is 90% *mathematics*, and only 10% *electronics*.

Fortunately, the design of high-pass filters using Sallen-Key stages doesn't require this mathematical manipulation. The "*1/s*" for "*s*" replacement in the s-domain corresponds to swapping the resistors and capacitors in the circuit. In the s-plane, this swap places the poles at a new position, and adds two zeros directly at the origin. This results in the frequency response having a value of zero at DC (zero frequency), just as you would expect for a high-pass filter. This brings the Sallen-Key circuit to its full potential: the implementation of two poles *and* two zeros.

The z-Transform

Just as analog filters are designed using the Laplace transform, recursive digital filters are developed with a parallel technique called the z-transform. The overall strategy of these two transforms is the same: probe the impulse response with sinusoids and exponentials to find the system's poles and zeros. The Laplace transform deals with differential equations, the s-domain, and the s-plane. Correspondingly, the z-transform deals with difference equations, the z-domain, and the z-plane. However, the two techniques are not a mirror image of each other; the s-plane is arranged in a rectangular coordinate system, while the z-plane uses a polar format. Recursive digital filters are often designed by starting with one of the classic analog filters, such as the Butterworth, Chebyshev, or elliptic. A series of mathematical conversions are then used to obtain the desired digital filter. The z-transform provides the framework for this mathematics. The Chebyshev filter design program presented in Chapter 20 uses this approach, and is discussed in detail in this chapter.

The Nature of the z-Domain

To reinforce that the Laplace and z-transforms are parallel techniques, we will start with the Laplace transform and show how it can be changed into the z-transform. From the last chapter, the Laplace transform is defined by the relationship between the time domain and s-domain signals:

$$X(s) = \int_{t=-\infty}^{\infty} x(t)\, e^{-st}\, dt$$

where $x(t)$ and $X(s)$ are the time domain and s-domain representation of the signal, respectively. As discussed in the last chapter, this equation analyzes the time domain signal in terms of sine and cosine waves that have an exponentially changing amplitude. This can be understood by replacing the complex variable, s, with its equivalent expression, $\sigma + j\omega$. Using this alternate notation, the Laplace transform becomes:

$$X(\sigma,\omega) = \int_{t=-\infty}^{\infty} x(t)\, e^{-\sigma t} e^{-j\omega t}\, dt$$

If we are only concerned with *real* time domain signals (the usual case), the top and bottom halves of the s-plane are mirror images of each other, and the term, $e^{-j\omega t}$, reduces to simple cosine and sine waves. This equation identifies each *location* in the s-plane by the two parameters, σ and ω. The *value* at each location is a complex number, consisting of a real part and an imaginary part. To find the real part, the time domain signal is multiplied by a cosine wave with a frequency of ω, and an amplitude that changes exponentially according to the decay parameter, σ. The value of the real part of $X(\sigma,\omega)$ is then equal to the integral of the resulting waveform. The value of the imaginary part of $X(\sigma,\omega)$ is found in a similar way, except using a sine wave. If this doesn't sound very familiar, you need to review the previous chapter before continuing.

The Laplace transform can be changed into the z-transform in three steps. The first step is the most obvious: change from continuous to discrete signals. This is done by replacing the time variable, t, with the sample number, n, and changing the integral into a summation:

$$X(\sigma,\omega) = \sum_{n=-\infty}^{\infty} x[n]\, e^{-\sigma n} e^{-j\omega n}$$

Notice that $X(\sigma,\omega)$ uses parentheses, indicating it is *continuous*, not discrete. Even though we are now dealing with a discrete time domain signal, $x[n]$, the parameters σ and ω can still take on a continuous range of values. The second step is to rewrite the exponential term. An exponential signal can be mathematically represented in either of two ways:

$$y[n] = e^{-\sigma n} \qquad \text{or} \qquad y[n] = r^n$$

As illustrated in Fig. 31-1, both these equations generate an exponential curve. The first expression controls the decay of the signal through the parameter, σ. If σ is positive, the waveform will *decrease* in value as the sample number, n, becomes larger. Likewise, the curve will progressively *increase* if σ is negative. If σ is exactly zero, the signal will have a constant value of *one*.

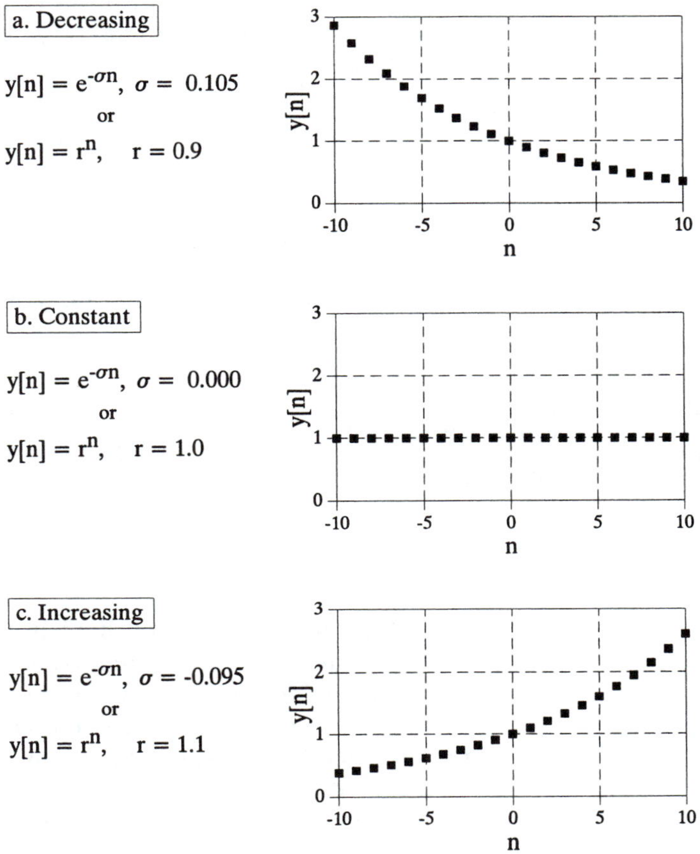

a. Decreasing

$$y[n] = e^{-\sigma n}, \; \sigma = 0.105$$

or

$$y[n] = r^n, \quad r = 0.9$$

FIGURE 31-1
Exponential signals. Exponentials can be represented in two different mathematical forms. The Laplace transform uses one way, while the z-transform uses the other.

b. Constant

$$y[n] = e^{-\sigma n}, \; \sigma = 0.000$$

or

$$y[n] = r^n, \quad r = 1.0$$

c. Increasing

$$y[n] = e^{-\sigma n}, \; \sigma = -0.095$$

or

$$y[n] = r^n, \quad r = 1.1$$

The second expression uses the parameter, r, to control the decay of the waveform. The waveform will decrease if $r < 1$, and increase if $r > 1$. The signal will have a constant value when $r = 1$. These two equations are just different ways of expressing the same thing. One method can be swapped for the other by using the relation:

$$r^n = \left[e^{\ln(r)} \right]^n = e^{n \ln(r)} = e^{-\sigma n}$$

where: $\sigma = -\ln(r)$

The second step of converting the Laplace transform into the z-transform is completed by using the *other* exponential form:

$$X(r,\omega) = \sum_{n=-\infty}^{\infty} x[n] \, r^n e^{-j\omega n}$$

While this is a perfectly correct expression of the z-transform, it is not in the most compact form for complex notation. This problem was overcome in the Laplace transform by introducing a new complex variable, s, defined to be: $s = \sigma + j\omega$. In this same way, we will define a new variable for the z-transform:

$$z = re^{-j\omega}$$

This is defining the complex variable, z, as the polar notation combination of the two real variables, r and ω. The third step in deriving the z-transform is to replace: r and ω, with z. This produces the standard form of the z-transform:

EQUATION 31-1
The z-transform. The z-transform defines the relationship between the time domain signal, $x[n]$, and the z-domain signal, $X(z)$.

$$X(z) = \sum_{n=-\infty}^{\infty} x[n]z^{-n}$$

Why does the z-transform use r^n instead of $e^{-\sigma n}$, and z instead of s? As described in Chapter 19, recursive filters are implemented by a set of *recursion coefficients*. To analyze these systems in the z-domain, we must be able to convert these recursion coefficients into the z-domain *transfer function*, and back again. As we will show shortly, defining the z-transform is this manner (r^n and z) provides the simplest means of moving between these two important representations. In fact, defining the z-domain in this way makes it *trivial* to move from one representation to the other.

Figure 31-2 illustrates the difference between the Laplace transform's s-plane, and the z-transform's z-plane. Locations in the s-plane are identified by two parameters: σ, the exponential decay variable along the horizontal axis, and ω, the frequency variable along the vertical axis. In other words, these two real parameters are arranged in a *rectangular* coordinate system. This geometry results from defining s, the complex variable representing position in the s-plane, by the relation: $s = \sigma + j\omega$.

In comparison, the z-domain uses the variables: r and ω, arranged in *polar* coordinates. The distance from the origin, r, is the value of the exponential decay. The angular distance measured from the positive horizontal axis, ω, is the frequency. This geometry results from defining z by: $z = re^{-j\omega}$. In other words, the complex variable representing position in the z-plane is formed by combining the two real parameters in a polar form.

These differences result in vertical *lines* in the s-plane matching *circles* in the z-plane. For example, the s-plane in Fig. 31-2 shows a pole-zero pattern where all of the poles & zeros lie on vertical lines. The equivalent poles & zeros in the z-plane lie on circles concentric with the origin. This can be understood by examining the relation presented earlier: $\sigma = -\ln(r)$. For instance, the s-plane's vertical axis (i.e., σ=0) corresponds to the z-plane's

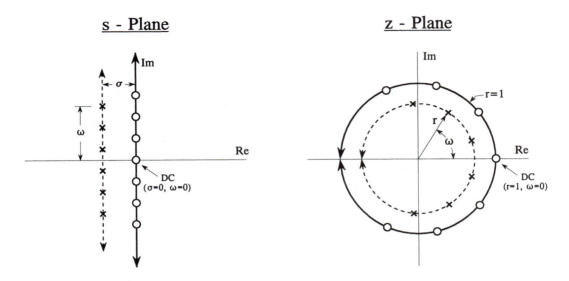

FIGURE 31-2
Relationship between the s-plane and the z-plane. The s-plane is a rectangular coordinate system with σ expressing the distance along the real (horizontal) axis, and ω the distance along the imaginary (vertical) axis. In comparison, the z-plane is in polar form, with r being the distance to the origin, and ω the angle measured to the positive horizontal axis. Vertical lines in the s-plane, such as illustrated by the example poles and zeros in this figure, correspond to circles in the z-plane.

unit circle (that is, $r=1$). Vertical lines in the left half of the s-plane correspond to circles inside the z-plane's unit circle. Likewise, vertical lines in the right half of the s-plane match with circles on the outside of the z-plane's unit circle. In other words, the left and right sides of the s-plane correspond to the interior and the exterior of the unit circle, respectively. For instance, a continuous system is unstable when poles occupy the *right half* of the s-plane. In this same way, a discrete system is unstable when poles are *outside* the unit circle in the z-plane. When the time domain signal is completely real (the most common case), the upper and lower halves of the z-plane are mirror images of each other, just as with the s-domain.

Pay particular attention to how the frequency variable, ω, is used in the two transforms. A *continuous* sinusoid can have any frequency between DC and infinity. This means that the s-plane must allow ω to run from negative to positive infinity. In comparison, a *discrete* sinusoid can only have a frequency between DC and one-half the sampling rate. That is, the frequency must between 0 and 0.5 when expressed as a fraction of the sampling rate, or between 0 and π when expressed as a natural frequency (i.e., $\omega = 2\pi f$). This matches the geometry of the z-plane when we interpret ω to be an angle expressed in *radians*. That is, the positive frequencies correspond to angles of 0 to π radians, while the negative frequencies correspond to 0 to $-\pi$ radians. Since the z-plane express frequency in a different way than the s-plane, some authors use different symbols to distinguish the two. A common notation is to use Ω (an upper case omega) to represent frequency in the z-domain, and ω (a lower case omega) for frequency in the s-domain. In this book we will use ω to represent both types of frequency, but look for this in other DSP material.

In the s-plane, the values that lie along the vertical axis are equal to the frequency response of the system. That is, the Laplace transform, evaluated at $\sigma = 0$, is equal to the Fourier transform. In an analogous manner, the frequency response in the z-domain is found along the unit circle. This can be seen by evaluating the z-transform (Eq. 31-1) at $r = 1$, resulting in the equation reducing to the Discrete Time Fourier Transform (DTFT). This places zero frequency (DC) at a value of *one* on the horizontal axis in the s-plane. The spectrum's positive frequencies are positioned in a counter-clockwise pattern from this DC position, occupying the upper semicircle. Likewise the negative frequencies are arranged from the DC position along the clockwise path, forming the lower semicircle. The positive and negative frequencies in the spectrum meet at the common point of $\omega = \pi$ and $\omega = -\pi$. This circular geometry also corresponds to the frequency spectrum of a discrete signal being *periodic*. That is, when the frequency angle is increased beyond π, the same values are encountered as between 0 and π. When you run around in a circle, you see the same scenery over and over.

Analysis of Recursive Systems

As outlined in Chapter 19, a recursive filter is described by a **difference equation**:

EQUATION 31-2
Difference equation. See Chapter
19 for details.

$$y[n] \;=\; a_0 x[n] + a_1 x[n-1] + a_2 x[n-2] + \cdots + b_1 y[n-1] + b_2 y[n-2] + b_3 y[n-3] + \cdots$$

where $x[\,]$ and $y[\,]$ are the input and output signals, respectively, and the "a" and "b" terms are the **recursion coefficients**. An obvious use of this equation is to describe how a programmer would implement the filter. An equally important aspect is that it represents a mathematical relationship between the input and output that must be continually satisfied. Just as continuous systems are controlled by *differential* equations, recursive discrete systems operate in accordance with this *difference* equation. From this relationship we can derive the key characteristics of the system: the impulse response, step response, frequency response, pole-zero plot, etc.

We start the analysis by taking the z-transform (Eq. 31-1) of both sides of Eq. 31-2. In other words, we want to see what this controlling relationship looks like in the z-domain. With a fair amount of algebra, we can separate the relation into: $Y[z]/X[z]$, that is, the z-domain representation of the output signal divided by the z-domain representation of the input signal. Just as with the Laplace transform, this is called the **system's transfer function**, and designate it by $H[z]$. Here is what we find:

EQUATION 31-3
Transfer function in polynomial form. The recursion coefficients are directly identifiable in this relation.

$$H[z] = \frac{a_0 + a_1 z^{-1} + a_2 z^{-2} + a_3 z^{-3} + \cdots}{1 - b_1 z^{-1} - b_2 z^{-2} - b_3 z^{-3} - \cdots}$$

This is one of two ways that the transfer function can be written. This form is important because it directly contains the recursion coefficients. For example, suppose we know the recursion coefficients of a digital filter, such as might be provided from a design table:

$$a_0 = 0.389$$
$$a_1 = -1.558 \qquad b_1 = 2.161$$
$$a_2 = 2.338 \qquad b_2 = -2.033$$
$$a_3 = -1.558 \qquad b_3 = 0.878$$
$$a_4 = 0.389 \qquad b_4 = -0.161$$

Without having to worry about nasty complex algebra, we can directly write down the system's transfer function:

$$H[z] = \frac{0.389 - 1.558 z^{-1} + 2.338 z^{-2} - 1.558 z^{-3} + 0.389 z^{-4}}{1 - 2.161 z^{-1} + 2.033 z^{-2} - 0.878 z^{-3} + 0.161 z^{-4}}$$

Notice that the "b" coefficients enter the transfer function with a *negative* sign in front of them. Alternatively, some authors write this equation using additions, but change the sign of all the "b" coefficients. Here's the problem. If you are given a set of recursion coefficients (such as from a table or filter design program), there is a 50-50 chance that the "b" coefficients will have the opposite sign from what you expect. If you don't catch this discrepancy, the filter will be grossly unstable.

Equation 31-3 expresses the transfer function using *negative* powers of z, such as: z^{-1}, z^{-2}, z^{-3}, etc. After an actual set of recursion coefficients have been plugged in, we can convert the transfer function into a more conventional form that uses *positive* powers: i.e., $z, z^2, z^3, \cdots$. By multiplying both the numerator and denominator of our example by z^4, we obtain:

$$H[z] = \frac{0.389 z^4 - 1.558 z^3 + 2.338 z^2 - 1.558 z + 0.389}{z^4 - 2.161 z^3 + 2.033 z^2 - 0.878 z + 0.161}$$

Positive powers are often easier to use, and they are *required* by some z-domain techniques. Why not just rewrite Eq. 31-3 using positive powers and forget about negative powers entirely? We can't! The trick of dividing the numerator and denominator by the highest power of z (such as z^4 in our example) can only be used if the number of recursion coefficients is already known. Equation 31-3 is written for an *arbitrary* number of coefficients. The point is, both positive and negative powers are routinely used in DSP and you need to know how to convert between the two forms.

The transfer function of a recursive system is useful because it can be manipulated in ways that the recursion coefficients cannot. This includes such tasks as: combining cascade and parallel stages into a single system, designing filters by specifying the pole and zero locations, converting analog filters into digital, etc. These operations are carried out by algebra performed in the s-domain, such as: multiplication, addition, and factoring. After these operations are completed, the transfer function is placed in the form of Eq. 31-3, allowing the new recursion coefficients to be identified.

Just as with the s-domain, an important feature of the z-domain is that the transfer function can be expressed as **poles** and **zeros**. This provides the second general form of the z-domain:

EQUATION 31-4
Transfer function in pole-zero form.

$$H[z] = \frac{(z-z_1)(z-z_2)(z-z_3)\cdots}{(z-p_1)(z-p_2)(z-p_3)\cdots}$$

Each of the poles $(p_1, p_2, p_3, \cdots)$ and zeros $(z_1, z_2, z_3 \cdots)$ is a complex number. To move from Eq. 31-4 to 31-3, multiply out the expressions and collect like terms. While this can involve a tremendous amount of algebra, it is straightforward in principle and can easily be written into a computer routine. Moving from Eq. 31-3 to 31-4 is more difficult because it requires *factoring* of the polynomials. As discussed in Chapter 30, the quadratic equation can be used for the factoring if the transfer function is second order or less (i.e., there are no powers of z higher than z^2). Algebraic methods cannot be used to factor systems greater than second order and numerical methods must be employed. Fortunately, this is seldom needed; digital filter design *starts* with the pole-zero locations (Eq. 31-4) and *ends* with the recursion coefficients (Eq. 31-3), not the other way around.

As with all complex numbers, the pole and zero locations can be represented in either polar or rectangular form. Polar notation has the advantage of being more consistent with the natural organization of the z-plane. In comparison, rectangular form is generally preferred for mathematical work, that is, it is usually easier to manipulate: $\sigma + j\omega$, as compared with: $re^{j\omega}$.

As an example of using these equations, we will design a notch filter by the following steps: (1) specify the pole-zero placement in the z-plane, (2)

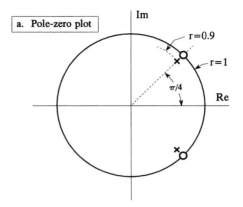

FIGURE 31-3
Notch filter designed in the z-domain. The design starts by locating two poles and two zeros in the z-plane, as shown in (a). The resulting impulse and frequency response are shown in (b) and (c), respectively. The sharpness of the notch is controlled by the distance of the poles from the zeros.

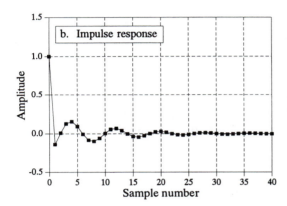

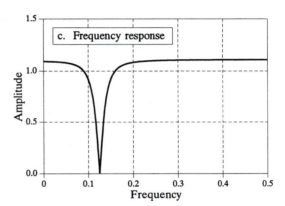

write down the transfer function in the form of Eq. 31-4, (3) rearrange the transfer function into the form of Eq. 31-3, and (4) identify the recursion coefficients needed to implement the filter. Fig. 31-3 shows the example we will use: a notch filter formed from two poles and two zeros located at

In polar form:

$$z_1 = 1.00\, e^{\,j(\pi/4)}$$
$$z_2 = 1.00\, e^{\,j(-\pi/4)}$$
$$p_1 = 0.90\, e^{\,j(\pi/4)}$$
$$p_2 = 0.90\, e^{\,j(-\pi/4)}$$

In rectangular form:

$$z_1 = 0.7071 + j\,0.7071$$
$$z_2 = 0.7071 - j\,0.7071$$
$$p_1 = 0.6364 + j\,0.6364$$
$$p_2 = 0.6364 - j\,0.6364$$

To understand why this is a notch filter, compare this pole-zero plot with Fig. 30-6, a notch filter in the s-plane. The only difference is that we are moving along the *unit circle* to find the frequency response from the z-plane, as opposed to moving along the *vertical axis* to find the frequency response from the s-plane. From the polar form of the poles and zeros, it can be seen that the notch will occur at a natural frequency of $\pi/4$, corresponding to 0.125 of the sampling rate.

Since the pole and zero locations are known, the transfer function can be written in the form of Eq. 31-4 by simply plugging in the values:

$$H(z) = \frac{\left(z - (0.7071 + j0.7071)\right)\left(z - (0.7071 - j0.7071)\right)}{\left(z - (0.6364 + j0.6364)\right)\left(z - (0.6364 - j0.6364)\right)}$$

To find the recursion coefficients that implement this filter, the transfer function must be rearranged into the form of Eq. 31-3. To start, expand the expression by multiplying out the terms:

$$H(z) = \frac{z^2 - 0.7071z + j0.7071z - 0.7071z + 0.7071^2 - j0.7071^2 - j0.7071z + j0.7071^2 - j^2 0.7071^2}{z^2 - 0.6364z + j0.6364z - 0.6364z + 0.6364^2 - j0.6364^2 - j0.6364z + j0.6364^2 - j^2 0.6364^2}$$

Next, we collect like terms and reduce. As long as the upper half of the z-plane is a mirror image of the lower half (which is always the case if we are dealing with a *real* impulse response), all of the terms containing a "j" will cancel out of the expression:

$$H[z] = \frac{1.000 - 1.414\,z + 1.000\,z^2}{0.810 - 1.273\,z + 1.000\,z^2}$$

While this is in the form of one polynomial divided by another, it does not use negative exponents of z, as required by Eq. 31-3. This can be changed by dividing both the numerator and denominator by the highest power of z in the expression, in this case, z^2:

$$H[z] = \frac{1.000 - 1.414\,z^{-1} + 1.000\,z^{-2}}{1.000 - 1.273\,z^{-1} + 0.810\,z^{-2}}$$

Since the transfer function is now in the form of Eq. 31-3, the recursive coefficients can be directly extracted by inspection:

$$a_0 = 1.000$$
$$a_1 = -1.414 \qquad b_1 = 1.273$$
$$a_2 = 1.000 \qquad b_2 = -0.810$$

This example provides the general strategy for obtaining the recursion coefficients from a pole-zero plot. In specific cases, it is possible to derive simpler equations directly relating the pole-zero positions to the recursion coefficients. For example, a system containing two poles and two zeros, called as **biquad**, has the following relations:

EQUATION 31-5
Biquad design equations. These equations give the recursion coefficients, a_0, a_1, a_2, b_1, b_2, from the position of the poles: r_p & ω_p, and the zeros: r_0 & ω_0.

$$a_0 = 1$$
$$a_1 = -2r_0 \cos(\omega_0)$$
$$a_2 = r_0^2$$

$$b_1 = 2r_p \cos(\omega_p)$$
$$b_2 = -r_p^2$$

After the transfer function has been specified, how do we find the frequency response? There are three methods: one is mathematical and two are computational (programming). The mathematical method is based on finding the values in the z-plane that lie on the unit circle. This is done by evaluating the transfer function, $H(z)$, at $r = 1$. Specifically, we start by writing down the transfer function in the form of either Eq. 31-3 or 31-4. We then replace each z with $e^{-j\omega}$ (that is, $re^{-j\omega}$ with $r = 1$). This provides a mathematical equation of the frequency response, $H(\omega)$. The problem is, the resulting expression is in a very inconvenient form. A significant amount of algebra is usually required to obtain something recognizable, such as the magnitude and phase. While this method provides an exact equation for the frequency response, it is difficult to automate in computer programs, such as needed in filter design packages.

The second method for finding the frequency response also uses the approach of evaluating the z-plane on the unit circle. The difference is that we only calculate *samples* of the frequency response, not a mathematical solution for the entire curve. A computer program loops through, perhaps, 1000 equally spaced frequencies between $\omega = 0$ and $\omega = \pi$. Think of an ant moving between 1000 discrete points on the upper half of the z-plane's unit circle. The magnitude and phase of the frequency response are found at each of these location by evaluating the transfer function.

This method works well and is often used in filter design packages. Its major limitation is that it does not account for *round-off noise* affecting the system's characteristics. Even if the frequency response found by this method looks perfect, the implemented system can be completely unstable!

This brings up the third method: find the frequency response from the recursion coefficients that are actually used to implement the filter. To start, we find the impulse response of the filter by passing an impulse through the system. In the second step, we take the DFT of the impulse response (using the FFT, of course) to find the system's frequency response. The only critical item to remember with this procedure is that enough samples must be taken of the impulse response so that the discarded samples are *insignificant*. While books could be written on the theoretical criteria for this, the practical rules are much simpler. Use as many samples as you *think* are necessary. After finding the frequency response, go back and repeat the procedure using twice as many samples. If the two frequency responses are adequately similar, you can be assured that the truncation of the impulse response hasn't fooled you in some way.

Cascade and Parallel Stages

Sophisticated recursive filters are usually designed in stages to simplify the tedious algebra of the z-domain. Figure 31-4 illustrates the two common ways that individual stages can be arranged: cascaded stages and parallel stages with added outputs. For example, a low-pass and high-pass stage can be cascaded to form a *band-pass* filter. Likewise, a parallel combination of low-pass and high-pass stages can form a *band-reject* filter. We will call the two stages being combined *system 1* and *system 2*, with their recursion coefficients being called: a_0, a_1, a_2, b_1, b_2 and A_0, A_1, A_2, B_1, B_2, respectively. Our goal is to combine these stages (in cascade or parallel) into a single recursive filter, which we will call *system 3*, with recursion coefficients given by: $a_0, a_1, a_2, a_3, a_4, b_1, b_2, b_3, b_4$.

As you recall from previous chapters, the frequency responses of systems in a cascade are combined by *multiplication*. Also, the frequency responses of systems in parallel are combined by *addition*. These same rules are followed by the z-domain transfer functions. This allows recursive systems to be combined by moving the problem into the z-domain, performing the required multiplication or addition, and then returning to the recursion coefficients of the final system.

As an example of this method, we will work out the algebra for combining two biquad stages in a cascade. The transfer function of each stage is found by writing Eq. 31-3 using the appropriate recursion coefficients. The transfer function of the entire system, $H[z]$, is then found by multiplying the transfer functions of the two stage:

$$H[z] = \frac{a_0 + a_1 z^{-1} + a_2 z^{-2}}{1 - b_1 z^{-1} - b_2 z^{-2}} \times \frac{A_0 + A_1 z^{-1} + A_2 z^{-2}}{1 - B_1 z^{-1} - B_2 z^{-2}}$$

Multiplying out the polynomials and collecting like terms:

$$H[z] = \frac{a_0 A_0 + (a_0 A_1 + a_1 A_0) z^{-1} + (a_0 A_2 + a_1 A_1 + a_2 A_0) z^{-2} + (a_1 A_2 + a_2 A_1) z^{-3} + (a_2 A_2) z^{-4}}{1 - (b_1 + B_1) z^{-1} - (b_2 + B_2 - b_1 B_1) z^{-2} - (-b_1 B_2 - b_2 B_1) z^{-3} - (-b_2 B_2) z^{-4}}$$

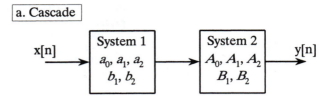

a. Cascade

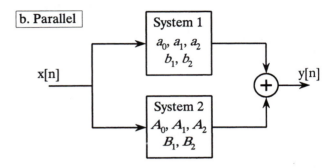

FIGURE 31-4
Combining cascade and parallel stages.
The z-domain allows recursive stages in
a cascade, (a), or in parallel, (b), to be
combined into a single system, (c).

b. Parallel

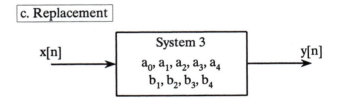

c. Replacement

Since this is in the form of Eq. 31-3, we can directly extract the recursion coefficients that implement the cascaded system:

$$a_0 = a_0 A_0$$
$$a_1 = a_0 A_1 + a_1 A_0 \qquad\qquad b_1 = b_1 + B_1$$
$$a_2 = a_0 A_2 + a_1 A_1 + a_2 A_0 \qquad b_2 = b_2 + B_2 - b_1 B_1$$
$$a_3 = a_1 A_2 + a_2 A_1 \qquad\qquad b_3 = -b_1 B_2 - b_2 B_1$$
$$a_4 = a_2 A_2 \qquad\qquad b_4 = -b_2 B_2$$

The obvious problem with this technique is the large amount of algebra needed to multiply and rearrange the polynomial terms. Fortunately, the entire algorithm can be expressed in a short computer program, shown in Table 31-1. Although the cascade and parallel combinations require different mathematics, they use nearly the same program. In particular, only one line of code is different between the two algorithms, allowing both to be combined into a single program.

```
100 'COMBINING RECURSION COEFFICIENTS OF CASCADE AND PARALLEL STAGES
110 '
120 '                              'INITIALIZE VARIABLES
130 DIM A1[8], B1[8]              'a and b coefficients for system 1, one of the stages
140 DIM A2[8], B2[8]              'a and b coefficients for system 2, one of the stages
150 DIM A3[16], B3[16]           'a and b coefficients for system 3, the combined system
160 '
170                               'Indicate cascade or parallel combination
180 INPUT "Enter 0 for cascade, 1 for parallel: ", CP%
190 '
200 GOSUB XXXX                    'Mythical subroutine to load: A1[ ], B1[ ], A2[ ], B2[ ]
210 '
220 FOR I% = 0 TO 8              'Convert the recursion coefficients into transfer functions
230    B2[I%] = -B2[I%]
240    B1[I%] = -B1[I%]
250 NEXT I%
260 B1[0] = 1
270 B2[0] = 1
280 '
290 FOR I% = 0 TO 16             'Multiply the polynomials by convolving
300    A3[I%] = 0
310    B3[I%] = 0
320    FOR J% = 0 TO 8
330      IF I%-J% < 0  OR I%-J% > 8 THEN GOTO 370
340      IF CP% = 0 THEN A3[I%] = A3[I%] + A1[J%] * A2[I%-J%]
350      IF CP% = 1 THEN A3[I%] = A3[I%] + A1[J%] * B2[I%-J%] + A2[J%] * B1[I%-J%]
360      B3[I%] = B3[I%] + B1[J%] * B2[I%-J%]
370    NEXT J%
380 NEXT I%
390 '
400 FOR I% = 0 TO 16             'Convert the transfer function into recursion coefficients.
410    B3[I%] = -B3[I%]
420 NEXT I%
430 B3[0] = 0
440 '                            'The recursion coefficients of the combined system now
450 END                          'reside in A3[ ] & B3[ ]
```

TABLE 31-1
Combining cascade and parallel stages. This program combines the recursion coefficients of stages in cascade or parallel. The recursive coefficients for the two stages being combined enter the program in the arrays: A1[ ], B1[ ], & A2[ ], B2[ ]. The recursion coefficients that implement the entire system leave the program in the arrays: A3[ ], B3[ ].

This program operates by changing the recursive coefficients from each of the individual stages into transfer functions in the form of Eq. 31-3 (lines 220-270). After combining these transfer functions in the appropriate manner (lines 290-380), the information is moved back to being recursive coefficients (lines 400 to 430).

The heart of this program is how the transfer function polynomials are represented and combined. For example, the numerator of the first stage being combined is: $a_0 + a_1 z^{-1} + a_2 z^{-2} + a_3 z^{-3} \cdots$. This polynomial is represented in the program by storing the coefficients: $a_0, a_1, a_2, a_3 \cdots$, in the array: A1[0], A1[1], A1[2], A1[3] $\cdots$. Likewise, the numerator for the second stage is represented by the values stored in: A2[0], A2[1], A2[2], A2[3] $\cdots$, and the numerator for the combined system in: A3[0], A3[1], A3[2], A3[3] $\cdots$. The idea is to represent and manipulate *polynomials* by only referring to their *coefficients*. The question is, how do we calculate A3[ ], given that A1[ ], A2[ ], and A3[ ] all represent polynomials? The answer is that when two polynomials are multiplied, their coefficients are *convolved*. In equation form: A1[ ] * A2[ ] = A3[ ]. This allows a standard convolution algorithm to find the transfer function of cascaded stages by convolving the two numerator arrays and the two denominator arrays.

The procedure for combining parallel stages is slightly more complicated. In algebra, fractions are added according to:

$$\frac{W}{X} + \frac{y}{Z} = \frac{W \cdot Z + y \cdot Z}{X \cdot Z}$$

Since each of the transfer functions is a fraction (one polynomial divided by another polynomial), we combine stages in parallel by multiplying the denominators, and adding the cross products in the numerators. This means that the denominator is calculated in the same way as for cascaded stages, but the numerator calculation is more elaborate. In line 340, the numerators of *cascaded* stages are convolved to find the numerator of the combined transfer function. In line 350, the numerator of the *parallel* stage combination is calculated as the sum of the two numerators convolved with the two denominators. Line 360 handles the denominator calculation for both cases.

Spectral Inversion

Chapter 14 describes an FIR filter technique called *spectral inversion*. This is a way of changing the filter kernel such that the frequency response is flipped top-for-bottom. All the passbands are changed into stopbands, and vice versa. For example, a low-pass filter is changed into high-pass, a band-pass filter into band-reject, etc. A similar procedure can be done with recursive filters, although it is far less successful.

As illustrated in Fig. 31-5, spectral inversion is accomplished by subtracting the output of the system from the original signal. This procedure can be

FIGURE 31-5
Spectral inversion. This procedure is the same as subtracting the output of the system from the original signal.

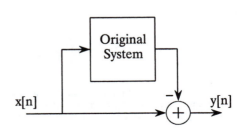

viewed as combining two stages in parallel, where one of the stages happens to be the *identity system* (the output is identical to the input). Using this approach, it can be shown that the "b" coefficients are left unchanged, and the modified "a" coefficients are given by:

EQUATION 31-6
Spectral inversion. The frequency response of a recursive filter can be flipped top-for-bottom by modifying the "a" coefficients according to these equations. The original coefficients are shown in italics, and the modified coefficients in roman. The "b" coefficients are not changed. This method usually provides poor results.

$$a_0 = 1 - a_0$$
$$a_1 = -a_1 - b_1$$
$$a_2 = -a_2 - b_2$$
$$a_3 = -a_3 - b_3$$
$$\vdots$$

Figure 31-6 shows spectral inversion for two common frequency responses: a low-pass filter, (a), and a notch filter, (c). This results in a high-pass filter, (b), and a band-pass filter, (d), respectively. How do the resulting frequency responses look? The high-pass filter is absolutely terrible! While

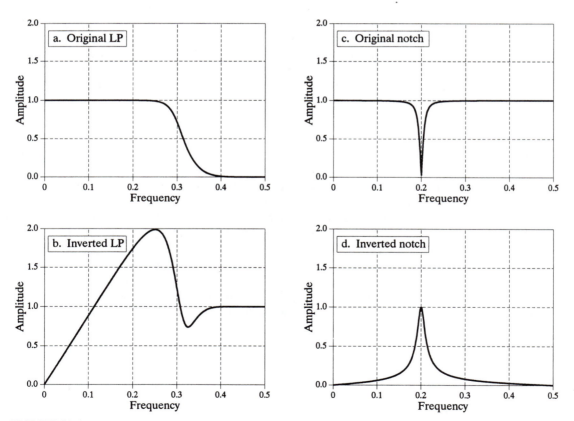

FIGURE 31-6
Examples of spectral inversion. Figure (a) shows the frequency response of a 6 pole low-pass Butterworth filter. Figure (b) shows the corresponding high-pass filter obtained by spectral inversion; its a mess! A more successful case is shown in (c) and (d) where a notch filter is transformed in to a band-pass frequency response.

the band-pass is better, the peak is not as sharp as the notch filter from which it was derived. These mediocre results are especially disappointing in comparison to the excellent performance seen in Chapter 14. Why the difference? The answer lies in something that is often forgotten in filter design: the *phase response*.

To illustrate how phase is the culprit, consider a system called the **Hilbert transformer**. The Hilbert transformer is not a specific device, but any system that has the frequency response: Magnitude = 1 and phase = 90 degrees, for all frequencies. This means that any sinusoid passing through a Hilbert transformer will be unaffected in amplitude, but changed in phase by one-quarter of a cycle. Hilbert transformers can be analog or discrete (that is, hardware or software), and are commonly used in communications for various modulation and demodulation techniques.

Now, suppose we spectrally invert the Hilbert transformer by subtracting its output from the original signal. Looking only at the *magnitude* of the frequency responses, we would conclude that the entire system would have an output of *zero*. That is, the magnitude of the Hilbert transformer's output is identical to the magnitude of the original signal, and the two will cancel. This, of course, is completely incorrect. Two sinusoids will exactly cancel only if they have the same magnitude *and* phase. In reality, the frequency response of this composite system has a magnitude of $\sqrt{2}$, and a phase shift of -45 degrees. Rather than being zero (our naive guess), the output is *larger* in amplitude than the input!

Spectral inversion works well in Chapter 14 because of the specific kind of filter used: *zero phase*. That is, the filter kernels have a left-right symmetry. When there is no phase shift introduced by a system, the subtraction of the output from the input is dictated solely by the magnitudes. Since recursive filters are plagued with phase shift, spectral inversion generally produces unsatisfactory filters.

Gain Changes

Suppose we have a recursive filter and need to modify the recursion coefficients such that the output signal is changed in amplitude. This might be needed, for example, to insure that a filter has unity gain in the passband. The method to achieve this is very simple: multiply the "a" coefficients by whatever factor we want the gain to change by, and leave the "b" coefficients alone.

Before adjusting the gain, we would probably like to know its current value. Since the gain must be specified at a frequency in the *passband*, the procedure depends on the type of filter being used. Low-pass filters have their gain measured at a frequency of *zero*, while high-pass filters use a frequency of 0.5, the maximum frequency allowable. It is quite simple to derive expressions for the gain at both these special frequencies. Here's how it is done.

First, we will derive an equation for the gain at zero frequency. The idea is to force each of the input samples to have a value of *one*, resulting in each of the output samples having a value of G, the gain of the system we are trying to find. We will start by writing the recursion equation, the mathematical relationship between the input and output signals:

$$y[n] = a_0 x[n] + a_1 x[n-1] + a_2 x[n-2] + \cdots + b_1 y[n-1] + b_2 y[n-2] + b_3 y[n-3] + \cdots$$

Next, we plug in *one* for each input sample, and G for each output sample. In other words, we force the system to operate at zero frequency. The equation becomes:

$$G = a_0 + a_1 + a_2 + a_3 + \cdots + b_1 G + b_2 G + b_3 G + b_4 G \cdots$$

Solving for G provides the gain of the system at zero frequency, based on its recursion coefficients:

EQUATION 31-7
DC gain of recursive filters. This relation provides the DC gain from the recursion coefficients.

$$G = \frac{a_0 + a_1 + a_2 + a_3 \cdots}{1 - (b_1 + b_2 + b_3 \cdots)}$$

To make a filter have a gain of *one* at DC, calculate the existing gain by using this relation, and then divide all the "a" coefficients by G.

The gain at a frequency of 0.5 is found in a similar way: we force the input and output signals to operate at this frequency, and see how the system responds. At a frequency of 0.5, the samples in the input signal alternate between -1 and 1. That is, successive samples are: 1, -1, 1, -1, 1, -1, 1, etc. The corresponding output signal also alternates in sign, with an amplitude equal to the gain of the system: G, $-G$, G, $-G$, G, $-G$, etc. Plugging these signals into the recursion equation:

$$G = a_0 - a_1 + a_2 - a_3 + \cdots - b_1 G + b_2 G - b_3 G + b_4 G \cdots$$

Solving for G provides the gain of the system at a frequency of 0.5, using its recursion coefficients:

EQUATION 31-8
Gain at maximum frequency. This relation gives the recursive filter's gain at a frequency of 0.5, based on the system's recursion coefficients.

$$G = \frac{a_0 - a_1 + a_2 - a_3 + a_4 \cdots}{1 - (-b_1 + b_2 - b_3 + b_4 \cdots)}$$

Just as before, a filter can be normalized for unity gain by dividing all of the "a" coefficients by this calculated value of G. Calculation of Eq. 31-8 in a computer program requires a method for generating negative signs for the odd coefficients, and positive signs for the even coefficients. The most common method is to multiply each coefficient by $(-1)^k$, where k is the index of the coefficient being worked on. That is, as k runs through the values: 0,1,2,3,4,5,6 etc., the expression, $(-1)^k$, takes on the values: 1, -1, 1, -1, 1, -1, 1 etc.

Chebyshev-Butterworth Filter Design

A common method of designing recursive digital filters is shown by the Chebyshev-Butterworth program presented in Chapter 20. It starts with a pole-zero diagram of an *analog* filter in the s-plane, and converts it into the desired *digital* filter through several mathematical *transforms*. To reduce the complexity of the algebra, the filter is designed as a cascade of several stages, with each stage implementing one pair of poles. The recursive coefficients for each stage are then combined into the recursive coefficients for the entire filter. This is a very sophisticated and complicated algorithm; a fitting way to end this book. Here's how it works.

Loop Control
Figure 31-7 shows the program and flowchart for the method, duplicated from Chapter 20. After initialization and parameter entry, the main portion of the program is a loop that runs through each pole-pair in the filter. This loop is controlled by block 11 in the flowchart, and the FOR-NEXT loop in lines 320 & 460 of the program. For example, the loop will be executed three times for a 6 pole filter, with the loop index, P%, taking on the values 1,2,3. That is, a 6 pole filter is implemented in three stages, with two poles per stage.

Combining Coefficients
During each loop, subroutine 1000 (listed in Fig. 31-8) calculates the recursive coefficients *for that stage*. These are returned from the subroutine in the five variables: $A0, A1, A2, B1, B2$. In step 10 of the flowchart (lines 360-440), these coefficients are combined with the coefficients of all the previous stages, held in the arrays: A[ ] and B[ ]. At the end of the first loop, A[ ] and B[ ] hold the coefficients for stage one. At the end of the second loop, A[ ] and B[ ] hold the coefficients of the cascade of stage one and stage two. When all the loops have been completed, A[ ] and B[ ] hold the coefficients needed to implement the entire filter.

The coefficients are combined as previously outlined in Table 31-1, with a few modifications to make the code more compact. First, the index of the arrays, A[ ] and B[ ], is shifted by *two* during the loop. For example, a_0 is held in A[2], a_1 & b_1 are held in A[3] & B[3], etc. This is done to prevent the program from trying to access values outside the defined arrays. This shift is removed in block 12 (lines 480-520), such that the final recursion coefficients reside in A[ ] and B[ ] without an index offset.

Second, A[ ] and B[ ] must be initialized with coefficients corresponding to the *identity* system, not all zeros. This is done in lines 180 to 240. During the first loop, the coefficients for the first stage are combined with the information initially present in these arrays. If all zeros were initially present, the arrays would always remain zero. Third, two temporary arrays are used, TA[ ] and TB[ ]. These hold the old values of A[ ] and B[ ] during the convolution, freeing A[ ] and B[ ] to hold the new values.

To finish the program, block 13 (lines 540-670) adjusts the filter to have a unity gain in the passband. This operates as previously described: calculate the existing gain with Eq. 31-7 or 31-8, and divide all the "a" coefficients to normalize. The intermediate variables, SA and SB, are the sums of the "a" and "b" coefficients, respectively.

Calculate Pole Locations in the s-Plane

Regardless of the type of filter being designed, this program begins with a Butterworth low-pass filter in the *s-plane*, with a cutoff frequency of $\omega = 1$. As described in the last chapter, Butterworth filters have poles that are equally spaced around a circle in the s-plane. Since the filter is low-pass, no zeros are used. The radius of the circle is *one*, corresponding to the cutoff frequency of $\omega = 1$. Block 3 of the flowchart (lines 1080 & 1090) calculate the location of each pole-pair in rectangular coordinates. The program variables, RP and IP, are the real and imaginary parts of the pole location, respectively. These program variables correspond to σ and ω, where the pole-pair is located at $\sigma \pm j\omega$. This pole location is calculated from the number of poles in the filter and the stage being worked on, the program variables: NP and P%, respectively.

Warp from Circle to Ellipse

To implement a Chebyshev filter, this *circular* pattern of poles must be transformed into an *elliptical* pattern. The relative flatness of the ellipse determines how much ripple will be present in the passband of the filter. If the pole location on the circle is given by: σ and ω, the corresponding location on the ellipse, σ' and ω', is given by:

EQUATION 31-9
Circular to elliptical transform. These equations change the pole location on a circle to a corresponding location on an ellipse. The variables, *NP* and *PR*, are the number of poles in the filter, and the percent ripple in the passband, respectively. The location on the circle is given by σ and ω, and the location on the ellipse by σ' and ω'. The variables ϵ, v, and k, are used only to make the equations shorter.

$$\sigma' = \sigma \sinh(v)/k$$

$$\omega' = \omega \cosh(v)/k$$

where:

$$v = \frac{\sinh^{-1}(1/\epsilon)}{NP}$$

$$k = \cosh\left(\frac{1}{NP}\cosh^{-1}\frac{1}{\epsilon}\right)$$

$$\epsilon = \left[\left(\frac{100}{100-PR}\right)^2 - 1\right]^{1/2}$$

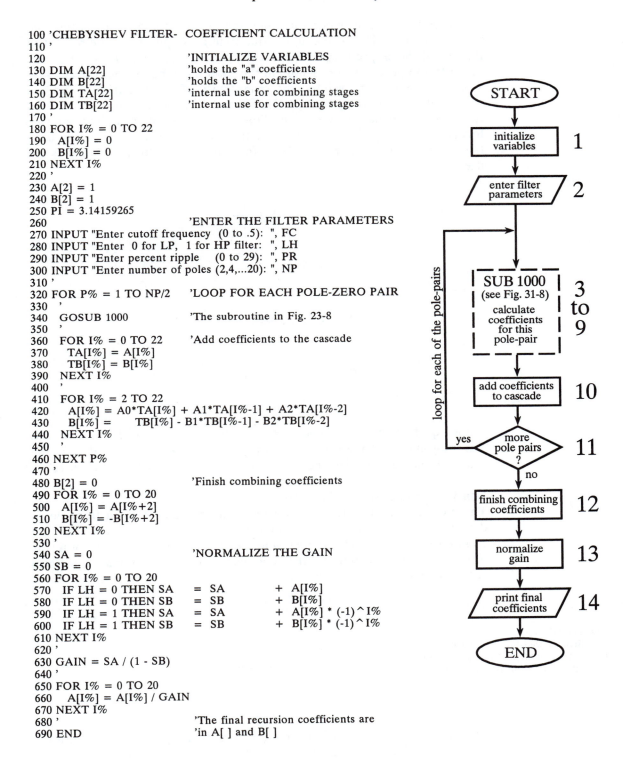

```
100 'CHEBYSHEV FILTER- COEFFICIENT CALCULATION
110 '
120                          'INITIALIZE VARIABLES
130 DIM A[22]                'holds the "a" coefficients
140 DIM B[22]                'holds the "b" coefficients
150 DIM TA[22]               'internal use for combining stages
160 DIM TB[22]               'internal use for combining stages
170 '
180 FOR I% = 0 TO 22
190   A[I%] = 0
200   B[I%] = 0
210 NEXT I%
220 '
230 A[2] = 1
240 B[2] = 1
250 PI = 3.14159265
260                          'ENTER THE FILTER PARAMETERS
270 INPUT "Enter cutoff frequency  (0 to .5): ", FC
280 INPUT "Enter  0 for LP,  1 for HP filter: ", LH
290 INPUT "Enter percent ripple    (0 to 29): ", PR
300 INPUT "Enter number of poles (2,4,...20): ", NP
310 '
320 FOR P% = 1 TO NP/2       'LOOP FOR EACH POLE-ZERO PAIR
330   '
340   GOSUB 1000             'The subroutine in Fig. 23-8
350   '
360   FOR I% = 0 TO 22       'Add coefficients to the cascade
370     TA[I%] = A[I%]
380     TB[I%] = B[I%]
390   NEXT I%
400   '
410   FOR I% = 2 TO 22
420     A[I%] = A0*TA[I%] + A1*TA[I%-1] + A2*TA[I%-2]
430     B[I%] =      TB[I%] - B1*TB[I%-1] - B2*TB[I%-2]
440   NEXT I%
450   '
460 NEXT P%
470 '
480 B[2] = 0                 'Finish combining coefficients
490 FOR I% = 0 TO 20
500   A[I%] = A[I%+2]
510   B[I%] = -B[I%+2]
520 NEXT I%
530 '
540 SA = 0                   'NORMALIZE THE GAIN
550 SB = 0
560 FOR I% = 0 TO 20
570   IF LH = 0 THEN SA    = SA      + A[I%]
580   IF LH = 0 THEN SB    = SB      + B[I%]
590   IF LH = 1 THEN SA    = SA      + A[I%] * (-1) ^ I%
600   IF LH = 1 THEN SB    = SB      + B[I%] * (-1) ^ I%
610 NEXT I%
620 '
630 GAIN = SA / (1 - SB)
640 '
650 FOR I% = 0 TO 20
660   A[I%] = A[I%] / GAIN
670 NEXT I%
680 '                        'The final recursion coefficients are
690 END                      'in A[ ] and B[ ]
```

FIGURE 31-7

Chebyshev-Butterworth filter design. This program was previously presented as Table 20-4 and Table 20-5 in Chapter 20. Figure 31-8 shows the program and flowchart for subroutine 1000, called from line 340 of this main program.

These equations use hyperbolic sine and cosine functions to define the ellipse, just as ordinary sine and cosine functions operate on a circle. The flatness of the ellipse is controlled by the variable: PR, which is numerically equal to the percentage of ripple in the filter's passband. The variables: ϵ, v and k are used to reduce the complexity of the equations, and are represented in the program by: ES, VX and KX, respectively. In addition to converting from a circle to an ellipse, these equations correct the pole locations to keep a unity cutoff frequency. Since many programming languages do not support hyperbolic functions, the following identities are used:

$$\sinh(x) = \frac{e^x - e^{-x}}{2}$$

$$\cosh(x) = \frac{e^x + e^{-x}}{2}$$

$$\sinh^{-1}(x) = \log_e\left(x + (x^2+1)^{1/2}\right)$$

$$\cosh^{-1}(x) = \log_e\left(x + (x^2-1)^{1/2}\right)$$

These equations produce illegal operations for $PR \geq 30$ and $PR = 0$. To use this program to calculate Butterworth filters (i.e., zero ripple, $PR = 0$), the program lines that implement these equations must be bypassed (line 1120).

Continuous to Discrete Conversion
The most common method of converting a pole-zero pattern from the s-domain into the z-domain is the **bilinear transform**. This is a mathematical technique of *conformal mapping*, where one complex plane is algebraically distorted or warped into another complex plane. The bilinear transform changes $H(s)$, into $H(z)$, by the substitution:

EQUATION 31-10
The Bilinear transform. This substitution maps every point in the s-plane into a corresponding piont in the z-plane.

$$s \rightarrow \frac{2(1 - z^{-1})}{T(1 + z^{-1})}$$

That is, we write an equation for $H(s)$, and then replaced each s with the above expression. In most cases, $T = 2\tan(1/2) = 1.093$ is used. This results in the s-domain's frequency range of 0 to π radians/second, being mapped to the z-domain's frequency range of 0 to π radians. Without going into more detail, the bilinear transform has the desired properties to convert

```
1000 'THIS SUBROUTINE IS CALLED FROM FIG. 31-7,  LINE 340
1010 '
1020 'Variables entering subroutine: PI, FC, LH, PR, HP, P%
1030 'Variables exiting subroutine:  A0, A1, A2, B1, B2
1040 'Variables used internally:     RP, IP, ES, VX, KX, T, W, M, D, K,
1050 '                               X0, X1, X2, Y1, Y2
1060 '
1070 '                               'Calculate pole location on unit circle
1080 RP = -COS(PI/(NP*2) + (P%-1) * PI/NP)
1090 IP =  SIN(PI/(NP*2) + (P%-1) * PI/NP)
1100 '
1110 '                               'Warp from a circle to an ellipse
1120 IF PR = 0 THEN GOTO 1210
1130 ES = SQR( (100 / (100-PR))^2 -1 )
1140 VX = (1/NP) * LOG( (1/ES) + SQR( (1/ES^2) + 1) )
1150 KX = (1/NP) * LOG( (1/ES) + SQR( (1/ES^2) - 1) )
1160 KX = (EXP(KX) + EXP(-KX))/2
1170 RP = RP * ( (EXP(VX) - EXP(-VX) ) /2 ) / KX
1180 IP = IP * ( (EXP(VX) + EXP(-VX) ) /2 ) / KX
1190 '
1200 '                               's-domain to z-domain conversion
1210 T  = 2 * TAN(1/2)
1220 W  = 2*PI*FC
1230 M  = RP^2 + IP^2
1240 D  = 4 - 4*RP*T + M*T^2
1250 X0 = T^2/D
1260 X1 = 2*T^2/D
1270 X2 = T^2/D
1280 Y1 = (8 - 2*M*T^2)/D
1290 Y2 = (-4 - 4*RP*T - M*T^2)/D
1300 '
1310 '                               'LP TO LP, or LP TO HP
1320 IF LH = 1 THEN K = -COS(W/2 + 1/2) / COS(W/2 - 1/2)
1330 IF LH = 0 THEN K =  SIN(1/2 - W/2) / SIN(1/2 + W/2)
1340 D = 1 + Y1*K - Y2*K^2
1350 A0 = (X0 - X1*K + X2*K^2)/D
1360 A1 = (-2*X0*K + X1 + X1*K^2 - 2*X2*K)/D
1370 A2 = (X0*K^2 - X1*K + X2)/D
1380 B1 = (2*K + Y1 + Y1*K^2 - 2*Y2*K)/D
1390 B2 = (-K^2 - Y1*K + Y2)/D
1400 IF LH = 1 THEN A1 = -A1
1410 IF LH = 1 THEN B1 = -B1
1420 '
1430 RETURN
```

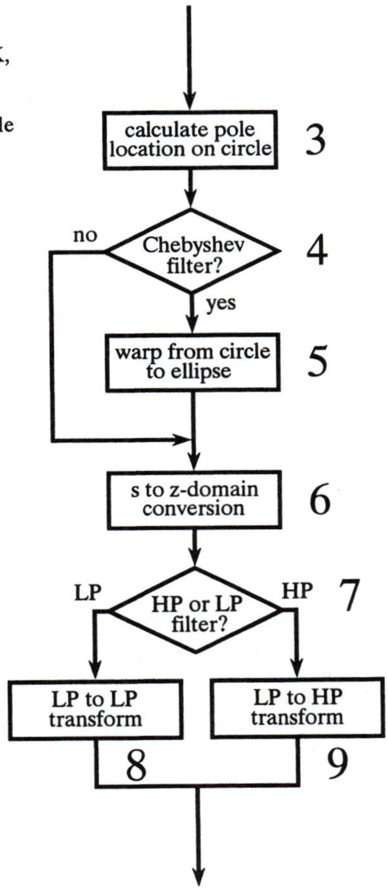

FIGURE 31-8
Subroutine called from Figure 31-7.

from the s-plane to the z-plane, such as vertical lines being mapped into circles. Here is an example of how it works. For a continuous system with a single pole-pair located at $p_1 = \sigma + j\omega$ and $p_2 = \sigma - j\omega$, the s-domain transfer function is given by:

$$H(s) = \frac{1}{(s-p_1)(s-p_2)}$$

The bilinear transform converts this into a discrete system by replacing each *s* with the expression given in Eq. 31-10. This creates a z-domain transfer function also containing two poles. The problem is, the substitution leaves the transfer function in a very unfriendly form:

$$H(z) = \cfrac{1}{\left(\cfrac{2(1-z^{-1})}{T(1+z^{-1})} - (\sigma+j\omega)\right)\left(\cfrac{2(1-z^{-1})}{T(1+z^{-1})} - (\sigma-j\omega)\right)}$$

Working through the long and tedious algebra, this expression can be placed in the standard form of Eq. 31-3, and the recursion coefficients identified as:

EQUATION 31-11
Bilinear transform for two poles. The pole-pair is located at $\sigma\pm\omega$ in the s-plane, and a_0, a_1, a_2, b_1, b_2 are the recursion coefficients for the discrete system.

$$a_0 = T^2/D$$
$$a_1 = 2T^2/D$$
$$a_2 = T^2/D$$

$$b_1 = (8 - 2MT^2)/D$$
$$b_2 = (-4 - 4\sigma T - MT^2)/D$$

where:
$$M = \sigma^2 + \omega^2$$
$$T = 2\tan(1/2)$$
$$D = 4 - 4\sigma T + MT^2$$

The variables M, T, and D have no physical meaning; they are simply used to make the equations shorter.

Lines 1200-1290 use these equations to convert the location of the s-domain pole-pair, held in the variables, RP and IP, directly into the recursive coefficients, held in the variables, X0, X1, X2, Y1, Y2. In other words, we have calculated an intermediate result: the recursion coefficients for one stage of a *low-pass* filter with a cutoff frequency of *one*.

Low-pass to Low-pass Frequency Change
Changing the frequency of the recursive filter is also accomplished with a conformal mapping technique. Suppose we know the transfer function of a recursive low-pass filter with a unity cutoff frequency. The transfer function of a similar low-pass filter with a new cutoff frequency, W, is obtained by using a **low-pass to low-pass transform**. This is also carried out

by *substituting variables*, just as with the bilinear transform. We start by writing the transfer function of the unity cutoff filter, and then replace each z^{-1} with the following:

EQUATION 31-12
Low-pass to low-pass transform. This is a method of changing the cutoff frequency of low-pass filters. The original filter has a cutoff frequency of unity, while the new filter has a cutoff frequency of W, in the range of 0 to π.

$$z^{-1} \rightarrow \frac{z^{-1} - k}{1 - kz^{-1}}$$

where:
$$k = \frac{\sin(1/2 - W/2)}{\sin(1/2 + W/2)}$$

This provides the transfer function of the filter with the new cutoff frequency. The following design equations result from applying this substitution to the biquad, i.e., no more than two poles and two zeros:

EQUATION 31-13
Low-pass to low-pass conversion. The recursion coefficients of the filter with unity cutoff are shown in italics. The coefficients of the low-pass filter with a cutoff frequency of W are in roman.

$$a_0 = (a_0 - a_1 k + a_2 k^2)/D$$

$$a_1 = (-2a_0 k + a_1 + a_1 k^2 - 2a_2 k)/D$$

$$a_2 = (a_0 k^2 - a_1 k + a_2)/D$$

$$b_1 = (2k + b_1 + b_1 k^2 - 2b_2 k)/D$$

$$b_2 = (-k^2 - b_1 k + b_2)/D$$

where:
$$D = 1 + b_1 k - b_2 k^2$$
$$k = \frac{\sin(1/2 - W/2)}{\sin(1/2 + W/2)}$$

Low-pass to High-pass Frequency Change

The above transform can be modified to change the response of the system from low-pass to high-pass while simultaneously changing the cutoff frequency. This is accomplished by using a **low-pass to high-pass transform**, via the substitution:

EQUATION 31-14
Low-pass to high-pass transform. This substitution changes a low-pass filter into a high-pass filter. The cutoff frequency of the low-pass filter is *one*, while the cutoff frequency of the high-pass filter is W.

$$z^{-1} \rightarrow \frac{-z^{-1} - k}{1 + kz^{-1}}$$

where:
$$k = -\frac{\cos(W/2 + 1/2)}{\cos(W/2 - 1/2)}$$

As before, this can be reduced to design equations for changing the coefficients of a biquad stage. As it turns out, the equations are identical to those of Eq. 31-13, with only two minor changes. The value of k is different (as given in Eq. 31-14), and two coefficients, a_1 and b_1, are negated in value. These equations are carried out in lines 1310 to 1410 in the program, providing the desired cutoff frequency, and the choice of a high-pass or low-pass response.

The Best and Worst of DSP

This book is based on a simple premise: **most DSP techniques can be used and understood with a minimum of mathematics.** The idea is to provide scientists and engineers *tools* for solving the DSP problems that arise in their non-DSP research or design activities.

These last four chapters are the other side of the coin: DSP techniques that can *only* be understood through extensive math. For example, consider the Chebyshev-Butterworth filter just described. This is the *best* of DSP, a series of elegant mathematical steps leading to an optimal solution. However, it is also the *worst* of DSP, a design method so complicated that most scientists and engineers will look for another alternative.

Where do you fit into this scheme? This depends on *who* your are and *what* you plan on using DSP for. The material in the last four chapters provides the theoretical basis for signal processing. If you plan on pursuing a *career* in DSP, you need to have a detailed understanding of this mathematics. On the other hand, specialists in other areas of science and engineering only need to know how DSP is *used*, not how it is *derived*. To this group, the theoretical material is more of a background, rather than a central topic.

Study Guide

FOUNDATIONS

Chapter 1- The Breadth and Depth of DSP

Digital Signal Processing is used in a *broad* range of fields. Each of these areas has developed a *deep* DSP technology. This combination of breadth and depth makes it impossible for any one individual to master all the techniques that have been developed.

Much of DSP was developed in the 1960s and 1970s from work in radar & sonar, oil exploration, space exploration, and medical imaging.

Most DSP literature is mathematically intensive and very difficult for nonspecialists to understand.

Digital Signal Processing has produced revolutionary changes in many diverse fields, including: telecommunications, audio processing, echo location, and image processing.

KEY WORDS AND TERMS: artificial reverberation, computed tomography, data compression, echo control, multiplexing, mix-down, radar, seismology, speech generation, speech recognition, sonar

Chapter 2- Statistics, Probability and Noise

A *signal* is a description of how one parameter (called the dependent variable) depends on another parameter (called the independent variable).

The *domain* of a signal is determined by the independent variable. For instance, a voltage that varies with time is said to be in the "time domain."

With *continuous* signals, the two parameters can take on a continuous range of values. With *discrete* (digitized) signals, both parameters are quantized. For the most part, continuous signals exist in nature, while discrete signals exist inside computers.

Statistics is used to characterize acquired signals. *Probability* is used to characterize the underlying processes that generate the signals.

The variable, N, is often used to represent the number of samples in a signal. The indexes of a signal usually run from 0 to N-1; however, 1 to N can also be used. Watch out for this difference, it can be very confusing.

The *mean* is the average value of a signal. The *standard deviation* describes how the samples in a signal differ from the mean. These are the same as the DC and AC values in electronics. The mean and standard deviation can be calculated in several different ways.

The *histogram, probability mass function* (pmf), and *probability distribution function* (pdf) all describe how frequently a variable takes on specific values. The histogram is used with acquired data, while the pdf or pmf is used to describe the underlying process.

From the *Central Limit Theorem*, a sum of random numbers has a *Gaussian* distribution, regardless of the distribution of the individual random numbers. The Gaussian distribution is also called a *normal* distribution.

Random number generators produce a uniform distribution between zero and one. Simple steps allow this to be turned into a Gaussian distribution.

Accuracy describes the systematic errors in a measurement, and can be improved by better calibration. *Precision* describes the random errors in a measurement, and can be improved by averaging measurements.

KEY WORDS AND TERMS: AC, accuracy, binning, Central Limit Theorem, coefficient-of-variation (CV), continuous signal, dependent variable, discrete signal, DC, domain, frequency domain, Gaussian, histogram, independent variable, mean, normal distribution, precision, probability, probability distribution function (pdf), probability mass function (pmf), random errors, random number generator, range, root-mean-square, running statistics, signal, signal-to-noise ratio (SNR), spatial domain, standard deviation, stationary process, statistical noise, statistics, systematic errors, time domain, underlying process, variance

Chapter 3- ADC and DAC

Analog-to-digital conversion (ADC) can be modeled as a two step process: *sampling* and *quantization*.

The error due to quantization usually acts like added random noise.

The *sampling theorem* describes what information is lost during sampling. Sampled data can only contain frequencies up to one-half the sampling rate. To avoid losing information during sampling, the sampling rate must be *twice* the highest frequency present in the signal.

If frequencies greater than one-half the sampling rate are present in a signal being sampled, they will be *aliased* to lower frequencies in the digitized signal. This "change of frequency" (aliasing) is generally a loss of information, and should usually be avoided.

An *antialias filter* is a low-pass analog filter placed before the ADC. Its purpose is to remove frequencies above one-half the sampling rate. Antialias filters usually have a Butterworth or Chebyshev response. Butterworth filters have the fastest possible roll-off while keeping a flat passband. Chebyshev filters achieve a faster roll-off than the Butterworth by allowing a small amount of ripple in the passband.

Signals with *frequency domain encoding* usually use an antialias filter. This is because aliasing directly destroys information encoded in the frequency domain.

The sampling theorem is little help in understanding how to digitize signals with *time domain encoded* information. An antialias filter is generally not used, because the distortion it introduces is often worse than the aliasing it prevents. For these signals, the required sampling rate is determined by understanding the shape of the waveform to be captured.

In some cases, the *Bessel filter* is used to filter time domain encoded signals before ADC. It smoothes the signal to remove high-frequency noise and interference. The Bessel filter produces minimal distortion of the waveform because the step response has no overshoot or ringing, and the rising and falling edges look similar.

Digital-to-analog conversion (DAC) is usually accomplished by a *zeroth-order hold* followed by a *reconstruction filter*, an analog low-pass filter at one-half the sampling rate. However, the high frequency components in this reproduced signal have too low of amplitude (by up to 36%). To correct this, a high frequency boost can be applied in the digital data before the DAC, or in the analog signal after.

Multirate techniques for ADC and DAC replace *analog circuitry* with *digital algorithms*. This can have tremendous cost and performance advantages. A very high sampling rate is used during ADC, followed by decimation to reduce the sampling rate to whatever is desired. During DAC, the digital signals are interpolated to a higher sampling rate before being converted to an analog signal.

Single bit ADC and DAC are common multirate techniques used in telecommunications and high fidelity music reproduction. There are several variations, including: delta modulation, CVSD, and delta-sigma conversion.

KEY WORDS AND TERMS: aliasing, antialias filter, Bessel filter, Butterworth filter, Chebyshev filter, CVSD, decimation, delta modulation, delta-sigma, dithering, elliptic filter, frequency domain encoding, impulse train, interpolation, least significant bit (LSB), multirate, Nyquist frequency, Nyquist rate, overshoot, passband ripple, passband, proper sampling, quantization error, reconstruction filter, ringing, roll-off, sampling theorem, sidebands, sinc function, step response, stopband, switched capacitor filter, time domain encoding, zeroth-order hold

Chapter 4- DSP Software

Computer numbers are stored in two common formats: *fixed point* (also called integers) and *floating point* (also called real numbers).

In *fixed point*, the bit patterns can be arranged as: unsigned integer, offset binary, sign and magnitude, one's complement, or two's complement. Two's complement is the most common because it is the easiest way to design hardware.

When 16 bits are used to store a two's complement number, all of the integers from -32,768 to 32,767 can be represented.

Integers are ideal for loop indexes, counters, and other applications where round-off noise must be avoided.

The encoding scheme for *floating point* is the same as scientific notation, a *mantissa* is multiplied by a base raised to an *exponent*. This allows incredibly small and large numbers to be represented, but the spacing between adjacent numbers is unequal.

Single precision floating point uses 32 bits to represent numbers, while double precision uses 64 bits. Double precision is only required in the most demanding applications.

Expect that every single precision number will have a round-off error of about one part in *forty million*, multiplied by the number of mathematical operations it has been through. For double precision, this increases to one part in *forty quadrillion*.

DSP uses the same programming languages as other scientific and engineering applications. The most common language is C; however, assembly can achieve a higher execution speed and the simplicity of BASIC makes it attractive for the infrequent programmer.

Computer speed doubles about every two years. During the last 15 years it has increased by nearly one-thousand times. Learning about the current technology is not enough; you must understand and track how it is continually evolving.

To make your programs execute faster, learn what is *fast* and what is *slow*. Integers generally run about 10-20 times faster than floating point numbers. Multiplication and division are slower than addition and subtraction. Transcendental functions such as $\sin(x)$, $\log(x)$, and e^x require about 10 time longer to calculate than a single multiplication. I/O and graphics operations are notoriously slow.

KEY WORDS AND TERMS: assembly, BASIC, C, compiled language, compiler, double precision, DSP microprocessor, fixed point, floating point, high-level language, integers, long integer, machine code, memory cache, offset binary, one's complement, pipeline, RISC, round-off error, sign and magnitude, sign bit, single precision, source code, two's complement, unsigned integer

FUNDAMENTALS

Chapter 5- Linear Systems

A *signal* is a description of how one parameter depends on another parameter. A *system* is any process that produces an output signal in response to an input signal.

In this book, parentheses are used to denote continuous signals, such as $x(t)$, while brackets are used for discrete signals, such as $x[n]$.

If a more descriptive name is not available, the input signal to a system is often called $x[n]$ or $x(t)$, while the output signal is called $y[n]$ or $y(t)$.

There are an infinite number of different systems. Fortunately, many useful ones fall into a category called *linear systems*. This is extremely important. Without the linear system concept, we would be forced to examine the individual characteristics of many unrelated systems. With this approach, we can focus on the traits of the linear system category as a whole.

A system is linear if it has two mathematical properties: *homogeneity* and *additivity*. A third property, *shift invariance*, is not a strict requirement for linearity, but is mandatory for most DSP techniques.

Homogeneity means that a change in the amplitude of the input signal results in the same change in the amplitude of the output signal. That is, if $x[n]$ produces $y[n]$, then $kx[n]$ produces $ky[n]$, where k is a constant.

When a system is *additive*, signals that are added in the input produce signals that are added in the output. In mathematical form, if $a[n]$ produces $y[n]$, and if $b[n]$ produces $z[n]$, then the combined input $a[n]+b[n]$ produces $x[n]+z[n]$.

Shift invariance means that the system does not change with time (or whatever independent variable is used). That is, if $x[n]$ produces $y[n]$, then a shift in the input signal, $x[n+k]$ produces nothing more than a shift in the output signal, $y[n+k]$, where k is a constant.

All linear systems have *static linearity*; when the input is a constant value, the output is simply equal to the input multiplied by a constant.

All linear systems have *sinusoidal fidelity*; when the input is a sinusoid, the output will be a sinusoid at exactly the same frequency, but may have a different amplitude and phase.

The fundamental concept in DSP: A signal can be decomposed into additive components, each of the components passed through a linear system, and the resulting output components synthesized into an output signal. The output signal found in this manner is identical to that found by passing the original signal through the system. The allows a *difficult* problem to be replaced with many *simple* problems.

Several different decompositions are used in DSP, but two are the most important: *Impulse decomposition* (leading to the technique of convolution) and *Fourier decomposition*.

KEY WORDS AND TERMS: additivity, cascade, decomposition, even/odd decomposition, Fourier decomposition, homogeneity, impulse decomposition, interlaced decomposition, linear system, memoryless, shift invariance, signal, sinusoidal fidelity, static linearity, step decomposition, superposition, system

Chapter 6- Convolution

An *impulse* is a signal consisting of all zeros, except for a single nonzero value. The *delta function* is a normalized impulse, that is, the nonzero value occurs at sample number zero, and has an amplitude of one. The delta function is represented by $\delta[n]$.

The *impulse response* of a system is the output produced by a delta function input. The impulse response contains complete information about a linear system. If the system is a filter, the impulse response is also called the *filter kernel*, the *convolution kernel*, or the *kernel*. In image processing, the impulse response is also called the *point spread function*. The impulse response is often represented by $h[n]$.

Convolution is a mathematical operation that combines two signals to form a third signal. It is important in linear systems because it describes how the input signal and impulse response combine to form the output signal.

A star is used to indicate convolution. That is, $x[n] * h[n] = y[n]$ means that $x[n]$ and $h[n]$ are convolved to produce $y[n]$.

Convolution can be understood in two different ways, from the viewpoint of the input signal or the viewpoint of the output signal.

From the *input side*, each sample in the input signal is viewed as contributing a scaled and shifted version of the impulse response to the output signal. From the *output side*, each sample in the output signal is viewed as the sum of samples in the input signal weighted by the impulse response flipped left-for-right.

When a signal of length N is convolved with a signal of length M, the resulting signal has a length of $N+M$-1.

KEY WORDS AND TERMS: convolution kernel, convolution sum, delta function, end effects, filter kernel, immersed impulse response, impulse, impulse response, kernel, left-for-right flip, point spread function, unit impulse, weighing coefficients

Chapter 7- Properties of Convolution

The delta function is the *identity* for convolution; any signal convolved with the delta function is left unchanged: $x[n] * \delta[n] = x[n]$.

An impulse response consisting of a scaled delta function is an *amplifier* or *attenuator*. Echoes have an impulse response containing a delta function plus a scaled and shifted delta function.

The *first difference* and *running sum* are the discrete versions of the first derivative and integral, respectively. These can be implemented by convolution or recursion equations.

Low-pass filter kernels are formed from a group of adjacent positive points. This smoothes the signal thereby removing high frequencies. *High-pass filter kernels* can be formed by subtracting a low-pass filter kernel from a delta function.

In a *causal* signal, all of the negative indexed samples have a value of zero.

In a *linear phase* signal, the left half of the signal is symmetrical with the right half. If the point of symmetry is at sample zero, the signal is additionally said to be *zero phase*. If this symmetry is not present, the signal is said to be *nonlinear phase*.

The *commutative* property of convolution is stated: $x[n] * h[n] = h[n] * x[n]$. This means that the input signal and impulse response can be exchange without affecting the output; however, this usually has no physical meaning.

The *associate* property of convolution is stated: $(a[n] * b[n]) * c[n] = a[n] * (b[n] * c[n])$. This describes how cascaded systems behave. When two or more stages are in a cascade, the order of the stages can be rearranged without changing the output signal. Further, any number of cascaded stages can be replaced with a new system. The impulse response of the replacement system is found by convolving the impulse response of all the original stages.

The *distributive* property of convolution is stated: $a[n] * b[n] + a[n] * c[n] = a[n] * (b[n] + c[n])$. This shows that parallel stages with added outputs can be replaced with a single system. The impulse response of the replacement system is found by adding the impulse responses of all the original stages.

By the *Central Limit Theorem*, a pulse-like signal convolved with itself many times becomes Gaussian.

Correlation is used to locate a known waveform in another signal. It is carried out the same as convolution, except the left-for-right flip is not used. This is the basis of a technique called *matched filtering*. When two signals are correlated, the resulting signal is called the *cross-correlation*. If a signal is correlated with itself, the resulting signal is call the *autocorrelation*.

KEY WORDS AND TERMS: associative property, autocorrelation, cascaded systems, causal, commutative property, correlation, cross-correlation, difference equation, discrete integral, discrete derivative, first difference, identity, linear phase, matched filtering, noncausal signal, nonlinear phase, parallel systems, recursion equation, running sum, zero phase

Chapter 8- The Discrete Fourier Transform (DFT)

Fourier analysis is a family of mathematical techniques based on decomposing signals into sinusoids. Sinusoids are used because of *sinusoidal fidelity*: a sinusoid entering a linear system produces a sinusoidal output. Other waveforms do not have this property.

A signal can be either *discrete* or *continuous*, and *periodic* or *aperiodic*. This results in four categories of signals, each with its own version of the Fourier transform:

> *Fourier transform* (aperiodic-continuous signals)
> *Fourier series* (periodic-continuous signals)
> *Discrete time Fourier transform, or DTFT* (aperiodic-discrete signals)
> *Discrete Fourier transform, or DFT* (periodic-discrete signals)

If a signal is discrete in one domain, it will be periodic in the other, and vice versa. If a signal is continuous in one domain, it will be aperiodic in the other, and vice versa.

The *DFT* is the most important for DSP, since this is the only one of the four that can be represented in a digital computer. The other three must be carried out by mathematical equations.

A *function* is an algorithm or procedure for changing one or more variables into another variable. A *transform* is an extension of this, changing one *group of data* into another *group of data*.

Each of the four Fourier transforms can be subdivided into *real* and *complex versions*. The real version is the simplest, using only ordinary numbers. The complex versions are much more complicated, requiring the use of *complex numbers* (numbers with the term: $j = \sqrt{-1}$). Chapters 28-31 discuss the *complex Fourier transforms*, while the earlier chapters only deal with the *real Fourier transforms*.

The real DFT transforms an N sample signal in the time domain into two $N/2+1$ sample signals in the frequency domain. This is called the *forward transform* or *decomposition*. Changing the frequency domain back into the time domain is called the *inverse transform* or *synthesis*. In most applications, N is selected to be a power of two between 32 and 4096.

Lower case letters are used to represent time domain signals, such as $x[n]$. Upper case is used to represent the corresponding frequency domain signals. One of the two frequency domain signals is called the *real part*, written $Re\,X[k]$. The other is called the *imaginary part*, and is written $Im\,X[k]$. The real part holds the amplitudes of the cosine waves contained in the time domain signal, while the imaginary part holds the amplitudes of the sine waves (neglecting scaling factors).

The frequency domain's independent variable can be expressed in four different forms: (1) a sample index, such as the variable k, that runs from 0 to $N/2+1$, (2) a frequency, such as represented by the variable f, that runs from 0 to 0.5, (3) a natural frequency, usually denoted by ω, that runs between 0 and π, (4) the analog frequencies of the particular system, expressed in hertz, running from zero to one-half of the actual sampling rate.

The *basis functions* of the Fourier transform are cosine and sine waves with unity amplitude. To *synthesize* a time domain signal, the frequency domain values are assigned to the basis functions, scaling factors are applied, and the resulting sinusoids added.

Calculating the frequency domain from the time domain (*analysis*) can be accomplished in several ways. This includes: *simultaneous equations*, *correlation* of the time domain signal with the basis functions, and the *fast Fourier transform*.

The frequency domain can be expressed in *rectangular form* (real and imaginary parts) or in polar form (*magnitude and phase*). Rectangular form is usually used during calculations, while polar form is usually best for displaying the signals to humans.

There are several *nuisances* associated with polar notation. Watch out for these; they will confuse you!

KEY WORDS AND TERMS: analysis, basis functions, complex Fourier transform, decomposition, discrete time Fourier transform (DTFT), discrete Fourier transform (DFT), forward transform, Fourier transform, Fourier series, imaginary part, inverse transform, magnitude, natural frequency, phase, polar form, real Fourier transform, real part, rectangular form, synthesis, transform

Chapter 9- Applications of the DFT

This chapter presents three uses for the DFT: (1) spectral analysis, (2) calculating the frequency response of systems, and (3) convolution via the frequency domain.

Spectral analysis is used to examine signals with information encoded in the frequency domain.

The most common spectral analysis procedure: (1) the acquired signal is broken into segments, (2) each segment is multiplied by a smooth window (Blackman or Hamming), (3) the DFT is taken of each segment, (4) the frequency spectrum of each segment is converted to polar form, and (4) the magnitudes are averaged.

Longer segments provide better resolution in the frequency spectrum. The noise in the final spectrum is determined by the number of segments averaged.

The *frequency response* of a system is the Fourier Transform of its impulse response.

To find the frequency response of a system: (1) the impulse response is padded with zeros to make its length a large power of two, such as 512 samples, (2) the DFT is taken of the padded impulse response, using the FFT algorithm, (3) the frequency spectrum is converted to polar form to be displayed.

The longer the DFT used (i.e., the larger the value of N), the more samples that will appear in the frequency response between 0 and 0.5. If the time domain could be padded with an infinite number of zeros, the frequency domain would become a continuous line. This means the frequency response of a system *exists* as a continuous line between 0 and 0.5. However, when an N point DFT is used, only $N/2+1$ samples of this continuous curve are calculated.

Frequency domain convolution is a way to bypass convolving signals in the time domain. This can eliminate two problems. First, convolution is mathematically difficult to deal with; it is not a simple operation such as addition or multiplication. Second, convolution requires many calculations, resulting in a long execution time.

In the time domain, the input signal is convolved with the impulse response to produce the output signal. In the frequency domain, this corresponds to multiplying the spectrum of the input signal with the system's frequency response, creating the spectrum of the output signal.

To carry out frequency domain convolution, the DFT is taken of both signals to be convolved. The frequency spectra of these two signals are multiplied, resulting in the frequency spectrum of the output signal. The Inverse DFT is then used find the output signal from its spectrum.

Care must be taken during frequency domain convolution to avoid *circular convolution*.

KEY WORDS AND TERMS: 1/f noise, Blackman window, circular convolution, deconvolution, fast Fourier transform, FFT, frequency resolution, frequency domain convolution, frequency response, fundamental frequency, Hamming window, harmonics, microphonics, passive sonar, rectangular window, spectral analysis, spectral tails, spectral leakage, white noise

Chapter 10- Fourier Transform Properties

Every operation in the time domain has a corresponding operation in the frequency domain and vice versa. For instance, convolution in one domain corresponds to multiplication in the other domain. These relationships are called Fourier Transform *properties*.

The Fourier transform is *linear*, meaning it is *additive* and *homogeneous*.

Additivity means that signals added in one domain are also added in the other domain. In mathematical form, if the Fourier transform of $a[n]$ is $A[k]$, and if the Fourier transform of $b[n]$ is $B[k]$, then the Fourier transform of $a[n]+b[n]$ is $A[k]+B[k]$.

Homogeneity means that a change of amplitude in one domain results in the same change of amplitude in the other domain. That is, if the Fourier transform of $x[n]$ is $X[k]$, then the Fourier transform of $kx[n]$ is $kX[k]$, where k is a constant.

In spite of being linear, the Fourier transform is not *shift invariant*; a shift in the time domain does *not* correspond to a shift in the frequency domain.

When a time domain signal is *shifted*, the frequency domain's magnitude remains the same, but the slope of the phase changes.

Left-to-right symmetry around sample zero in the time domain corresponds to a zero phase in the frequency domain. This is true for all four members of the Fourier transform family.

For the DFT, a signal that is symmetrical around sample $N/2$ will also be symmetrical around sample zero, and therefore have a zero phase. (This assumes the indexes run from 0 to $N-1$, and may need manipulation of the π and 2π ambiguities).

When a signal is flipped left-for-right in the time domain, the magnitude is left unchanged, but the phase is reversed in sign. In rectangular form, this corresponds to leaving the real part unchanged, and reversing the sign of the imaginary part. The name given this operation is *complex conjugation*.

This explains why a symmetrical signal has a zero phase; the phase of its left half is exactly opposite that of its right half, cancelling to zero.

The DFT views both the time and frequency domains as being *periodic*. When only a single period is considered, the left side of the signal seems to be connected to its right side, that is, the signal appears to be *circular*.

An *N* point DFT views the time domain as being periodic with a period of *N*. This can be confusing since most of the signals used with the DFT are *not* periodic; they are simply *N* samples acquired from some scientific experiment or engineering application. This periodicity is *forced* upon the nonperiodic data in order to use the DFT.

A *N* point DFT also views the frequency domain as being periodic with a period of *N*, but the pattern is more complicated since negative frequencies must be taken into account.

Mathematics allows any sinusoid to be expressed as a *positive* or as a *negative* frequency. Many science and engineering applications only need to deal with the positive frequencies, and the negative frequencies can be ignored. Two examples were presented in the last chapter: spectral analysis of signals and the frequency response of systems.

Negative frequencies are important for two reasons. First, they describe how signals behave in certain operations, such as amplitude modulation. These are cases where the negative frequencies are shifted to the positive frequency portion of the spectrum. Second, negative frequencies are an inherent part of the *complex* Fourier transform, discussed in Chapter 29.

Aliasing is a result of the periodic nature of the time and frequency domains. It occurs when the information contained in one period becomes too long to be held within that period. The information expands into adjacent periods, mixing with what is already there.

Frequency domain aliasing occurs when an operation in the time domain causes the periods in the frequency domain to overflow. Likewise, *time domain aliasing* occurs when an operation in the frequency domain causes the periods in the time domain to overflow.

The *discrete time Fourier transform* (DTFT) is the member of the Fourier Transform family that operates on aperiodic discrete signals. It can be understood by starting with the DFT and making *N* large and larger, resulting in more and more samples between 0 and 0.5 in the frequency domain. As *N* approaches infinity, the time domain becomes aperiodic, and the frequency domain becomes a continuous signal. Since a continuous signal is involved, the DTFT must be calculated in equation form, not with a digital computer.

KEY WORDS AND TERMS: additivity, amplitude modulation, carrier wave, circularity, complex conjugation, decimation, frequency domain multiplexing, frequency domain aliasing, homogeneity, interpolation, linear phase, multirate, negative frequencies, nonlinear phase, Parseval's relation, shift invariance, sidebands, time domain aliasing, zero phase

Chapter 11- Fourier Transform Pairs

A waveform in one domain corresponds to another waveform in the opposite domain. For instance, a rectangular pulse in the time domain corresponds to a sinc function in the frequency domain. By duality, the reverse is also true; a rectangular pulse in the frequency domain corresponds to a sinc function in the time domain. These matching waveforms are called *Fourier transform pairs*.

The most common Fourier transform pairs are:

delta function	↔	*constant value*
shifted delta function	↔	*sinusoid*
rectangular pulse	↔	*sinc function*
triangular pulse	↔	*sinc function squared*
Gaussian	↔	*Gaussian*

When a time domain signal is synthesized from only a portion of the component sinusoids, edges in the reconstructed signal show overshoot and ringing. This is called the *Gibbs effect*. In more general terms, an abrupt truncation of a signal in one domain results in overshoot and ringing at discontinuities in the other domain.

The *fundamental frequency* is the frequency that a periodic waveform repeats itself. Any periodic waveform contains only the fundamental frequency plus integer multiples of the fundamental frequency. These frequencies are called *harmonics*, and are numbered starting with the fundamental being called the *first* harmonic. If the periodic waveform is symmetrical between the top and bottom, it only contains even numbered harmonics.

Harmonics explain why a *periodic signal* in one domain corresponds to a *discrete signal* in the other domain, and vice versa. The samples in the discrete domain are the harmonics of the periodicity of the other domain.

Chirp signals are used in echo location equipment such as radar and sonar. A chirp system expands an impulse into a longer length signal, while an antichirp system changes the longer signal back into an impulse.

KEY WORDS AND TERMS: chirp signal and system, Fourier transform pairs, fundamental frequency, Gibbs effect, harmonics, overshoot, ringing, sinc function

Chapter 12- The Fast Fourier Transform (FFT)

The FFT is an efficient way of calculating the *discrete Fourier transform* (DFT).

The FFT inherently calculates the *complex DFT*, however, it can also be used to calculate the simpler *real DFT*.

To calculate the *forward* real DFT using the FFT, make the imaginary part of the time domain signal contain all zeros, calculate the FFT, and then ignore the negative frequencies in the frequency domain.

To calculate the *inverse* real DFT using the FFT, make the negative frequencies a mirror image of the positive frequencies, change the sign of the imaginary part of the frequency domain, calculate the FFT, and divide all the samples in the time domain signal by N. All the values in the imaginary part of the time domain will be zero (within round-off noise).

The FFT operates by *decomposing* an N point time domain signal into N signals each containing one point. This is done with repeated use of the interlaced decomposition. Each of these single point signals is then *transformed* into the frequency domain. These N frequency domain signals are then *recombined* to produce the frequency domain of the original signal. (Since the FFT uses complex math, each of these "points" consists of two numbers, the real and imaginary parts).

Although the FFT can be carried out in only a few dozen lines of code, it is one of the most sophisticated and complicated algorithms in DSP. This is usually a routine you copy from a book or find in a software library, not one you write from scratch.

Calculation of the DFT by correlation (the standard method) requires an execution time proportional to N^2. The FFT requires a time proportional to $N \log N$. This makes the FFT faster whenever N is greater than about 16 points, which is nearly always the case in DSP.

Since the FFT performs the calculation faster, it also carries it out with greater precision (lower round-off noise).

There are several methods for making the FFT even faster, but the improvements are only about 30%. The most common is called the *real FFT* (or a similar name). It is used when the signal being passed through the FFT is real (i.e., it has no imaginary part, the most common case). The program code is about twice as long as the standard FFT.

KEY WORDS AND TERMS: bit reversal sorting, butterfly, complex DFT, Cooley and Tukey, decimation in time, decimation in frequency, FFT, in-place computation, interlaced decomposition, negative frequencies, real DFT, real FFT

Chapter 13- Continuous Signal Processing

Continuous signals are processed in the same general way as discrete signals; convolution and Fourier analysis are the two key techniques. Since continuous signals cannot be handled in computers, calculus is the primary tool used.

The continuous *delta function* is defined by three characteristics: (1) the pulse must be infinitesimally brief, (2) the pulse must occur at time zero, and (3) the pulse must have an area of one.

The continuous delta function is a mathematical abstraction, not a signal that can actually exist in an electronic circuit. However, if a signal in an electronic circuit is *brief* compared to the response of the circuit, it will *act* as a delta function.

Continuous *convolution* can be viewed from the *input side* or the *output side*, just as with discrete convolution. The mathematics is the same, except that the summations are replaced by integrals.

The *Fourier transform* (one of the four members of the Fourier transform family) is used with signals that are continuous and aperiodic. The *Fourier series* (another of the four members) is used with signals that are continuous and periodic.

KEY WORDS AND TERMS: continuous impulse response, continuous delta function, convolution integral, Fourier transform, Fourier series, fundamental frequency, harmonics

DIGITAL FILTERS

Chapter 14- Introduction to Digital Filters

Filters are used for two general purposes: signal *separation* and signal *restoration*.

Digital filters can be carried out in two different ways: *convolution* and *recursion*.

Filters carried out by convolution have impulse responses that are a fixed number of samples long. This accounts for these filters being called *Finite Impulse Response (FIR) filters*. An FIR filter is designed by specifying the *filter kernel* (the name given to the impulse response).

In comparison, filters carried out by recursion have impulse responses containing decaying exponentials, and are thus called *Infinite Impulse Response (IIR) filters*. These filters are designed by specifying a set of recursion coefficients. The impulse response of these filters can be found by passing an impulse through the system.

The *step response* can be found by passing a step function through the filter. Alternatively, the step response can be calculated as the running sum of the impulse response.

The *frequency response* of a filter is found by taking the DFT of the impulse response, usually by means of the FFT algorithm. The frequency response is often displayed in *decibels*.

Information is represented in signals in many ways. The two most common are *time domain encoding* and *frequency domain encoding*. Time domain encoding represents the information in the shape of the waveform. Frequency domain encoding represents the information in the amplitude, frequency and phase of the component sinusoids.

When time domain encoding is used, the *step response* provides the key information about the filter's performance, such as: risetime, overshoot, and phase linearity.

When frequency domain encoding is used, the *frequency response* contains the important parameters, such as: roll-off, passband ripple, and stopband attenuation.

There are two method of turning a low-pass filter into a high-pass filter. *Spectral inversion* flips the frequency response top-for-bottom. *Spectral reversal* flips the frequency response left-for-right.

Band-pass and *band-reject* filters are formed by combining low-pass and high-pass stages in cascade or parallel.

KEY WORDS AND TERMS: band-pass, band-reject, cutoff frequency, finite impulse response (FIR), frequency response, high-pass, impulse response, infinite impulse response (IIR), linear phase, low-pass, nonlinear phase, passband, roll-off, signal separation, signal restoration, spectral inversion, spectral reversal, step response, step response risetime, step response overshoot, stopband, stopband attenuation, transition band

Chapter 15- Moving Average Filters

The moving average filter and its relatives are used with time domain encoded signals. They *smooth* the signal to reduce noise and change the shape of the waveform.

The moving average filter is *optimal* for reducing white (random) noise while maintaining a sharp step response. However, relatives of the moving average are virtually as good in this respect.

When carried out by *convolution*, the moving average filter is very easy to program, but may require a long execution time. When carried out by *recursion*, the execution is extremely fast, but the programming is more complicated.

The frequency response of the moving average is very bumpy (a sinc function) and has little ability to separate one band of frequencies from another

Relatives of the moving average filter include: the multiple pass moving average, the Gaussian, and the Blackman window. These filters have better stopband attenuation and a less abrupt step response; however, they are slower to execute.

KEY WORDS AND TERMS: Blackman window, Gaussian, moving average filter, multiple pass moving average, noise reduction vs. step response sharpness

Chapter 16- Windowed-Sinc Filters

Windowed-sinc filters are nearly ideal for separating one band of frequencies from another. They come in the four common responses: high-pass, low-pass, band-pass and band-reject.

Derivation of the windowed-sinc: The perfect low-pass filter kernel is an infinitely long sinc function. When this is truncated to a usable length, there is undesirable ripple and overshoot in the frequency response. This is corrected by multiplying the truncated sinc by a smooth window.

These filters are carried out by convolution. This results in their main limitation: a slow execution speed due to the large number of calculations required.

Two parameters must be specified to design a windowed-sinc filter, the desired cut-off frequency and the number of samples in the filter kernel.

The length of the filter kernel determines the roll-off of the frequency response.

The passband ripple and stopband attenuation are determined by the window used. Several windows are available, but only two are considered in most applications: the Blackman and the Hamming.

The frequency response is symmetrical between the passband and stopband. The shape of the frequency response does not depend on the cutoff frequency used.

The windowed-sinc can be pushed to incredible performance levels, such as extremely fast roll-offs and high stopband attenuations.

KEY WORDS AND TERMS: Blackman window, Gibbs effect, Hamming window, ideal low-pass filter, passband ripple, rectangular window, roll-off, sinc function, spectral inversion, stopband attenuation

Chapter 17- Custom Filters

Custom filters have an arbitrary frequency response. The strategy is similar to that used with the windowed-sinc filter.

The desired frequency response is represented by data in an array. The IDFT is used to find the corresponding impulse response. Before using this calculated impulse response as a filter kernel, it must be multiplied by a smooth window (the Hamming or Blackman) to reduce the effects of aliasing.

These filters are carried out by convolution, or FFT convolution. This results in their main limitation: a slow execution speed due to the large number of calculations required.

The length of the filter kernel determines how closely the actual frequency response matches the desired frequency response.

Custom filtering is used in two important DSP applications: *deconvolution* and *optimal filtering*.

Deconvolution is the process of counteracting an unwanted convolution. Two examples of this are: shortening the output pulse of a gamma ray detector, and reducing the resonances in old audio recordings.

Three optimal filters are presented. The *moving average filter* is optimal at reducing white noise while retaining the sharpest step response. The *matched filter* is optimal at detecting a known pattern in the presence of white noise. The *Wiener filter* is optimal at maximizing the signal-to-noise ratio of a signal.

KEY WORDS AND TERMS: aliasing, Blackman window, deconvolution, Hamming window, matched filter, moving average filter, optimal filters, Wiener filter, white noise

Chapter 18- FFT Convolution

The *overlap add* method is used to process a signal in *segments*. This may be needed for several reasons, such as: the signal may be too long to fit in the computer's memory, the processing must be done in real time, the filtering algorithm could require it (such as with FFT convolution), etc.

FFT convolution is a technique for filtering signals at a much faster rate than conventional convolution, while producing exactly the same result.

FFT convolution uses the overlap add method to break the signal into segments. The segments are moved into the frequency domain via the FFT, where they are multiplied by the frequency response of the system. The Inverse FFT moves the processed segments back into the time domain, where they are recombined into the final output signal.

Care must be taken to pad the signals with zeros such that *circular convolution* does not take place.

The *speed improvement* depends on the length of the filter kernel. Filter kernels shorter than about 60 points can be implemented faster by conventional convolution, while FFT convolution is faster for longer filter kernels. For filter kernels that are thousands of points long, FFT convolution can be hundreds of times faster than conventional convolution.

KEY WORDS AND TERMS: FFT convolution, real time processing, high-speed convolution, circular convolution, overlap add method

Chapter 19- Recursive Filters

Recursive filters are carried out by a *recursion equation* rather than by convolution. This means they are defined by a set of *recursion coefficients*, rather than a filter kernel.

The impulse response of a recursive filter is infinitely long, consisting of a combination of exponentially decaying sinusoids. For this reason, they are also called *infinite impulse response (IIR) filters*. In comparison, filters carried out by convolution are called finite impulse response (FIR) filters.

Recursive filters are useful because they have a long impulse response, but require relatively few calculations to implement. In other words, they have good performance while executing very quickly.

The z-transform (Chapter 31) provides the mathematics for designing recursive filters. There are three ways to design an IIR filter without having to use this detailed math: (1) simple filters can be created with the design equations presented in this chapter, (2) a "cookbook" program is presented in Chapter 20 for designing sophisticated Chebyshev filters, (3) Chapter 26 illustrates an iterative method for designing filters with a custom frequency response.

Single pole recursive filters (low-pass and high-pass) can be used for the same processing as RC networks in electronics: DC removal, high-frequency noise suppression, waveshaping, smoothing etc. *Narrow-band* filters (band-pass and band-reject) can also be designed. All these filters are easy to program, fast to execute, and produce few surprises.

Linear phase filters have filter kernels with left-to-right symmetry. If the point of symmetry is at sample zero, the filter is additionally said to be *zero phase*. Filter kernels that do not have this left-to-right symmetry are *nonlinear* phase.

Linear phase filters are desirable for processing signals when information is encoded in the time domain. This is because the left and right sides of pulses are modified in the same way.

It is easy to make an FIR filter have linear phase, since the impulse response (filter kernel) is directly designed.

Recursive filters can only achieve a linear phase when *bidirectional filtering* is used.

Analog filters cannot have a linear phase, because their impulse responses cannot have left-to-right symmetry. The *Bessel* filter is designed to have as linear phase as possible.

KEY WORDS AND TERMS: -3dB cutoff frequency, analog RC filters, bidirectional filtering, Chebyshev filter, finite impulse response (FIR), infinite impulse response (IIR), linear phase, narrow-band filters, nonlinear phase, notch filter, pulse response, reverse filtering, single-pole filters, time constant, z-transform, zero phase

Chapter 20- Chebyshev Filters

Chebyshev filters are a sophisticated IIR design used to separate one band of frequencies from another. They can be designed for a high-pass or low-pass response.

The Chebyshev response is an optimal tradeoff between *roll-off* and *passband ripple*. When a Chebyshev filter is designed for zero passband ripple, it is called a *maximally flat* or *Butterworth* filter.

Chebyshev filters cannot match the performance of the windowed-sinc filters; however, they are adequate for many applications and execute about an order of magnitude faster.

Four parameters must be specified to design a Chebyshev filter: the cutoff frequency, a high-pass or low-pass response, the percent ripple in the passband, and the number of poles. From these four parameters, the recursion coefficients can be found from a table or by a "cookbook" computer program.

The more ripple allowed in the passband, the faster the roll-off. A passband ripple of 0.5% is often a good choice.

When more poles are used, the filter's performance is better, but the execution is slower because more recursion coefficients are used.

The performance of recursive filters is limited by round-off noise. This can be extended by using double precision, or carrying out the filter in stages.

KEY WORDS AND TERMS: Butterworth response, cascade of stages, Chebyshev response, double precision, maximally flat, passband ripple, poles, recursion coefficients, roll-off, round-off noise, stability, z-transform

Chapter 21- Filter Comparison

Match #1: Analog vs. Digital. Digital filters can beat analog filters in virtually every filter parameter, such as: passband flatness, roll-off sharpness, stopband attenuation and phase linearity. However, analog circuits have a much larger dynamic range in both amplitude and frequency, and are much faster.

Match #2: Windowed-sinc vs. Chebyshev. This compares the FIR and IIR filters used for *frequency domain processing*. The windowed-sinc (FIR) has far better performance but is about an order of magnitude slower to execute,

Match #3: Moving average vs. Single pole recursive. This is a comparison between the FIR and IIR filters used for *time domain processing*. These filters are very similar and the choice is usually based on personal preference.

KEY WORDS AND TERMS: amplitude dynamic range, analog vs. digital, FIR vs. IIR, frequency dynamic range, moving average vs. single pole recursive, Windowed-sinc vs. Chebyshev

APPLICATIONS

Chapter 22- Audio Processing

The human ear acts as a *spectrum analyzer*; individual fibers in the cochlear nerve respond to specific frequencies.

The *phase* of the component frequencies is not detected in hearing (with minor exceptions). This means the shape of the audio waveform is not important, and two different waveforms can sound exactly the same. In other words, the ear is optimized to detect information encoded in the *frequency domain*, not the *time domain*.

Human hearing is generally considered to cover 20 Hz to 20 kHz, but is far more sensitive to sounds between 1 kHz and 4 kHz.

The perception of a continuous sound can be divided into three parts: *loudness*, a measure of the amplitude of the audio waveform; *pitch*, the fundamental frequency of the sound; and *timbre*, determined by the harmonic content of the signal.

Since the ear is accustomed to hearing a fundamental plus harmonics, the frequency range of 20 Hz to 20 kHz is perceived on a *logarithmic* scale. That is, as much information is contained between 100 and 200 Hz, as between 1 kHz and 2 kHz, and as between 10 kHz and 20 kHz.

Telephone quality speech only needs to be sampled at about 8 kHz, so that frequencies up to 3.2 kHz can be contained in the digitized signal. A quantization of at least 12 bits is needed if the levels are equally spaced, but only 8 bits if logarithmic spacing (companding) is used.

High fidelity audio samples at 44.1 kHz, allowing the full 20 Hz to 20 kHz band to be retained. A quantization of 16 bits per sample is the standard.

Human speech can be modeled as an *excitation* followed by a *filter*. The excitation is one of two types: *voiced*, consisting of periodic pulses of air from the vocal cords, or *fricative*, formed from random noise caused by air turbulence. The filter is formed by the resonances of the acoustic cavities of the throat, mouth and nose. The type of excitation and the characteristics of the filter can change about every 25 milliseconds. *Speech generation* and *speech recognition* are based on this model.

Homomorphic processing is used for separating signals that have been combined in a nonlinear way, such as by multiplication or convolution. The strategy is to pass the combined signal through a function that makes the components *additive*, allowing standard linear techniques to be used.

KEY WORDS AND TERMS: "A" law, basilar membrane, cepstrum, cochlea, cochlear nerve, companding, decibel SPL, format frequencies, fricative, high fidelity, homomorphic, linear predictive coding, loudness, mu law, octave, oval window, pitch, spectrogram, speech recognition, speech synthesis, timbre, tympanic membrane, voiced, voiceprint

Chapter 23- Image Formation & Display

Images have their information encoded in the *spatial domain*, the two-dimensional equivalent of the *time domain*. This means the relevant information is contained in the shape of the waveform, not in the component frequencies.

Typical digital images have 500 to 1000 rows and columns. Black and white images usually represent each pixel with a single byte, allowing 256 gray levels between pure black and pure white. In color images, it is common to use one byte for each of the primary colors, blue, green and red, resulting in three bytes per pixel.

Science and engineering frequently display nonvisual parameters as visual images, such as the temperature of an integrated circuit, or the x-ray emission from a distant galaxy. This is using the human visual system as an analysis *tool*. Black and white is usually sufficient for these applications.

The eye and camera are very similar in operation. Both use a lens to focus an image onto a light sensor. The most common electronic light sensor is the *charge coupled device* (CCD), a two-dimensional array of light sensitive regions called *wells*. The light sensor in the eye is the *retina*, a two-dimensional array of light sensitive cells called *rods* and *cones*.

The *fovea* is a small region in the center of the retina with a high density of cones, resulting in exceptionally good vision. At normal reading distance, the fovea only sees about a 1 mm diameter area, with a resolution of about 20×20 pixels. This small region is scanned over the entire visual scene in jerky eye movements called *saccades*.

Brightness refers to the overall lightness or darkness of an image. *Contrast* refers to how the brightness of an object differs from the brightness of its background.

Grayscale transforms are used to map the values of the stored pixels into the shades of gray displayed in an image. This can be as simple as adjusting the brightness and contrast, or as complicated as *histogram equalization* and manually generated curves.

Warping is used to change the viewing angle of an image, such as a region on a planet being viewed obliquely instead of from above. It is carried out by copying each of the pixel values in the original image to a different location in the processed image.

KEY WORDS AND TERMS: aspect ratio, brightness, charge coupled device (CCD), composite video, contrast, field, focus, fovea, frame grabber, frame, gamma curve, gray levels, grayscale stretch, grayscale, halftone, histogram equalization, interlaced, iris, lens, line scanning, morphing, NTSC, output transform, PAL, pixel, retina, RGB encoding, rods and cones, row major order, saccades, sampling grid, sampling aperture, SECAM, spatial domain, warping, well

Chapter 24- Linear Image Processing

The two-dimensional *delta function* is a image composed of all zeros, except for the pixel at the origin which has a value of one. When the delta function image is the input to an imaging system, the output image is the system's impulse response, also called the *point spread function* or *PSF*.

Within the eye, the PSF of the retina produces *edge sharpening*, providing greater distinction between objects and the background. The Mach bands are an optical illusion resulting from this PSF.

Image convolution can be viewed from the input side or the output side, just as with one-dimensional convolution. From the input side, each pixel in the input image contributes a scaled and shifted version of the PSF to the output image. From the output side, each pixel in the output image is the sum of pixels in the input image multiplied by the PSF flipped left-for-right and top-for-bottom.

The pillbox, Gaussian, and square PSF are common low-pass filter kernels. High-pass filter kernels are formed by subtracting a low-pass filter kernel from a delta function.

Image convolution requires a long execution time. Three strategies are often used to improve the speed: (1) use a *very small PSF*, often only 3×3 pixels; (2) use a PSF that is *separable*, allowing the two-dimensional convolution to be carried out in one-dimension; or (3) use *FFT convolution*.

A 3×3 pixel filter kernel can provide significant image processing because it is large enough to affect the edges in an image. Algorithms include: shift and subtract (an embossed effect), smoothing, sharpening, and edge detection.

Convolution by separability is used when a large filter kernel is needed, but its exact shape is not important. An example is *illumination flattening*, where the illumination component of an image is suppressed to allow the reflectance component to dominate.

The DFT of an image is calculated by taking the one-dimensional DFT of each of the rows, followed by taking the one-dimensional DFT of each of the resulting columns.

Since images have their information encoded in the spatial domain, the frequency domain cannot usually be used to understand the information in an image. Likewise, the frequency domain it is generally no help in designing image filters.

Two areas where the Fourier transform is useful in image processing are: *FFT convolution*, used to reduce the execution time of image convolution, and *computed tomography*, the reconstruction of an image from its projections.

KEY WORDS AND TERMS: edge detection, edge enhancement, FFT convolution, Mach bands, pillbox, point spread function, PSF, separable, sharpening, shift and subtract

Chapter 25- Special Imaging Techniques

The chapter presents four specific aspects of image processing: *spatial resolution*, *signal-to noise ratio*, *morphological techniques*, and *computed tomography*.

Spatial resolution describes how small of an object can be seen in an image. It is limited by two distinct but interrelated effects: *sample spacing* and *sampling aperture*. If there is less than 1 sample along the rising portion of the edge response, the sample spacing is the limiting factor. If more than 3 samples are present, the sampling aperture is dominate. For 1-3 samples, both factors limit the resolution.

Three methods are commonly used to describe the spatial resolution of a system: (1) the *modulation transfer function* or *MTF* (the imaging jargon for the *frequency response*), (2) the *limiting resolution* observable on a *line pair gauge*, and (3) the width of the *edge response*, often measured from the 10%-90% amplitudes.

The *signal-to-noise* ratio is important because an object is visible in an image when its *contrast* (the *signal*) is distinguishable from the *noise*. The larger the object, the easier it is to detect in a noisy background.

The human eye can detect a minimum contrast of 0.5% to 5%, depending on the conditions.

Image noise comes in two common forms. If the noise originates in the electronics of the camera, it will have a constant amplitude regardless of the signal level. In comparison, noise that results from the image having been represented by a small number of particles (such as x-rays or light photons) will increase as the square-root of the signal level.

Morphological processing can refer to a wide range of techniques; however, it is generally used to describe simple nonlinear operations applied to binary images. The four "standard" operations are: erosion, dilation, opening, and closing. These four are seldom used by themselves; custom algorithms are developed by trial and error for the particular task at hand.

Computed tomography was developed to produce high quality images of the living human body from x-ray projections. X-rays are only allowed to pass through the *slice* of the body being imaged. The profile of the detected x-rays at each of the possible angles is called a *view*. From the concept of *simultaneous linear equations*, the number of points in each view, multiplied by the number of views, must be greater than the number of pixels in the image being reconstructed.

Three methods can be used to reconstruct an image from its views: (1) *iterative techniques*, where the final image is obtained by gradually adjusting each of the pixel values in the proper direction until convergence is obtained, (2) *filtered backprojection*, where each of the views is convolved with a one-dimensional filter kernel, and the filtered views backprojected through the reconstructed image, and (3) *Fourier reconstruction*, where the reconstruction takes place in the frequency domain, and then converted into the spatial domain.

KEY WORDS AND TERMS: backprojection, closing, computed tomography (CT), dilation, edge response, erosion, filtered backprojection, Fourier reconstruction, Full-width-at-half-maximum (FWHM), integrated profile, iterative, limiting resolution, line pair, line spread function (LSF), line pair gauge, modulation transfer function, morphological, MTF, opening, Poisson statistics, quantum sink, reconstruction algorithms, sample spacing, sampling aperture

Chapter 26- Neural Networks (and more!)

Most DSP techniques need both *algorithms* and *parameters* to operate. Neural networks are technique that uses very simple algorithms, but many highly optimized parameters. An important lesson of this chapter is that there is not a sharp dividing line between neural networks and more conventional techniques.

The goal of *target detection* is to determine if a condition is present or not present based on the value of a given parameter or parameters. For a single parameter system, this involves examining the overlap of target and nontarget distributions, and selecting a threshold.

There are four possible outcomes for each target detection trial: *true-positive* (target was present and correctly identified), *true-negative* (target was not present and its absence was correctly identified), *false-positive* (target was not present but erroneously said to be present), and *false-negative* (target was present but erroneously said to be not present).

Changing the threshold on a single parameter system allows the rate of false-negatives to be traded for the rate of false-positives. An ROC curve is used to show this effect of changing the threshold.

In multiple parameter systems the problem is more difficult to understand and solve. This is because it involves separating regions in a parameter *hyperspace*. Neural networks operate by using *examples* of the problem to find and implement an optimal separation of the hyperspace.

The most common neural network is the three-layer fully-interconnected structure. This simple structure is capable of carrying out a wide variety of operations, all controlled by the selection of the *weights*.

A *learning algorithm* is a procedure for determining the appropriate weights, based on examples of how the network *should* operate. The most common learning algorithm is called *steepest decent*. This is an iterative technique; each weight is slightly improved as the examples are repeatedly examined, until an optimal set of weights is obtained.

Neural networks are very good at finding patterns in abstract and complicated data. However, the solutions are not in a form that *humans* can understand. A certain set of weights can be *observed* to work, but *why* they work is nearly always a mystery.

Neural network techniques can also be combined with conventional DSP methods to form superb algorithms. For example, recursive filters can be designed with a *custom response*. In this method, the steepest decent learning algorithm is applied to the recursion coefficients until an optimal solution is reached.

KEY WORDS AND TERMS: classifiers, convergence, evolution, false-negative, false-positive, feature extraction, hidden layer, hyperspace, input layer, iterative, learning algorithm, node, output layer, parameter space, ROC curve, sigmoid, steepest descent, target detection, threshold, true-negative, true-positive, weights

Chapter 27- Data Compression

Data compression algorithms can be *lossless* or *lossy*.

Lossless algorithms are used when the uncompressed data must be identical with the original, such as with executable code and word processing files. Lossy algorithms can be used on acquired signals that already contain a certain amount of noise. If the error resulting from lossy data compression resembles a small amount of additional noise, no harm is done.

Lossless algorithms generally achieve a compression ratio of about 2:1. In comparison, lossy algorithms can routinely achieve a 10:1 compression with little degradation of the data.

Three simple lossless encoding schemes are *run-length encoding, Huffman encoding* and *delta encoding*. Run length encoding is useful when the same character is repeated many times. Huffman encoding is used when some characters are repeated more often than other characters. Delta encoding is a useful data compression strategy when the signals change very slowly.

The most successful lossless compression method is *LZW encoding*. This is based on storing strings of characters in a code table, and then transmitting the code table's address of the string, rather than the string itself. The LZW algorithm cleverly handles two problems: how to determine what strings should be in the code table, and how to provide the uncompression algorithm with a copy of the code table. LZW encoding is always used with GIF images, and as an option in TIFF and PostScript images.

Transform compression is the most common approach to lossy compression. The idea is to pass the data through a Fourier (or similar) transform, and then eliminate some of the bits used to represent the high frequencies. The distortion resulting from the discarded bits resembles random noise in the time (or spatial) domain. Many different transforms have been investigated for data compression. The most commonly used is the *discrete cosine transform* (DCT), a relative of the Fourier transform.

JPEG image compression involves several steps. First, the image is broken into 8×8 pixel groups, and the DCT taken of each group. Bits are then removed from the highest frequencies of each group according to a predetermined *quantization table*. The data are then converted into a serial data stream, and then further compressed with run-length and Huffman encoding.

MPEG is a compression scheme for digital video. The JPEG algorithm is used to compress some of the individual frames in the sequence (called I-pictures in the jargon of MPEG). The other frames are encoded by reference these I-pictures by a sophisticated form of delta encoding.

KEY WORDS AND TERMS: ASCII, compression, delta encoding, discrete cosine transform (DCT), GIF, Huffman encoding, JPEG, lossless compression, lossy compression, LZW compression, MPEG, quantization table, run-length encoding, TIFF, transform compression, uncompression

COMPLEX TECHNIQUES

Chapter 28- Complex Numbers

Complex numbers are formed from two groups, the *real numbers* (used in everyday life), and the *imaginary numbers* (numbers that contain the term: $j = \sqrt{-1}$). Complex numbers often arise in common algebra problems, such as in the solutions of polynomial equations.

Complex numbers are expressed in one of two forms, $a + bj$ (rectangular) or $Me^{j\theta}$ (polar). The conversion between these two forms is based on *Euler's relation*, a key equation in complex mathematics. Using either notation, a single complex number contains two separate pieces of information, either a & b, or M & θ.

When they naturally arise in science and engineering problems, complex numbers usually represent solutions that cannot exist in the world as we know it. In spite of this elusive nature, complex numbers follow mathematical laws that are similar (or identical) to those governing ordinary numbers.

Complex numbers are used in science and engineering in two ways. This chapter discusses the simpler method of *substitution*, while the next chapter discusses the more advanced method of *mathematical equivalence*. In both methods, a physical problem is changed into in a complex form, the complex numbers are then manipulated, and the results converted back into a physical meaning. While the *middle* of this procedure is complex, both *ends* are completely real.

Complex numbers are popular in electronics and DSP because they are a compact way to represent sine and cosine waves. The simplest way to do this is by *substitution*: in rectangular form: $A\cos(\omega t) + B\sin(\omega t) \Rightarrow a + jb$, and in polar form: $M\cos(\omega t + \phi) \Rightarrow Me^{j\theta}$. It is important to understand that these are not equations, but merely ways of letting complex numbers *represent* sinusoids.

This method of substitution is also called the *phasor transform* (although a more sophisticated definition is often used).

A sinusoid passing through a linear system remains a sinusoid; the amplitude and phase may change, but the frequency will remain the same. When the input and output sinusoids are represented by complex numbers, the system can also be represented by a complex number. The complex number representing the input, multiplied by the complex number representing the system, is equal to the complex number representing the output.

The phasor transform is particularly useful for analyzing circuits composed of resistors, capacitors and inductors. Each of these individual components can be viewed as a system, with a sinusoidal *current* being the input signal and a sinusoidal *voltage* being the output signal. The complex number representing the system (the component) is called the *impedance*. This allows the component to use the relation: *current* × *impedance* = *voltage*, where all three are represented by complex numbers. This can be thought of as "Ohm's law for sinusoids."

The impedance of the resistor, capacitor and inductor are: R, $-j/\omega C$, and $j\omega L$, respectively. Circuits composed of combinations of these components can also be analyzed with complex numbers. This is done by treating the *impedance* the same as simple *resistance* in a DC circuit. This allows all the basic analysis techniques to be used, such as resistors in series, resistors in parallel, voltage dividers, etc.

KEY WORDS AND TERMS: complex exponential, complex plane, complex number, Euler's relation, impedance, mathematical equivalence, notch filter, phasor transform, substitution

Chapter 29- The Complex Fourier Transform

The real Fourier transform can be written with complex numbers by using *substitution*; however, this is *not* the complex Fourier transform. The complex Fourier transform is a more sophisticated technique that uses complex numbers by *mathematical equivalence*.

The complex Fourier transform has three advantages over the real Fourier transform: (1) the complex numbers are mathematically *equal* to the original problem, rather than just *representing* the problem (as in substitution), (2) the negative frequency portion of the spectrum is inherently included in the analysis, and (3), there is no special handling of $Re\,X[0]$ and $Re\,X[N/2]$ for the inverse DFT.

Mathematical equivalence is based on *Euler's relation*, expressing sine and cosine waves as complex exponentials: $\cos(\omega t) = \tfrac{1}{2}e^{j(-\omega)t} + \tfrac{1}{2}e^{j\omega t}$ and $\sin(\omega t) = \tfrac{1}{2}je^{j(-\omega)t} - \tfrac{1}{2}je^{j\omega t}$. That is, complex numbers can be made *equal* to the sine and cosine waves when the negative frequencies are taken into account.

The complex Fourier transform is formed by replacing the cosine and sine terms in the real Fourier transform with the above expressions. Here are five subtle details:

(1) The real Fourier transform relates a real time domain signal with two real frequency domains signals, while the complex Fourier transform relates a complex time domain signal with a complex frequency domain signal.

(2) Since the real DFT only uses positive frequencies, the frequency domain index only runs from: $k = 0$ to $N/2$, or $f = 0$ to 0.5, or $\omega = 0$ to π. The frequency index for the complex DFT also includes negative frequencies: $k = -(N/2-1)$ to $N/2$, or $f = -0.5$ to 0.5, or $\omega = -\pi$ to π. Because the DFT is periodic, this is the same as: $k = 0$ to $N-1$, or $f = 0$ to 1, or $\omega = 0$ to 2π.

(3) In substitution, j terms are added to move from a physical problem to a complex number form, and then deleted to move in the reverse direction. Since the complex Fourier transform uses complex numbers in exact mathematical expressions, the j terms cannot be added or deleted any more than any other variable.   A physical problem is converted to complex numbers by using *Euler's relation*. At the end of the mathematical manipulation, the answer to the physical problem is found by having all the imaginary terms cancel, providing an entirely real solution.

(4) The real Fourier transform uses a scaling factor of *two* that is not found in the complex Fourier transform.   This is because the complex Fourier transform uses both negative and positive frequencies to form a sinusoid, while the real Fourier transform only use a positive frequency.

(5) The complex DFT does not require $Re\,X[0]$ and $Re\,X[N/2]$ to be divided by two during the inverse DFT, as does the real DFT. These two frequencies have no matching negative frequency, automatically correcting for the factor of two.

When the time domain is completely real (i.e., the imaginary part of the time domain is zero), the real part of the frequency domain will have *even symmetry* and the imaginary part will have *odd symmetry*.   Likewise, when the time domain is completely imaginary (i.e., the real part of the time domain is zero), the real part of the frequency domain will have *odd symmetry* and the imaginary part will have *even symmetry*. When the time domain contains both real and imaginary signals, *additivity* can be used to understand the frequency domain as a combination of these two types of symmetry.

The family of Fourier transforms should be understood as a well organized and symmetrical *group*, rather than many unrelated equations.   There are *four versions* of the Fourier transform, corresponding to the time domain signal being discrete or continuous, and periodic or aperiodic. All four can be carried out using either *real* or *complex* math. Each of these four transforms involves a *synthesis* and an *analysis* equation (i.e., a forward and inverse transform).

KEY WORDS AND TERMS: complex DFT, mathematical equivalence, negative frequency, real DFT, substitution

Chapter 30- The Laplace Transform

Many physical systems are governed by *differential equations*. The impulse response of these systems must be a solution to these differential equations, and can therefore only consist of sinusoids and decaying exponentials. The *Laplace transform* is a technique for understanding these continuous systems. The *z-transform* is a similar technique used with discrete systems.

The Laplace transform converts a signal in the time domain into a signal in the s-domain (also called the s-plane). It is carried out by multiplying the time domain signal with a decaying exponential, and then taking the complex Fourier transform. This can be thought of as *probing* the time domain signal to identity its key properties: the *frequencies* of the sinusoids and the *decay constants* of the exponentials.

The vertical or *imaginary axis* of the s-plane represents *frequency*, using the variable ω. The upper half contains the positive frequencies and the lower half the negative frequencies. Since the time domain signal is real, the upper half will always be a mirror image of the lower half. The horizontal or *real axis* of the s-plane represents *exponential decay*, using the variable σ.

A location in the s-plane is specified by the values of ω and σ. This can be placed in a complex number form by defining: $s = \sigma + j\omega$. That is, s specifies the location in the s-plane, while $X(s)$ gives the value at s. Just as with the Fourier transform, time domain signals use lower case letters, while s-domain signals use upper case. For example, the Laplace transform of $x(t)$ is $X(s)$, the Laplace transform of $h(t)$ is $H(s)$, etc.

The values in the s-domain are not important in themselves, only the *locations* where certain values appear. Locations where the value is *zero* are called *zeros* of the system. Locations where the value is just *barely infinite* are called *poles*. The locations of the poles and zeros define the characteristics of the system. This is because they are related to the key parameters of the impulse response: the frequencies of the sinusoids and the decay constants of the exponentials.

The phasor transform allows the frequency response of a physical system (such as an electric circuit) to be calculated. Alternatively, the Laplace transform finds the location of the poles and zeros of the physical system. From these poles and zeros the frequency response can be found. While both methods provide the same end result, the intermediate step of the pole-zero diagram provides *insight* into why the frequency response behaves as it does.

The method for finding the poles and zeros of an electric circuit is similar to that of the phasor transform. Resistors, capacitors and inductors become R, $1/sC$, and sL in the s-domain. The resulting transfer function is then arranged in the form of one polynomial divided by another polynomial. The roots of the numerator polynomial are the zeros of the system, while the roots of the denominator polynomial are the poles. The frequency response of the system is found by evaluating the s-domain at $\sigma = 0$, that is, along the imaginary axis.

The most powerful use of the Laplace transform is the design of systems by directly specifying the location of the poles and zeros. The Butterworth filter is designed by placing poles on a *circle* in the s-plane. Chebyshev filters are designed by placing poles on an *ellipse*.

KEY WORDS AND TERMS: biquad, Butterworth filter, Chebyshev filter, differential equation, elliptic filter, even order filter, Laplace transform, odd order filter, pole-zero diagram, poles, region-of-convergence, s-domain, s-plane, stable, transfer function, zeros

Chapter 31- The z-Transform

The z-transform is used with discrete systems in the same way the Laplace transform is used with continuous systems. Just as a continuous system can be controlled by *differential equations,* discrete systems can be controlled by *difference equations*.

The Laplace transform contains an exponential of the form: $e^{-\sigma t}$, where σ is the parameter controlling the decay rate. In comparison, the z-transform uses the exponential term: r^n, where r is the decay parameter. This makes the s-domain a rectangular coordinate system defined by: $s = \sigma + j\omega$, and the z-domain a polar coordinate system defined by: $z = re^{-j\omega}$.

The left and right halves of the s-plane corresponds to the interior and exterior of the unit circle in the z-plane, respectively. Just as the frequency response is along the imaginary axis in the s-plane, it is along the unit circle in the z-plane.

The z-domain transfer function can be represented in two forms: as one polynomial divided by another polynomial, such as: $(a_0 + a_1 z^{-1} + a_2 z^{-2})/(1 - b_1 z^{-1} + b_2 z^{-2})$, or as a *factored* polynomial divided by another *factored* polynomial: $(z - z_1)(z - z_2)/(z - p_1)(z - p_2)$. The first form is important because it directly provides the recursion coefficients needed to implement the filter. The factored form is important because it directly provides the location of the poles and zeros of the system. Converting between a system's recursion coefficients and its poles & zeros is accomplished by algebraically manipulating the transfer function from one form to the other.

The z-transform has two primary uses in DSP. First, filters can be designed by directly selecting the location of the poles and zeros in the z-domain. The recursion coefficients for the desired pole-zero pattern are then calculated and used to carry out the filter. Second, the z-transform allows cascaded and parallel stages of recursive filters to be combined into a single system. The recursion coefficients of the individual stages are converted into the z-domain, combined in the desired manner, and then converted into the recursion coefficients of the combined system.

Several steps are required to design Chebyshev recursive filters, with each step requiring mathematical manipulation of the s-domain or z-domain.: (1) An analog low-pass Butterworth filter with a unity cutoff frequency is designed. This is done by specifying the pole locations on a circle in the s-plane. (2) The Butterworth filter is changed to a Chebyshev filter by warping the locations of the poles from a circle to an ellipse. (3) The analog filter is converted to a digital design by mapping the pole locations in the s-plane into the pole locations in the z-plane. This is done with the *bilinear transform*. (4) The desired cutoff frequency is set by using a *low-pass to low-pass* transform. (5) If a high-pass filter is desired, a *low-pass to high-pass* transform is used instead. (6) Steps 1-5 are repeated for each 2 pole stage of the filter. The final filter can be implemented as a cascade of these stages, or the stages can be combined into a single system.

KEY WORDS AND TERMS: bilinear transform, Butterworth, cascade stages, Chebyshev, difference equation, Hilbert transformer, parallel stages, pole, recursion coefficients, spectral inversion, transfer function, unit circle, z-domain, z-plane, z-transform, zero

Glossary

1/f noise: A type of random noise that increases in amplitude at lower frequencies. It is widely observable in physical systems, but not well understood. See *white noise* for comparison.

-3dB cutoff frequency: The division between a filter's passband and transition band. Defined as the frequency where the frequency response is reduced to -3dB (0.707 in amplitude).

"A" law: Companding standard used in Europe. Allows digital voice signals to be represented with only 8 bits instead of 12 bits by making the quantization levels unequal. See *mu law* for comparison.

AC: Alternating Current. Electrical term for the portion of a signal that fluctuates around the average (DC) value.

Accuracy: The error in a measurement (or a prediction) that is repeatable from trial to trial. Accuracy is limited by systematic (repeatable) errors. See *precision* for comparison.

Additivity: A mathematical property that is necessary for linear systems. If input a produces output p, and if input b produces output q, then an input of $a+b$ produces an output of $p+q$.

Aliasing: The process where a sinusoid changes from one frequency to another as a result of sampling or other nonlinear action. Usually results in a loss of the signal's information.

Amplitude modulation: Method used in radio communication for combining an information carrying signal (such as audio) with a carrier wave. Usually done by multiplying the signals.

Analysis: The forward Fourier transform; calculating the frequency domain from the time domain. See *synthesis* for comparison.

Antialias filter: Low-pass analog filter placed before an analog-to-digital converter. Removes frequencies above one-half the sampling rate that would alias during conversion.

ASCII: A method of representing letters and numbers in binary form. Each character is assigned a number between 0 and 127. Very widely used in computers and communication.

Aspect ratio: The ratio of an image's width to its height. Standard television has an aspect ratio of 4:3, while motion pictures have an aspect ratio of 16:9.

Assembly: Low-level programming language that directly manipulates the registers and internal hardware of a microprocessor. See *high-level language* for comparison.

Associative property of convolution: Written as: $(a[n] * b[n]) * c[n] = a[n] * (b[n] * c[n])$. This is important in signal processing because it describes how cascaded stages behave.

Autocorrelation: A signal correlated with itself. The Fourier transform of the autocorrelation is the power spectrum of the original signal.

Backprojection: A technique used in computed tomography for reconstructing an image from its views. Results in poor image quality unless used with a more advanced method.

BASIC: A high-level programming language known for its simplicity, but also for its many weaknesses. Most of the programs in this book are in BASIC.

Basilar membrane: Small organ in the ear that acts as a spectrum analyzer. It allows different fibers in the cochlear nerve to be stimulated by different frequencies.

Basis functions: The set of waveforms that a decomposition uses. For instance, the basis functions for the Fourier decomposition are unity amplitude sine and cosine waves.

Bessel filter: Analog filter optimized for linear phase. It has almost no overshoot in the step response and similar rising and falling edges. Used to smooth time domain encoded signals.

Bidirectional filtering: Recursive method used to produce a zero phase filter. The signal is first filtered from left-to-right, then the intermediate signal is filtered from right-to-left.

Bilinear transform: Technique used to map the s-plane into the z-plane. Allows analog filters to be converted into equivalent digital filters.

Binning: Method of forming a histogram when the data (or signal) has numerous quantization levels, such as in floating point numbers.

Biquad: An analog or digital system with two poles and up to two zeros. Often cascaded to create a more sophisticated filter design.

Bit reversal sorting: Algorithm used in the FFT to achieve an interlaced decomposition of the signal. Carried out by counting in binary with the bits flipped left-for-right.

Blackman window: A smooth curve used in the design of filters and spectral analysis, calculated from: $0.42 - 0.5\cos(2\pi n/M) + 0.08\cos(4\pi n/M)$, where n runs from 0 to M.

Brightness: The overall lightness or darkness of an image. See *contrast* for comparison.

Butterfly: The basic computation used in the FFT. Changes two complex numbers into two other complex numbers.

Butterworth filter: Separates one band of frequencies from another; fastest roll-off while keeping the passband flat; can be analog or digital. Also called a *maximally flat* filter.

C: Common programming language used in science, engineering and DSP. Also comes in the more advanced $C++$.

Carrier wave: Term used in amplitude modulation of radio signals. Refers to the high frequency sine wave that is combined with a lower frequency information carrying signal.

Cascade: A combination of two or more stages where the output of one stage becomes the input for the next.

Causal signal: Any signal that has a value of zero for all negative numbered samples.

Causal system: A system that has a zero output until a nonzero value has appeared on its input (i.e., the input *causes* the output). The impulse response of a causal system is a causal signal.

Central Limit Theorem: Important theorem in statistics. In one form: a sum of many random numbers will have a Gaussian pdf, regardless of the pdf of the individual random numbers.

Cepstrum: A rearrangement of "spectrum." Used in homomorphic processing to describe the spectrum when the time and frequency domains are switched.

Charge coupled device (CCD): The light sensor in electronic cameras. Formed from a thin sheet of silicon containing a two-dimensional array of light sensitive regions called *wells*.

Chebyshev filter: Used for separating one band of frequencies from another. Achieves a faster roll-off than the Butterworth by allowing ripple in the passband. Can be analog or digital.

Chirp system: Used in radar and sonar. An impulse is converted into a longer duration signal before transmission, and compressed back into an impulse after reception.

Circular convolution: Aliasing that can occur in the time domain when frequency domain signals are multiplied. Each period in the time domain overflows into adjacent periods.

Circularity: The appearance that the end of a signal is connected to its beginning. This arises when considering only a single period of a periodic signal.

Classifiers: A parameter extracted from and representing a larger data set. For example: size of a region, amplitude of a peak, sharpness of an edge, etc. Used in pattern recognition.

Closing: A morphological operation defined as an erosion operation followed by a dilation operation.

Cochlea: Organ in the ear where sound in converted into a neural signal.

Cochlear nerve: Nerve that transmits audio information from the ear to the brain.

Coefficient-of-variation (CV): Common way of stating the variation (noise) in data. Defined as: $100\% \times$ standard deviation / mean.

Commutative property of convolution: Written as: $a[n] * b[n] = b[n] * a[n]$.

Companding: An "s" shaped nonlinearity allows voice signals to be digitized using only 8 bits instead of 12 bits. Europe uses "*A*" law, while the United States uses the *mu law* version.

Complex conjugation: Changing the sign of the imaginary part of a complex number. Often denoted by a star placed next to the variable. Example: if $A = 3+2j$, then $A^* = 3-2j$.

Complex DFT: The discrete Fourier transform using complex numbers. A more complicated and powerful technique than the *real* DFT.

Complex exponential: A complex number of the form: e^{a+bj}. They are useful in engineering and science because Euler's relation allows them to represent sinusoids.

Complex Fourier transform: Any of the four members of the Fourier transform family written using complex numbers. See *real Fourier transform* for comparison.

Complex numbers: The *real numbers* (used in everyday math) plus the *imaginary numbers* (numbers containing the term j, where $j = \sqrt{-1}$). Example: $3+2j$.

Complex plane: A graphical interpretation of complex numbers, with the real part on the x-axis and the imaginary part on the y-axis. This is analogous to the *number line* used with ordinary numbers.

Composite video: An analog television signal that contains synchronization pulses to separate the fields or frames.

Computed tomography (CT): A method used to reconstruct an image of the interior of an object from its x-ray projections. Widely used in medicine. Old name: CAT scanner.

Continuous signal: A signal formed from continuous (as opposed to discrete) variables. Example: a *voltage* that varies with *time*. Often used interchangeably with *analog signal*.

Contrast: The difference between the brightness of an object and the brightness of the background. See *brightness* for comparison.

Converge: Term used in iterative methods to indicate that progress is being made toward a solution ("The algorithm is converging") or that a solution has been reached ("The algorithm has converged").

Convolution integral: Mathematical equation that defines convolution in continuous systems;

analogous to the *convolution sum* for discrete systems.

Convolution kernel: The impulse response of a filter implemented by convolution. Also known as the *filter kernel* and the *kernel*.

Convolution sum: Mathematical equation defining convolution for discrete systems.

Cooley and Tukey: J.W. Cooley and J.W. Tukey, given credit for bringing the FFT to the world in a paper they published in 1965.

Correlation: Mathematical operation carried out the same as convolution, except a left-for-right flip of one signal. This is an optimal way to detect a known waveform in a signal.

Cross-correlation: The signal formed when one signal is correlated with another signal. Peaks in this signal indicate a similarity between the original signals. See also *autocorrelation*.

Cutoff frequency: In analog and digital filters, the frequency separating the passband from the transition band. Often measured where the amplitude is reduced to 0.707 (-3dB).

CVSD: Continuously Variable Slope Delta modulation, a technique used to convert a voice signal into a continuous binary stream.

DC: Direct Current. Electrical term for the portion of the signal that does not change with time; the average value or mean. See *AC* for comparison.

Decibel SPL: Sound Pressure Level. Log scale used to express the intensity of a sound wave: 0 dB SPL is barely detectable; 60 dB SPL is normal speech, and 140 dB SPL causes ear damage.

Decimation: Reducing the sampling rate of a digitized signal. Generally involves low-pass filtering followed by discarding samples. See *interpolation* for comparison.

Decomposition: The process of breaking a signal into two or more additive components. Often refers specifically to the *forward Fourier transform*, breaking a signal into sinusoids.

Deconvolution: The inverse operation of convolution: if $x[n] * h[n] = y[n]$, find $x[n]$ given only $h[n]$ and $y[n]$. Deconvolution is usually carried out by dividing the frequency spectra of the signals.

Delta encoding: A broad term referring to techniques that store data as the difference between adjacent samples. Used in ADC, data compression and many other applications.

Delta function: A normalized impulse. The discrete delta function is a signal composed of all zeros, except the sample at zero that has a value of one. The continuous delta function is similar, but more abstract.

Delta-sigma: Analog-to-digital conversion method popular in voice and music processing. Uses a very high sampling rate with only a single bit per sample, followed by decimation.

Dependent variable: In a signal, the dependent variable depends on the value of the independent variable. Example: when a voltage changes over time, time is the independent variable and voltage is the dependent variable.

Difference equation: Equation relating the past and present samples of the output signal with past and present samples of the input signal. Also called a *recursion equation*.

Dilation: A morphological operation. When applied to binary images, dilation makes the objects larger and can combine disconnected objects into a single object.

Discrete cosine transform (DCT): A relative of the Fourier transform. Decomposes a signal into cosine waves. Used in data compression.

Discrete derivative: An operation for discrete signals that is analogous to the derivative for continuous signals. A better name is the *first difference*.

Discrete Fourier transform (DFT): Member of the Fourier transform family dealing with time domain signals that are *discrete* and *periodic*.

Discrete integral: Operation on discrete signals that is analogous to the integral for continuous signals. A better name is the *running sum*.

Discrete signal: A signal that uses quantized variables, such as a digitized signal residing in a computer.

Discrete time Fourier transform (DTFT): Member of the Fourier transform family dealing with time domain signals that are *discrete* and *aperiodic*

Dithering: Adding noise to an analog signal before analog-to-digital conversion to prevent the digitized signal from becoming "stuck" on one value.

Domain: The independent variable of a signal. For example, a voltage that varies with time is in the *time domain*. Other common domains are the *spatial domain* (such as images) and the *frequency domain* (the output of the Fourier transform).

Double precision: A standard for floating point notation that used 64 bits to represent each number. See *single precision* for comparison.

DSP microprocessor: A type of microprocessor designed for rapid math calculations. Often has a pipeline and/or Harvard architecture. Also called a RISC.

Dynamic range: The largest amplitude a system can deal with divided by the inherent noise of the system. Also used to indicate the number of bits used in an ADC. Can also be used with parameters other than amplitude; see *frequency dynamic range*.

Edge enhancement: Any image processing algorithm that makes the edges more obvious. Also called a *sharpening* operation.

Edge response: In image processing, the output of a system when the input is an edge. The sharpness of the edge response is often used as a measure of the resolution of the system.

Elliptic filter: Used to separate one band of frequencies from another. Achieves a fast roll-off by allowing ripple in the passband and the stopband. Can be used in both analog and digital designs.

End effects: The poorly behaved ends of a filtered signal resulting from the filter kernel not being completely immersed in the input signal.

Erosion: A morphological operation. When applied to binary images, erosion makes the objects smaller and can break objects into two or more pieces.

Euler's relation: The most important equation in complex math, relating sine and cosine waves with complex exponentials.

Even/odd decomposition: A way of breaking a signal into two other signals, one having even symmetry, and the other having odd symmetry.

Even order filter: An analog or digital filter having an even number of poles.

False-negative: One of four possible outcomes of a target detection trial. The target is present, but incorrectly indicated to be not present.

False-positive: One of four possible outcomes of a target detection trial. The target is not present, but incorrectly indicated to be present.

Fast Fourier transform (FFT): An efficient algorithm for calculating the discrete Fourier transform (DFT). Reduces the execution time by *hundreds* in some cases.

FFT convolution: A method of convolving signals by multiplying their frequency spectra. So named because the FFT is used to efficiently move between the time and frequency domains.

Field: Interlaced television displays the even lines of each frame (image) followed by the odd lines. The even lines are called the *even field*, and the odd lines the *odd field*.

Filter kernel: The impulse response of a filter implemented by convolution. Also known as the *convolution kernel* and the *kernel*.

Filtered backprojection: A technique used in computed tomography for reconstructing an image from its views. The views are *filtered* and then *backprojected*.

Finite impulse response (FIR): An impulse response that has a finite number of nonzero values. Often used to indicate that a filter is carried out by using convolution, rather than recursion.

First difference: An operation for discrete signals that mimics the first derivative for continuous signals. Also called the *discrete derivative*.

Fixed point: One of two common ways that computers store numbers; usually used to store integers. See *floating point* for comparison.

Floating point: One of the two common ways that computers store numbers. Floating point uses a form of scientific notation, where a mantissa is raised to an exponent. See *fixed point* for comparison.

Forward transform: The analysis equation of the Fourier transform, calculating the frequency domain from the time domain. See *inverse transform* for comparison.

Fourier reconstruction: One of the methods used in computed tomography to calculate an image from its views.

Fourier series: The member of the Fourier transform family that deals with time domain signals that are *continuous* and *periodic*.

Fourier transform: A family of mathematical techniques based on decomposing signals into sinusoids. In the complex version, signals are decomposed into complex exponentials.

Fourier transform pair: Waveforms in the time and frequency domains that correspond to each other. For example, the rectangular pulse and the sinc function.

Fovea: A small region in the retina of the eye that is optimized for high-resolution vision.

Frame: An individual image in a television signal. The NTSC television standard uses 30 frames per second.

Frame grabber: A analog-to-digital converter used to digitize and store a frame (image) from a television signal.

Frequency domain: A signal having frequency as the independent variable. The output of the Fourier transform.

Frequency domain aliasing: Aliasing that occurs occurring in the frequency domain in response to an action taken in the time domain. Aliasing during sampling is an example.

Frequency domain convolution: Convolution carried out by multiplying the frequency spectra of the signals.

Frequency domain encoding: One of the two main ways that information can be encoded in a signal. The information is contained in the amplitude, frequency, and phase of the signal's component sinusoids. Audio signals are the best example. See *time domain encoding* for comparison.

Frequency domain multiplexing: A method of combining signals for simultaneous transmission by shifting them to different parts of the frequency spectrum.

Frequency dynamic range: The ratio of the largest to the lowest frequency a system can deal with. Analog systems usually have a much larger frequency dynamic range than digital systems.

Frequency resolution: The ability to distinguish or separate closely spaced frequencies.

Frequency response: The magnitude and phase changes that sinusoids experience when passing through a linear system. Usually expressed as a function of frequency. Often found by taking the Fourier transform of the impulse response.

Fricative: Human speech sound that originates as random noise from air turbulence, such as: *s, f, sh, z, v* and *th*. See *voiced* for comparison.

Full-width-at-half-maximum (FWHM): A common way of measuring the width of a peak in a signal. The width of the peak is measured at one-half of the peak's maximum amplitude.

Fundamental frequency: The frequency that a periodic waveform repeats itself. See *harmonic* for comparison.

Gamma curve: The mathematical function or look-up table relating a stored pixel value and the brightness it appears in a displayed image. Also called a *grayscale transform*.

Gaussian: A bell shaped curve of the general form: e^{x^2}. The Gaussian has many unique properties. Also called the *normal distribution*.

Gibbs effect: When a signal is truncated in one domain, ringing and overshoot appear at edges and corners in the other domain.

GIF: A common image file format using LZW (lossless) compression. Widely used on the world wide web for graphics. See *TIFF* and *JPEG* for comparison.

Grayscale: image A digital image where each pixel is displayed in shades of gray between black and white; also called a black and white image.

Grayscale stretch: Greatly increasing the contrast of a digital image to allow the detailed examination of a small range of quantization levels. Quantization levels outside of this range are displayed as saturated black or white.

Grayscale transform: The conversion function between a stored pixel value and the brightness that appears in a displayed image. Also called a *gamma curve*.

Halftone: A common method of printing images on paper. Shades of gray are created by various patterns of small black dots. Color halftones use dots of red, green and blue.

Hamming window: A smooth curve used in the design of filters and spectral analysis, calculated from: $0.54 - 0.46\cos(2\pi n/M)$, where n runs from 0 to M.

Harmonics: The frequency components of a periodic signal, always consisting of integer multiples of the fundamental frequency. The fundamental is the first harmonic, twice this frequency is the second harmonic, etc.

High fidelity: High quality music reproduction, such as provided by CD players.

High-level language: Programming languages such as C, BASIC and FORTRAN.

High-speed convolution: Another name for FFT convolution.

Hilbert transformer: A system having the frequency response: Mag = 1, Phase = 90°, for all frequencies. Used in communications systems for modulation. Can be analog or digital.

Histogram equalization: Processing an image by using the integrated histogram of the image as the grayscale transform. Works by giving large areas of the image higher contrast than the small areas.

Histogram: Displays the distribution of values in a signal. The x-axis show the possible values the samples can take on; the y-axis indicates the number of samples having each value.

Homogeneity: A mathematical property of all linear systems. If an input $x[n]$ produces an output of $y[n]$, then an input $kx[n]$ produces an output of $ky[n]$, for any constant k.

Homomorphic: DSP technique for separating signals combined in a nonlinear way, such as by multiplication or convolution. The nonlinear problem is converted to a linear one by an appropriate transform.

Huffman encoding: Data compression method that assigns frequently encountered characters fewer bits than seldom used characters.

Hyperspace: Term used in target detection and neural network analysis. One parameter can be graphically interpreted as a *line*, two parameters a *plane*, three parameters a *space*, and more than three parameters a *hyperspace*.

Imaginary part: The portion of a complex number that has a *j* term, such as 2 in 3+2*j*. In the real Fourier transform, the *imaginary part* also refers to the portion of the frequency domain that holds the amplitudes of the sine waves, even though *j* terms are not used.

Impulse: A signal composed of all zeros except for a very brief pulse. For discrete signals, the pulse consists of a single nonzero sample. For continuous signals, the width of the pulse must be much shorter than the inherent response of any system the signal is used with.

Impulse decomposition: Breaking an *N* point signal into *N* signals, each containing a single sample from the original signal, with all the other samples being zero. This is the basis of convolution.

Impulse response: The output of a system when the input is a normalized impulse (a delta function).

Impulse train: A signal consisting of a series of equally spaced impulses.

Independent variable: In a signal, the dependent variable depends on the value of the independent variable. Example: when a voltage changes over time, time is the independent variable and voltage is the dependent variable.

Infinite impulse response (IIR): An impulse response that has an infinite number of nonzero values, such as a decaying exponential. Often used to indicate that a filter is carried out by using recursion, rather than convolution.

Integers: Whole numbers: ···–2, –1, 0, 1, 2, ···. Also refers to numbers stored in fixed point notation. See *floating point* for comparison.

Interlaced decomposition: Breaking a signal into its even numbered and odd numbered samples. Used in the FFT.

Interlaced video: A video signal that displays the even lines of each image followed by the odd lines. Used in television; developed to reduce flicker.

Interpolation: Increasing the sampling rate of a digitized signal. Generally done by placing zeros between the original samples and using a low-pass filter. See *decimation* for comparison.

Inverse transform: The synthesis equation of the Fourier transform, calculating the time domain from the frequency domain. See *forward transform* for comparison.

Iterative: Method of finding a solution by gradually adjusting the variables in the right direction until convergence is achieved. Used in CT reconstruction and neural networks.

JPEG: A common image file format using transform (lossy) compression. Widely used on the world wide web for graphics. See *GIF* and *TIFF* for comparison.

Kernel: The impulse response of a filter implemented by convolution. Also known as the *convolution kernel* and the *filter kernel*.

Laplace transform: Mathematical method of analyzing systems controlled by differential equations. A main tool in the design of electric circuits, such as analog filters. Changes a signal in the time domain into the s-domain

Learning algorithm: The procedure used to find a set of neural network weights based on examples of how the network should operate.

Line pair: Imaging term for *cycle*. For example, 5 cycles per mm is the same as 5 line pairs per mm.

Line pair gauge: A device used to measure the resolution of an imaging system. Contains a series of light and dark lines that move closer together at one end.

Line spread function (LSF): The response of an imaging system to a thin line in the input image.

Linear phase: A system with a phase that is a straight line. Usually important because it means the impulse response has left-to-right symmetry, making rising edges in the output signal look the same as falling edges. See also *zero phase*.

Linear system: By definition, a system that has the properties of additivity and homogeneity.

Lossless compression: Data compression technique that exactly reconstructs the original data, such as LZW compression.

Lossy compression: Data compression methods that only reconstruct an approximation to the original data. This allows higher compression ratios to be achieved. JPEG is an example.

Matched filtering: Method used to determine where, or if, a know pattern occurs in a signal. Matched filtering is based on correlation, but implemented by convolution.

Mathematical equivalence: A way of using complex numbers to represent real problems. Based on Euler's relation equating sinusoids with complex exponentials. See *substitution* for comparison.

Mean: The average value of a signal or other group of data.

Memoryless: Systems where the current value of the output depends only on the current value of the input, and not past values.

Modulation transfer function (MTF): Imaging jargon for the *frequency response*.

Morphing: Gradually warping an image from one form to another. Used for special effects, such as a man turning into a werewolf.

Morphological: Usually refers to simple non-linear operations performed on binary images, such as erosion and dilation.

Moving average filter: Each sample in the output signal is the average of many adjacent samples in the input signal. Can be carried out by convolution or recursion.

MPEG: Compression standard for video, such as digital television.

Mu law: Companding standard used in the United States. Allows digital voice signals to be represented with only 8 bits instead of 12 bits by making the quantization levels unequal. See *"A" law* for comparison.

Multiplexing: Combining two or move signals together for transmission. This can be carried out in many different ways.

Multirate: Systems that use more than one sampling rate. Often used in ADC and DAC to obtain better performance with less electronics.

Natural frequency: A frequency expressed in radians per second, as compared to cycles per second (hertz). To convert frequency (in hertz) to natural frequency, multiply by 2π.

Negative frequencies: Sinusoids can be written

as a positive frequency: $\cos(\omega t)$, or a negative frequency: $\cos(-\omega t)$. Negative frequencies are included in the complex Fourier transform, making it more powerful.

Normal distribution: A bell shaped curve of the form: e^{x^2}. Also called a *Gaussian*.

NTSC: Television standard used in the United States, Japan, and other countries. See *PAL* and *SECAM* for comparison.

Nyquist frequency, Nyquist rate: These terms refer to the sampling theorem, but are used in different ways by different authors. They can be used to mean four different things: the highest frequency contained in a signal, twice this frequency, the sampling rate, or one-half the sampling rate.

Octave: A factor of two in frequency.

Odd order filter: An analog or digital filter having an odd number of poles.

Opening: A morphological operation defined as a dilation operation followed by an erosion operation.

Optimal filter: A filter that is "best" in some specific way. For example, Wiener filters produce an optimal signal-to-noise ratio and matched filters are optimal for target detection.

Overlap add: Method used to break long signals into segments for processing.

PAL: Television standard used in Europe. See *NTSC* for comparison.

Parallel stages: A combination of two or more stages with the same input and added outputs.

Parameter space: Target detection jargon. One parameter can be graphically interpreted as a *line*, two parameters a *plane*, three parameters a *space*, and more than three parameters a *hyperspace*.

Parseval's relation: Equation relating the energy in the time domain to the energy in the frequency domain.

Passband: The band of frequencies a filter is designed to pass unaltered.

Passive sonar: Detection of submarines and other undersea objects by the sounds they produce. Used for covert surveillance.

Phasor transform: Method of using complex numbers to find the frequency response of *RLC* circuits. Resistors, capacitors and inductors become R, $-j/\omega C$, and $j\omega L$, respectively.

Pillbox: Shape of a filter kernel used in image processing: circular region of a constant value surrounded by zeros.

Pitch: Human perception of the fundamental frequency of an continuous tone. See *timbre* for comparison.

Pixel: A contraction of "picture element." An individual sample in a digital image.

Point spread function (PSF): Imaging jargon for the impulse response.

Poisson statistics: Variations in a signal's value resulting from it being represented by a finite number of particles, such as: x-rays, light photons or electrons. Also called *Poisson noise* and *statistical noise*.

Polar form: Representing sinusoids by their magnitude and phase: $M\cos(\omega t + \phi)$, where M is the magnitude and ϕ is the phase. See *rectangular form* for comparison.

Pole: Term used in the Laplace transform and z-transform. When the s-domain or z-domain transfer function is written as one polynomial divided by another polynomial, the roots of the denominator are the *poles* of the system, while the roots of the numerator are the *zeros*.

Pole-zero diagram: Term used in the Laplace and z-transforms. A graphical display of the location of the poles and zeros in the s-plane or z-plane.

Precision: The error in a measurement or prediction that is not repeatable from trial to trial. Precision is determined by random errors. See *accuracy* for comparison.

Probability distribution function (pdf): Gives the probability that a *continuous* variable will take on a certain value.

Probability mass function (pmf): Gives the probability that a *discrete* variable will take on a certain value. See *pdf* for comparison.

Pulse response: The output of a system when the input is a pulse.

Quantization error: The error introduced when a signal is quantized. In most cases, this results in a maximum error of $\pm\frac{1}{2}$ LSB, and an rms error of $1/\sqrt{12}$ LSB. Also called *quantization noise*.

Random error: Errors in a measurement or prediction that are not repeatable from trial to trial. Determines *precision*. See *systematic error* for comparison.

Radar: Radio Detection And Ranging. Echo location technique using radio waves to detect aircraft.

Real DFT: The discrete Fourier transform using only real (ordinary) numbers. A less powerful technique than the complex DFT, but simpler. See *complex DFT* for comparison.

Real FFT: A modified version of the FFT. About 30% faster than the standard FFT when the time domain is completely real (i.e., the imaginary part of the time domain is zero).

Real Fourier transform: Any of the members of the Fourier transform family using only real (as opposed to imaginary or complex) numbers. See *complex Fourier transform* for comparison.

Real part: The portion of a complex number that does not have the j term, such as 3 in $3+2j$. In the real Fourier transform, the *real* part refers to the part of the frequency domain that holds the amplitudes of the cosine waves, even though no j terms are present.

Real time processing: Processing data as it is acquired, rather than storing it for later use. Example: DSP algorithms for controlling echoes in long distance telephone calls.

Reconstruction filter: A low-pass analog filter placed after a digital-to-analog converter. Smoothes the stepped waveform by removing frequencies above one-half the sampling rate.

Rectangular form: Representing a sinusoid by the form: $A\cos(\omega t) + B\sin(\omega t)$, where A is called the *real part* and B is called the *imaginary part* (even though these are not imaginary numbers).

Rectangular window: A signal with a group of adjacent points having unity value, and zero elsewhere. Usually multiplied by another signal to select a section of the signal to be processed.

Recursion coefficients: The weighing values used in a recursion equation. The recursion coefficients determine the characteristics of a recursive (IIR) filter.

Recursion equation: Equation relating the past and present samples of the output signal with the past and present values of the input signal. Also called a *difference equation*.

Region-of-convergence: The term used in the Laplace and z-transforms. Those regions in the s-plane and z-planes that have a defined value.

RGB encoding: Representing a color image by specifying the amount of red, green, and blue for each pixel.

RISC: Reduced Instruction Set Computer, also called a DSP microprocessor. A fewer number of programming commands allows much higher speed math calculations. The opposite is the Complex Instruction Set Computer, such as the Pentium.

ROC curve: A graphical display showing how threshold selection affects the performance of a target detection problem.

Roll-off: Jargon used to describe the sharpness of the transition between a filter's passband and stopband. A *fast* roll-off means the transition is sharp; a *slow* roll-off means it is gradual.

Root-mean-square (rms): Used to express the fluctuation of a signal around *zero*. Often used in electronics. Defined as the square-root of the mean of the squares. See *standard deviation* for comparison.

Round-off noise: The error introduced by rounding the result of a math calculation to the nearest quantization level.

Row major order: A pattern for converting an image to serial form. Operates the same as English writing: left-to-right on the first line, left-to-right on the second line, etc.

Run-length encoding: Simple data compression technique with many variations. Characters that are repeated many times in succession are replaced by codes indicating the character and the length of the run.

Running sum: An operation used with discrete signals that mimics integration of continuous signals. Also called the *discrete integral*.

s-domain: The domain defined by the Laplace transform. Also called the *s-plane*.

Sample spacing: The spacing between samples when a continuous image is digitized. Defined as the center-to-center distance between pixels.

Sampling aperture: The region in a continuous image that contributes to an individual pixel during digitization. Generally about the same size as the sample spacing.

Sampling theorem: If a continuous signal composed of frequencies less than f is sampled at $2f$, all of the information contained in the continuous signal will be present in the sampled signal. Frequently called the *Shannon* sampling theorem or the *Nyquist* sampling theorem.

SECAM: Television standard used in Europe. See NTSC for comparison.

Seismology: Branch of geophysics dealing with the mechanical properties of the earth.

Separable: An image that can be represented as the product of its vertical and horizontal profiles. Used to improve the speed of image convolution.

Sharpening: Image processing operation that makes edges more abrupt.

Shift and subtract: Image processing operation that creates a 3D or embossed effect.

Shift invariance: A property of many systems. A shift in the input signal produces nothing more than a shift in the output signal. Means that the characteristics of the system do not changing with time (or other independent variable).

Sigmoid: An "s" shaped curve used in neural networks.

Signal: A description of how one parameter varies with another parameter. Example: a *voltage* that varies with *time*.

Signal restoration: Returning a signal to its original form after it has been changed or degraded in some way. One of the main uses of *filtering*.

Sinc function: Formally defined by the relation: $sinc(a) = sin(\pi a)/\pi a$. The π terms are often hidden in other variables, making it in the general form: $sin(x)/x$. Important because it is the Fourier transform of the rectangular pulse.

Single precision: A floating point notation that

used 32 bits to represent each number. See *double precision* for comparison.

Single-pole digital filters: Simple recursive filters that mimic *RC* high-pass and low-pass filters in electronics.

Sinusoidal fidelity: An important property of linear systems. A sinusoidal input can only produce a sinusoidal output; the amplitude and phase may change, but the frequency will remain the same.

Sonar: <u>So</u>und <u>N</u>avigation <u>A</u>nd <u>R</u>anging. The use of sound to detect submarines and other underwater objects. *Active sonar* uses echo location, while *passive sonar* only listens.

Source code: A computer program in the form written by the programmer; distinguished from *executable code*, a form that can be directly run on a computer.

Spatial domain: A signal having distance (space) as the independent variable. Images are signals in the spatial domain.

Spectral analysis: Understanding a signal by examining the amplitude, frequency, and phase of its component sinusoids. The primary tool of spectral analysis is the Fourier transform.

Spectral inversion: Method of changing a filter kernel such that the corresponding frequency response is flipped top-for-bottom. This can change low-pass filters to high-pass, band-pass to band-reject, etc.

Spectral leakage: Term used in spectral analysis. Since the DFT can only be taken of a finite length signal, the frequency spectrum of a sinusoid is a peak with tails. These tails are referred to as *leakage* from the main peak.

Spectral reversal: Technique for changing a filter kernel such that the corresponding frequency response is flipped left-for-right. This changes low-pass filters into high-pass filters.

Spectrogram: Measurement of how an audio frequency spectrum changes over time. Usually displayed as an image. Also called a *voiceprint*.

Standard deviation: A way of expressing the fluctuation of a signal around its average value. Defined as the square-root of the average of the deviations squared, where the deviation is the difference between a sample and the mean. See *root-mean-square* for comparison.

Static linearity: Refers to how a linear system acts when the signals are not changing (i.e., they are *DC* or *static*). In this case, the output is equal to the input multiplied by a constant.

Statistical noise: Variations in a signal's value resulting from it being represented by a finite number of particles, such as: x-rays, electrons, or light photons. Also called *Poisson statistics* and *Poisson noise*.

Steepest descent: Strategy used in designing iterative algorithms. Analogous to finding the bottom of a valley by always moving in the down-hill direction.

Step response: The output of a system when the input is a step function.

Stopband: The band of frequencies that a filter is designed to block.

Stopband attenuation: The amount by which frequencies in the stopband are reduced in amplitude, usually expressed in decibels. Used to describe a filter's performance.

Substitution: A way of using complex numbers to represent a physical problem, such as electric circuit design. In this method, *j* terms are added to change the physical problem to a complex form, and then removed to move back again. See *mathematical equivalence* for comparison.

Superposition: The fundamental concept in signal processing. A signal can be broken into additive components, each of the components passed through a linear system, and the outputs added. The signal found in this manner is the same as if the original signal had been passed through the system. This allows complicated signals to be processed by breaking them into simpler signals.

Switched capacitor filter: Analog filter that uses rapid switching to replace resistors. Made as easy-to-use integrated circuits. Often used as antialias filters for ADC and reconstruction filters for DAC.

Synthesis: The inverse Fourier transform, calculating the time domain from the frequency domain. See *analysis* for comparison.

System: Any process that produces an output signal in response to an input signal.

Systematic error: Errors in a measurement or prediction that are repeatable from trial to trial.

Systematic errors determines *accuracy*. See *random error* for comparison.

Target detection: Deciding if an object or condition is present based on measured values.

TIFF: A common image file format used in word processing and similar programs. Usually not compressed, although LZW compression is an option. See *GIF* and *JPEG* for comparison.

Timbre: The human perception of harmonics in sound. See *pitch* for comparison.

Time domain: A signal having time as the independent variable. Also used as a general reference to any domain that data has been acquired in.

Time domain aliasing: Aliasing occurring in the time domain when an action is taken in the frequency domain. Circular convolution is an example.

Time domain encoding: Signal information contained in the shape of the waveform. See *frequency domain encoding* for comparison.

Transfer function: The output signal divided by the input signal. This comes in several different forms, depending on how the signals are represented. For instance, in the s-domain and z-domain, this will be one polynomial divided by another polynomial, and can be expressed as *poles* and *zeros*.

Transform: A procedure, equation or algorithm that changes one group of data into another group of data.

Transform compression: Data compression technique based on assigning fewer bits to the high frequencies. *JPEG* is the best example.

Transition band: Filter jargon; the band of frequencies between the passband and stopband where the roll-off occurs.

True-negative: One of four possible outcomes of a target detection trial. The target is not present, and is correctly indicated to be not present.

True-positive: One of four possible outcomes of a target detection trial. The target is present, and correctly indicated to be present.

Unit circle: The circle in the z-plane at $r = 1$. The values along this circle are the frequency

response of the system.

Unit impulse: Another name for *delta function*.

Voiced: Human speech sound that originates as pulses of air passing the vocal cords. Vowels are an example of voiced sounds. See *fricative* for comparison.

Well: Short for *potential well*; the region in a CCD that is sensitive to light.

White noise: Random noise that has a flat frequency spectrum. Occurs when each sample in the time domain contains no information about the other samples. See *1/f noise* for comparison.

Wiener filter: Optimal filter for increasing the signal-to-noise ratio based on the frequency spectra of the signal and noise.

Windowed-sinc: Digital filter used to separate one band of frequencies from another.

z-domain: The domain defined by the z-transform. Also called the *z-plane*.

z-transform: Mathematical method used to analyze discrete systems that are controlled by difference equations, such as recursive (IIR) filters. Changes a signal in the time domain into a signal in the z-domain.

Zero: A term used in both the Laplace and z-transforms. When the s-domain or z-domain transfer function is written as one polynomial divided by another polynomial, the roots of the numerator are the *zeros* of the system. See also *pole*.

Zero phase: A system with a phase that is entirely zero. Occurs only when the impulse response has left-to-right symmetry around the origin. See also *linear phase*.

Zeroth-order hold: A term used in DAC to describe that the analog signal is maintained at a constant value between conversions, resulting in a staircase appearance.

Index